Analytical Applications of Lasers

Edited by

EDWARD H. PIEPMEIER

Department of Chemistry
Oregon State University
Corvallis, Oregon

A WILEY-INTERSCIENCE PUBLICATION

JOHN WILEY & SONS

New York / Chichester / Brisbane / Toronto / Singapore

Copyright © 1986 by John Wiley & Sons, Inc.

All rights reserved. Published simultaneously in Canada.

Reproduction or translation of any part of this work
beyond that permitted by Section 107 or 108 of the
1976 United States Copyright Act without the permission
of the copyright owner is unlawful. Requests for
permission or further information should be addressed to
the Permissions Department, John Wiley & Sons, Inc.

Library of Congress Cataloging in Publication Data:

Analytical applications of lasers.

  (Chemical analysis ; v. 87)
  ''A Wiley-Interscience publication.''
  Includes index.
  1. Laser spectroscopy.   I. Piepmeier, Edward H.
II. Series.

QD96.L3A48 1986   543'.0858   86-4088
ISBN 0-471-87023-4

Printed in the United States of America

10 9 8 7 6 5 4 3 2 1

CONTRIBUTORS

C. Th. J. Alkemade, Physical Laboratory of the State University, Utrecht, The Netherlands

G. J. Blanchard, Bell Communications Research Inc., Red Bank, New Jersey

Robert B. Green, Naval Weapons Center, China Lake, California

Joel M. Harris, Department of Chemistry, University of Utah, Salt Lake City, Utah

R. J. Haskell, Department of Chemistry, University of Wisconsin, Madison, Wisconsin

Scott J. Hein, Department of Chemistry, Oregon State University, Corvallis, Oregon

Robert S. Houk, Ames Laboratory-USDOE and Department of Chemistry, Iowa State University, Ames, Iowa

A. C. Koskelo, Department of Chemistry, University of Oregon, Eugene, Oregon

R. M. Measures, University of Toronto, Institute for Aerospace Studies, Toronto, Ontario, Canada

D. C. Nguyen, Department of Chemistry, University of Wisconsin, Madison, Wisconsin

N. Omenetto, CCR, European Community Center, Stabilimento di Ispra, Analytical Chemistry Division, Ispra, Italy

Edward H. Piepmeier, Department of Chemistry, Oregon State University, Corvallis, Oregon

J. K. Steehler, Department of Chemistry, University of Wisconsin, Madison, Wisconsin

Andrew C. Tam, IBM Research Laboratory, San Jose, California

Lawrence C. Thomas, Department of Chemistry, Seattle University, Seattle, Washington

M. A. Valentini, Department of Chemistry, University of Wisconsin, Madison, Wisconsin

Bennie R. Ware, Department of Chemistry, Syracuse University, Syracuse, New York

E. L. Wehry, Department of Chemistry, University of Tennessee, Knoxville, Tennessee

J. D. Winefordner, Department of Chemistry, University of Florida, Gainesville, Florida

M. J. Wirth, Lawrence Livermore National Laboratory, Livermore, California

J. C. Wright, Department of Chemistry, University of Wisconsin, Madison, Wisconsin

Edward S. Yeung, Department of Chemistry and Ames Laboratory, Iowa State University, Ames, Iowa

To my wife Karen

and

my parents

PREFACE

This book is written for a person who has a scientific background and who is interested in learning about the many different ways lasers are used for determining concentrations of a wide variety of chemicals in many types of samples. The person who is familiar with lasers also will find this book helpful in learning more about how lasers are applied in analytical chemistry to solve both unique and common measurement problems.

Lasers make possible the determination of concentrations of chemical species spread over a range of many kilometers, or the counting of individual molecules passing the focal point of a tightly focused beam of light. The chemical species may range from small atoms to very large molecules; from those that are free in a vacuum to those that are bound onto a surface. Samples may be gases, liquids, or solids. Analyte concentrations range from those of major constituents to ultra trace levels.

Lasers allow absorption measurements that are limited only by shot noise over a long period of time, or the observation of transient species on a sub-picosecond time scale. High spectral resolution is achieved for species that are otherwise overlapped. Multiphoton transitions populate excited states that would not otherwise be conveniently attainable and thereby improve the selectivity of an analytical determination.

Some laser measurements provide information that is unique and complementary to that obtained by other methods. Some methods take advantage of a special characteristic of a laser and simply substitute the laser as a component in an ordinary type of measurement system. Still other measurements, envisioned decades ago, have now become practical because of the use of lasers.

The book begins with an introductory chapter on the basic principles of lasers, as well as nonlinear optical effects, which are fundamental to some laser methods. The person familiar with lasers and nonlinear effects may skip this chapter or use it as a brief consolidating review. With so many laser methods to cover, the book was further subdivided into several parts, based on prominent characteristics of the methods. This is purely a convenience, and the reader will quickly recognize that the subdivisions are not exclusive; some methods could reside in more than one place.

To preserve the individual styles of the authors, I made few editorial changes in the content of any chapter. Some overlap of material remains in order to help the reader who wishes to concentrate on selected chapters, without having to study preceding chapters (except perhaps for Chapter 1).

It is intended that this book be educational, and that it present some of the most recent advances in the field. Chapters generally include discussions of the principles of the methods, diagrams of instrumental configurations, typical figures of merit, examples of applications to analytical problems, and speculation about the future. I anticipate that this book will find use in graduate level courses for students who are interested in learning more about and using laser methods in analytical chemistry. I also hope that it will attract people to this expanding field, who will bring new ideas to further advance the applications of lasers in analytical chemistry.

EDWARD H. PIEPMEIER

Corvallis, Oregon
April 1986

CONTENTS

PART I
Introduction

PART II
Selected Methods that Use Various Detection Schemes

xi

PART III
Methods With Improved Spectral Resolution

PART IV
Selected Multiphoton and Multiwavelength Methods

PART V
Methods Based on Special Characteristics of Lasers

Analytical Applications
of Lasers

PART

I

INTRODUCTION

CHAPTER

1

BASIC PRINCIPLES OF LASERS AND NONLINEAR OPTICAL EFFECTS

EDWARD H. PIEPMEIER

Department of Chemistry
Oregon State University
Corvallis, Oregon

3

1. INTRODUCTION

This chapter provides the background for the reader to understand the properties of lasers that are important in the analytical applications discussed in the rest of the book. Lasers emit radiation that has unique properties, which are employed to solve both routine and special problems in analytical chemistry. Because all lasers do not exhibit these properties to the same extent, this chapter will also help in understanding compromises that are often necessary.

We shall start by mentioning important properties and then continue with a brief discussion of the theory of operation of a laser and laser components, which control beam properties. With this background, we shall discuss laser beam properties in sufficient detail to satisfy the requirements for the chapters that follow. Additional details are included in the chapters when appropriate.

Well-known properties of laser beams are high intensity, directionality, and monochromaticity. Coherence is also an important property because it influences the cross-sectional spatial profile, the temporal profile, and the spectral profile of the laser beam. In addition to these properties are characteristics that are important in analytical applications, although one might not immediately think of them when the word *laser* is mentioned. These include polarization and tunability over a wavelength region. And, of course, daily operating costs and serviceability are important in routine applications.

One of the most desirable characteristics of a laser beam for analytical applications is reproducibility. Unfortunately, obtaining a laser beam that has excellent reproducibility in all of the desired characteristics may be expensive, and not always practical. When the unique properties of a particular laser are necessary to help solve a problem, it is important to consider the influence that reproducibility of those properties may have on the final results.

2. STIMULATED EMISSION

The laser is made possible by stimulated emission, an effect which is inherently related to the well-known effects of spontaneous emission and absorption. All three of these effects are related to each other by the Einstein probability coefficients, which occur in the rate constants for the equations that give the rate at which atoms in one energy level make transitions to another energy level. For an atomic system with energy levels E_1 and E_2, the rate of spontaneous decay (per unit volume) to level 1 of the population in the higher energy level 2 is given by

$$\frac{dn_2}{dt} = -A_{21}n_2 \tag{1.1}$$

where n_2 represents the number of atoms per unit volume in level 2 and A_{21} (s^{-1}) is the Einstein probability coefficient for spontaneous emission for that transition. Its reciprocal $\tau_{sp} = 1/A_{12}$ is the radiative lifetime and is a characteristic parameter of that transition. The other two Einstein coefficients apply when the atoms are irradiated by a plane electromagnetic wave, when the radiation beam is polarized and the atoms are randomly oriented, or when the beam is unpolarized. Then the rate of transitions from level i to level j is given by

$$\frac{dn_i}{dt} = -W_{ij}n_i \tag{1.2}$$

where the rate constants W_{ij} (in s^{-1}) are given by

$$W_{ij} = B_{ij}EL(\lambda - \lambda_0) \tag{1.3}$$

In this equation E (W cm^{-2}) is the irradiance of the incident monochromatic radiation, at wavelength λ, and $L(\lambda - \lambda_0)$ is the spectral lineshape function with a central wavelength λ_0 and normalized so that its integral over all wavelengths λ is unity. B_{ij} is the Einstein probability coefficient for stimulated emission when the transition is from the upper level 2 to the lower level 1, and is the Einstein probability coefficient for (stimulated) absorption when the transition is from the lower level 1 to the upper level 2. Equation (1.3) shows that the stimulated transition rates are proportional to the irradiance of the external radiation, and also depend on how far the wavelength λ of the radiation is from the central wavelength λ_0 of the line.

When the spectral width of the laser beam is large relative to the atomic linewidth, the rate constants are given by

$$W_{ij} = B_{ij}E_{\lambda_0} \tag{1.4}$$

where E_{λ_0} is the spectral irradiance (irradiance per unit of wavelength interval, $\text{W cm}^{-2} \text{nm}^{-1}$) at the atomic line center λ_0.

The Einstein coefficients are directly proportional to each other:

$$A_{21} = \frac{8\pi h c^2}{\lambda_0^5} B_{21} \qquad (1.5a)$$

$$B_{12} = B_{21}\frac{g_2}{g_1} \qquad (1.5b)$$

where h is Planck's constant, c is the speed of light, and each degeneracy factor, g_1 and g_2 (for energy levels 1 and 2, respectively), is the number of different physical configurations of the atom that happen to have the same energy level in the absence of a magnetic field. It should be emphasized that B_{ij} has been defined in the wavelength rather than the frequency domain, and in terms of irradiance rather than power density or radiant energy density.

2.1. Saturation and Bleaching

Notice in Eq. (1.5) that the probability coefficients for stimulated emission and absorption are equal, except for the factor g_2/g_1, which is unity in the simplest case and is usually not very far from unity in real systems. Consequently, when a population of atoms is strongly irradiated so that the stimulated transitions are much more frequent than spontaneous transitions or transitions caused by collisions (e.g., quenching), the population approaches a condition of saturation, where there is an equal number of atoms in levels 1 and 2. Then it is equally probable that a photon will be absorbed or that a photon will be emitted. Photons that are emitted by stimulated emission have the same wavelength and travel in the same direction as the incident photons. Therefore, when a beam passes through a saturated population of atoms, there is no net gain or loss of photons and the population appears to be transparent or bleached. This effect makes it impossible to invert a two-energy-level population by *optical* pumping, and it has important consequences when exciting an analyte population with a laser beam to obtain a signal.

Because the stimulated transitions (in both directions) in a saturated population are much more frequent than other transitions such as quenching, the fraction of atoms in the excited state in a population in a steady-state condition is not influenced by changes in quenching; a loss by quenching is quickly replaced by an excited state formed by a stimulated transition. Consequently, fluorescence emission, which is proportional to the excited-state population, is independent of irradiance or quenching rates in a saturated population. This contrasts with an unsaturated population in which fluorescence may be dramatically decreased by quenching collisions.

In addition to its advantages, saturation has a disadvantage: it decreases spectral selectivity because it broadens the absorption, or excitation, spectral line profile. Consider the case where a monochromatic excitation beam scans across a spectral transition. At low irradiances the ordinary spectral profile of the transition can be observed by absorption. As the irradiance caused by the excitation beam is increased, the center of the spectral line profile eventually approaches complete transparency, its peak value, while the wings of the profile still absorb. After the center has become transparent, the wings continue to approach transparency (the same peak value) as the irradiance continues to increase. The effect is a profile that becomes broader and broader as the irradiance of the excitation beam is increased. The result is reduced spectral resolution for adjacent spectral lines.

2.2. Beam Amplification

Ordinarily, most atoms in a population are in the lower energy level, and absorption occurs much more frequently than stimulated emission. If the population of atoms were inverted so that most of the atoms were in the upper energy level, then stimulated emission would occur more frequently than absorption. When this happens, the beam gains photons at the same wavelength, all of which travel in the same direction, and the beam is said to be amplified. Beam amplification is basic to the operation of a laser.

3. BASIC LASER COMPONENTS

A laser has three basic components: an optical amplifier, an excited-state pump, and an optical resonator or cavity. The optical amplifier has a population or set of atoms, molecules, or ions with a larger fraction of the population in higher energy levels than in the corresponding lower energy levels. Because an ordinary population has a larger fraction in lower energy levels, the population in the amplifier is said to be inverted. The excited-state pump is the means by which this population inversion is maintained. The resonator is a tuned feedback component that converts the amplifier into an optical oscillator, in much the same way that tuned feedback components are used to convert an electronic amplifier into an oscillator. The characteristics of a particular laser beam depend on the details of the energy levels in the amplifier, the mechanism used to pump the amplifier, and the resonator design. These will be described in more detail.

3.1. Optical Amplifier

An optical amplifier has an active medium containing an inverted population that causes a beam of radiation passing through it to gain photons if the wavelength of the radiation corresponds to the transition between the inverted energy level pair. This is not a sufficient condition for amplification to occur, however.

3.1.1. Amplifier Losses

Losses in the amplifier must also be considered when determining the actual increase (or decrease) in beam intensity that occurs when a beam passes through an amplifier. Amplifier losses may be classified into (a) fixed fractional losses, such as those caused by imperfect reflections at the end mirrors and imperfect transmissions and undesired reflections of internal optical components, and (b) those that increase with the length of the path through the amplifier, such as losses caused by diffraction, scattering, and absorption (by other particles in the amplifier medium). In order for amplification to occur, the amplification produced by the inverted population must exceed the amplifier losses as the beam passes through the amplifier. If the loss per unit length exceeds the gain per unit length, then amplification cannot occur, no matter how long the path is. If the gain per unit length exceeds the loss per unit length, then amplification will occur when the length is long enough to more than compensate for the fixed fractional losses.

3.1.2. Amplifier Saturation

If the beam intensity is not too high, so that it does not deplete the inverted population faster than the pumping rate can replenish it, then the beam will have little influence on the amplification factor of the amplifier medium; any beam passing through the amplifier will continue to be amplified by the same factor. However, in some cases, after much amplification, the intensity of a beam may become so high that the leading edge of the beam reduces the inverted population to zero in each region of the amplifier medium through which the beam passes. In this case the gain is said to be saturated. Later parts of the beam will experience little if any amplification in these regions, until the pump has a chance to reestablish an inversion in the population. This effect can be used to shorten a laser pulse by causing its leading edge to grow faster in intensity than the trailing edge. It also causes a laser that has a relatively weak excited-state pump to produce trains of pulses, rather than a continuously operating (continuous-wave or cw) laser beam.

3.2. Excited-State Pump

3.2.1. Pumping Methods

The type of excited-state pump used to invert the population of an amplifier depends on the amplifier medium that supports the population of particles, and upon the pattern of the energy levels and their lifetimes, that are involved in the pumping and amplifying processes. In a gas medium a chemical reaction or an electrical discharge may be used as a pump, whereas optical pumping is used for a solid or liquid medium. A ruby laser rod can be pumped with a relatively long millisecond pulsed flashlamp, whereas an organic dye laser requires a short microsecond pulse, the difference being due to differences in energy level lifetimes.

3.2.2. Pumping Requirements

If a two-level system were pumped from the ground state to an excited state with an external light source, an inverted population could never be achieved. As soon as the photon density of the external source was sufficient to excite almost one-half of the population to the excited state, the system would become transparent and no more pumping photons would be absorbed. This is because the rate at which transitions from the upper state to the lower state were induced by the pumping light source would equal the rate of induced transitions from the lower to the upper state; there would be no net absorption or gain of photons from the pumping source.

This problem is overcome when a multilevel system is used. Consider the three-level system in Fig. 1.1 where the relative transition rates between energy levels have been chosen so that energy level 2 can be adequately pumped. In this case the pump photons are chosen to populate energy level 3, and the amplifier amplifies photons corresponding to the transition from level 2 to 1. Particles that are pumped into energy level 3 are quickly converted (by collisions) to particles with an energy at level 2. Because the transition from level 2 to 1 in this case is relatively infrequent, level 2 achieves a higher population than level 1. When this happens, the system is inverted and amplification can occur (if the pumping rate exceeds the threshold set by losses). For a system with a nanosecond lifetime for level 2, an auxillary laser beam would be required as the pump to achieve a pumping rate that would produce a sufficiently high inversion. For a millisecond lifetime, the required pumping rate could be achieved by a gas discharge lamp.

A four-level system eliminates a major disadvantage of a three-level system:

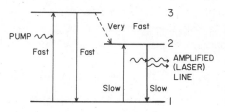

Fig. 1.1. A laser amplifier medium with three energy levels. The medium is pumped from the lowest level to the highest level. Populating the middle from the highest level is very fast. Inversion occurs when the middle level population exceeds the population of the lowest level.

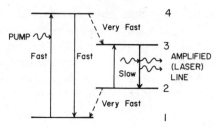

Fig. 1.2. A laser amplifier medium with four energy levels. Pumping occurs from level 1 to level 4. Conversion to level 3 is very fast. Level 2 is populated from level 3 by photon emission at the lasing wavelength (the three wavy arrows symbolize the stimulated emission process). Quick conversion of the population in level 2 to level 1 helps to maintain an inversion between levels 2 and 3. Not shown is a possible transition from level 3 to 1, which also must be slow.

the need to pump essentially half of all of the particles into level 2 before inversion is achieved. This results in an inefficient transfer of energy from the pump to the amplified beam. In Fig. 1.2 the transition rates for a four-level system have been chosen so that inversion can be achieved between levels 2 and 3 by pumping level 4 from level 1. Just as in the case of the three-level system, the spontaneous and collisional transitions between the two energy levels that amplify the beam (levels 3 and 2 in this case) must be relatively slow compared to other transitions. That is, the lifetime of level 3 must be relatively long compared to the lifetimes of levels 4 and 2. As mentioned above, if saturation of the gain is to be avoided and steady amplification is to occur, the circulation of particles from levels 2 to 1 to 4 to 3 must be fast compared to the transition rate from level 3 to 2 caused by the beam that is being amplified. Therefore, the pump must be strong enough to cause a high transition rate from level 1 to level 4. In the limiting case of a very strong pump, which produces transitions between levels 1 and 4 at an extremely rapid rate, the steady-state amplification factor is limited by the transition rates from level 4 to 3 and from level 2 to 1.

4. PROPERTIES OF A LASER BEAM

The resonant cavity produces a laser beam with the following properties: (a) directionality, (b) high radiance, (c) monochromaticity, and (d) coherence. After introductory comments about each of these properties, we will discuss in more detail how the cavity controls them. These sections provide the basis for understanding the important temporal properties of a laser beam, which are discussed in Section 6.

Typical values for some properties of particular types of lasers are given in Table 2.4 and Tables 17.2 and 17.3.

4.1. Directionality

A laser beam is confined to a narrow cone of angles, from a few tenths of a milliradian to a few milliradians. This directionality is the result of the mirrors of the resonant cavity specifically defining the path for the beam to take, Fig. 1.3. As we will later see, directionality is limited by optical diffraction and the presence of undesired off-axis modes.

4.2. Radiance

Radiance is the radiant power (W) per unit solid angle (sr) per unit surface area (cm^2) that is emitted from an infinitesimal area of a surface into a cone perpendicular to the surface. Because of its high directionality and radiant power, the radiance of a laser is typically 1 to 10 orders of magnitude higher than the radiance of the most intense incoherent sources.

4.3. Monochromaticity

Monochromaticity means that the laser beam covers only a narrow range of frequencies or wavelengths. The bandwidth of a laser beam is typically many times to several orders of magnitude narrower than the bandwidth observed by spontaneous emission. Monochromaticity is due mainly to the higher gain of the optical amplifier at the central wavelength of the active transition of the amplifier, and to the wavelength selectability of the resonant cavity. Lasing occurs at wavelengths for which there are an integral number of half wavelengths between the two mirrors of the cavity. Additional wavelength selection devices (gratings, prisms, filters) are often added to the resonant cavity to cause high losses for undesirable wavelengths, as shown in Fig. 1.4.

4.4. Coherence

Coherence is the orderly relationship of one part of the beam to another part. Two types of coherence can be distinguished, spatial and temporal. Consider

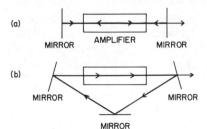

Fig. 1.3. Laser cavities with (a) a conventional two-mirror design and (b) a three-mirror ring design.

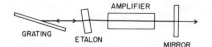

Fig. 1.4. A laser cavity with wavelength selection components. The grating acts as one mirror of the cavity.

any two points in a beam. If the electromagnetic fields at both points oscillate in the same manner (i.e., have a fixed phase relationship to each other), then the fields at the two points are coherent. If the fields at all points in the beam are coherent, then the beam is said to be spatially coherent.

If at any point in space, the electromagnetic field oscillates in a stable, predictable manner from one point in time to any other point in time, then there is perfect temporal coherence. Temporal and spatial coherence are interrelated for a traveling laser beam because there will be temporal coherence at any point that is passed by a beam that is spatially coherent. If coherence is valid only for a limited time interval τ_c, then the beam is *partially* coherent, with a coherence time τ_c. Coherence is related to monochromaticity; a wave with a bandwidth $\Delta\nu$ has a coherence time $\tau_c \simeq 1/\Delta\nu$.

5. INFLUENCE OF THE CAVITY UPON BEAM PROPERTIES

The spatial, spectral, and temporal properties of a laser beam are ultimately related to each other. A beam with a greater divergence caused by its mode structure will have a wider spectral bandwidth. Ultrashort pulses require a laser capable of lasing over a wide spectral bandwidth. The influence that the cavity has upon these relationships will now be considered.

5.1. Spatial Modes

The spatial structure influences the divergence angle and the focal properties of the beam. The laser cavity supports certain stable spatial configurations, called spatial modes, of the electromagnetic field of the laser beam. The spatial configurations can be described mathematically by solving the fundamental Maxwell equations with boundary conditions imposed by the cavity. The electromagnetic field distribution in the cavity must replicate itself upon round trip reflection by the mirrors. The simplest mode is a standing wave formed by the superposition of two waves that propagate in the cavity in opposite directions. When the end mirrors of the cavity are electrically conducting, the standing wave must satisfy the boundary condition of zero electric field at the mirror surfaces. Consequently, the resonator frequencies are determined by how many half wavelengths there are between the two mirrors. For a cavity with two plane parallel mirrors, the resonant frequencies ν are given by (1)

$$v = \frac{nc}{2d}\left[m^2 + \left(\frac{dj_{q,m}}{\pi a}\right)^2\right]^{1/2} \qquad (1.6a)$$

$$\doteq \frac{mnc}{2d} \qquad (1.6b)$$

where n is the refractive index of laser material, c is the velocity of light, d is the distance between the mirrors, m is a large integer (equal to the number of half wavelengths of the light contained between the mirrors), a is the radius of the cylindrical cavity and $j_{q,m}$ is the mth root of the first derivative of the ordinary Bessel function of the first kind and order q. In practice, m^2 is usually much larger than the term in parentheses and the approximation given by Eq. (1.6b) is usually used. The term in parentheses in Eq. (1.6a) accounts for off-axis modes that propagate at slight angles to the optical axis and increase the divergence of the beam.

5.2. Concave Mirror Cavities

Because of diffraction, an electromagnetic wave spreads out as it travels through the cavity, causing some of the energy to leave the cavity. Diffraction losses can be reduced by using spherical concave mirrors to focus the spreading beam back into the cavity. A cavity made with spherical mirrors is said to be *stable* if rays are refocused so that their energy remains in the cavity. When the spherical mirrors have radii R_1 and R_2 and are spaced a distance d apart, the cavity is stable if

$$0 < \left(1 - \frac{d}{R_1}\right)\left(1 - \frac{d}{R_2}\right) < 1$$

A stable cavity is needed if the amplifier gain is low, as it is with some gas lasers. An unstable cavity (a misleading term) will lase if the amplifier gain is high enough to compensate for diffraction losses. Unstable cavities have been used to improve beam quality because diffraction trims away the outer ragged edges of the beam, causing the center of the beam to be emphasized, and it helps to overlap heterogeneous regions of the beam.

5.3. TEM Modes

When the spherical mirrors are in a confocal arrangement (where the surface of each mirror is located at the center of the radius of curvature of the opposite mirror), the resonant frequencies are given by

$$\nu_{ijk} = \frac{nc}{4d} [2(k + 1) + i + j + 1] \tag{1.7}$$

where i and j are small integers related to the x and y rectangular coordinates perpendicular to the optical axis, and k is the number of field zeros of the standing-wave pattern along the optical axis between the end mirrors ($k + 1$ is the number of half wavelengths). The value of k is generally not given in the designation for the modes, which are termed TEM_{ijk} or simply TEM_{ij} modes (transverse electromagnetic). The axial modes correspond to $i = j = 0$. When i or j is greater than zero, the spatial cross section of the beam has more than one intensity maximum and the values of i and j equal the number of intensity minima that occur in the direction of their corresponding rectangular coordinate axes.

5.4. Mode Spacing and Mode Pulling

The spacing $\Delta \nu_s$ between the resonant frequencies of the axial modes of a cavity is obtained by subtracting the frequency of one mode from that of its adjacent mode. For a plane mirror cavity Eq. (1.6b) gives

$$\Delta \nu_s = \frac{c}{2d} [n_{m+1} + m(n_{m+1} - n_m)] \tag{1.8}$$

where n_{m+1} and n_m are the refractive indices for the respective modes. Notice that mode spacing depends on the separation of the mirrors and is independent of the large integer m when the refractive indices are the same.

The refractive index varies rapidly near the spectral peak of an atomic or molecular transition in such a way that n_m (for the longer wavelength mode) is greater than n_{m+1}. Therefore, Eq. (1.8) shows that the modes near the peak of the amplifying transition will be more closely spaced than the modes farther from the peak. This phenomenon is called mode pulling because the modes near the peak appear to be pulled toward the center of the peak rather than being equally spaced.

5.5. Real Cavities

Equations (1.6) and (1.7) were derived for ideal cases. In practice, the real mode structure of the beam is complicated by imperfections in the reflecting surfaces and by optical inhomogeneities in the active medium, which may cause further complications by varying during a pump pulse, or from one pulse to the next.

5.5.1. Pumping Homogeneity

The quality of a laser beam is influenced by the way that the active amplifier medium is pumped. Without additional apertures or mode selection devices, the cross-sectional energy distribution of the laser beam will roughly correspond to the cross-sectional distribution of the pumping rates within the amplifier medium. The pumping rate for an optically pumped rod depends on the concentration of absorbing species (active sites) within the rod as well as the spatial distribution of the pumping light. Even when embedded in a region of uniform radiation density, the pumping within a cylindrical rod is not uniform because of refraction of the light at the surface of the rod. Pump light rays that are nearly tangent to the surface are bent toward the center of the rod by refraction so that the center is pumped more strongly than the outer region (2). For pulsed operation the most highly pumped regions will be the first to lase. As the pumping increases during the pulse, the other regions will lase and more off-axis mode structure may occur. Placing an aperture in the laser cavity helps to minimize this effect.

A low-absorption rod can be uniformly pumped if the rod is clad with an appropriate nonlasing medium so that there is no change in refractive index as the pumping light enters the active region of the clad laser rod. Another way to produce more uniform pumping is to roughen the rod surface by rubbing it with a diamond powder. This technique reduces pumping efficiency because of diffuse reflection of the pumping light.

5.5.2. Concentration Effects

When the concentration of lasing sites is high, absorption becomes significant, and a larger fraction of the pumping radiation is absorbed near the edge of an optically pumped amplifier medium. Eventually, as the concentration is increased, the ring near the circumference of a cylindrical rod is pumped above lasing threshold so that the outer ring appears in the laser beam with a dark ring between it and the central lasing region. Pumping efficiency in the center of the rod is reduced so that the laser beam has a weak center surrounded in turn by a bright ring, a dark ring, and another bright ring.

5.5.3. Other Considerations

When an amplifier medium has an inhomogeneous concentration of lasing sites, or has optical imperfections, yet is uniformly illuminated, the medium cross section breaks up into separate lasing filaments, where the optical gain is highest. These filaments produce a heterogeneous beam.

Homogeneous optical pumping is important to minimize thermal distortions of the amplifier medium since most of the pump light ends up as thermal energy. Some thermal distortions can be compensated. Those that cannot will cause optical deterioration of the laser cavity and the quality of the resulting laser beam.

Additional modes of oscillation when the amplifier gain is high may result from secondary cavities caused by reflections from extra surfaces within the main cavity. Feedback from extra surfaces can be reduced by designing the surfaces so that they are at a slight angle with respect to the optical axis, or by using antireflection coatings (which must be designed to withstand the power of the beam). Reflections can be prevented for a polarized beam by designing the surfaces to be at the Brewster angle (the critical angle at which a beam of proper polarization is not reflected).

5.6. Near- and Far-Field Profiles

The diameter of the beam inside a cavity at some distance from concave mirrors is smaller than the diameter at the mirrors, and reaches a minimum called the waist. The minimum angle of divergence of the beam in radians is equal to a number between 0.6 and 1.3 times the ratio of the wavelength divided by the dimensions of the waist. The number depends on the shape of the beam and the definition of the radius.

After the output beam leaves any cavity, it will continue to spread because of diffraction. Fresnel diffraction describes the pattern of the cross section of the beam up to some critical distance away from the output aperture of the laser. This is called the near-field region. The region beyond about 10 times the critical distance is the far-field region, where Fraunhofer diffraction describes the intensity distribution. The Fraunhofer diffraction pattern for a cylindrical beam is the familiar Airy circle, which has a second-order Bessel function irradiance distribution.

The critical distance is given approximately by the ratio of the area of the output aperture of the beam to the wavelength of the light. For the general case, and even for a beam that starts out with a uniform intensity distribution, the intermediate region between the near-field and far-field regions is more complex to describe than can be done using the limiting cases of Fresnel and Fraunhofer diffraction. However, for the special case of a Gaussian beam, the cross-sectional irradiance distribution is greatly simplified. Although a Gaussian beam increases in diameter as it travels, its cross-sectional profile remains Gaussian at all distances from its origin. A beam with a Gaussian profile may be obtained by passing a beam through a long series of coaxial circular apertures of the same diameter.

5.7. Focal Spot

A Gaussian beam can be focused to a very sharp spot, which will also have a Gaussian profile if optics that are limited only by diffraction are used to focus the beam. If the radius of a Gaussian beam is defined as the radius where the irradiance of the beam has fallen to $1/e^2$ (13.5%) times the peak axial value, then the radius R_0 of the focal spot will be

$$R_0 = \frac{\lambda L}{\pi R} \tag{1.9}$$

where R is the radius of the original beam, L the focal length of the lens, and λ the wavelength of the beam. The size of the focal spot will increase in proportion to the number of axial modes that are lasing. Off-axial modes will produce even further increases in the focal spot size. For a beam of arbitrary intensity pattern, the focal spot pattern will be approximately the same as the far-field pattern of the beam.

6. TEMPORAL MODES

The temporal operating modes of lasers can be subdivided into five groups: (a) continuous wave (cw), (b) normal pulsed (or simply, pulsed), (c) Q switched, (d) cavity dumped, and (e) mode locked. Continuous-wave operation can be achieved only if the lifetime of the upper laser energy level is longer than that of the lower level. If it is not, the lasing transition is called self-terminating, and it can produce amplification and lasing only in a pulsed mode where the pump pulse is shorter than the upper-level lifetime. Lasers that could, in principle, operate in a cw mode are sometimes operated in a small pulsed mode. A normal pulsed laser is one that has a pulsed pumping system and a cavity with no special devices to vary its losses. The shape of the envelope of the laser pulse (or series of shorter pulses) approximately follows the shape of the pump pulse. A normal pulse may contain a series of shorter pulses if the cavity gain reaches a value during the pulse that is so high that the beam quickly depopulates the excited states each time lasing threshold is reestablished during the pump cycle.

6.1. Q Switching

High-power pulses with a duration of only a few nanoseconds are made possible with a Q-switched laser. The resonator quality (Q) of the cavity is reduced by introducing an intensity-dependent or time-varying loss so that lasing will not take place until after the gain of the inverted population has been pumped to a very high value. When a large population inversion is reached, the losses of the cavity are suddenly reduced, thereby switching the Q of the cavity to a high value. Then the laser pulse grows quickly to a value that rapidly depopulates the excited state to the point where lasing is no longer sustained. The growth and termination of the pulse takes only a few nanoseconds.

Q switching is accomplished by several methods including a synchronized rotating prism or mirror used as one of the cavity mirrors, an electronically controlled optical switch such as a Pockels cell or an acousto-optical deflector, or simply a saturable absorber. A Pockels cell requires a polarized beam whereas an acousto-optical deflector does not. The Pockels cell rotates the polarization of the beam in proportion to an applied voltage. An acousto-optical deflector sets up an acoustic standing wave in a crystal when a radio-frequency signal is applied to a transducer attached to the crystal. The wave causes the refractive index of the crystal to vary in a grating-like pattern, which causes the beam to deflect when the radio-frequency signal is present. An example of a saturable absorber is a dye that has a strong absorption band in the lasing region. A cell containing the dye at a concentration that allows say 50% of the light to pass through is placed in the cavity. As the intensity in the cavity builds up, more excited states of the dye quickly become filled, leaving fewer absorbing states, until the dye is saturated or bleached. The cavity has then been switched to full gain and the laser pulse grows quickly to a value that rapidly depopulates the excited states of the amplifier to the point where lasing is no longer sustained.

6.2. Cavity Dumping

In a cavity-dumped laser, the beam energy is allowed to build up within a laser cavity that has 100% reflecting mirrors, and then the beam is suddenly diverted out of the cavity to the experiment. One method of diverting the beam out of the cavity is to suddenly switch on the diffraction grating formed in a standing acoustic wave Bragg cell located in the cavity (3). All of the energy in the beam can be extracted in a pulse of duration equal to the round-trip time for light in the cavity (about 6 ns in a 100-cm cavity). The repetition rate depends on the recovery time for the laser amplifier, and typically approaches 1 MHz.

6.3. Mode Locking

Mode phase locking, or just *mode locking*, generates trains of the shortest possible laser pulses (femtosecond pulses have been achieved) with very high power (more than 100 MW). The active medium of the amplifier must have a broad enough spectrum to amplify many modes (frequencies) in a laser cavity. The output of the laser is the result of the superposition of the electromagnetic waves of each of the lasing modes. In a non-mode-locked laser, the frequencies, phases, and amplitudes of these modes vary with time, and the output of the laser fluctuates quasirandomly. If a cavity perturbation forces the various mode oscillations to maintain fixed phases relative to one another, then the output becomes a repetitive train of pulses, and the laser is said to be mode locked. Figure 1.5 illustrates the formation of mode-locked pulses (bottom trace) by showing the result of the superposition of nine sine waves of equal amplitude (only every other one of the nine sine waves is shown to minimize cluttering of the figure; additional examples are shown in Chapter 14, Fig. 14.3). The resultant electric field wave (not shown) is formed by adding the electric fields of the individual waves together at each point in time. The emission intensity shown at the bottom of Fig. 1.5 is the square of the resultant electric field wave. By studying the equation for the resultant wave in a laser cavity, one finds that the pulses are spaced apart by the round-trip time for light in the cavity. The laser oscillation can therefore be interpreted as an ultrashort pulse propagating back and forth in the cavity, the output beam being a train of short pulses spaced in time by the round-trip time. As expected from Fourier transform theory, the width of the pulse is approximately equal to the inverse of the difference in frequency be-

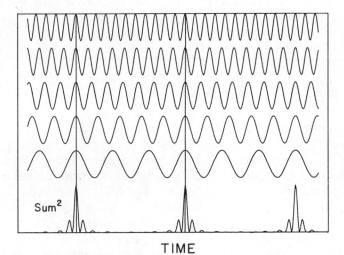

TIME

Fig. 1.5. Computer simulation of mode-locked sine waves. The maximum intensity (square of the sum of the sine waves), bottom trace, occurs periodically when the peaks of all sine waves coincide. For clarity, the relative range of frequencies in this example is much larger than in a mode-locked laser, and only every other one of the nine sine waves used to generate the bottom trace is shown.

tween the highest frequency and lowest frequency modes whose phases are locked together. The shortest pulses therefore require a laser capable of lasing over a broad spectral band.

The cavity peturbation required to force the phases of modes to lock together can be obtained in several ways discussed in the literature (4). Usually the losses or gain of the laser cavity are modulated at a frequency equal to the mode spacing $\Delta\nu$. This is the beat frequency of adjacent modes, and the rhythm of the cavity at this frequency causes the modes to become phase locked. Loss modulation is achieved, for example, by a saturable absorber that has a fast recovery time, or by an electronically driven device such an acoustic wave Bragg cell. Loss modulation is less effective than gain modulation when a laser is operated far above threshold (5). Gain modulation or synchronous pumping is achieved by pumping the amplifier with a mode-locked laser that has a mode spacing equal to the mode spacing of the gain-modulated laser, as described in more detail in Chapter 14. An example is the use of a mode-locked argon ion laser to pump a dye laser, in order to achieve a shorter pulse. Since the spectral bandwidth of the argon laser is realtively narrow, the mode-locked pulse width is relatively long. However, a dye laser has a wide spectral bandwidth, and can generate subpicosecond pulses.

Single pulses can be extracted from a mode-locked train of pulses with an electro-optic or acousto-optic modulator outside of the cavity. Cavity dumping, however, is a more efficient way to extract single pulses because nearly all of the energy that has built up in the amplifier can be extracted in each pulse. An example of a synchronously pumped, cavity-dumped dye laser is found in Harris et al. (6). Pulse rates up to 500 kHz were obtained without loss of pulse energy.

7. WAVELENGTH SELECTION

The wavelength range over which a laser operates depends on the spectral range over which the laser amplifier medium and cavity has a net gain, and the optical length of the cavity, which determines the wavelengths of the longitudinal modes. When it is desired to restrict the lasing wavelength of a laser that can lase over a broad spectral bandwidth, wavelength selection devices are placed in the cavity to reduce the gain of the cavity below lasing threshold for all the desired wavelengths. For example a grating may be used as one of the end mirrors of the cavity, as shown in Fig. 1.4, or a prism or birefringent filter may be placed within the cavity. These methods reduce the spectral bandpass to about 0.1 nm. Further reductions are obtained by placing a Fabry–Perot interferometer (or etalon) in the cavity. This etalon has two dielectric reflecting surfaces usually spaced less than 1 cm apart. As with the laser cavity, this etalon has a longitudinal mode spacing (free spectral range) that allows only those wavelengths to pass through that can fit an integral number of half wavelengths between the reflecting surfaces. Since other wavelengths are reflected rather than passed, the etalon is tilted slightly to divert the reflected rays out of the cavity. Figure 1.6 shows schematically how the lasing region is first restricted by a coarse tuning element (top of figure), and how further restriction is obtained by an inner cavity etalon. The vertical axis indicates the transmittance of the wavelength selection devices. The many wavelengths that the laser mirror spacing allows to lase are indicated by the multiple peaks in the center of Fig. 1.6. The bottom of the figure shows the mode that is allowed to lase at λ_1 when the laser threshold is at the level indicated by A. Reducing the lasing threshold to B, for example by increasing the pumping rate, allows two modes to lase, at λ_1 and λ_2.

Even when the wavelength selection devices and pumping rate are carefully chosen to encourage single-mode laser operation, optical imperfections in the laser cavity will allow the laser to lase not only at exactly the wavelength of a longitudinal mode but at wavelengths nearby. Typical imperfections include imperfect optical surfaces, variations in the refractive index of the lasing medium and other material in the cavity (including air) and mechanical vibrations. A

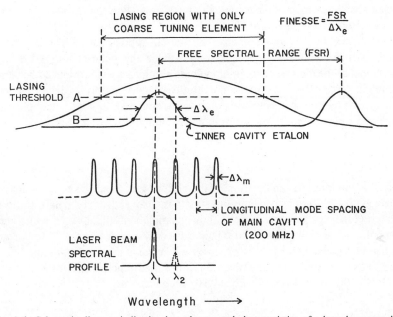

Fig. 1.6. Schematic diagram indicating how the spectral characteristics of a laser beam are determined by the lasing threshold and the bandpass characteristics of a laser cavity containing an internal etalon and a coarse tuning element such as a birefringent filter. From Piepmeier (11); reproduced by permission.

vibration isolation table may be necessary, as well as a means of obtaining thermal stability to reduce slow mechanical and refractive index changes.

8. NONLINEAR OPTICAL EFFECTS

Nonlinear optical effects have important applications in both analytical measurements and in generating laser wavelengths. Laser beams with wavelengths that are not present in the amplifier medium can be generated with nonlinear optical effects. For example, crystals that respond nonlinearly to the electric field in a laser cavity are commonly used to generate laser beams with two and three times the frequency of the fundamental laser beam. This is called frequency doubling or harmonic generation. Two laser beams can be mixed together in a nonlinear medium to generate laser beams with frequencies that are sums and differences of the frequencies of the original beams. (See Chapter 6, Section 2.1 for the parametric oscillator, where the frequency is halved.)

These same nonlinear effects are used to make analytical measurements that are only possible with the high irradiances that lasers provide. We shall briefly discuss here the basis for nonlinear effects and the production of new laser wavelengths. Uses of nonlinear optical effects in analytical measurements will be considered in later chapters.

If an electric field is applied across a material, a polarization of the electric charges will be induced. Although at low fields the polarization is linearly proportional to the electric field, at high fields nonlinearity is introduced because the binding force of the electron is not strictly proportional to its displacement. Although the relationship between polarization and electric field is complex, it is common to describe the relationship as an infinite power series:

$$P = \chi^{(1)}E + \chi^{(2)}E^2 + \chi^{(3)}E^3 + \ldots \qquad (1.10)$$

where P is the induced polarization, E the electric field, and the coefficients $\chi^{(i)}$ are called the ith-order susceptibilities ($\chi^{(1)}$ is also called the polarizability). Equation (1.10) can be used to reveal how a simple sinusoidal wave is distorted by a nonlinear medium. The electric field caused by a laser beam with frequency of ω can be represented by $E = E_0 \sin \omega t$ where E_0 is the maximum amplitude of the sine wave and t is time. When this sinusoidal function is substituted into Eq. (1.10), the second term in Eq. (1.10) becomes

$$\frac{\chi^{(2)} E_0^2}{2}(1 - \cos 2\omega t) \qquad (1.11)$$

which shows that the induced polarization is the sum of a constant term and a term oscillating at frequency 2ω. The constant term indicates that a dc voltage is produced across the crystal by the incident laser beam. Because an oscillating polarization corresponds to oscillating electric charge, the oscillating term in Eq. (1.11) reveals the existence of an electromagnetic field radiating at the second-harmonic frequency 2ω of the incident laser beam.

Constructive interference causes the second-harmonic wave to have the same directional properties as the incident beam because the induced polarization oscillations are in phase with each other throughout the region that is irradiated by the laser beam. Phase matching is not obtained automatically, because the refractive indices of the crystal are usually different for the two wavelengths. The methods used to achieve phase matching (such as using the different refractive indices for beams having different polarizations in a birefringent crystal) usually result in a small angle between the two beams.

In a similar manner, the third-order susceptibility term of Eq. (1.10) generates a wave at a frequency 3ω with similar directionality properties as the incident laser beam.

When two laser beams of different frequencies are mixed together in a crystal that has good nonlinear properties, the polarization oscillates with frequency components that are sums and differences of the two frequencies. Laser beams can be generated at these new frequencies when the medium allows phase matching at the frequencies involved.

In a medium in which the polarizability is changing at some frequency ω_v (e.g., because of a molecular vibration), and the polarization is caused to oscillate at the frequency ω_L of a laser beam, the polarization, which is the product of the electric field of the laser and the polarizability, will vibrate with frequency components at the sum and difference frequencies $\omega_L + \omega_v$ and $\omega_L - \omega_v$. These frequencies are the anti-Stokes (sum) ω_A and Stokes (difference) ω_S Raman scattering frequencies. When the laser is intense enough, the electric field produced by the photons at ω_S, as well as at ω_L, will be strong enough to synchronously drive the molecular vibrations and the photons at ω_S will produce a coherent laser beam. This new beam can produce another new beam at the second-order Stokes frequency, $\omega_S - \omega_v$, and so forth, so that laser beams with frequencies $\omega_L - n\omega_v$ can be produced, where n is an integer. In a similar manner, the anti-Stokes lines produce laser beams with frequencies $\omega_L + n\omega_v$.

These nonlinear methods are useful in helping to produce coherent beams in spectral regions where laser lines are otherwise unavailable.

9. EXAMPLES OF LASERS

9.1. Solid-State Lasers

Ruby is an example of an ionic crystal laser, which uses a crystal of Al_2O_3 doped with 0.05% (by weight) Cr_2O_3. The chromium ion is the active species and has an absorption band with wavelengths from about 500 to 600 nm that can be used by an optical pump. The energy level system is essentially a three-level system. The ions that are excited by the pump quickly decay to a fluorescing energy level that has a lifetime of 3 ms and causes amplification at 694.3 nm $\pm$ 0.2 nm. The 3-ms lifetime facilitates pumping with flashlamps. Millisecond trains of many submicrosecond pulses are obtained when millisecond flashlamp pump pulses are used and the cavity is not Q switched.

A neodymium laser is a common solid-state ion laser that uses the Nd^{3+} ion in a suitable glass or a host crystal such as yttrium aluminum garnet (YAG), $Y_3Al_5O_{12}$. Other rare-earth ions have also been used. The absorption bands for pumping the Nd^{3+} lasers are in the red region of the spectrum, while the wavelength that is amplified is in the near infrared, 1064 nm, with a lifetime of about 0.5 ms. The energy levels involved form a four-level system, which facilitates continuous amplification and lasing with a gas discharge lamp pump.

An F-center laser uses as the amplifier medium an alkali halide crystal containing F_2^+ lattice defects (where two adjacent anion vacancies share one electron). They lase in the near IR region, where three different crystals are required to cover the range from 1200 to 3500 nm. They must be cooled below 50 K for high quantum yields and to prevent deionization of the F-centers by diffusion of electrons from other locations. The crystals are pumped by visible lasers, although the longer wavelength lasers can be pumped by the 1064-nm line of neodynium laser. The lasing bandwidth is broad because there are many lattice vibrations that couple to the F-center transition. Prisms and gratings are used for wavelength selection.

Semiconductor lasers made with *pn* junctions provide narrow bandwidths for high-resolution spectroscopy in the infrared region from 4 to 32 μm, as well as the red and near infrared regions (650–840 nm). Gallium arsenide salts are used in the red and near infrared regions. Lead salts at cryogenic temperatures are used in the infrared region. The position in the infrared region of the rather broad spectral band over which lasing can occur is controlled by the salt composition, which is usually not stoichiometric. Semiconductors have been made to lase in the ultraviolet region, but require pumping with an external source such as bombardment with a beam of high-energy (>20 keV) electrons.

The energy of the lasing photons is close to the energy bandgap that lies between the bottom of the conduction band and the top of the valance band of the semiconductor material. When a *pn* junction is forward biased with a high enough voltage, electrons from one side of the junction are injected fast into the conduction band of the material on the other side at a rate fast enough to produce a population inversion. The highest optical gain over a reasonable distance is in the plane of the junction. The opposite ends of the semiconductor perpendicular to the junction are polished flat and parallel to make the laser cavity. The refractive index of the semiconductor is often high enough so that the Fresnel reflection at the ends permits lasing to occur without additional coatings on the ends. The spectral bandwidth of the emitting amplifier medium is narrow so that wavelength selection devices may not be required. Tuning is achieved by changing the frequency of the narrow emitting band. Tuning over a broad spectral range is possible and is commonly done by changing the applied voltage, which changes the energy gap, and by changing the temperature, which causes the wavelength of a mode to shift, mainly because of the resulting change in refractive index. Changes in pressure, up to 15 kbar with the help of a compact high-pressure liquid cell, will also change the lasing frequency.

9.2. Dye Lasers

Dye lasers use suitable dyes, usually in liquid solvents, as the active amplifier material. They are effectively four-level systems that use as their lasing transitions the fluorescence transitions from the first excited singlet states to the upper ground states. Nanosecond fluorescence lifetimes are typical. The molecule is pumped into upper singlet states, which decay in picosecond times to the first excited singlet state. The band of wavelengths that can lase is essentially as broad as the spectral fluorescence band of the dye. Transitions from the triplet state, phosphorescence, are not used because these strongly forbidden transitions would usually require too high a concentration of dye to provide adequate gain for lasing. Triplet states must be considered, however, because lasing can be stopped by transitions (intersystem crossings) from the singlet state to the triplet states.

Lifetimes of the triplet states are so long that in a short time after pumping begins, a significant fraction of the total population can end up in the triplet states, causing the population of the singlet state to fall below that needed for lasing. A strong pump, such as an argon ion laser, is needed to maintain sufficient inversion for cw operation. For pulsed operation submicrosecond rise times of the pumping pulse helps to ensure that lasing begins before the triplet state steals too much of the population. Flashlamp-pumped dye lasers typically are pumped with submicrosecond flashlamp pulses. Other lasers are commonly used to pump dye lasers. Typical characteristics of laser-pumped dye lasers are given in Table 2.4 .

The detrimental effect of the triplet state can be considerably reduced by the addition of triplet-state quenchers such as molecular oxygen (7), provided they are not also as effective at quenching the singlet state (8). Quenchers quickly return triplet-state molecules to the ground state where they again become active in the lasing cycle. Cyclooctatetraene is particularly effective as a triplet-state quencher for rhodamine 6G dye molecules (9).

9.3. Gas Lasers

Gases have narrow absorption bands or lines because line-broadening mechanisms are less effective than in condensed phases. Therefore optical pumping with flashlamps is not efficient except in very special cases where the pump has narrow lines that happen to overlap the narrow absorption peaks. Electrical discharges are therefore commonly used to pump the atoms, ions, or molecules in a gas optical amplifier. An electrical discharge can be sustained by a strong electric field that accelerates ions and free electrons. Excitation of the amplifying particles occurs via energy transferred during one or more collisions. For example, inverted populations for amplifying some infrared lines can be established in noble-gas atoms by collisions of the *first kind*, where energy is transferred to an atom when it collides with a free electron that has been accelerated by the electric field:

$$e^- + X \rightarrow X^* + e^-$$

where X^* represents the particle in the excited state. When the gas has more than one constituent, the amplifying particle can be excited by energy transfer during a collision with another excited species:

$$A^* + X \rightarrow A + X^*$$

The most effective examples are those where the energy levels of the two particles are about the same and when the excited state of A is metastable, that is, it has a long lifetime. An important deexcitation mechanism in a gas laser is collisions with the walls of the container. Consequently, the gas pressure and the diameter of the container are important design considerations.

In the popular helium–neon laser, He atoms are excited to metastable states by collisions with electrons. They transfer their energy by collisions to the lasing neon atoms to produce a population inversion. The lower state of the visible 632.8-nm lasing transition is rapidly depopulated to the ground state by wall collisions.

The noble-gas ion lasers are the most powerful visible lasers that operate in the cw mode. Ions are generated by collisions with electrons, and the ions are excited by further electron collisions. The discharge current densities must be very high to achieve the necessary inverted population of excited ions. The argon ion laser is the most common, with two strong lines at 488.0 and 514.5 nm, and others at 457.9, 476.5, 496.5, 501.7, 514.5, 528.7, and 1092.3 nm. Krypton is sometimes used because its transitions are spread across the visible spectrum from 468.0 to 647.1 nm.

The He–Cd and He–Se lasers are examples of metal vapor lasers. The He–Cd laser has lines at 325.0 and 441.6 nm and the He–Se laser has a number of visible lines. In the He–Cd laser an electrical discharge is formed in a mixture of He gas and Cd vapor. Metastable He atoms transfer their energy to Cd atoms by collisions. The result of the collision is an excited Cd ion. A population inversion is maintained because the ions decay faster from the lower levels than from the higher levels. In the He–Se laser helium ions collide with Se atoms and exchange their charge, leaving the Se ions in excited states. In both lasers wall collisions aid the return of the ions to the atomic ground states.

The CO_2 molecular laser has a high power efficiency (up to 30%) and high power in the infrared region of the spectrum at 10.6 μm. A competing band occurs at 9.6 μm. Pumping occurs via electron collisions or by collisions with excited nitrogen molecules that are purposely added to the gas. Lasing occurs between vibrational transitions, and the spectral range over which lasing can occur is increased by rotational energy levels associated with the vibrational energy levels. Helium is added to reduce the lifetime of the lower levels of the laser transition. Pulsed and cw operation are common.

The N_2 and H_2 lasers use vibrational transitions of the molecules for their lasing transitions. The N_2 laser emits at 337.1 nm in the ultraviolet and is commonly used to pump dye lasers. The H_2 laser emits in the vacuum ultraviolet at 161 and 116 nm. Since the lifetime of the upper level is shorter than that of the lower level, these lasers can only be operated in the pulsed mode.

Excimer lasers use excited noble-gas halides, which are reasonably stable while in the excited state. But because their ground-state potential energy surface is repulsive, they have a spectral bandwidth of about 2500 cm^{-1} (10). Excimer lasers are notable for their ultraviolet lines. Examples are XeCl, which lases at 308 nm, KrF, which lases at 249 nm, and ArF, which lases in the vacuum ultraviolet at 193 nm.

The development of lasers has been rapid, with commercial units not far behind the introduction of new concepts. New beam characteristics open up new measurement possibilities, not only to analytical chemistry but to many other fields. A laser built for one purpose may not be appropriate for another. In such a rapidly developing field, the analytical chemist must ensure that the choice of a laser meets all of the many requirements of the analytical application, including cost and reliability. When confronted with a list of specifications for a particular laser, it is important to know which specifications are met simultaneously. For example, does the maximum output power correspond to a single-mode beam structure or (more likely) to a multimode beam structure? Understandably, a new beam characteristic that opens up new measurement possibilities may excite both the manufacturer and the analytical chemist. However, both must work together to ensure that the *reproducibility* of the many characteristics of the laser is adequate to make reliable analytical measurements.

ACKNOWLEDGMENT

The author is very grateful to Professor C. Th. J. Alkemade and Professor E. S. Yeung for their careful reading of the original manuscript and their many helpful suggestions.

REFERENCES

1. S. Ramo and J. R. Whinnery, *Fields and Waves in Modern Radio*, Wiley, New York, 1953.
2. C. H. Cooke, J. McKenna, and J. G. Skinner, *Appl. Opt.*, **3**, 957 (1964).
3. D. Maydan, *J. Appl. Phys.*, **41**, 1552 (1970).
4. P. W. Smith, M. A. Duguay, and E. P. Ippen, "Mode-locking of Lasers," in *Progress in Quantum Electronics*, Vol. 1, Part 2, J. H. Sanders and S. S. Stenholm, Eds., Pergamon Press, London, 1974.
5. J. M. Harris, R. W. Chrisman, F. E. Lytle, and R. S. Tobias, *Anal. Chem.*, **48**, 1937 (1976).
6. J. M. Harris, L. M. Gray, M. J. Pelletier, and F. E. Lytle, *Molec. Photochem.*, **8**, 161 (1977).
7. F. P. Schafer and L. Ringwelski, *Z. Naturforsch.*, **28A**, 792 (1973).
8. J. B. Marling, D. W. Gregg, and S. J. Thomas, *IEEE J. Quant. Electr.*, **QE-6**, 570 (1970).
9. R. Pappalardo, H. Samelson, and A. Lempicki, *IEEE J. Quant. Electr.*, **QE-6**, 716 (1970).
10. J. J. Ewing and C. A. Bran, *Tunable Lasers and Applications*, A. Mooradian, et al., Eds., Springer-Verlag, Berlin, 1976, Chapter 2.
11. E. H. Piepmeier, "Atomic Absorption Spectroscopy With Laser Primary Sources," in *Analytical Laser Spectroscopy*, N. Omenetto, Ed., Wiley, New York, 1979.

BIBLIOGRAPHY

Alkemade, C. Th. J., Tj. Hollander, W. Snelleman, and P. J. Th. Zeegers, *Metal Vapours in Flames*, Pergamon Press, Oxford, 1982. Contains an excellent summary of the structures and spectra of atoms and molecules, as well as an advanced discussion of radiative transitions including line intensities, saturation, and bleaching.

Lengyel, B. A., *Lasers*, Wiley-Interscience, New York, 1971. An excellent introductory textbook, particularly good on laser principles.

O'Shea, D. C., W. R. Callen, and W. T. Rhodes, *Introduction to Lasers and Their Applications*, Addison-Wesley, Reading, MA, 1978. An introductory textbook with good diagrams.

Ross, D. *Laser Light Amplifiers and Oscillators*, Academic Press, London, 1969. An intermediate-level textbook with a compendium of laser transitions.

Siegman, A. E., *An Introduction to Lasers and Masers*, McGraw-Hill, New York, 1971. An excellent textbook written from an electrical engineering viewpoint.

Stitch, M. L., Ed., *Laser Handbook*, Vol. 3, North-Holland, Amsterdam, 1979. An excellent book on special topics such as excimer lasers, unstable resonators, pulsed dye lasers, and picosecond spectroscopy.

Verdeyn, J. T., *Laser Electronics*, Prentice-Hall, Englewood Cliffs, NJ, 1981. An interesting textbook written from an electrical engineering viewpoint.

PART

II

SELECTED METHODS THAT USE VARIOUS DETECTION SCHEMES

CHAPTER

2

LASER-EXCITED ATOMIC AND IONIC FLUORESCENCE IN FLAMES AND PLASMAS

J. D. WINEFORDNER

Department of Chemistry
University of Florida
Gainesville, Florida

N. OMENETTO

CCR, European Community Center
Stabilimento di Ispra
Analytical Chemistry Division
Ispra, Italy

Work supported by AFOSR-F49620-80-C-0005.

1. INTRODUCTION

Atomic fluorescence spectrometry (AFS) is based upon radiational excitation of atoms (produced in a suitable atomizer) and measurement of the consequent radiational deactivation (called fluorescence). Laser-excited AFS (LEAFS) involves the same processes except that the source of excitation is a laser, and in most cases of analytical interest, a dye laser. Atomic fluorescence spectrometry has been of considerable interest to researchers in atomic spectrometry because of its use for analytical and diagnostical purposes. However, no commercial AFS instruments are currently available except for the Baird hollow cathode lamp-excited inductively coupled plasma (HCL–ICP–AFS) system. There are no commercial analytical LEAFS instruments.

This chapter is concerned strictly with LEAFS. To facilitate the discussion of principles, instrumentation, analytical figures of merit, and applications of LEAFS and to conform to rather limited space requirements, the authors will use primarily figures and tables with minimal textual discussions. The authors have attempted to give a rather comprehensive survey of the literature available (up to about January 1, 1984). The references are divided into three main groups: Reviews (R); Fundamentals (F) of fluorescence; and Instrumentation and Methodologies (I) of LEAFS.

The reader is referred to the previous reviews (see references) on LEAFS. The considerable interest in LEAFS is apparent when one considers that it is within the realm of possibility to detect *selectively* single atoms in complex real-world samples. Alkemade (F52) in a *classic* paper has given an overview of the principles and achievements of various laser-based methods for single-atom detection (called SAD). He showed the ultimate detection limit is an intrinsic value dependent on the statistical fluctuations of the atoms within the observation volume. The intrinsic detection limit is ultimately the limit to quantitative trace analysis. Various experimentalists have approached, and in fact exceeded, single-atom detection within an observation volume that consists of either a beam oven or a pure vapor vessel, that is, a cell system often not of analytical importance. In more analytically significant cells such as flames, but especially electrothermal atomizers, detection limits of under 100 atoms/cm^3 are obtained and have been predicted by Omenetto and Winefordner (F33) and by Falk (F57).

LEAFS is not only a technique with high detection power but also one with high *spectral selectivity*. The high spectral selectivity is achieved because it is possible to excite the atomic vapor with a single, narrow-bandwidth line, thus minimizing problems associated with spectral interferences and stray light that often plague other atomic methods, for example, plasma atomic emission methods. LEAFS is certainly a technique worthy of consideration by analytical chemists, and it is hoped that this chapter will provide an impetus to others to work in this area of research.

In Tables 2.1 and 2.2, a brief review is given of the considerations concerning

Table 2.1. Why Are Dye Lasers of Interest in Atomic Fluorescence Spectrometry?

1. High spectral irradiance

   Optical saturation possible (especially in the fundamental region, >350 nm), and consequently:
   a. Improved limits of detection (assuming noise is not related to source intensity)
   b. Increased linear dynamic range (because of reduced self-absorption and prefilter effects)
   c. Increased precision (because of increased signal levels)
   d. Freedom from quenching effects

2. Narrow spectral bandwidth

   a. Stray light is minimal
   b. Spectral interferences are minimal since it is easy to excite *only* the wavelength of the element of interest and also to select the fluorescence wavelength.

3. Small beam size, beam collimation, narrow temporal pulses

   a. Optimal for small atom reservoirs, e.g., furnace
   b. Optimal for spatial profile of atoms in flames, plasmas, etc.

Table 2.2. Why Are No Commercial Laser-excited Atomic Fluorescence Spectrometers Available?

1. Complexity of dye lasers
   Frequency doubling, frequency narrowing, frequency mixing, Raman shifting, spatial matching, temporal matching, etc.
2. Expense of dye laser
   Equipment costs of ~$50,000–$120,000

   Operational costs of ~$1,000–$20,000/year depending on amount and kind of dye used, part replacement such as thyratrons, capacitors, etc.
3. Stability (amplitude) of dye lasers (pulsed)
   Pulse to pulse stability can be poor.
4. Wavelength scan difficulties
   Difficult to scan wavelength over more than one dye
5. Low UV spectral irradiances, E_λ
   In the ultraviolet region, where doubling techniques (for Raman shifting) must be used

   E_λ is in general less than that necessary to reach optical saturation
6. Size and special conditions
   Dye laser systems are often bulky, nonmobile, and demanding in terms of utilities (cooling and power requirements)

Table 2.3. Selected Projects on Fundamental Aspects of Laser-excited Fluorescence Spectrometry

Reference	Intensity Expressions	Source[a]	Ys and ks	τs and ks	n_Ts	Line Broadening	Other[b]
F1	—	Both	—	—	—	—	Fundamentals
F2	—	Both	—	—	—	—	Fundamentals
F3	—	Both	—	—	—	Yes	Fundamentals
F4	Yes	Both	Yes	Yes	Yes	Yes	Fundamentals
F5	Yes	Both	Yes	Yes	Yes	Yes	Edited book
F6	—	Both	—	—	—	—	Fundamentals
F7	Yes	Both	Yes	Yes	Yes	Yes	Edited book
F8	—	Both	—	—	—	—	Fundamentals
F9	Yes	Both	Yes	Yes	Yes	Yes	Key paper
F10	Yes	Cont	Yes	Yes	Yes	No	Key paper
F11	Yes	Cont	Yes	Yes	Yes	No	—
F12	Yes	Line	Yes	No	Yes	No	Key paper
F13	Yes	Line	Yes	No	Yes	No	Key paper
F14	No	No	No	No	No	No	Designation of AF transitions
F15	Yes	Cont	Yes	No	Yes	No	Key paper
F16	Yes	Cont	Yes	No	Yes	No	Shapes of curves of growth
F17	Yes	Line	Yes	No	Yes	No	—
F18	Yes	Cont	Yes	No	Yes	No	—
F19	No	Both	No	Yes	Yes	Yes	Key paper
F20	Yes	Cont	Yes	No	Yes	No	—
F21	Yes	Cont	Yes	Yes	Yes	No	Key paper
F22	Yes	Cont	Yes	No	Yes	No	Exptal verification
F23	Yes	Cont	Yes	No	Yes	No	Exptal study
F24	No	Cont	No	No	No	Yes	Excitation profile
F25	No	None	No	No	No	No	S/N expressions for AFS

Ref							Description
F26	No	Yes	No	Yes	Cont	Yes	—
F27	No	Yes	No	Yes	Cont	Yes	General paper
F28	No	Yes	No	Yes	Cont	Yes	—
F29	No	No	No	No	Cont	No	—
F30	No	Yes	No	Yes	Cont	Yes	Derivation of two-level/three-level intensities
F31	Yes	No	No	No	Line	No	Laser saturation broadening
F32	No	Yes	No	Yes	Cont	Yes	—
F33	Yes	Yes	Yes	Yes	Both	Yes	General treatment of AFS
F34	—	Yes	—	Yes	Both	Yes	Saturated absorption profiles
F35	Yes	No	No	Yes	Cont	No	—
F36	No	Yes	Yes	Yes	Cont	Yes	Saturation curves
F37	No	Yes	No	Yes	Cont	Yes	Exptal study on Ne microwave discharge
F38	No	Yes	No	Yes	Cont	Yes	Ionization in flames
F39	No	Yes	No	Yes	Cont	Yes	Ionization in flames
F40	No	Yes	No	Yes	Cont	Yes	—
F41	No	Yes	No	Yes	Cont	Yes	—
F42	No	No	Yes	Yes	Cont	Yes	Exptal verification
F43	No	No	No	No	Cont	Yes	Fluorescence ratio of Na-D lines
F44	No	Yes	No	Yes	Cont	Yes	Compound formation in excited state
F45	No	Yes	No	Yes	Cont	Yes	OH studies
F46	No	Yes	No	Yes	Cont	Yes	OH studies
F47	Yes	No	No	Yes	Cont	Yes	Laser saturation broadening

Table 2.3. (*Continued*)

Reference	Intensity Expressions	Source[a]	Ys and ks	τs and ks	n_Ts	Line Broadening	Other[b]
F48	Yes	Cont	No	No	No	No	Ionization by one and two photons
F49	Yes	Cont	Yes	No	No	No	Extended model for saturation in two-level atom
F50	Yes	Cont	Yes	No	Yes	No	Experimental verification
F51	Yes	Both	Yes	No	Yes	No	Ionization of vapors
F52	No	Both	No	No	Yes	No	Single-atom detection Key paper
F53	No	Cont	Yes	No	No	No	Response time to steplike excitation
F54	No	Cont	Yes	No	Yes	No	Experimental verification
F55	Yes	Cont	Yes	No	No	No	Saturation parameter in two- and three-level atoms
F56, F58	No	Cont	Yes	Yes	No	No	Lifetimes and Y's as function of height in ICP
F57	Yes	Both	Yes	No	Yes	No	Comparison of methods

[a]Cont = Continuum source, Line = Line source, Both = Line or continuum source.
[b]Fundamental means that principles of excitation, deexcitation, broadening, etc., are given. Key paper means papers that have stimulated considerable *analytical* interest (opinion of authors).

why dye lasers are of interest in AFS and why no LEAFS system are commercially available. These considerations will be discussed at numerous points throughout this chapter.

Table 2.3, in addition to a few selected books and chapters, lists a number of papers that have had, in the author's opinion, a significant impact in the analytical development of the technique.

2. PRINCIPLES

2.1. Atomic Fluorescence Transitions

The possible types of AFS transitions with the appropriate designations (F14) have been described several times in the literature and are shown again in Fig. 2.1 for the sake of completeness. Essentially, all processes shown in Fig. 2.1 have been observed in LEAFS. It should be stressed that the analytical interest in nonresonance transitions (either direct line fluorescence, DLF, or stepwise line fluorescence, SLF) exists because of the freedom from scattering problems (background, background noise). Scatter can be either *elastic*, for example, Rayleigh scatter from atoms and molecules or Mie scatter from particles much larger than the excitation wavelength, or *inelastic*, for example, atomic and molecular fluorescence and Raman scatter. It should be emphasized here that the use of laser excitation for resonance fluorescence, assuming optical saturation occurs, will degrade analytical figures of merit since the fluorescence signal will reach a plateau with an increase in source intensity, but the scatter signal *and* scatter noise will continue to increase.

2.2. Comparison of Sources of Excitation for AFS

In Table 2.4, a comparison is given of several of the characteristics (figures of merit) of three major types of dye lasers (excimer laser-pumped dye laser, nitrogen laser-pumped dye laser, synchronously pumped, mode-locked dye laser), a conventional light (line) source, and an Eimac xenon arc lamp (300 W). Footnotes are given to describe the characteristics in the right side of the table. It is apparent that dye lasers have much higher spectral irradiances (peak values) than conventional sources, by factors of 10^5–10^{12}, which will allow optical saturation in the visible (fundamental region of the dye laser) region and in some cases in the UV (doubled fundamental of the dye laser). In addition, their use will result in far fewer spectral interferences than a conventional continuum source (Eimac xenon arc lamp) or a conventional line source (electrodeless lamp or hollow cathode discharge lamp). On the other hand, dye lasers are

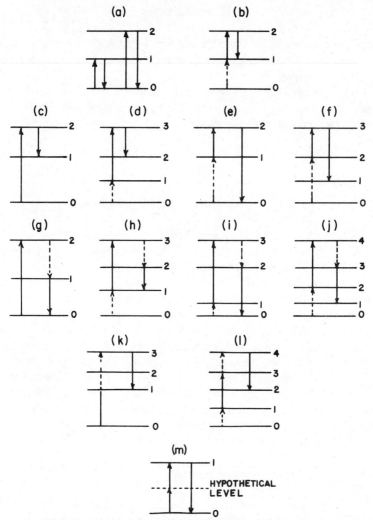

Fig. 2.1. Types of atomic fluorescence transitions (the spacing between atomic levels is not indicative of any specific atom). (a) RF resonance fluorescence (either process); (b) ERF, excited-state resonance fluorescence; (c) SDLF, Stokes direct line fluorescence; (d) ESDLF, excited-state Stokes direct line fluorescence; (e) ASDLF, anti-Stokes direct line fluorescence; (f) EASDLF, excited-state anti-Stokes direct line fluorescence; (g) SSLF, Stokes stepwise line fluorescence; (h) ESSLF, excited-state Stokes stepwise line fluorescence; (i) ASSLF, anti-Stokes stepwise line fluorescence; (j) EASSLF or ESSLF, excited-state anti-Stokes or Stokes stepwise line fluorescence; (k) TASSLF or TAASSLF, thermally assisted Stokes or anti-Stokes stepwise line fluorescence (depending on whether the absorbed radiation has shorter or longer wavelengths, respectively, than the fluorescent radiation); (l) ETASSLF or ETASSSLF, excited-state thermally assisted Stokes or anti-Stokes stepwise line fluorescence (depending on whether the absorbed radiation has shorter or longer wavelengths, respectively, than the fluorescence radiation); (m) TPF, two-photon excitation fluorescence (multiphoton processes involving more than two identical photons are even less probable than the two-photon processes).

more expensive to purchase and operate, more difficult to use, and of poorer stability than conventinal sources.

2.3. Atomic Fluorescence Radiance

Expressions for the atomic fluorescence radiance, B_F, in $J\ s^{-1}\ cm^{-2}\ sr^{-1}$, in the limiting cases of low source intensity (said to be the *linear case*) and of high source intensity (called *optical saturation*) are given in Table 2.5, with definitions of symbols. The reader is referred to numerous articles (F4, F9, F15, F27, F30, F33) where the above expressions have been derived and discussed.

Several conclusions or comments can be made based upon the radiance expressions in Table 2.5:

1. In all cases, B_F is proportional to the analyte concentration.
2. In linear AFS (low-intensity source cases), B_F is proportional to the source spectral irradiance, $E_\nu(\nu_0)$ and to the quantum efficiency, Y, of the fluorescence process.
3. In nonlinear AFS (high-intensity source cases), B_F is independent of source spectral irradiance, $E_\nu(\nu_0)$.
4. For two-level atoms and for nonlinear AFS, B_F is also independent of Y and, a simple expression results.
5. For a three-level Na-like atom and for nonlinear AFS, B_F will be independent of the collisional rate coefficients provided that the mixing between levels 2 and 3 is very effective, that is, k_{32} or $k_{23} \gg (A_{21} + k_{21})$ for $1 \to 3$ excitation and k_{32} (or k_{23}) $\gg (A_{31} + k_{31})$ for $1 \to 2$ excitation.
6. For a three-level Tl-like atom and for nonlinear AFS, B_F will depend also on the gas temperature, T, although the dependency is small for the $1 \to 3$ excitation process since $\exp(-E_{12}/kT) \ll 1$, but is significant for the $2 \to 3$ excitation since $\exp(E_{12}/kT) \gg 1$.
7. It is possible to determine *absolute concentrations of atoms* (atoms/cm^3) via saturated fluorescence (F21–23, F29–31, F37–39, F40, F44, F46) as long as the atoms within the observation volume are optically saturated, and B_F is measured in absolute units (use of a calibrated spectrometer-detector). However, in the case of a three-level atom, the collisional rate coefficients should be known or their effect should be negligible. In addition, such measurements require critical attention to the spatial, spectral, and temporal characteristics of the laser beam and the excited-state processes, including compound formation, ionization, and so on (see Table 2.3, especially references F4, F29, F32, F35, F36, F38, F39, F44).

Table 2.4. Comparison of Excitation Sources

| | Peak Spectral Irradiance $(W\ cm^{-2}\ nm^{-1})$ | | Pulse Width | Optical Saturation |
	Fundamental	Doubled	(ns)	Vis, UV
Source				
Excimer dye laser	10^9-10^{10}	10^7-10^9	5–20	Yes, ?
N_2 dye laser	10^8-10^{10}	10^6-10^8	3–10	Yes, ?
Synch pump dye laser	10^5-10^7	10^4-10^5	0.005–0.05	?, ?
Eimac arc	$10^{-1}-10^0$	$10^{-2}-10^{-1}$	cw	No, No
Electrodeless discharge lamp	$10^{-3}-10^{-5}$	$10^{-3}-10^{-5}$	cw	No, No

[a] E = easy, F_1 = difficult, F_2 = more difficult, F_3 = most difficult, Go = good, Ex = excellent, VL = very low, Gr = great, L = low, Hi = high, Mo_2 = moderate but still high, Mo_1 = moderate but low; Lo_1 = lowest, Lo_2 = low (higher than Lo_1).
[b] Spectral interferences correspond to those expected when the source is used to execute atoms in a flame or an ICP

2.4. Saturation Steady-State Spectral Irradiance, E_ν^s

The saturation steady state-spectral irradiance E_ν^s is the value of the source spectral irradiance, E_ν, where $B_F = 50\%$ of the maximum possible value (optically saturated) plateau value. Expressions for E_λ^s are given in Table 2.6. The saturation spectral irradiance, E_ν^s $(J\ s^{-1}\ cm^{-2}\ nm^{-1})$ is related to E_ν^s $(J\ s^{-1}\ cm^{-2}\ Hz^{-1})$ by

$$E_\lambda^s(\lambda_0) = E_\nu^s(\nu_0) \frac{c}{\lambda_0^2} 10^{-7} = \left(\frac{3 \times 10^3}{\lambda_0^2}\right) E_\nu^s$$

where λ_0 is the wavelength of excitation (in centimeters). In Table 2.7, saturation spectral irradiance values (E_λ^s) for a two-level atom are given for excitation wavelengths from 200 to 800 nm in 100-nm intervals. It is apparent that optical saturation occurs more readily as the quantum efficiency increases (toward unity) and the wavelength increases (E_λ^s is proportional to $\lambda_0^{-5} Y^{-1}$).

The importance of the saturation parameter as well as the pitfalls associated with its determination from the experimental saturation curves have received much attention (see especially references F4, F19, F36, F49 and F55).

2.5. Signal-to-Noise Ratio in AFS

The signal-to-noise (S/N) ratios in AFS have been described in detail by Omenetto and Winefordner (R14 and R17) and so will be treated only briefly here.

in Atomic Fluorescence Spectrometry[a]

Wavelength Range		Ease of Use		Stability	Spectral Interferences[b]	Cost	
Vis, UV		Vis, UV				Initial	Operation
Yes, $\lesssim 215$ nm		F_1, F_2		Go	VL	Mo_2	Mo_2
Yes, $\lesssim 215$ nm		F_1, F_2		Go	VL	Mo_1	Mo_1
Yes, 285–315 nm		F_3, F_3		Go	VL	Hi	Hi
Yes, Yes		E, E		Ex	Gr	Lo_1	Lo_1
Yes, Yes		E, E		Ex	L	Lo_2	Lo_2

The noise in LEAFS is determined by the following sources: background (flame, ICP, furnace) shot and flicker noise; source scatter (resonance case only) shot and flicker noise; background radical fluorescence shot and flicker noise; analyte fluorescence (and emission) shot and flicker noises; and interferent fluorescence (and emission) shot and flicker noises. In Table 2.8, some of the flame gas radicals and molecular analyte species that have been found to fluoresce in flames are given. As an example, Fig. 2.2 shows a typical molecular fluorescence spectrum of manganese in an air–acetylene flame. If the analyte excitation–emission combination is chosen properly, then interferent fluorescence and molecular fluorescence of flame radicals can generally be avoided. If the analyte concentration is low (as near the limit of detection), then the analyte-related noises are minimal. If nonresonance fluorescence is measured, then the scatter noises (except in cases where stray light could cause problems) are negligible. If the source is modulated or pulsed and a phase-sensitive or boxcar detection system used, the background flicker noises can be minimized. Therefore, the ultimate limiting noise at the powers of detection experimentally measured in LEAFS (and in AFS, in general) is cell background emission shot noise. In Table 2.9, detection limits and other analytical figures of merit of LEAFS and several other atomic methods are given. The detection limits were estimated as discussed in the footnotes and based upon general calculations by Omenetto and Winefordner (F33) and by Falk (F57).

Source pulsing and gated detection will often lead to improved detection limits (and S/N ratios at concentrations above the detection limits) compared to continuous-wave (cw) operation of the source and detector. Based on work described in references (R1) and (F25), the expressions and gain factors reported

Table 2.5. Atomic Fluorescence Steady-State Radiance Expression (F30)

General Expression

$$B_F = \frac{\ell}{4\pi} h\nu_{ul} A_{ul} n_u$$

Two-Level Atom

Low-excitation irradiance

$$B_F = \frac{\ell}{4\pi} Y_{21} E_\nu(\nu_0) \int k(\nu)\, d\nu$$

High-excitation irradiance

$$B_F = \frac{\ell}{4\pi} h\nu_{ul} A_{21} n_T \frac{g_u}{g_u + g_\ell}$$

Three-Level Atom

The three levels are indicated here as 1, 2, and 3 in order of increasing energy, 1 being the ground state. The levels not coupled by radiation are levels 2 and 3 for sodiumlike atoms, and levels 2 and 1 for thalliumlike atoms. The subscripts for B_F refer to the fluorescence and absorption transitions, e.g., $B_{F_{2\to1}}_{1\to3}$ means that laser excitation is set at λ_{13} and fluorescence is measured at λ_{21}.

Na atom type

Low-excitation irradiance (resonance fluorescence)

$$B_{F_{3\to1}}_{1\to3} = \frac{\ell}{4\pi} A_{31} h\nu_{31} n_T \frac{g_3}{g_1} \frac{E_\nu(\nu_{31})}{E^*_\nu(\nu_{31})}$$

High-excitation irradiance (resonance fluorescence)

$$B_{F_{3\to1}}_{1\to3} = \frac{\ell}{4\pi} A_{31} h\nu_{31} n_T \left(1 + \frac{g_1}{g_3} + \frac{k_{32}}{A_{21} + k_{21} + k_{23}} \right)^{-1}$$

High-excitation irradiance (Stokes stepwise line fluorescence)

$$B_{F_{2\to1}}_{1\to3} = \frac{\ell}{4\pi} A_{21} h\nu_{21} n_T \frac{g_3 k_{32} E_\nu(\nu_{31})}{g_1 (A_{21} + k_{23} + k_{21}) E^*_\nu(\nu_{31})}$$

High-excitation irradiance (stepwise line fluorescence)

$$B_{F_{2\to1}}_{1\to3} = \frac{\ell}{4\pi} A_{21} h\nu_{21} n_T \left[1 + \frac{A_{21} + k_{21} + k_{23}}{k_{32}} + \left(1 + \frac{g_1}{g_3} \right) \right]^{-1}$$

[Note: for the opposite cases, where excitation is at $1 \to 2$, replace all 2's by 3's and all 3's by 2's.]

42

Table 2.5 *(Continued)*

Three-Level Atom

Tl atom type
  Low-excitation irradiance (resonance fluorescence)

$$B_{F_{3 \to 1} \atop 1 \to 3} = \frac{\ell}{4\pi} A_{31} h\nu_{31} n_T \frac{g_3(E_\nu(\nu_{31}))}{g_1 E_\nu^*(\nu_{31}) [1 + (g_2/g_1) e^{-E_{12}/kT}]}$$

High-excitation irradiance (resonance fluorescence)

$$B_{F_{3 \to 1} \atop 1 \to 3} = \frac{\ell}{4\pi} A_{31} h\nu_{31} n_T \left(1 + \frac{g_1}{g_3} + \frac{g_2}{g_3} e^{-E_{12}/kT} + \frac{A_{32} + k_2}{k_{21}}\right)^{-1}$$

Low-excitation irradiance (Stokes direct line fluorescence)

$$B_{F_{3 \to 2} \atop 1 \to 3} = \frac{\ell}{4\pi} A_{32} h\nu_{32} n_T \frac{g_3 E_\nu(\nu_{31})}{g_1 E_\nu^*(\nu_{31}) [1 + (g_2/g_1) e^{-E_{12}/kT}]}$$

High-excitation irradiance (Stokes direct line fluorescence)

$$B_{F_{3 \to 2} \atop 1 \to 3} = \frac{\ell}{4\pi} A_{32} h\nu_{32} n_T \left(1 + \frac{g_1}{g_3} + \frac{g_2}{g_3} e^{-E_{12}/kT} + \frac{A_{32} + k_{32}}{k_{21}}\right)^{-1}$$

[Note: for the opposite cases, where excitation is at $2 \to 3$, replace all 2's by 1's and all 1's by 2's and replace E_{12} by $-E_{12}$.]

Key to Symbols

ℓ = emission (fluorescence path length), m
$h\nu_{ul}$ = energy of emission photon, J
n_u = upper (radiative) level population density, m^{-3}
n_T = total population density of *all* electronic states of same atom, m^{-3}
g_l, g_u = statistical weights of lower and upper levels, respectively
$k(\nu)$ = atomic absorption coefficient at frequency ν, m
$E_\nu(\nu_{ul})$ = source spectral irradiance at ν_{ul}, J s^{-1} m^{-2} Hz^{-1}
$E_\nu^*(\nu_{ul})$ = modified saturation spectral irradiance at ν_{ul}, J s^{-1} m^{-2} Hz^{-1}

$$= \frac{cA_{ul}}{B_{ul}Y_{ul}} = \frac{8\pi h\nu_{ul}^3}{c^2 Y_{ul}}$$

B_{ul} = Einstein coefficient of induced emission, s^{-1} (J/m^3 Hz)$^{-1}$
k = pseudo-first-order radiationless rate coefficient between levels shown, s^{-1}
A = Einstein coefficient of spontaneous emission between levels shown, s^{-1}
E_{12} = energy separation between levels 1 and 2, J
k = Boltzmann constant, J K^{-1} atom^{-1}
T = temperature of gas, K
c = velocity of light in vacuum, m s^{-1}

Table 2.5 (*Continued*)

Assumptions Necessary to Obtain Expressions

1. All expressions are for low optical densities of analyte (no self-absorption).
2. Prefilter and postfilter (self-reversal) effects are absent.
3. Rate equations approach is valid (no coherent effects).
4. Analyte atom concentration and cell temperature are constant over the cell (e.g., flame) dimensions.
5. Source of excitation is a continuum with respect to the absorption line width $(\delta\lambda_s > \delta\lambda_a)$.

Quantum Efficiencies, Y

Two-level atom

$$Y_{21} = \frac{A_{21}}{k_{21} + A_{21}}$$

Three-level atom Na-like

$$Y_{31} = \frac{A_{31}}{(A_{31} + k_{31} + k_{32})\left(1 - \dfrac{k_{32}k_{23}}{(A_{31} + k_{31} + k_{32})(A_{21} + k_{21} + k_{23})}\right)}$$

$$Y_{21} = \frac{A_{21}}{(A_{21} + k_{21} + k_{23})\left(1 - \dfrac{k_{32}k_{23}}{(A_{32} + k_{31} + k_{32})(A_{21} + k_{21} + k_{23})}\right)}$$

Three-level atom Tl-like

$$Y_{31} = \frac{A_{31}}{A_{31} + A_{32} + k_{32} + k_{31}} \qquad Y_{32} = \frac{A_{32}}{A_{31} + A_{32} + k_{32} + k_{31}}$$

Table 2.6. Saturation Steady-State Spectral Irradiance $E_{\nu}^{s}(\nu_0)^a$

Two-level atom

$$E_{\nu_{12}}^{s} = \left(\frac{c}{B_{21}}\right)\left(\frac{g_1}{g_1 + g_2}\right)(k_{21} + A_{21}) = \left(\frac{g_1}{g_1 + g_3}\right)E_{\nu_{12}}^{*}$$

Three-level Na-like atom

$$E_{\nu_{13}}^{s} = E_{\nu_{13}}^{*}\frac{g_1}{g_3}\left(1 + \frac{g_1}{g_3} + \frac{k_{32}}{k_{23} + k_{21} + A_{21}}\right)^{-1}$$

Three-level Tl-like atom

$$E_{\nu_{13}}^{s} = E_{\nu_{13}}^{*}\frac{g_1}{g_3}\left(1 + \frac{g_1}{g_3} + \frac{A_{32} + k_{32}}{k_{21}}\right)^{-1}$$

^{a}See Table 2.5 for definition of all terms and units.

Table 2.7. Saturation Steady-State Spectral Irradiance Values for a Two-level Atom ($g_1 = g_2$)[a]

λ_0 (nm)	E_λ^s (λ_0) (J s^{-1}/ cm^{-2}/ nm^{-1})
200	$2.5 \times 10^5/Y_{21}$
300	$3.0 \times 10^4/Y_{21}$
400	$7.0 \times 10^3/Y_{21}$
500	$2.5 \times 10^3/Y_{21}$
600	$1.0 \times 10^3/Y_{21}$
700	$4.5 \times 10^2/Y_{21}$
800	$2.0 \times 10^2/Y_{21}$

[a] Y_{21} = fluorescence quantum efficiency.

Table 2.8. Molecular Fluorescence Observed in Flames with Laser Excitation

Species	Excitation (nm)	Fluorescence (nm)	Transition	Reference
C_2	473.7, 516.5, 563.4	500–670	$A^3\Pi \leftarrow X^3\Pi$	a
CH	387.2, 431.2	390–450	$B^2\Sigma \leftarrow X^2\Pi$	a
	387.2, 431.2	420–500	$A^2\Delta \leftarrow X^2\Pi$	a
CN	388.3, 421.6	300–700	$B^2\Sigma \leftarrow X^2\Pi$	a
OH	306.4	280–350	$A^2\Sigma \leftarrow X^2\Pi$	a
BaCl	506–532	506–532	$C^2\Pi \leftarrow X^2\Sigma$	b
BaO	480–540	480–560	$A^1\Sigma \leftarrow X^1\Sigma$	b
BaOH	480–540	470–550		b
		700–850		
CaOH	540–560	540–660	$B^2X \leftarrow X^2\Sigma$	c
	590–650	540–660	$A^2\Pi \leftarrow X^2\Sigma$	
SrOH	620–630	600–700	$B^2X \leftarrow X^2\Sigma$	c
			$A^2\Pi \leftarrow X^2\Sigma$	
CrO	570–605	605–680	$A^5\Pi \leftarrow X^5\Sigma$	c
MnO	585–591	585–591		c
YO	$A^2\Pi_{3/2, 1/2} \leftarrow X^2\Sigma$ Q (1, 1) and Q (0, 0)	594–628	$A^2\Pi \rightarrow X^2\Sigma$	d

[a] K. Fujiwara, N. Omenetto, J. D. Bradshaw, J. N. Bower, S. Nikdel, and J. D. Winefordner, *Spectrochim. Acta,* **34B,** 317 (1979), and references therein cited.
[b] H. Haraguchi, S. J. Weeks, and J. D. Winefordner, *Spectrochim. Acta,* **35A,** 391 (1979), and references therein cited.
[c] M. B. Blackburn, J. M. Mermet, and J. D. Winefordner, *Spectrochim. Acta,* **34A,** 847 (1978), and references therein cited.
[d] T. Wijchers, H. A. Dijkerman, P. J. Th. Zeegers, and C. Th. J. Alkemade, *Spectrochim. Acta,* **35B,** 271 (1980).

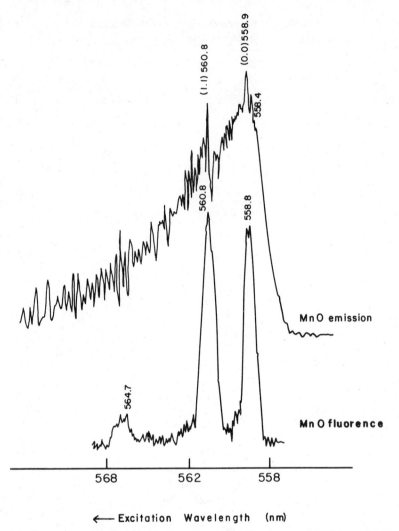

Fig. 2.2. Comparison of molecular emission and molecular fluorescence of MnO in an Ar–O_2–H_2 flame (Table 2.8, ref. *c*). MnO flame emission; MnO flame fluorescence (laser excitation at 587.9 nm).

in Table 2.10 have been derived. In general, in those cases where the noise is determined predominately by the cell background, it is possible to obtain an appreciable gain by source pulsing and detector gating if one assumes that the same average source power as in cw operation can be obtained and if no deleterious effects such as those described in footnote *f* of Table 2.10 are present.

Table 2.9. Comparison of Laser-excited Atomic Fluorescence Spectrometry with Other Atomic Methods

Method[a]	Cell	LOD[b] (atoms/cm^3)	LDR[c]	Spectral[d] Interferences	Matrix[e] Interferences	Comments
LEAFS	Flame	$\sim 10^5$	$\sim 10^6 X$	+	++	Background
	ICP	$\sim 10^5$	$\sim 10^6 X$	+	+	shot noise
	furnace	$\sim 10^5$	$\sim 10^5 X$	+	++++	
XEAFS	Flame	$\sim 10^9$	$\sim 10^5 X$	+++	++	Background shot noise
HCAFS[f]	ICP	$\sim 10^7$	$\sim 10^5 X$	++	+	Background shot noise
LEIS	Flame	$\sim 10^5$	$\sim 10^6 X$	++	+++	Background current shot noise
AAS	Flame	$\sim 10^8$	$\sim 10^2 X$	++	++	Source Shot noise
	Furnace	$\sim 10^8$	$\sim 10^2 X$	++	++++	Source shot noise
AES	ICP	$\sim 10^5$	$\sim 10^6 X$	++++	+	Background shot noise

[a]LEAFS = Laser-excited atomic fluorescence spectrometry with flame or ICP as cell

XEAFS = Xenon-excited atomic fluorescence spectrometry with flame cell

HCAFS = Pulsed hollow cathode lamp AFS with ICP cell

LEIS = Laser-enhanced ionization spectroscopy in flame cell

AAS = Atomic absorption spectrometry with flame or furnace cell; source could be hollow cathode lamp or xenon arc with high-resolution spectrometer (echelle)

AES = Atomic emission spectrometry with ICP source/cell.

[b]LOD = Limit of detection in atoms/cm^3. Values calculated by Omenetto and Winefordner (F33) and by Falk (F57). Values are averages of those obtained for typical experimental conditions.

[c]LDR = Linear dynamic range, i.e., range of linearity of calibration curve = C_u/C_1 where C_u = upper concentration which is within 5% of linearity and C_1 = limit of detection. The LDR values represent experimental values.

[d]Spectral interferences = selectivity of method to spectral interferences. ++++ = many spectral interferences and + = few spectral interferences.

[e]Matrix interferences = selectivity of method to matrix interferences. ++++ = many matrix interferences and + = few matrix interferences. In case of furnace AAS, some of the newer methods (platform vaporization, probe vaporization, and capacitative discharge represent ways to minimize such intereferences).

[f]The LOD was estimated by assuming the same conditions as in (F33) for the electrodeless discharge lamp excitation except that the duty cycle was taken as $\frac{1}{12}$ and the average hollow cathode irradiance was assumed to be the same as in the dc operated lamp used for the calculations in (F33).

Table 2.10. Benefit of Using Pulsed Sources and Various Detection in Atomic Fluorescence Spectrometry

Case	Source Type[a]	Detector Type[b]	Noise Type[c]	S/N[d]	Gain Factor[e]	Assumptions[f]
I	cw	cw	Source/bkgnd-shot	$K_1\bar{E}_{s,cw}/\sqrt{\bar{E}_{s,cw}+\bar{E}_b}$	1 (Refn)	—
I	pu	ga	Source/bkgnd-shot	$K_1\bar{E}_{s,pu}/\sqrt{\bar{E}_{s,pu}+f_{t_p}\bar{E}_b}$	$\bar{E}_{s,pu}/\bar{E}_{s,cw}\,\sqrt{f_{t_p}} \to \dfrac{1}{\sqrt{f_{t_p}}}$	$\bar{E}_b > \bar{E}_{s,cw}$ or $\bar{E}_{s,pu}$; $\bar{E}_{s,cw}\approx\bar{E}_{s,pu}$
I	pu	ga	Source/bkgnd-shot	$K_1\bar{E}_{s,pu}/\sqrt{\bar{E}_{s,pu}+f_{t_p}\bar{E}_b}$	$\sqrt{\bar{E}_{s,pu}/\bar{E}_{s,cw}} \to 1$	$\bar{E}_b < \bar{E}_{s,cw}$ or $\bar{E}_{s,pu}$; $\bar{E}_{s,cw}\approx\bar{E}_{s,pu}$
II	cw	cw	Bkgnd-flicker	$K_2\bar{E}_{s,cw}/\xi_b\bar{E}_b$	1 (Refn)	—
II	pu	ga	Bkgnd-flicker	$K_2\bar{E}_{s,pu}/\xi_b\,f_{t_p}\bar{E}_b$	$\bar{E}_{s,pu}/\bar{E}_{s,cw}f_{t_p} \to \dfrac{1}{f_{t_p}}$	$\bar{E}_{s,cw}\approx\bar{E}_{s,pu}$
III	cw	cw	Source-flicker	$\bar{E}_{s,cw}/\xi_s\bar{E}_{s,cw}$	1 (Refn)	—
III	pu	ga	Source-flicker	$\bar{E}_{s,pu}/\xi\bar{E}_{s,pu}$	1	—

[a] cw = continuous wave source (on all of the time); pu = pulsed source.

[b] cw = continuous wave detector (on all of the time); ga = gated detector.

[c] source/bkgnd-shot = source and background shot noise; source-flicker = source flicker noise; bkgnd-flicker = background flicker noise.

[d] $\bar{E}_{s,cw}$ = average excitation irradiance of cw source; $\bar{E}_{s,pu}$ = average excitation irradiance of pulsed source = $f_{t_p}\,\bar{E}_{s,peak}$ (assuming a rectangular pulse of width t_p seconds, and a repetition rate of f, Hz); $\bar{E}_{s,peak}$ = peak pulsed source irradiance; $\bar{E}_b$ = average background irradiance; ξ_b = flicker noise factor for background; ξ_s = flicker noise factor for source related noise; K = proportionality factors accounting for optical throughput.

[e] Gain factor is always with respect to the cw/cw case for each type of noise.

[f] There are other inherent assumptions concerning pulsing: (i) the pulsed source with a duty cycle of f_{t_p} can be pulsed at sufficient power so that $E_{s,pu} = E_{s,cw}$; (ii) the pulsing does not cause self-reversal, shift in the spectral line or assymetry; (iii) saturation is not reached or else source scatter would increase faster with $\bar{E}_s$ than the fluorescence. Also, the value of $\bar{E}_{s,pu}$ necessary to achieve saturation represents the maximum value to be substituted in the above equations.

No gain (enhancement) of S/N ratio occurs for those cases where the noise is associated with the analyte fluorescence and/or with the source intensity (e.g., analyte fluorescence, scatter, interferent fluorescence, flame radical fluorescence).

2.6. Fluorescence Growth Curves

Although no expressions were given in the fluorescence radiance section for high optical densities of the analyte, it is well known (F4, F9, F33) that, because of self-absorption, a plot of log B_F versus log n_T (where n_T is the total atomic population density, Table 2.5) will have a slope of zero at high optical densities if the excitation source is a spectral continuum with regards to the absorption profile, which is the case of the dye laser considered here (and of most dye lasers). It is clear that the linearity of the analytical fluorescence growth curve will be extended when using a laser compared to a conventional source because the detection limit, when the noise is not source related, will be reduced, and because, if optical saturation can be achieved, self-absorption is minimized. Therefore, it is not unreasonable for an analytical growth curve to extend over 6–8 orders of magnitude for LEAFS as compared to 4–7 orders of magnitude for the same excitation wavelength and the same element in the case of conventional (linear) source AFS.

Of course, experimental analytical growth curves are plots of S_F, that is, the fluorescence signal due to B_F versus C, which is the analyte concentration introduced as a nebulized aerosol into the atomization cell (flame or an ICP). The log S_F versus log C curves generally do not extend over as many decades (perhaps 5–7) as the log B_F versus log n_T curves because: (i) detector saturation may occur at large B_F values and (ii) the analyte concentration in the cell, n_T, may not be linear with the sample concentration C in the solution. In the case of flame and ICP atomization, this is due to a decrease in nebulization efficiency, solution transport rate, and solute vaporization efficiency with increased concentration of sample solution. For electrothermal atomizers, it is due to a decreased atomization time and decreased vaporization and atomization efficiencies.

3. INSTRUMENTAL AND METHODOLOGICAL CONSIDERATIONS

3.1. Instrumentation

A LEAFS instrument is basically an AFS system with a dye laser source. There are two major types of LEAFS instruments: a pulsed dye laser, cell, emission monochromator–photomultiplier tube with boxcar averager (see Figs. 2.3a and 2.3b) and a modulated cw dye laser (usually Ar laser-pumped dye laser), cell,

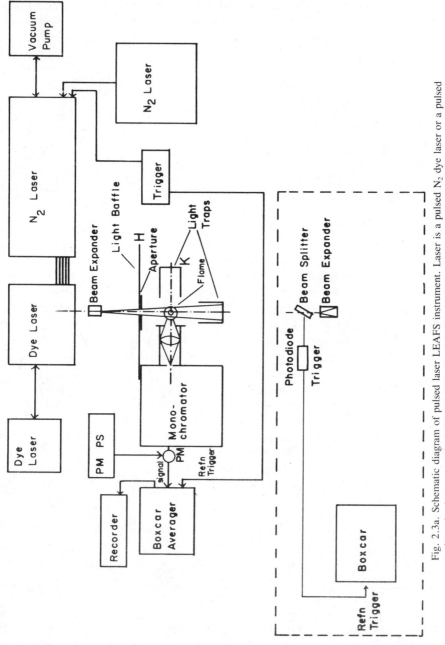

Fig. 2.3a. Schematic diagram of pulsed laser LEAFS instrument. Laser is a pulsed N_2 dye laser or a pulsed excimer dye laser or a pulsed Nd : YAG dye laser. Both external and optical triggering (with a photodiode) are indicated.

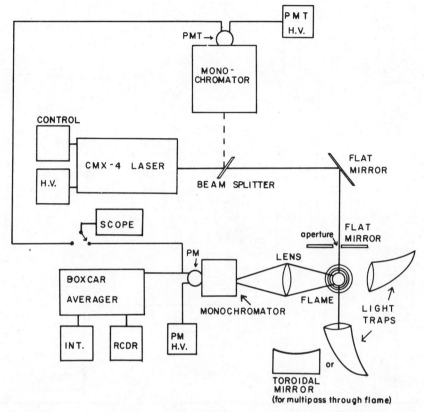

Fig. 2.3b. Schematic diagram of pulsed laser LEAFS instrument. Laser is a flashlamp dye laser.

emission monochromator–photomultiplier tube with phase-sensitive detection (see Fig. 2.4). No attempt will be made here to describe the operation of pulsed or modulation systems, or of detection systems; the reader is referred to other references.

3.2. Analytical Figures of Merit LEAFS

Table 2.11 collects several experimental papers on LEAFS. In addition to the instrumental components used in each reference, the types of analytically significant results, such as the number of elements measured, the presence of limits of detection, linear dynamic ranges (calibration curves), and the application to real samples are also given.

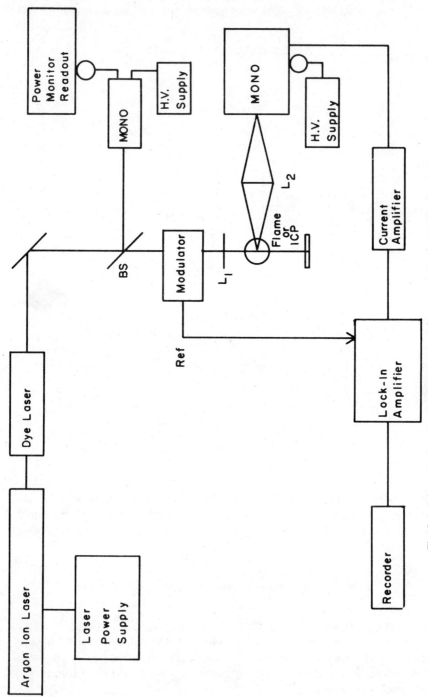

Fig. 2.4. Schematic diagram of an amplitude-modulated cw dye laser LEAFS instrument.

Table 2.11. Experimental Studies in Laser-excited Analytical Atomic Fluorescence Spectrometry

References	Laser System[a]	Cell	Modulation–Detection	Elements	Analytical Results			Comments
					LOD	LDR	Applic.	
I1	Ruby-DL	Flame	Amp-Osc	Ba	Yes	Yes	No	First paper
I2	N$_2$-DL	Flame	Boxcar	Many	Yes	Yes	No	First paper
I3	FDL	Na cell	Lock-in	Na	Yes	Yes	No	—
I4	N$_2$-DL	Flame	Boxcar	Many	Yes	Yes	No	—
I5	N$_2$-DL	Flame	Boxcar	Rare earth	Yes	Yes	No	Ionic fluorescence
I6	N$_2$-DL	Flame	Boxcar	Trans. elem.	Yes	No	No	Nonresonance fluorescence
I7	FDL	Flame	Boxcar	Na	Yes	Yes	No	—
I8	FDL	Flame	Boxcar	Mg, Ni, Pb	Yes	Yes	No	—
I9	FDL	Graphite rod	Lock-in	Pb	Yes	Yes	No	—
I10	NDL	Quartz cell	Photon counting	Alkali metals	No	No	No	Compared sensitized to resonance fluorescence
I11	ADL	Quartz cell	Lock-in	Na	Yes	Yes	No	—
I12	ADL	Quartz cell	Lock-in	Na	Yes	Yes	No	—
I13	FDL	Quartz cell	Filter + osc	Na	Yes	Yes	No	Fundamental study
I14	ADL	Flame	Lock-in	Ba, Na	Yes	Yes	No	—
I15	FDL	Quartz cell	Amp-Osc	Na	Yes	Yes	No	—
I16	ADL	Quartz cell	Lock-in	Na	Yes	No	No	—
I17	FDL	Sputtering cell	Boxcar	Fe	No	No	No	—
I18	YAG-DL	Graphite furnace	Filter + osc	Fe, Pb	Yes	Yes	No	—
I19	FDL	Flame	Boxcar	Na	Yes	No	No	Nonresonance fluorescence
I20	NDL	Flame	Boxcar	Sr, Mn, Na, Ca, In	No	No	No	Microprocessor-controlled scanning

53

Table 2.11. (*Continued*)

References	Laser System[a]	Cell	Modulation–Detection	Elements	Analytical Results			Comments
					LOD	LDR	Applic.	
I21	NDL	Flame	Boxcar	Na	No	No	No	Saturation study
I22	ADL	Flame	Lock-in	Na, V, B, Sr, Mo	Yes	Yes	No	—
I23	YAG-DL	Graphite furnace	Filter + osc	Pb	No	No	No	Fundamental studies
I24	FDL	Quartz cell	Boxcar	Na	Yes	Yes	No	Single-atom detector
I25	FDL	Flame	Boxcar	Na, MgO	Yes	Yes	No	Dye laser
I26	NDL	Flame	Boxcar	Many	Yes	Yes	No	—
I27	NDL	Graphite furnace	Boxcar	Tl	Yes	Yes	No	—
I28	ADL	Flame	Lock-in	Ba	Yes	Yes	No	Used wavelength modulator
I29	YAG-DL	Flame	Boxcar	Na	Yes	Yes	No	—
I30	YAG-DL	Flame	Boxcar	Na	Yes	Yes	No	—
I31	YAG-DL	Flame	Boxcar	Na	Yes	No	No	Comparison with photoionization
I32	ADL	Interactivity ABS cell	Fluorescence detector	I_2	Yes	Yes	No	Isotope detect.
I33	ADL	ICP	Lock-in	Na, Ba, Li, V	Yes	Yes	No	—
I34	NDL	Ablation chamber	Boxcar	Cr	Yes	Yes	Yes	Ruby laser used to ablate sample
I35	NDL	Ablation chamber	Boxcar	Cr	Yes	Yes	Yes	Ruby laser used to ablate sample
I36	?	Quartz cell	?	Na	Yes	?	No	Single-atom detector
I37	NDL	U-oven	Transient recorder	U	No	No	No	Spectra only
I38	FDL	Flame	Boxcar	Fe	Yes	Yes	Yes	Multipass cell

54

I39	FDL or NDL	ICP	Boxcar	Fe, Sn, Ba, In	Yes	No	No	—
I40	FDL	Flame	Boxcar	Sn, Ni	Yes	Yes	Yes	—
I41	ADL	Low-pressure cell	Lock-in	Na	Yes	No	No	Single-atom detector
I42	Resonance line	Flame	Boxcar	Tl, Na	Yes	No	No	Resonance line lasers, photoionization detection
I43	ADL	Flame	Lock-in	Na	Yes	No	No	Scatter correction
I44	NDL or ADL	Flame	Boxcar/lock-in	Na	No	No	No	Chemical rx's in exc. state
I45	YAG-DL	Furnace	Amp/disc/counter	Pb	Yes	Yes	No	—
I46	YAG-DL	Quartz cell	Amp/disc/counter	Hg	Yes	No	No	Multiphoton ioniz. & recomb.
I47	NDL	Flame	Boxcar	Ca, Sr, Mg, Pb, Fe	Yes	No	Yes	—
I48	YAG-DL	Quartz cell/furnace	Boxcar	Pb	Yes	Yes	No	—
I49	YAG-DL	Graphite furnace	Boxcar	Pb, Fe, Na, Pt, Ir, Eu, Ca, Ag, Co, Mn	Yes	Yes	Yes	—
I50	YAG-DL	Low-pressure discharge	Boxcar	Cl	No	No	No	—
I51	YAG-DL	Flame	Boxcar	?	Yes	No	No	Single-atom detector
I52	YAG-DL	Quartz cell	Photon counting	Hg	Yes	Yes	No	Projected use for atmosphere
I53	NDL	ICP	Boxcar	Tl, Ba, Ca, V, Y, Mo, Pb	Yes	Yes	No	Short torch
I54	NDL	ICP	Boxcar	Many	Yes	Yes	No	Long Torch

Table 2.11. (*Continued*)

References	Laser System[a]	Cell	Modulation–Detection	Elements	Analytical Results				Comments
					LOD	LDR	Applic.		
155	Exc-DL	Flame, furnace ICP	Boxcar	—					General considerations
156	Exc-DL	ICP, flame	Boxcar	Al, B, Ba, Ga, Mo, Pb, Si, Sn, Ti, Tl, V, Y, Zr	Yes	No	No		Short torch
157	Exc-DL	ICP	Boxcar						
158	Exc-DL	Flame	Boxcar	Pb	Yes	No	Yes		Direct determination in blood
159	ADL	Flame	Lock-in	Na	No	No	No		Scattering correction
160	NDL	Graphite cuvette	Boxcar	Many	Yes	No	No		
161	ADL	Flame	Lock-in	Na	No	No	No		Scattering correction
162	ADL	Flame	Lock-in	Na	No	No	No		

[a]NDL = Nitrogen laser-pumped dye laser; FDL = flashlamp-pumped laser; ADL = argon ion laser-pumped dye laser; Exc-DL = excimer laser-pumped dye laser.

56

3.2.1. Limits of Detection with LEAFS

Table 2.12 collects the best detection limits, in concentration units (ng/mL) so far achieved by LEAFS in either flame or plasma atomizers, together with the linear dynamic ranges obtained and the best combination for the excitation and fluorescence lines. In Table 2.13, these limits are compared with limits of detection by several other atomic methods: Atomic fluorescence flame spectrometry with conventional line sources (AFS-line); atomic fluorescence flame spectrometry with conventional continuum source (AFS-continuum); atomic absorption flame spectrometry (AAS); atomic emission spectrometry with inductively coupled plasma (ICPAES); and laser-enhanced ionization spectrometry in flames (LEIS). In Table 2.14, the best limits of detection in absolute limits (in picograms) for LEAFS with a graphite filament electrothermal atomizer are compared with EAAAS (AAS with an electrothermal atomizer), which has been the "star" of analytical methods for absolute elemental detection limits.

Although current concentration and absolute detection limits by LEAFS have not yet reached those *predicted* (F33, F57), they are nevertheless highly competitive with the best values achieved by other atomic (elemental) methods. The concentration limits of detection in LEAFS approach the values calculated, assuming background (e.g., flame) emission shot noise. If the background can be reduced (as in electrothermal atomizers), then it should be possible to lower the detection limits by $\sim 10^5 \times$, assuming the ultimate limiting noise level, that is, detector shot noise, can be reached. Experimentally, with carbon rod atomizers, $\sim 10^3$ atoms of lead per cubic centimeter have indeed been detected (I48, I56).

3.2.2. Analytical Calibration Curves

In Table 2.12, the linear dynamic ranges of LEAFS (flame) are listed along with the limits of detection. The linear dynamic ranges in LEAFS (ICP) are similar. As an example, Fig. 2.5 shows two typical analytical calibration curves for Sn and Ni.

3.2.3. Precision

The precision (percent relative standard deviation) in LEAFS is generally slightly poorer than that achieved by AAS, ICPAES, and conventional source AFS, mainly because of the pulse-to-pulse stability of pulsed dye lasers assuming no optical saturation. The precision is limited to ~ 3–5% for a 1-s integration time. In Fig. 2.6, a typical precision curve for Fe is given (I38).

Table 2.12. Concentration Detection Limits (Aqueous Solution) Obtained by Laser-excited Atomic Fluorescence Spectrometry

Element	$\lambda_{ex}/\lambda_{fl}$ [a]	Laser-Excited AFS[c-f]		
		Type of AF[b]	LOD(ng/mL)	LDR
Ag	328.1/328.1	RF	4	4.2
Al	394.4/396.1	S–DLF	0.4	—
Au	267.6/242.8	TA–AS–SLF	4	5.5
B	249.6/249.7	S–DLF	4	—
Ba	455.4/614.2(i)	S–DLF	0.7	—
Bi	306.8/306.8	RF	3	5.2
Ca	422.7/422.7	RF	0.01	5.0
Cd	228.8/228.8	RF	8	3.5
Ce	371.6/294.1(i)	TA–AS–SLF	500	—
Co	357.5/347.4	AS–DLF	19	5.0
Cr	359.3/359.3	RF	1	5.5
Cu	324.7/324.7	RF	1	5.0
Dy	364.5/356.3(i)	TA–AS–SLF	300	—
Er	400.8/400.8	RF	500	—
Eu	459.4/462.7	S–DLF	20	—
Fe	296.7/373.5	S–DLF	30	6.0
Ga	403.3/417.2	S–DLF	0.9	5.4
Gd	336.2/376.8(i)	E–S–SLF	800	—
Ho	405.4/410.4	S–DLF	100	—
In	410.4/451.1	S–DLF	0.2	6.2
Ir	266.5/406.9	S–DLF	9	4.5
Li	670.8/670.8	RF	0.5	4.3
Lu	465.8/513.5	S–DLF	3,000	—
Mg	285.2/285.2	RF	0.2	6.0
Mn	279.5/279.5	RF	0.4	5.4
Mo	313.2/317.0	S–SLF	5	—
Na	589.0/589.0	RF	0.1	5.7
Nb	405.9/408.0	E–S–SLF	1,400	—
Nd	463.4/489.7	S–SLF	2,000	—
Ni	311.2/342.0	S–DLF/S–SLF	2	6.0
Os	442.0/426.1	TA–AS–SLF	150,000	—
Pb	383.3/405.8	S–DLF	0.02	—
Pd	324.3/340.4	E–S–SLF	1	6.0
Pr	422.5/430.6(i)	S–DLF	1,000	—
Pt	264.7/270.2	S–DLF	0.7	6.0
Rh	369.2/350.2	TA–AS–SLF	100	—
Ru	287.5/366.3	S–DLF	2	6.0
Sc	391.2/402.4	E–S–SLF	10	—
Si	288.1/251.6	E–AS–SLF	1	—
Sm	367.4/373.9(i)	S–SLF	100	—

58

Table 2.12. (*Continued*)

Element	$\lambda_{ex}/\lambda_{fl}{}^a$	Laser-Excited AFS[c-f]		
		Type of AF[b]	LOD(ng/mL)	LDR
Sn	300.9/317.5	S–DLF	3	—
Sr	460.7/460.7	RF	0.3	5.0
Tb	370.3/350.9	?	500	—
Ti	307.8/316.2(i)	E–S–DLF	1	—
	307.8/316.8(i)	E–TA/S–SLF	1	—
Tl	276.7/352.9	S–DLF	0.8	—
U	409.0/385.9	AS–DLF	20	—
V	268.8/290.8(i)	S–DLF	3	—
Y	508.7/371.0(i)	E–AS–DLF	0.6	—
Yb	398.8/346.4	TA–AS–SLF	10	—
Zr	310.6/256.8(i)	E–AS–DLF	3	—
	310.6/257.1(i)	E–AS–SLF	3	—

[a] $\lambda_{ex}/\lambda_{fl}$ = excitation and fluorescence wavelengths, in nm. (i) means ionic fluorescence.
[b] RF = resonance fluorescence; DLF = direct line fluorescence; SLF = stepwise line fluorescence; S– = Stokes; AS– = anti-Stokes; TA = thermally assisted; E = excited states; LOD = limit of detection; LDR = linear dynamic ranges, number of decades.
[c] S. J. Weeks, H. Haraguchi, and J. D. Winefordner, *Anal. Chem.*, **50,** 360 (1978).
[d] N. Omenetto, N. N. Hatch, L. M. Fraser, and J. D. Winefordner, *Anal. Chem.*, **45,** 195 (1973); N. Omenetto, N. N. Hatch, L. M. Fraser, and J. D. Winefordner, *Spectrochim. Acta*, **28B,** 66 (1973). These values were obtained with one of the first laser experimental fluorescence setups and can be considered, for some elements, of historical interest.
[e] N. Omenetto, H. G. C. Human, P. Cavalli, and G. Rossi, *Spectrochim. Acta*, **39B,** 115 (1984).
[f] S. Kachin, B. W. Smith, and J. D. Winefordner, *Appl. Spectrosc.*, **39,** 587 (1985).

3.2.4. Spectral Selectivity

The spectral selectivity of LEAFS (also see Table 2.9 for a comparison with other atomic methods) is the highest of all atomic methods, perhaps *all* methods for analysis of elements. The high spectral selectivity rests with the extreme excitation selectivity based on single-line laser excitation *and* upon emission selectivity (selection of the fluorescence wavelength to observe the AFS signal). In Fig. 2.7, the excitation spectrum of Na is given (Fig. 2.7a), which indicates that a dye laser with a bandwidth (FWHM) of ~0.03 nm allows selection of either of the Na-*D* lines for excitation. In this case, as shown in Fig. 2.7b, the emission monochromator's bandpass was quite large, here 1.6 nm, to allow measurement of the fluorescence from both Na-*D* lines, since mixing collisions in the flame populate both Na-*D* states. Of course, the monochromator bandpass could be reduced to resolve the two lines, thus adding to the selectivity of the method.

Table 2.13. Limits of Detection (Aqueous Solution) Obtained by Laser-excited Atomic Fluorescence Spectrometry and by Several Other Methods (ng/mL)

Element	AFS Line[a]	AFS Continuum[b]	LEAFS (flame)[c]	AAS[d]	ICPAES[e]	LEIS[f]	LEAFS (ICP)[g]
Ag	0.1	1	4	2	0.2 (4)	1	—
Al	100	200	0.6	20	0.4	0.2	0.4
As	100	—	—	400	2	—	—
Au	1,000	150	4^ΔΔ	4^ΔΔ	—	—	—
Ba	—	—	8	20	0.01 (0.2)	0.2	0.7
Be	10	15	—	2	— (0.3)	—	—
Bi	10	—	3	30	— (50)	2	—
Ca	20	—	0.8	2	0.0001 (4)	0.1	2
Cd	0.001	6	8	1.5	0.07** (1)	0.1^≠	—
Ce	—	—	500*	—	0.4 (20)	—	—
Co	5	15	200*	15	0.1** (2)	0.08^≠	—
Cr	50	1.5	1	3	0.2 (4)	2	—
Cu	1	1.5	1	2	0.04** (2)	0.7^≠	—
Dy	—	—	300*	—	— (2)	—	—
Er	—	—	500*	—	— (—)	—	—
Eu	—	—	20*	—	— (1)	—	—
Fe	8	10	(0.06)† 30	10	0.09(2)	2	50*
Ga	10	—	0.9	—	0.6(40)	0.7	1
Gd	—	—	800*	—	0.5(8)	0.006	—
Ge	15,000	—	—	—	— (50)	—	—
Hf	—	—	—	200	— (—)	—	—
Hg	80	—	—	—	— (50)	—	—
Ho	—	—	100*	—	— (—)	—	—
In	100	25	0.2	—	— (40)	0.008	300*

60

	1	2	3	4	5	6	7
Li	—	—	0.5	1	—	0.001	—
Lu	—	—	3000*	—	(3)	0.1	—
Mg	1	0.1	0.2	0.2	(—)	0.3	—
Mn	6	2	0.4	3	0.003 (20)	—	—
Mo	500	100	12	20	0.02 (0.5)	0.05	5
Na	100,000	—	<0.1	0.5	0.4 (5)	—	—
Nb	—	—	1500*	—	0.02 (10)	0.08	—
Nd	3	25	2000*	10	0.2 (20)	—	—
Ni	—	—	0.5△	—	(10)	0.09#	—
Os	10	50	150,000*	15	0.2 (6)	—	1
Pb	1,000	100	0.02**	—	(200)	—	—
Pd	—	—	1△△	—	1** (20)	—	—
Pr	50,000	700	1000*	—	2 (40)	—	—
Pt	3,000	—	0.7△△	—	(30)	—	—
Rh	—	—	100*	—	(30)	—	—
Ru	—	—	2△△	—	(30)	—	—
Sb	40	—	50*	30	(60)	—	—
Sc	600	—	10*	250	(30)	—	—
Se	—	—	—	100	(1)	—	100
Si	30	150	—	—	1*** (20)	3	—
Sm	30	0.9	100*	70	(10)	—	1
Sn	—	—	3△	1	(10)	—	—
Sr	5	—	0.3	—	3 (6)	0.2	3
Tb	—	200	500*	70	0.003 (0.2)	0.09	0.5
Te	8	6	—	80	(—)	—	—
Ti	—	—	2	30	(20)	—	—
Tl	70	30	0.8**	—	0.03 (1)	—	1
Tm	—	—	100*	50	(75)	—	7
V	—	—	30	—	0.06 (2)	—	—
Yb	—	—	10*	—	—	—	3

61

Table 2.13 *(Continued)*

Element	AFS Line[a]	AFS Continuum[b]	LEAFS (flame)[c]	AAS[d]	ICPAES[e]	LEIS[f]	LEAFS (ICP)[g]
Y	—	—	—	200	0.08	—	0.6
Zn	0.2	15	—	1	0.1 (2)	—	—
Zr	—	—	—	1,000	0.3	—	3

[a] The values come from references within J. D. Winefordner, *J. Chem. Ed.*, **55**, 72 (1978).

[b] The values come from D. J. Johnson, F. W. Plankey, and J. D. Winefordner, *Anal. Chem.*, **46**, 1858 (1974).

[c] Values from S. J. Weeks, H. Haraguchi, and J. D. Winefordner, *Anal. Chem.*, **50**, 360 (1978), except those with * which were taken from references listed in Winefordner, *J. Chem. Ed.*, **55**, 72 (1978). The one with † was taken from M. S. Epstein et al., *Spectrochim. Acta*, **35B**, 233 (1980), that with △ was taken from M. S. Epstein et al., *Appl. Spectrosc.*, **34**, 372 (180), those with ** were taken from H. G. C. Human, et al., *Spectrochim. Acta*, **39B**, (1984), and those with △△ were taken from S. Kachin et al., *Appl. Spectrosc.*, **39**, 587 (1985).

[d] All values come from Perkin-Elmer atomic absorption commercial literature on the Model 460.

[e] All values from P. W. J. M. Boumans and F. J. De Boer, *Spectrochim. Acta*, **30B**, 309 (1975), except for those with ** and those in ( ). All values in ( ) come from commercial literature from Jarrell-Ash Division, and Fisher Scientific Co., Waltham, Mass. for their third-generation ICP Plasma Atom Comp. All values with ** come from K. W. Olson, J. W. Haas, and V. A. Fassel, *Anal. Chem.*, **49**, 632 (1977).

[f] Values taken from G. C. Turk, et al., *Anal. Chem.*, **51**, 1890 (1979), except * are dual laser-excited values taken from G. C. Turk, J. C. Travis, J. R. Dobbs, and T. C. O'Haver, *Anal. Chem.*, **54**, 643 (1982).

[g] Values taken from N. Omenetto et al., *Spectrochim. Acta*, **39B**, 115 (1984) except from those designated with *, which were taken from M. Epstein et al., *Anal. Chim. Acta*, **113**, 221 (1980).

Table 2.14. Absolute Limits of Detection (Aqueous Solution) by Laser-excited Atomic Fluorescence Spectrometry in Graphite Atomizer and by Electrothermal Atomizer Atomic Absorption Spectrometry (EAAAS)

Element	LOD (pg)	
	Laser AFS[a]	EAAS[b]
Ag	0.1	0.4
Co	0.06	3
Cu	0.15	2
Eu	300	2,000
Fe	0.1	5
Ir	6	2,000
Mn	0.2	1
Na	0.6†	50
Pb	0.0015	2
Pt	120	200
Sn	5†	0.6
Tl	0.1*	4

[a] All values from M. A. Bolshov, A. V. Zybin, and I. I. Smirenkina, *Spectrochim. Acta,* **36B,** 1143 (1981) except for those with † which are from P. Wittman, University of Florida, Ph.D. dissertation, 1982, and for those with *, which were taken from H. G. C. Human et al., *Spectrochim. Acta,* **39B,** (1984). It is worth mentioning that similar results were also obtained by J. Tilch et al., Proceedings of "Analytikteffen 1982," Neubrandenburg, DDR, November, 1982.
[b] Data from commercial literature (Perkin-Elmer, HGA-2100).

A good example of the much greater spectral selectivity of LEAFS (in the ICP) compared to ICPAES is given in Fig. 2.8. In the top system, the LEAFS spectrum of Y(II) in an iron matrix (100 μg/mL Y in 5000 μg/mL Fe) is given in the range ≈ 370–372 nm, and in the bottom spectrum, the ICPAES emission spectrum is given. It is quite obvious that in the fluorescence spectrum no problems arise with spectral interferences from Fe, whereas in the lower emission spectrum, Fe emission at 371.030 nm overlaps the 370.925 nm Y(II) line (I54), and one has therefore to search for another (perhaps less sensitive) emission line.

3.3. RF–ICP versus Flame Cells for Fluorescence

The flame has been used as an atom reservoir in most of the previous studies involving laser excitation. However, several publications (I53, I54, I56, I57)

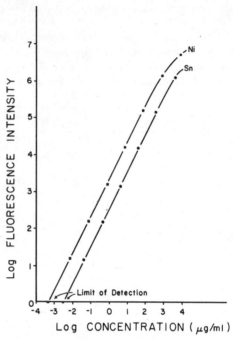

Fig. 2.5. Analytical calibration curves for nickel fluorescence excited at 300.249 nm and measured at approximately 342 nm and tin fluorescence excited at 300.914 nm and measured at 317.505 nm (I39).

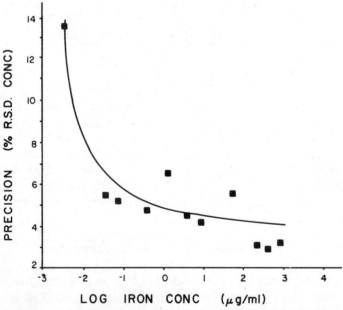

Fig. 2.6. Precision curve (% relative standard deviation vs concentration) for LEAFS measurement of Fe (I38). The Chromatix CMX-4 flashlamp dye laser in the single-pass configuration is used (see Fig. 2.3b).

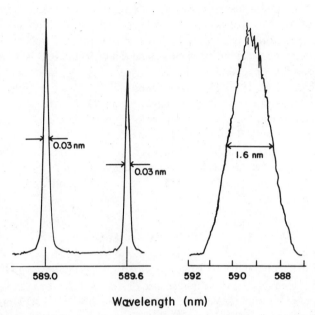

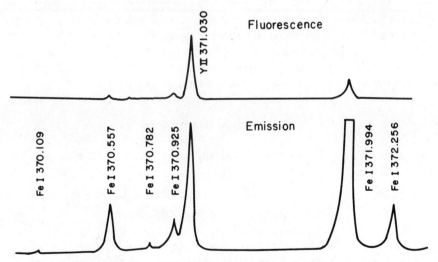

Fig. 2.7. Excitation and fluorescence profile of Na-*D* lines (at 589.0 nm and 589.6 nm) in an air–acetylene flame (I26). The monochromator slit width was 0.8 mm in both cases. (a) Profile observed by scanning the dye laser (excitation spectrum); (b) profile observed by scanning the emission monochromator (emission spectrum).

Fig. 2.8. Comparison of emission and fluorescence spectra of yttrium in an iron matrix. Conditions: 100 μg/mL Y in 5000 μg/mL Fe; RF power = 1 kW; 15 mm observation height above lead coil using conventional short torch; emission slit width = 0.02 mm (0.04 nm bandpass) and fluorescence slit width = 1 mm (2 nm bandpass) (I54).

have recently described the use of a radio-frequency Ar inductively coupled plasma as an atom–ion reservoir for laser-excited fluorescence. The following advantages of an ICP over a flame have been stressed: (i) both ion and atom lines can be excited; (ii) excitation of atoms and ions from levels up to ≈ 2 eV above the ground state can be efficiently done; and (iii) the ICP is a rather ideal atom–ion reservoir for several elements because of its high quantum efficiency. Indeed, Uchida et al. (I53, I54) have found that the quantum efficiencies of atoms with rather simple energy level diagrams, as Na, Li, and so on, approach unity low in the plasma, decreasing rather slowly with the increase in the height above the coil. Analytical studies with a pulsed N_2 dye laser ICP system (I53, I54) have so far been disappointing. However, the detection limits obtained with an excimer pumped dye laser (I56, I57) and a conventional ICP emission setup, were found to be superior to "standard" tabulated emission values for several of the elements investigated.

3.4. In Situ Elemental Analysis

Measures and Kwong (I34, I35) have reported on a technique based on laser ablation and selectively excited fluorescence (TAB LASER) for micro-ultratrace determination of in situ elements in solid samples. A block diagram of the instrumental approach is shown in Fig. 2.9. The authors were able to determine Cr in samples of NBS standard reference steel, doped skim milk, and doped flour. In their approach, a single pulse from an ablation laser (e.g., Q-switched ruby laser) strikes the sample target producing a plume that expands rapidly into a low-pressure region (10^{-3}–10^{-4} torr). After an appropriate delay to minimize Mie scatter from particulate debris and plasma background emission, the outward-streaming highly recombined plume of material is interrogated by a brief pulse (6 ns) of laser radiation that is tuned to one of the strong transitions of the analyte. The resulting saturated resonance fluorescence was directly related to the concentration of the trace element within the target.

3.5. Analytical Applications

Few applications of LEAFS to real analyses have been carried out. Several are listed in Table 2.15.

3.6. Diagnostics of Plasmas and Flames

The potentiality of LEAFS to the measurement of physical parameters in flames and plasmas with spatial ($\gtrsim 1$ mm^3), and temporal ($\gtrsim 1$ μs) resolution is apparent

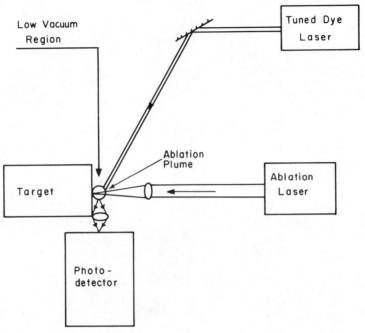

Fig. 2.9. Trace (element) analyzer based on laser ablation and selectivity excited radiation (redrawn from I34).

from the previous discussion. No attempt will be made to give a detailed evaluation of the use of LEAFS for average (multiple pulses) and single-pulse temperature, absolute concentration, rise velocity, and quantum efficiency measurements. The reader is referred to several excellent discussions of these parameters in the fundamental references (F4, F5, F9, F20, F21, F26, F28, F32, F33, F37, F40–42, F44–46, F50, F54, and other references in various journals, but especially *Appl. Optics*). The laser characteristics of vital importance for such studies include: high peak spectral irradiance (optical saturation possible); short pulse width (single pulse measurement possible); collimation and small beam size (spatial resolution possible).

4. EPILOGUE

The great excitement evident in the early workers on LEAFS is still present. However, the researchers of LEAFS have become more realistic in recent years on where LEAFS will find analytical usefulness (refer to Tables 2.1 and 2.2). In Table 2.16, a summary of the present status of LEAFS for analytical use is given.

Table 2.15. Determination of Various Elements in Different Materials by Laser-excited Atomic Fluorescence Spectrometry

Sample	Element	LEAFS Analysis	Certified Value	Reference
Trace element in waters (SRM-1643)	Fe	78 ng/g	75 ± 1 ng/g	138
	Ni	50 ± 2 ng/g	44 ± 1 ng/g	139
Unalloyed copper (SRM-394)	Fe	145 ± 6 μg/g	147 ± 8 μg/g	138
	Sn	66 ± 3 μg/g	65 ± 5 μg/g	139
Unalloyed copper (SRM-396)	Sn	0.7 ± 0.2 μg/g	0.8 ± 0.3 μg/g	138
	Ni	4.1 ± 0.1 μg/g	4.2 ± 0.1 μg/g	139
Fly ash (SRM-1633)	Fe	6.2 ± 0.2 %	6.2 ± 0.3 %	138
	Ni	99 ± 3 μg/g	98 ± 3 μg/g	139
Blood	Pb	8–40 μg/100 mL	—	158
Soil extracts	Fe, Co, Cu	10^{-9}–10^{-4} %	—	149

Table 2.16. Where Will Laser-excited Atomic Fluorescence Spectrometry Find Analytical Usefulness?

1. Specialized analyses where improved sensitivity, detection power, and spectral selectivity needed.
2. Optimization studies where combustion (plasma) diagnostics needed (temperature profiles, species profiles, deexcitation rate constants).
3. Spatial profile studies where sample is ablated by a separate means and plume is then excited.
4. Remote-sensing analysis, where the sample (solid, liquid, gas) can be directly addressed with a laser beam.
5. Semiquantitative absolute analysis, where calibration standards are not available.

REFERENCES

Reviews (R)

R1. N. Omenetto, L. M. Fraser, and J. D. Winefordner, *Appl. Spectrosc. Rev.*, **7**, 147 (1973).
R2. R. Browner, *Analyst*, **99**, 617 (1974).
R3. T. S. West, *Analyst*, **99**, 886 (1974).
R4. J. D. Winefordner, *Chem. Tech.*, **2**, 123 (1975).
R5. V. A. Razumov, *Zh. Anal. Khim.*, **32**, 596 (1977).
R6. K. Schofield, *J. Quant. Spectrosc. Radiat. Transfer*, **17**, 13 (1977).
R7. N. Omenetto and J. D. Winefordner, Chapter 4 in *Analytical Laser Spectroscopy*, N. Omenetto, Ed., Wiley, New York, 1979.
R8. J. D. Winefordner, *J. Chem. Ed.*, **55**, 72 (1978).
R9. J. D. Winefordner, in *New Applications of Lasers to Chemistry*, G. M. Hieftje, Ed., ACS Symposium Series No. 85, Washington, D.C., 1978.
R10. K. Sakurai, *Gendai Kagaku*, **92**, 29 (1978).
R11. C. G. Morgan. *Chem. Soc. Rev.*, **8**, 367 (1979).
R12. V. A. Razumov and A. M. Zvyagintsev, *Zh. Prikl. Spektrosk.*, **31**, 381 (1979).
R13: R. A. Keller and J. C. Travis, Chapter 7 in *Analytical Laser Spectroscopy*, N. Omenetto, Ed., Wiley, New York, 1979.
R14. N. Omenetto and J. D. Winefordner, *Prog. Anal. At. Spectrosc.*, **2**, 1 (1979).
R15. T-T, Lin, *Fen. Hsi. Hua. Hsueh.*, **8**, 567 (1980).
R16. S. J. Weeks and J. D. Winefordner, in *Lasers in Chemical Analysis*, G. M. Hieftje, J. C. Travis, and F. E. Lytle, Eds., Humana Press, Clifton, N.J., 1981.
R17. N. Omenetto and J. D. Winefordner, *CRC Crit. Rev. Anal. Chem.*, **13**, 59 (1981).
R18. J. D. Winefordner, in *Recent Advances in Analytical Chemistry*, K. Fuwa, Ed., Pergamon Press, Oxford, 1982.
R19. N. A. Narasimham, *Pure Appl. Chem.*, **54**, 841 (1982).

Fundamental Studies (F)

F1. L. Allen and J. H. Eberly, *Optical Resonances and Two Level Atoms*, Wiley, New York, 1975.

F2.   R. H. Pantell and H. E. Puthoff, *Fundamentals of Quantum Electronics*, Wiley, New York, 1969.

F3.   A. Yariv, *Introduction to Optical Electronics*, Holt, Rinehart, and Winston, New York, 1971.

F4.   C. Th. J. Alkemade, Tj. Hollander, W. Snelleman, and P. J. Th. Zeegers, *Metal Vapours in Flames*, Pergamon, Oxford, 1982.

F5.   D. R. Crosley, *Laser Probes for Combustion Chemistry*, ACS Symposium Series 134, Washington, D.C. 1979.

F6.   A. Corney, *Atomic and Laser Spectroscopy*, Clarendon Press, Oxford, 1977.

F7.   N. Omenetto, Ed., *Analytical Laser Spectroscopy*, Wiley, New York, 1979.

F8.   J. Steinfeld, Ed., *Laser and Coherence Spectroscopy*, Plenum, New York, 1978.

F9.   N. Omenetto and J. D. Winefordner, Chapter 4 in *Analytical Laser Spectroscopy*, N. Omenetto, Ed., Wiley, New York, 1979.

F10. R. M. Measures, *J. Appl. Phys.*, **39**, 5232 (1968).

F11. A. B. Rodrigo and R. M. Measures, *IEEE J. Quantum Electronics*, **QE-9**, 972 (1973).

F12. E. H. Piepmeier, *Spectrochim. Acta.*, **27B**, 431 (1972).

F13. E. H. Piepmeier, *Spectrochim. Acta.*, **27B**, 445 (1972).

F14. N. Omenetto and J. D. Winefordner, *Appl. Spectrosc.*, **26**, 555 (1972).

F15. N. Omenetto, P. Benetti, L. P. Hart, J. D. Winefordner, and C. Th. J. Alkemade, *Spectrochim. Acta*, **28B**, 289 (1973).

F16. N. Omenetto, L. P. Hart, P. Benetti, and J. D. Winefordner, *Spectrochim. Acta*, **28B**, 301 (1973).

F17. B. L. Sharp and A. Goldwasser, *Spectrochim. Acta*, **31B**, 431 (1976).

F18. J. Kuhl, S. Neumann, and M. Kriese, *Z. Naturforsch*, **28A**, 273 (1973).

F19. C. Th. J. Alkemade, Talk at 20th Colloquim Spectroscopicum Internationale, Prague, 1977.

F20. J. W. Daily, *Appl. Optics*, **16**, 2322 (1977).

F21. J. W. Daily, *Appl. Optics*, **16**, 568 (1977).

F22. B. Smith, J. D. Winefordner, and N. Omenetto, *J. Appl. Phys.*, **48**, 2676 (1977).

F23. M. A. Bolshov, A. V. Zybin, V. G. Koloshnikov, and K. N. Koshelev, *Spectrochim. Acta*, **32B**, 279 (1977).

F24. C. A. Van Dijk, *Optics Commun.*, **22**, 343 (1977).

F25. G. D. Boutilier, J. D. Bradshaw, S. J. Weeks, and J. D. Winefordner, *Appl. Spectrosc.*, **31**, 307 (1977).

F25a. N. Omenetto, G. D. Boutilier, S. J. Weeks, B. W. Smith, and J. D. Winefordner, *Anal. Chem.*, **49**, 1076 (1977).

F26. J. W. Daily, *Appl. Optics*, **16**, 568 (1977).

F27. D. R. DeOlivares and G. M. Hieftje, *Spectrochim. Acta*, **33B**, 79 (1978).

F28. J. W. Daily. *Appl. Optics.* **17**, 225 (1978).

F29. C. A. Van Dijk, P. J. Th. Zeegers, G. Nienhuis, and C. Th. J. Alkemade, *J. Quant. Spectrosc. Radiat. Transfer*, **20**, 55 (1978).

F30. G. D. Boutilier, M. B. Blackburn, J. M. Mermet, S. J. Weeks, H. Haraguchi, J. D. Winefordner, and N. Omenetto, *Appl. Optics*, **17**, 2291 (1978).

F31. J. S. Hosch and E. H. Piepmeier, *Appl. Spectrosc.*, **32**, 444 (1978).

F32. K. G. Muller and M. Stania, *J. Appl. Phys.*, **49**, 5801 (1978).

F33. N. Omenetto and J. D. Winefordner, *Prog. Anal. At. Spectrosc.*, **2**, 1 (1979).

F34. E. H. Piepmeier, Chapter 3 in *Analytical Laser Spectroscopy,* N. Omenetto, Ed., Wiley, New York, 1979.

F35. C. A. Van Dijk, P. J. Th. Zeegers, and C. Th. J. Alkemade, *J. Quant. Spectrosc. Radiat. Transfer,* **21,** 115 (1979).

F36. R. A. Van Calcar, M. J. M. Van de Ven, B. K. Van Uitert, K. J. Bienwenga, Tj. Hollander, and C. Th. J. Alkemade, *J. Quant. Spectrosc. Radiat. Transfer,* **21,** 11 (1979).

F37. J. D. Berg and W. L. Shackleford, *Appl. Optics,* **18,** 2093 (1979).

F38. G. C. Turk, J. L. Travis, J. R. DeVoe, and T. C. O'Haver, *Anal. Chem.,* **51,** 1890 (1979).

F39. J. C. Travis, P. K. Schenck, G. C. Turk, and W. G. Mallard, *Anal. Chem.,* **51,** 1516 (1979).

F40. J. W. Daily, in *Laser Probes for Combustion Chemistry,* D. Crosley, Ed., ACS Symposium Series 134, Washington, D.C., 1979.

F41. J. W. Daily, Paper at 6th Biennial Turbulance Symposium, University of Missouri, Rolla, 1979.

F42. M. B. Blackburn, J. M. Mermet, G. D. Boutilier, and J. D. Winefordner, *Appl. Optics,* **18,** 1804 (1979).

F43. N. Omenetto, M. S. Epstein, J. D. Bradshaw, S. Bayer, J. J. Horvath, and J. D. Winefordner, *J. Quant Spectrosc. Radiat. Transfer,* **22,** 287 (1979).

F44. C. H. Muller, K. Schofield, and M. Steinberg, *J. Chem. Phys.,* **72,** 6620 (1980).

F45. C. Chan and J. W. Daily, *Appl. Optics,* **19,** 1357 (1980).

F46. R. P. Lucht, D. W. Sweeney, and N. M. Laurendeau, *Appl. Optics,* **19,** 3295 (1980).

F47. N. Omenetto, J. Bower, J. Bradshaw, C. A. Van Dijk, and J. D. Winefordner, *J. Quant. Spectrosc. Radiat. Transfer,* **24,** 147 (1980).

F48. C. A. Van Dijk and C. Th. J. Alkemade, *Combust. Flame,* **38,** 37 (1980).

F49. C. A. Van Dijk, N. Omenetto, and J. D. Winefordner, *Appl. Spectrosc.,* **35,** 389 (1981).

F50. D. R. DeOlivares and G. M. Hieftje, *Spectrochim. Acta.,* **36B,** 1059 (1981).

F51. R. M. Measures, P. G. Cardinal, and G. W. Schinn, *J. Appl. Phys.,* **52,** 1269 (1981).

F52. C. Th. J. Alkemade, *Appl. Spectrosc.,* **35,** 1 (1981).

F53. N. Omenetto, *Spectrochim. Acta,* **37B,** 1009 (1982).

F54. B. T. Ahn, G. J. Bastiaans, and F. Albahodily, *Appl. Spectrosc.,* **36,** 106 (1982).

F55. N. Omenetto, C. A. Van Dijk and J. D. Winefordner, *Spectrochim. Acta,* **37B,** 703 (1982).

F56. H. Uchida, M. A. Kosinski, N. Omenetto, and J. D. Winefordner, *Spectrochim. Acta B,* **38B,** 529 (1983).

F57. H. Falk, *Prog. Anal. At. Spectrosc.,* **3,** 181 (1980).

F58. H. Uchida, M. A. Kosinski, N. Omenetto, and J. D. Winefordner, *Spectrochim. Acta,* **39B,** 63 (1984).

Instrumental/Methodology (I)

I1. M. B. Denton and H. V. Malmstadt, *Appl. Phys. Lett.,* **18,** 485 (1971).

I2. L. M. Fraser and J. D. Winefordner, *Anal. Chem.,* **43,** 1693 (1971).

I3. J. Kuhl and G. Marowsky, *Optics Commun.*, **4**, 125 (1971).

I4. L. M. Fraser and J. D. Winefordner, *Anal. Chem.*, **44**, 1444 (1972).

I5. N. Omenetto, N. N. Hatch, L. M. Fraser, and J. D. Winefordner, *Anal. Chem.*, **45**, 195 (1973).

I6. N. Omenetto, N. N. Hatch, L. M. Fraser, and J. D. Winefordner, *Spectrochim. Acta*, **28B**, 65 (1973).

I7. J. Kuhl, S. Neamann, and M. Kriese, *Z. Naturforsch.*, **28A**, 273 (1973).

I8. J. Kuhl and H. Spitschan, *Optics Commun.*, **7**, 256 (1973).

I9. S. Neumann and M. Kriese, *Spectrochim. Acta*, **29B**, 127 (1974).

I10. P. W. Pace and J. B. Atkinson, *J. Phys.*, **7E**, 556 (1974).

I11. F. C. M. Coolan and H. L. Hagedoorn, *J. Opt. Soc. Am.*, **65**, 952 (1975).

I12. W. M. Fairbank, T. W. Hansch, and A. L. Schawlow, *J. Opt. Soc. Am.*, **69**, 199 (1975).

I13. B. L. Sharp and A. Goldwasser, *Spectrochim. Acta*, **31B**, 431 (1976).

I14. R. B. Green, J. C. Travis, and R. A. Kellen, *Anal. Chem.*, **48**, 1954 (1976).

I15. H. L. Brod and E. S. Yeung, *Anal. Chem.*, **48**, 344 (1976).

I16. S. Mayo, R. A. Keller, J. C. Travis, and R. B. Green, *J. Appl. Phys.*, **47**, 4012 (1976).

I17. A. Elbern, Proc. Int. Symp. on Wall Interaction, Julich, RFG, 1976.

I18. M. A. Bolshov, A. V. Zybin, L. A. Zybina, V. G. Koloshnikov, and I. A. Majorov, *Spectrochim. Acta.*, **31B**, 493 (1976).

I19. J. A. Gelbwachs, C. F. Klein, and J. E. Wessel, *Appl. Phys. Lett.*, **30**, 489 (1977).

I20. J. A. Perry, M. F. Bryant, and H. V. Malmstadt, *Anal. Chem.*, **49**, 1702 (1977).

I21. B. Smith, J. D. Winefordner, N. Omenetto, *J. Appl. Phys.*, **48**, 2676 (1977).

I22. B. W. Smith, M. B. Blackburn and J. D. Winefordner, *Canad. J. Spectrosc.*, **22**, 57 (1977).

I22a. N. Omenetto, *Anal. Chem.*, **48**, 75A (1976).

I23. M. A. Bolshov, A. V. Zybin, V. G. Koloshnokov, and K. N. Koshelev, *Spectrochim. Acta*, **32B**, 279 (1977).

I24. M. Maeda, M. Matsumoto, and M. Yashushi, *Oyo Butsuri*, **46**, 1184 (1977).

I25. L. Paternack, P. A. Baronavski, and J. R. McDonald, *J. Chem. Phys.*, **69**, 4830 (1978).

I26. S. J. Weeks, H. Haraguchi, and J. D. Winefordner, *Anal. Chem.*, **50**, 360 (1978).

I27. J. P. Hohimer and P. J. Hargis, *Anal. Chim. Acta*, **97**, 43 (1978).

I28. D. A. Goff and E. S. Yeung, *Anal. Chem.*, **50**, 625 (1978).

I29. A. S. Gonchakov, N. B. Zorov, and Y. Y. Kuzakov, *J. Anal. Chem. (USSR)*, **34**, 2057 (1979).

I30. A. S. Gonchakov, N. B. Zorov, Y. Y. Kuzakov, and O. I. Matveev, *Anal. Lett.*, **12**, 1037 (1979).

I31. A. S. Gonchakov, N. B. Zorov, Y. Y. Kuzakov, and O. I. Matveev, *J. Anal. Chem. (USSR)*, **34**, 2312 (1979).

I32. J. P. Hohimer and P. J. Hargis, *Anal. Chem.*, **51**, 430 (1979).

I33. B. D. Pollard, M. B. Blackburn, S. Nikdel, A. Massoumi, and J. D. Winefordner, *Appl. Spectrosc.*, **33**, 5 (1979).

I34. R. M. Measures and H. S. Kwong, *Appl. Optics*, **18**, 281 (1979).

I35. H. S. Kwong and R. M. Measures, *Anal. Chem.*, **51**, 428 (1979).

136. V. I. Balykin, V. A. Letkohov, and V. I. Mishin, *Zh. Eksp. Teor. Fiz.*, **77**, 2221 (1979).

137. E. Miron, R. David, G. Erez, S. Lavi, and L. A. Levin, *J. Opt. Soc. Am.*, **69**, 256 (1979).

138. M. S. Epstein, S. Bayer, J. Bradshaw, E. Voigtman, and J. D. Winefordner, *Spectrochim. Acta,* **35B**, 233 (1980).

139. M. S. Epstein, J. Bradshaw, S. Bayer, J. Bower, E. Voitman, and J. D. Winefordner, *Appl. Spectrosc.*, **34**, 372 (1980).

140. M. S. Epstein, S. Nikel, J. D. Bradshaw, M. A. Kosinski, J. N. Bower, and J. D. Winefordner, *Anal. Chim. Acta,* **113**, 221 (1980).

141. C. L. Pan, J. V. Prodan, W. M. Fairbanks, and C. Y. She, *Optics Lett.*, **5**, 459 (1980).

142. D. J. Ehrlich, R. M. Osgood, G. C. Turk, and J. C. Travis, *Anal. Chem.*, **52**, 1354 (1980).

143. R. P. Frueholz and J. A. Gelbwachs, *Appl. Optics,* **19**, 2735 (1980).

144. M. Lino, H. Yano, Y. Takubo, and M. Shimzau, *J. Appl. Phys.*, **52**, 6025 (1981).

145. A. W. Miziolek and R. J. Willis, *Optics Lett.*, **6**, 528 (1981).

146. A. W. Miziolek, *Anal. Chem.*, **53**, 118 (1981).

147. J. J. Horvath, J. D. Bradshaw, J. N. Bower, M. S. Epstein, and J. D. Winefordner, *Anal. Chem.*, **53**, 6 (1981).

148. M. A. Bolshov, A. V. Zybin, V. G. Koloshnikov, *Spectrochim. Acta,* **36B**, 345 (1981).

149. M. A. Bolshov, A. V. Zybin and I. I. Smirenkina, *Spectrochim. Acta,* **36B**, 1143 (1981).

150. M. Heaven, T. A. Miller, R. R. Freeman, J. C. White, and J. Bokor, *Chem. Phys. Lett.*, **86**, 458 (1982).

151. O. I. Matveev, N. B. Zorov, and Y. Y. Kuzyakov, *Vestn. Mosk. Univ. Ser. Z. Khim.*, **19**, 537 (1982).

152. M. O. Rodgers, J. D. Bradshaw, K. Lin, and D. D. Davis, *Optics Lett.*, **7**, 359 (1982).

153. H. Uchida, M. A. Kosinski, and J. D. Winefordner, *Spectrochim. Acta.*, **38B**, 5 (1983).

154. M. A. Kosinski, H. Uchida, and J. D. Winefordner, *Talanta,* **30**, 339 (1983).

155. N. Omenetto and H. G. C. Human, *Spectrochim. Acta,* **39B**, 1333 (1984).

156. H. G. C. Human, N. Omenetto, P. Cavalli, and G. Rossi, *Spectrochim. Acta,* **39B**, 1345 (1984).

157. N. Omenetto, H. G. C. Human, P. Cavalli, and G. Rossi, *Spectrochim. Acta,* **39B**, 115 (1984).

158. N. Omenetto, H. G. C. Human, P. Cavalli, and G. Rossi, *Analyst,* (1984).

159. R. P. Frueholz and J. A. Gelbwachs, *Spectrochim. Acta,* **39B**, 807 (1984).

160. J. Tilch, H. J. Paetzold, H. Falk, and K. P. Schmidt, "Analytiktriffen 1982," Atomspektroskopie, Neubrandenburg, DDR, November 1982, Abstract N.DV55.

161. N. Omenetto, L. P. Hart, and J. D. Winefordner, *Appl. Spectrosc.*, **38**, 619 (1984).

162. L. P. Hart, C. Th. J. Alkemade, N. Omenetto, and J. D. Winefordner, *Appl. Spectrosc.*, **39**, 677 (1985).

CHAPTER

3

LASER-ENHANCED IONIZATION IN FLAMES

ROBERT B. GREEN

Department of Chemistry
University of Arkansas
Fayetteville, Arkansas

1. INTRODUCTION

The flame has occupied a prominent position in the methodology of analytical spectrometry since the earliest beginnings of the science. The flame's role as a simple, versatile, and inexpensive atom reservoir has contributed to a variety of analytical techniques since the late 1800s. In contrast, the use of lasers for analytical chemistry has been a relatively recent development. Although the first laser was experimentally demonstrated in 1960, it was the early 1970s before applications that derived from intrinsic laser properties began to appear in the analytical chemistry literature. In laser-enhanced ionization (LEI) spectrometry,

75

the old and new have been combined to provide a highly sensitive and selective method of analytical atomic spectrometry (1, 2).

As with other spectroscopic techniques, the flame desolvates and atomizes the sample in LEI spectrometry. Then a pulsed dye laser tuned to an absorption transition of the analyte promotes the atom to an excited state from which it is collisionally ionized (see Fig. 3.1). The rate of collisional ionization is enhanced by laser excitation because the effective ionization potential is lowered, that is, the energy difference between the state occupied by the atom and its ionization potential is reduced. The laser-related current that is detected with electrodes is a quantitative measure of the concentration of the absorbing species. Laser excitation is necessary to increase the rate of collisional ionization sufficiently to permit trace determinations. Stepwise LEI involving the absorption of two photons of different energies via a common intermediate excited state is used to produce the maximum sensitivity (3, 4). In a few cases, simultaneous two-photon absorption may be the best scheme for populating an excited state. The detection process is essentially nonselective but excellent resolution is provided by the laser excitation source. LEI may proceed by photoexcitation (via one or

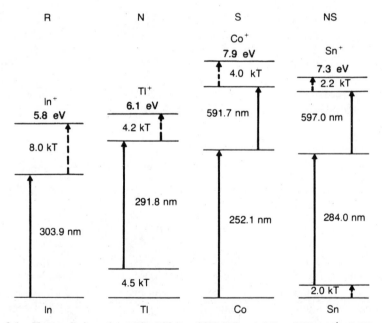

Fig. 3.1. Photoexcitation schemes for LEI in a 2500 K flame (kT = 1735 cm^{-1}). Solid arrows represent photoexcitation; dashed arrows are collisional excitation. Energy is the vertical axis. R = resonant, N = nonresonant, S = stepwise, resonant, NS = nonresonant, stepwise. From (2); reprinted by permission of the American Chemical Society.

more transitions) and collisional ionization or a combination of collisional excitation, photoexcitation, and collisional ionization.

A similar flame technique of more recent origin involves photoionization of the excited analyte atom. Direct laser ionization (DLI), sometimes referred to as dual-laser ionization (5), generally requires a minimum of two lasers. Since the second laser photoionizes the atom, this step may be accomplished by an off-resonant photon (see Fig. 3.2). Generally, a tunable dye laser excites the analyte atom to a bound state, and a fraction of the pump laser beam photoionizes it.

Therefore, DLI and LEI are similar in their implementation and methodology. Although single-wavelength LEI has produced the lowest detection limits in the one case where a comparison has been reported (5), LEI and DLI should be viewed as complementary techniques. A recent study demonstrated several cases where direct two-laser photoionization predominated over laser-enhanced collisional ionization although no detection limits were quoted (6). The enhancement of DLI signals over single-wavelength LEI depended on the energy difference between the laser-populated state and the analyte's ionization potential (i.e., the energy defect) and the energy match between the photoionizing laser and the energy defect. Since a nitrogen laser-pumped dye laser system was used for these experiments, the photoionizing laser wavelength was fixed. This condition led to considerable energy overshoots in some cases. Therefore, any improvement of the single-wavelength LEI signal was negated when the nitrogen laser beam was introduced into the excitation volume. In all cases, a hydrogen-oxygen–argon flame was the atom reservoir, so efficient collisional ionization was not favored. The predominance of DLI over LEI in a low-collisional environment has also been demonstrated by experiments in which Na was electrothermally atomized in a pure Ar atmosphere at room temperature (7). Under these conditions, the signal produced by laser excitation of Na atoms to a bound excited state was increased by addition of an ionizing laser beam.

When high-lying states are populated by the laser(s), collisional ionization efficiency approaches 100%. Two cases in which DLI was ultimately abandoned in favor of stepwise LEI illustrate the efficacy of collisional ionization when the energy defect is relatively small. The second harmonic of a Nd:YAG laser was

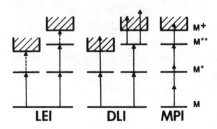

Fig. 3.2. Laser-induced ionization schemes in flames. LEI, laser-enhanced ionization; DLI, direct laser ionization; MPI, multiphoton ionization. Solid arrows represent photoexcitation; dashed arrows are collisional excitation. From (60); reprinted by permission of Springer Verlag, Berlin.

used to pump two dye lasers that excited sequential Na transitions (589.0 and 568.8 nm, respectively) with a common intermediate state (8). In addition, the 1064-nm YAG laser fundamental was introduced into the excitation volume in the flame to directly photoionize the excited Na atoms. The ionizing laser beam did not increase the signal. In a similar experiment, the YAG second harmonic was used to photoionize Li atoms that had been excited by two dye lasers tuned to stepwise transitions at 670.8 and 610.4 nm, respectively (9). As before, the photoionizing beam produced no additional signal, attesting to the efficiency of the collisional processes that are exploited in LEI spectrometry.

Single-wavelength LEI is the least complex and least expensive experiment, requiring only a flashlamp-pumped dye laser. At a minimum, DLI requires a laser-pumped dye laser that permits the use of a fraction of the pump laser beam for photoionizing the dye laser-excited atom. Stepwise LEI schemes require two tunable dye lasers, usually pumped by a third laser. Stepwise LEI provides additional selectivity over DLI that can be essential for the analysis of real samples. The additional complexity of a two-dye-laser system may be justified for DLI because it permits the optimization of the signal by tuning the second laser to the exact energy required for photoionization.

Resonance ionization spectrometry (RIS), which uses the same photoionization schemes as DLI, is another laser-induced ionization technique that has been applied to analytical determinations of trace metals (10–13). The difference, which is not indicated in the names, is that RIS is historically a nonflame method using specialized, low-background atom reservoirs. DLI may be referred to as RIS in flames; collisionally assisted RIS and stepwise LEI are synonomous (14). Although the similarities among these techniques are obvious, the distinction is in the significant differences in atom reservoirs. Whereas RIS is capable of single-atom detection under the proper experimental conditions (15), it would be unreasonable to expect similar sensitivity from DLI or LEI or any other techniques that utilized an atmospheric pressure flame as the sample reservoir.

Multiphoton ionization (MPI) may also occur in flames (16), but it has not been used for trace metal determinations. MPI has been used effectively for spectroscopy of organic molecules (17). MPI typically proceeds by the simultaneous absorption of three or more photons of the same wavelength via real and virtual states (see Fig. 3.2). A background signal may result from the MPI of added or native species in the flame. The noise carried by this additive signal may be a limiting factor since it is not as easily compensated for as the background itself.

As this introduction suggests, LEI and DLI share many of the properties of other atomic spectrometric methods while possessing unique qualities that may complement or supersede other techniques. No spectral dispersion is needed; the resolution is limited by the laser source. Scattered laser radiation, flame background emission, and ambient (room) light do not interfere with the electrical signal. The compatibility of the flame with these techniques improves the general utility of the methods. In addition, collisional processes in the flame are an asset rather than a liability for LEI. This chapter will focus primarily on LEI spectrometry because it is the more well-developed technique. Much of the work on LEI spectrometry applies equally well to DLI. Where DLI research is available, it has been integrated into the appropriate section to provide a more comprehensive view of laser-induced ionization spectrometry in flames. In combination, LEI and DLI have great potential for improving the methodology of analytical atomic spectrometry.

2. SIGNAL PRODUCTION AND COLLECTION

2.1. Ion Production

In its simplest form, LEI is a two-step process, photoexcitation and collisional ionization (see Fig. 3.1). It involves three quantum states: the atomic ground state, an atomic excited state, and an ionic ground state. For excited levels very near the ionization potential, ionization rates approach collision rates, giving ion yields near unity. The essential steps for LEI, photoexcitation and collisional ionization, are not the only processes occurring in an atmospheric pressure flame. An excited atom can also collisionally deactivate or fluoresce. A detailed description of signal production requires a complex expression involving several competing rate constants (18).

The probability of ionization of a given atom or molecule on collision is governed by the Arrhenius factor, $\exp\left[-(E_i - E_j)/kT\right]$ where E_i is the ionization potential and E_j is the energy level occupied by the atom or molecule, k is the Boltzmann constant, and T the flame temperature in degrees Kelvin. The collisional ionization probability for a low-lying atom may be increased by approximately two orders of magnitude by an electron volt of optical excitation, making LEI a viable approach for sensitive determinations of trace metals.

A helpful qualitative understanding of the dynamics of ion production in LEI may be gained by considering the hydrodynamic analogy illustrated in Fig. 3.3. The tubs represent the three energy levels, with the liquid levels indicating the atom or ion populations. (State multiplicities are ignored for simplicity, i.e., statistical weights are assumed to be equal.) The pumps representing the laser and thermal energy must have pumping rates proportional to the pressure head (population) as well as the rotational velocity of the pump rotor (laser power/ Arrhenius factor) for the analogy to be accurate.

In the absence of laser excitation, the fluid level in the top two tubs is negligible when compared with the bottom tub. Figure 3.3 illustrates the fluid levels some time after the laser pulse begins. The fluid levels (or populations)

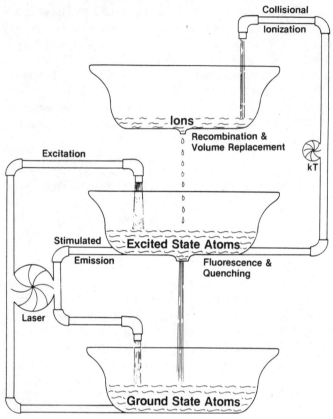

Fig. 3.3. Hydrodynamic analogy to LEI for a resonant photoexcitation scheme (R of Fig. 3.1).
From (2); reprinted by permission of the American Chemical Society. .

in the bottom two tubs have equilibrated under optical saturation. Collisional
ionization from the excited state relentlessly depletes the neutral population (the
bottom two tubs) while generating an ion population. In this example, the total
system has not yet equilibrated. Due to the relatively slow rate of ion–electron
recombination, ionization will approach completion if the laser remains on long
enough. The consequences of laser pulse duration and the ionization rate have
been examined, leading to the following rule of thumb: Unit ionization efficiency
may be approached with an optically saturating laser pulse whose duration significantly exceeds the reciprocal of the effective ionization rate of the laser-
populated excited state (2). The qualifier "effective" in the above statement
accommodates the effect of Boltzmann-populated states above the state in question (19–21).

2.2. Ion Collection

Once generated, the collection of analyte ions by applying an electric field is simple. Unfortunately, other ions in the flame will also be collected by this detection scheme. Therefore, the LEI signal can be influenced by high ion concentrations originating from the sample, the flame, and the analyte itself. The latter possibility is of little practical consequence because only IA and IIA elements have significant ion fractions in an acetylene–air flame, and samples containing these metals as analytes may be diluted if an electrical interference is a problem. Other instances of these electrical interferences with analyte signals have been the subject of many investigations (22–30). These studies have led to an evolution of the electrode design (Section 3.1.2) and a better understanding of the signal collection process.

The distribution of charge around a cylindrical probe extended into a flame may be divided into three regions (31a). If the probe is slightly negative, a region will form near the surface of the cathode where the electron concentration is much lower than the ion concentration. The excess positive charge at the cathode is a consequence of the large difference in the mobilities of ions and electrons. The velocity of electrons is 100–1000 times greater than positive ions at the same field strength (31b). This region has been referred to as the *sheath*. Just outside the sheath, a transition region exists where charge separation begins. In the bulk of the flame, the concentration of cations and electrons is essentially equal. The nonzero field necessary for LEI signal detection exists only in the sheath and the transition region. However, for large diameter or flat cathodes, with the order of -2000 V applied and typical flame ionization rates, the sheath may extend for 1 cm or more from the cathode.

As the concentration of ions in the flame increases, the collecting field withdraws toward the cathode surface. In the absence of a collecting field, no LEI signal will be recovered, accounting for the signal suppression observed at high ion concentrations with an external cathode (20–24). To a certain extent, this condition may be mitigated by increasing the applied voltage since the sheath expands with increased voltage. This approach is limited because when the applied voltage reaches a certain level, electrical breakdown (arcing) will occur through the flame.

Studies in which the voltage applied to the external split cathodes was pulsed have illustrated the formation of the capacitive double-layer responsible for signal suppression (28). The maximum LEI signal could be recovered only if the 1-μs laser pulse was delayed a minimum of 1.5 ms after the initiation of the 4-ms high-voltage pulse. Figure 3.4 shows the results of potential measurements made with a small-diameter rod replacing the laser beam while pulsing the applied voltage on the split cathodes. The burner head is used as the anode. These discharge time response curves indicate a time-dependent increase in the

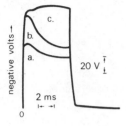

Fig. 3.4. Discharge time response curves in the flame measured at the laser beam position: (a) 100 μg/mL In, (b) 100 μg/mL In with 1 μg/mL K, (c) 100 μg/mL In with the cathodes moved into the flame. From (28); reprinted by permission of the American Chemical Society.

LEI signal collected by the external split cathodes. Addition of a low ionization potential concomitant further increases the time constant for field development. When the split cathodes are placed in contact with the flame, the maximum LEI signal is observed after only 200 μs. Using an immersed cathode, centered in the flame, and exciting near the electrode surface permitted the recovery of LEI signals after only 8 μs. These results strongly support the use of an immersed electrode (27). In addition to avoiding interferences by permitting excitation near the cathode surface where the field exists even at high ion concentrations, signal collection is improved by reducing the diffusion path in the flame. Observation of the LEI signal with laser delays less than 8 μs was not possible because the signal was obscured by a large voltage spike originating from the capacitor used to separate the signal from the background.

The voltage at which the sheath just extends to the anode is referred to as the *saturation voltage* (31c). At voltages higher than saturation, a nonzero field fills the region between the cathodes and the anode, and every ion and electron produced by thermal ionization is collected. For voltages above saturation, the *saturation current* is constant. Current versus voltage curves have been used to evaluate electrode designs and characterize interferences (27, 32).

Maps or images of LEI ions and electrons have been obtained by taking the signal from a small rod positioned between anode and cathode plates (32). Figure 3.5 shows the results of this experiment and the electrode configuration used. The rod is translated vertically to observe as a function of position the intercepted ions that were traveling to the electrode. At high voltage, the images for electrons and ions are centered at the same height above the burner head as the laser beam. Even at low voltages, although the images are shifted by the contributions of flame velocity and diffusion, essentially all of the signal is collected. Recombination, another possibility for loss of LEI signal, does not occur at a rate sufficient to deplete the ion concentration. Therefore, the probability of collecting 100% of the LEI signal is very high.

Because signal enhancement does not constitute as severe a problem for the analyst as suppression, it has not received as much attention. Although it may be advantageous in some cases, signal enhancement still qualifies as an inter-

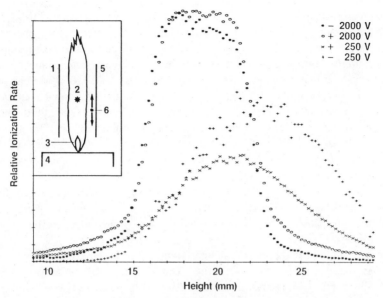

Fig. 3.5. Images of LEI ions and electrons, obtained by taking the LEI signal from a thin rod translated across the front of the normal collecting plate at the indicated high voltages (49). The experimental apparatus is shown in the inset: (1) high-voltage repelling plate, (2) laser beam, (3) flame reaction zone, (4) burner head, (5) low-voltage electrode plate, (6) vertically movable signal pick-off wire. From (49); reprinted by permission of the American Chemical Society.

ference. The immersed electrode largely eliminates electrical interferences, at least for acetylene–air flames, but signal enhancement still occurs when a high concentration of ions is present. Early speculation on the cause of LEI signal enhancement suggested that it might be related to the increasing field strength in the sheath as it compressed toward the cathode when a low ionization potential concomitant was added. This effect was predicted by electric field maps (27). A measured increase in the field strength produced by the addition of ions to the flame was observed experimentally (28).

More recent studies of pulse shape have added to the understanding of LEI (29) and provided an explanation for signal enhancement (30). Instead of the filtering and amplification normally used for analytical work, these studies used a faster, low-gain preamplifier to avoid distortion of the LEI pulse shape. A point charge model based on charge induction theory was developed to describe the time-resolved signal measurements, which consisted of an electron pulse followed by a relatively small ion pulse (30). The latter pulse is generally ignored by the gated integrator and has no analytical value because of its poor signal-to-noise ratio. Theoretical simulations and experimental measurements of current versus time curves for iron in sodium matrices were generated using the point charge model. These results demonstrate that the electron pulse sharpens as the matrix concentration increases. The higher peak currents are the result of sheath compression, which produces a higher local electric field and consequently faster moving charges. The slower ion pulse shifts to shorter arrival times with increasing sodium matrix concentrations. Under normal analytical measurement conditions with a 1-MHz frequency response preamplifier, most of the ion current is not measured for sodium-free samples. With 1000-μg/mL sodium matrix, both the electron and ion current are measured. Therefore, experimentally observed signal enhancements with sodium matrices are attributed to the increased percentage of the ion signal that is integrated. Further experimental verification, yielded large discrepancies for matrix concentrations of 1000 μg/mL, which suggested that other mechanisms of signal enhancement may prevail at high matrix concentrations. Field ionization may be a possible explanation (31d).

Limited studies of the electrical interferences for DLI have been reported (33), and these results are not definitive. The laser-induced ionization signals excited by a nitrogen laser-pumped dye laser were measured with two wire probes immersed in a hydrogen–oxygen–argon flame. In these experiments, the DLI signal results from a dye laser and a nitrogen laser photon; the LEI signal is produced by blocking the nitrogen laser beam. Under these conditions, it is necessary to increase the Na concentration by a factor of 10 to provide a LEI signal that is comparable to the DLI signal. When Cs is added as a concomitant, suppression of a Na LEI signal occurs well before suppression of the DLI signal. Generalization of these results might lead one to believe that DLI is more resistant to electrical interferences than LEI. On the contrary, these results are consistent with previous LEI studies (25, 26). The much higher concentration of Na required for the LEI measurement in this case contributes significantly to the suppression of its own signal. Other data also suggest the formation of a cation sheath, and it is concluded that the maximum signal recovery for DLI occurs when the point of ionization is near the cathode (33). Since electrical interferences are related to signal collection, not signal production, there is no reason to suspect that DLI and LEI spectrometry differ in this respect.

3. ANALYTICAL CONSIDERATIONS

3.1. Instrumentation

This section will discuss the components that comprise an LEI spectrometer. A schematic diagram of the typical instrumentation for LEI spectrometry is illustrated in Fig. 3.6.

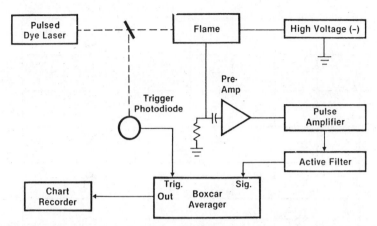

Fig. 3.6. Block diagram of a typical LEI spectrometer. From (60); reprinted by permission of Springer-Verlag, Berlin.

3.1.1. Burners

The bulk of LEI spectrometry and all of the detection limits reported in Fig. 3.7 except for Cs and Rb have been accomplished using a premixed burner. In a premixed burner, the nebulized sample is mixed with the fuel and oxidant prior to combustion. Premixing and the subsequent removal of large liquid droplets result in a quiet, laminar flow flame that accounts for the widespread use of the premixed burner for optically detected flame spectrometry. Long pathlength slot burner heads are commonly employed, but the choice of fuel–oxidant mixtures is limited if explosive flashback is to be avoided. A 10-cm slot burner head with an acetylene–air flame has been used predominantly for LEI spectrometry (22–28), but results for a 5-cm slot burner head with an acetylene–nitrous oxide flame have also been reported (34) and are included in Fig. 3.7. The acetylene–air flame is the sample reservoir of choice for a large number of metals except those that form refractory oxides in the flame. The higher temperature nitrous oxide flame decomposes these oxides, rendering them accessible to atomic spectrometry. Until the advent of the immersed electrode, the nitrous oxide flame was unusable for analytical determinations because of the high level of electrical noise generated by the flame.

A hydrogen–oxygen–argon mixture has been used for analytical DLI spectrometry (5). The sample is aspirated into a circular central flame, surrounded by a mantle, supported on a Meker burner. The hydrogen–oxygen–argon flame excludes molecular nitrogen and therefore should favor photoionization over collisional ionization.

Time resolution of LEI signals has been demonstrated as an approach for discriminating against electrical interferences using a premixed burner (28).

LEI PERIODIC TABLE

1																1	2
3 *Li* 0.001 R	4 Be											5 B	6	7	8	9	10
11 *Na* 0.03 R	12 *Mg* 0.1 R											13 *Al* 0.1 N	14 Si	15 P	16 S	17	18
19 *K* 0.1 R	20 *Ca* 0.1 N	21 *Sc* 0.2 N	22 *Ti* 600 N	23 *V* 2 N	24 *Cr* 2 N	25 *Mn* 0.02 S	26 *Fe* 2 R	27 *Co* 0.08 S	28 *Ni* 0.08 NS	29 *Cu* 0.07 S	30 Zn	31 *Ga* 0.07 R	32 Ge	33 As	34 Se	35	36
37 *Rb* 0.1 R	38 *Sr* 4 R	39 *Y* 10 N	40 Zr	41 Nb	42 Mo	43 Tc	44 Ru	45 Rh	46 Pd	47 *Ag* 1 R	48 *Cd* 0.1 S	49 *In* 0.006 R	50 *Sn* 0.3 NS	51 Sb	52 Te	53 I	54
55 *Cs* 0.004 R	56 *Ba* 0.2 R	57 La	72 Hf	73 Ta	74 W	75 Re	76 Os	77 Ir	78 Pt	79 *Au* 1 S	80 Hg	81 *Tl* 0.09 N	82 *Pb* 0.09 S	83 *Bi* 2 R	84	85	86
87	88	89	104														
				58 Ce	59 Pr	60 Nd	61	62 Sm	63 Eu	64 Gd	65 Tb	66 Dy	67 Ho	68 Er	69 *Tm* 200 R	70 Yb	71 *Lu* 0.2 N
				90 Th	91	92 U	93	94	95	96	97	98	99	100	101	102	103

Fig. 3.7. LEI periodic chart of the elements, indicating experimental limits of detection (in ng/mL) and the excitation scheme for the elements observed to date (see Fig. 3.1 for definitions of the excitation schemes.) Other elements shown are expected to yield LEI signals in flames. Omitted elements are not amenable to flame spectrometry. Detection limits were obtained from the following sources: Na, K (40); Al, Sc, Ti, V, Y, Tm, Lu (34); Rb, Cs (61); all other elements (2).

When a sample is aspirated into the premixing chamber, it is diluted prior to introduction into the flame. The sample concentration reaches a steady-state value in the flame a short time after aspiration begins. The sequence is reversed when sample aspiration is terminated. Figure 3.8 illustrates the sequence of events when 100 μg/mL In is the analyte and 5 μg/mL K is the concomitant. The In signal reaches a maximum a few milliseconds after aspiration begins and then is progressively suppressed as the K concentration in the flame increases. The great sensitivity of LEI permits observation of a signal from an extremely small flame concentration of the analyte before the concomitant concentration in the flame reaches a level where suppression of the analyte signal begins. This approach is equivalent to volumetric sample dilution to minimize concomitant interferences.

Recently, a total consumption (turbulent flow) burner has been examined for LEI spectrometry (35). Although excellent limits of detection have been reported for LEI spectrometry with premixed burners, a total consumption burner has several potential advantages. In the total consumption burner, the entire sample is aspirated into the flame. This should result in more analyte atoms in the flame, even considering unvaporized solution, producing a net gain in sensitivity. LEI detection is insensitive to scattered source light and so should not be limited by scattering from unvaporized solution droplets. Since the fuel and oxidant are mixed at the burner nozzle, there is no possibility of explosive flashback, regardless of the fuel and oxidant. Rapid-propagating, high-temperature, oxygen-based flames will be available for LEI spectrometry with the total consumption burner, permitting more effective utilization of atomic transitions originating in excited states. A total consumption burner with a hydrogen–oxygen or acetylene–oxygen flame presents an alternative for LEI spectrometry of metals that form refractory oxides.

Preliminary experiments with total consumption burners have resulted in the observation of substantial LEI signals in hydrogen–air, acetylene–air, hydrogen–oxygen, and acetylene–oxygen flames. Possible electrode configurations for LEI spectrometry with the total consumption burner have been evaluated using indium as the analyte and an acetylene–air flame. As shown in Table 3.1, the detection limits obtained for indium using the 2.5-cm pathlength total consumption burner with the optimum electrode configuration are comparable to

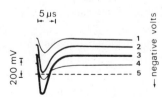

Fig. 3.8. Oscilloscope traces of sequential LEI signals for 100 μg/mL In with 5 μg/mL K. Note that while the transient signal increases to a maximum (1–3) and then decreases to zero with time (4, 5), the background increases linearly. From (28); reprinted by permission of the American Chemical Society.

Table 3.1. Comparison of the Best LEI Detection Limits for Indium, Copper, and Manganese Using Total Consumption and Premixed Burners

Element	Burner	Limits of Detection[a] (ng/mL)	
		Exptl	Adj
In	Total consumption[b]	0.12	0.3
	Premixed[b]	0.023	0.023
	Premixed[c]	0.006	0.006
Cu	Total consumption[b]	1.5	0.4
	Premixed[b]	2.2	2.2
	Premixed[d]	3	3
Mn	Total consumption[e]	0.09	0.02
	Premixed[e]	0.2	0.2
	Premixed[d]	0.3	0.3

[a] Detection limits for indium were determined using a flashlamp-pumped dye laser. Detection limits for copper and manganese were determined with a Nd:YAG laser-pumped dye laser. "Exptl" refers to experimental data while "Adj" is data adjusted for a 10-cm pathlength.
[b] Ref. 35.
[c] Ref. 2.
[d] Ref. 4.
[e] Ref. 36.

the values determined for a premixed burner with a 10-cm burner head if the pathlength difference is considered (35). For elements, such as indium, where extremely low LEI detection limits have been obtained, further improvements in sensitivity may be limited by the analyte atom fraction, which is determined by the fuel–oxidant combination, not sample introduction. The detection limits for copper and manganese were lowered using the total consumption burner (Table 3.1).

A background current has been observed when using both total consumption and premixed burners for LEI spectrometry with high-power Nd:YAG laser-pumped dye lasers (35). The background signal is related to water aspiration and has a two-photon dependence on laser power (36). This background may be due to direct two-photon photoionization (MPI) or two-photon excitation and LEI of a species in the flame. The background signal may be compensated for by subtraction, but the additional noise is detrimental to sensitivity. DLI also suffers from an electrical interference that is attributed to off-resonant multiphoton ionization of easily ionized matrix components (33).

In practice, the best LEI detection limits are obtained by reducing the incident laser irradiance until the background diminishes to zero (35). With a total consumption burner, a Nd:YAG laser-pumped dye laser can be operated at full power without inducing the background ionization signal by expanding the beam.

The larger excitation volume is also compatible with the total consumption burner's circular geometry and more widely distributed atom concentration. Therefore, beam expansion improves the total consumption burner detection limits significantly while having less effect on premixed burner results. When the detection limits are adjusted for the pathlength difference, the total consumption burner results are over a factor of 6 better for Cu and 10 times better for Mn than the best premixed burner detection limits (Table 3.1). The premixed burner detection limits reported in this study are also the best published single-wavelength LEI results for Cu and Mn using this burner.

The detection limits obtained with the total consumption burner may also be improved by confinement of the sample to a smaller volume and pathlength extension, provided the sample has been desolvated. Otherwise, the noise carried by the ionization background will overcome any improvements in signal. Samples have been desolvated electrothermally in a graphite rod furnace and then swept into the total consumption burner with an inert gas during a second heating cycle (36). The LEI signals for identical concentrations of desolvated and solvated Mn are compared in Fig. 3.9. The reduction in both the background and noise by sample desolvation is apparent. The improvement in Mn detection limits with desolvation is also illustrated in Table 3.2.

Once the sample is desolvated, strategies for lowering detection limits may be implemented. Although several methods have been investigated, the most successful approach involves using a 10-mm diameter Vycor tube to confine the

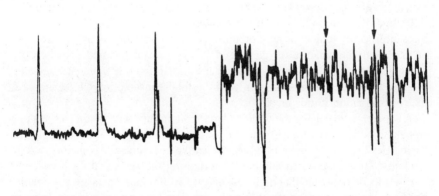

Fig. 3.9. Comparison of the signal to noise and background for desolvated and solvated Mn samples. The results of five experiments are shown. The arrows indicate the position of manganese signals buried in noise caused by MPI when a solvated sample is aspirated.

Table 3.2. LEI Detection Limits for Manganese Using a Total Consumption Burner with Graphite Furnace Desolvation and a Nd:YAG Laser-Pumped Dye Laser (36)

Water Present	Desolvated	Chimney Vertical, Desolvated	Chimney Horizontal, Desolvated
32 ng	5 ng	0.8 ng	0.4 ng

flame (see Table 3.2). The anode and cathode are positioned just above the 10-mm long Vycor chimney, which is flared at the bottom to accommodate the flame. The best detection limits are obtained by rotating the total consumption burner 90° to a horizontal position. A shorter Vycor tube was used to confine the flame. The streaming velocity of the flame gases is sufficient to maintain collimation for almost 8 cm beyond the Vycor tube. Approximately 2.5-cm long electrodes are positioned horizontally just beyond the tube, and the laser beam is directed axially between the electrodes, through the center of the flame.

Discrete sample introduction permits the use of microliter samples while continuous aspiration requires much larger volumes. The absolute detection limits for the total consumption burner coupled with graphite furnace desolvation reported as concentrational detection limits were not as good as those obtained by direct, continuous aspiration of the sample into the total consumption burner. No detailed study of the furnace-burner system was undertaken, but inefficiencies in the atomization and/or transfer system may account for the poorer limits of detection. Also there was no attempt to synchronize the laser pulse with the furnace atomization cycle. Further refinements of the present system await investigation of other desolvation approaches that may prove more satisfactory.

Routine desolvation also permits the maintenance of intrinsic flame temperatures afforded by total consumption burners. Total aspiration of a water solvent can cool high-temperature oxygen-based flames several hundred degrees Celsius, negating any advantages of these flames for refractory elements. Solvent stripping techniques for desolvation with the total consumption burner are also under investigation in this laboratory.

LEI spectrometry using the total consumption burner, with its greater sample throughput and a wider range of usable fuel-oxidant combinations, expands the possibilities for development of a more sensitive and versatile detection system for atomic spectrometry. In addition to furthering the analytical methodology, these results demonstrate that high-sensitivity LEI measurements are possible in an adverse sample environment where traditional methods of optical spectrometry have proven inadequate.

An electrothermal atomizer has been incorporated into a burner for introduction of the sample into a diffusion flame for LEI spectrometry (37). The sample is applied to a wire loop positioned just below the level of the burner head. The burner head is concentric with the tube that contains the sample loop. After the flame is ignited, the wire loop is electrically heated and the vaporized sample swept into the flame by a stream of argon. Programmed temperature control permits some discrimination against interferences from the sample matrix due to differences in the volatilization temperatures of the analyte and the interferent.

A graphite crucible and furnace have been used directly as atom reservoirs for LEI spectrometry (7). In the latter case, a 1-mm diameter tungsten rod electrode was inserted along the axis of the graphite tube. The tube itself serves as the other electrode. Less than 300 V were applied to the rod to avoid breakdown between the electrodes. Aqueous solutions of sodium, cesium, indium, and ytterbium were investigated with single-wavelength and stepwise excitation. Although the precision was poor, the analytical signal was 10–100 times greater when compared with flame ionization, suggesting that nonflame atomizers have promise for LEI spectrometry.

3.1.2. Electrodes

Many electrode configurations have been used with comparable success for determining the concentration of metals in pure, aqueous solutions. Early LEI measurements were made with a cathode in the flame (see Fig. 3.10a). The signal was taken off the burner head, which served as the anode. Soon thereafter the cathode was split and the electrodes were placed just outside the flame (Fig. 3.10b). This configuration was attractive because it was nonintrusive and the tungsten electrodes were not subject to deterioration in the flame. Unfortunately, this configuration was susceptible to severe electrical interferences.

Figure 3.11 illustrates the LEI signal behavior for the external split cathode when the aqueous samples contain different concomitants with low ionization potentials in addition to the indium analyte. The laser was tuned to the 303.9-nm indium line. Although electrical interferences are significant only when the concomitants are IA or some IIA elements, this interference makes the analysis of some real samples more difficult. For this reason, it is worthwhile to understand the cause of the interferences and discuss possible remedies.

Figure 3.12 demonstrates the improvement in LEI signal recovery that occurs when the diameter of cylindrical electrodes is increased. The use of planar (plate) cathodes considerably improves the tolerance of the LEI signal to high levels of ionization in the flame (26). The reduction in electrical interferences is due to the commensurate reduction in the field strength at the electrode surface. The high fields near small-diameter rods exacerbate the interference. As previously mentioned, more LEI signal may be recovered by increasing the applied voltage,

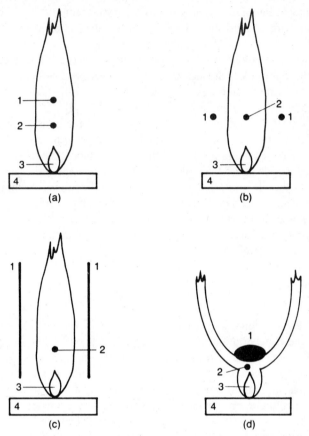

Fig. 3.10. Evolution of the electrode configurations used for LEI. 1 = high-voltage electrode, 2 = laser beam, 3 = reaction zone, 4 = burner head. (a) and (b) 1 is a rod cathode. (c) 1 is a plate cathode. (d) 1 is a water-cooled cathode. The burner head, 4, is the anode in all cases. From (2); reprinted by permission of the American Chemical Society.

but arc-over will occur at a voltage dictated by the ion concentration and electrode configuration.

As seen in Fig. 3.12, an immersed cathode (Fig. 3.10d) provides the most resistance to electrical interferences in an acetylene–air flame (27). This electrode may be constructed by slightly flattening a 0.25-in. diameter stainless steel tube. Water circulation through the electrode prevents degradation of the electrode surface. Since the cathode is positioned in the flame, the region near its surface is a suitable sampling volume. With external cathodes (Figs. 3.10b and 3.10c), the cathode surface is not adjacent to the region of maximum atom concentration, and the cathodes are separated from the flame by an air gap. With the immersed cathode, ions produced by laser enhancement remain within

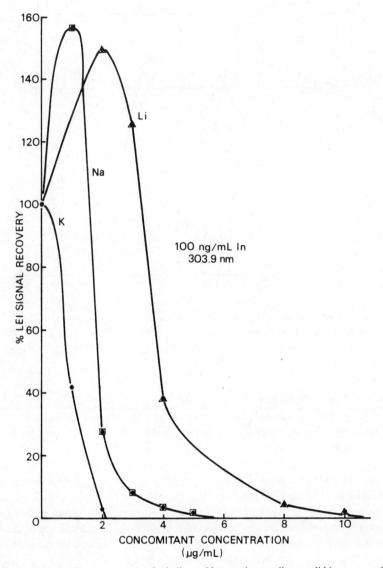

Fig. 3.11. LEI signal recovery curves for indium with potassium, sodium, or lithium concomitant.

the collecting field even when the sheath shrinks toward the cathode surface as a result of high flame ion concentrations. Small-diameter wires separated by a few millimeters and immersed in the flame have been used for DLI measurements (5). The laser beams are overlapped spatially and temporally between the electrodes, as they are for stepwise LEI (3, 4).

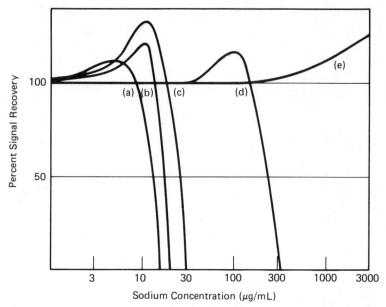

Fig. 3.12. The effect of electrode design on LEI signal recovery in the presence of varying sodium matrix concentrations: (a) 1-mm diameter rods, (b) 1.5-mm diameter rods, (c) 2.35-mm diameter rods, (d) 5-mm wide × 125-mm long plates, (e) water-cooled immersed cathode. From (2); reprinted by permission of the American Chemical Society.

Although it appears that LEI signal suppression for the analyte may be observed in the acetylene–air flame with the immersed electrode if the concomitant concentration exceeds 3000 μg/mL, other considerations are usually more important than signal suppression. At high salt concentrations, burner clogging and arc-over may be more relevant than signal suppression. When such high salt concentrations are present in samples, flame analysis by other spectroscopic techniques becomes problematic as well.

The best electrode configuration for a total consumption burner differs from that used for a premixed burner because of the difference in the maximum analyte concentration zones in the respective flames (35). In the total consumption burner, the analyte concentration is distributed throughout a much larger volume, higher in the flame. The best electrode configuration involves a horizontally opposed, water-cooled anode and cathode positioned 10 cm above the burner nozzle. Recovery of the maximum signal-to-noise ratios using a standard Beckman total consumption burner necessitates expanding the laser beam and using wider electrodes, separated at a greater distance to accommodate the laser beam. Insulation of the electrode surfaces that are not involved in signal collection improves signal-to-noise ratios. When the burner was tipped in a horizontal position and the flame was condensed and collimated with a Vycor tube, a longer pathlength was available that increased the signal collected (36).

Electrical interferences with the total consumption burner have not been examined in detail. Complete signal recovery for In in the presence of 1500 $\mu g/$ mL of Na has been observed (35). When the Na concentration exceeded this concentration, the electrodes were shorted and the power supply shut down. The same Na concomitant concentration aspirated into a premixed burner produces no shorting or LEI signal suppression because the actual flame concentration (which determines the onset of signal suppression) is a factor of 10 less. The use of interferent ion removal electrodes (38) or prior chromatography (39) should reduce electrical interferences from sample matrices for all electrodes and burner systems.

3.1.3. Lasers

An amplitude-modulated continuous wave (cw) dye laser was used as the source in the original LEI experiments (1). More recently, it was demonstrated that cw laser excitation provides immunity from the positional dependence of electrical interferences (40). The cw laser beam may be positioned at virtually any position in an acetylene–air flame without experiencing LEI signal suppression from low ionization potential concomitants, whereas pulsed laser excitation yields maximum signal recovery only near the immersed cathode surface. With cw excitation, convection or diffusion moves the analyte ion into the nonzero field region during the synchronization window for chopping rates less than 500 Hz. Unfortunately, cw dye lasers produce only low power emission in the ultraviolet (UV) spectral region due to the inefficiency of the frequency doubling process. Most metals have their strongest resonance lines in the UV. In addition, UV transitions terminate nearer the ionization limit, producing the highest LEI sensitivity in most cases.

Because of the high peak power available and the wide wavelength coverage, a pulsed dye laser is the most practical excitation source for LEI spectrometry. As implied in previous sections, a flashlamp-pumped dye laser should be the best source for LEI because of its relatively long (approximately 1 μs) pulse duration. LEI detection limits obtained with pulsed laser-pumped dye lasers suffer somewhat because of their relatively short pulsewidths (6–10 ns) which may be partially compensated for by their high peak powers. The efficiency of nonlinear second-harmonic generation is also improved by higher peak powers. Experience has shown that Nd:YAG laser-pumped dye lasers are capable of producing single-wavelength detection limits comparable to the best achieved with flashlamp-pumped dye lasers in many cases. Nitrogen laser-pumped dye lasers perform less well not only because of their narrow pulsewidths coupled with lower peak powers but also because of the radio frequency interference

(RFI) broadcast by the nitrogen laser. When stepwise excitation is required, dye lasers simultaneously pumped by either a nitrogen laser (18) or Nd:YAG (19) laser have been used. Again, the Nd:YAG system is preferred because of its superior peak pulse power, more complete wavelength coverage with the dye fundamental (using Nd:YAG second- and third-harmonic pumping), and low levels of RFI. Using the appropriate optics, one or both dye lasers may be frequency doubled for tunable ultraviolet operation. In addition, when a single laser is used to pump two dye lasers simultaneously, timing synchronization concerns are minimized. Excimer laser-pumped dye laser systems may prove to be as useful for LEI spectrometry as the Nd:YAG laser-pumped systems.

To date, most DLI experiments have utilized a nitrogen laser-pumped dye laser (5, 6). The relatively poor DLI detection limits that have been reported may be partially attributed to the low peak power and the high RFI associated with this laser. Unquestionably the wavelength flexibility provided by the ready availability of higher harmonics from a Nd:YAG laser for pumping a dye laser and providing a photoionizing beam would improve DLI results. The photoionization step becomes more efficient as the energy overshoot is reduced (6). A dual dye laser system may be appropriate for DLI for the same reason. An excimer laser-pumped dye laser system should also be considered for DLI spectrometry.

3.1.4. Signal Processing

The laser-related current pulses detected with electrodes may be simply monitored as a voltage drop across a resistor. More commonly, a high-gain preamplifier is used for analytical LEI (41). A resistor/capacitor network blocks the dc background current from the flame and passes the signal to the processing electronics. To achieve the optimum performance from the preamplifier, it should be mounted as close as possible to the anode where the signal is taken. This minimizes formation of ground loops, reduces cable capacitance, and permits the use of a short, shielded cable to avoid RFI pickup. Thorough grounding and shielding of the burner system are also effective in reducing environmental noise. This preamplifier will yield approximately 2-μs pulses regardless of the laser pulse-width because of intrinsic bandwidth limitations. This preamplifier is therefore conveniently resistant to 10–100-ns changes in electron transit times due to matrix and geometry effects (30).

A boxcar averager in the single-point mode may be used for processing the LEI signal (25, 26). The variable-width electronic gate is positioned over the LEI pulse. A trigger signal from a photodiode that monitors the laser beam opens the gate for a specified time after the laser pulse. The signal recovered on successive laser pulses is averaged depending on the instrumental time constant. The average signal amplitude may be read out on a strip chart recorder or converted into a digital signal for further processing with a microcomputer (42). Generally, the signal pulse is displayed on an oscilloscope simultaneously. Recent work has indicated that the total signal (electron and ion pulses) should be integrated to avoid the effects of electrical interferences (30). According to the "point charge model," the total LEI charge is matrix independent. Longer integration times (i.e., longer boxcar aperture durations) also reduce electron shot noise in the flame background current. In practice, noise due to the background current may reduce signal-to-noise ratios if longer boxcar aperture durations are used, particularly with alkali metals.

3.2. Methodology

A brief discussion of current methodology drawn from the cumulative research should be helpful in crystallizing the best approach to practical LEI spectrometry. At this point, one would choose a premixed burner with a water-cooled, immersed electrode for the reasons discusssed in previous sections. Because of the great sensitivity of LEI measurements, sample dilution should be considered as a viable option to reduce or eliminate matrix interferences.

When using a premixed burner, it is important to acidify all blank, standard, and sample solutions (43). This procedure prevents adsorption of solutions on the interior of the burner premixing chamber and contributes to larger LEI signals. Otherwise very large transient LEI signals result when the aspiration of an acid solution is alternated with a neutral solution containing the analyte. LEI is affected by sample adsorption in the premixed burner to a much greater extent than other flame techniques because of its greater sensitivity. Addition of nitric acid may be preferred for routine work because it provides constant signal enhancement over a relatively wide concentration range, making the amount added to samples less critical (43).

A flashlamp-pumped, nitrogen laser-pumped, or Nd:YAG laser-pumped dye laser system may be used for LEI, keeping in mind the trade-offs. Excimer laser-pumped dye laser systems are also suitable. Realistically, availability may be the most compelling factor in choosing a laser.

The selection of the best analytical line has not been discussed explicitly and deserves attention. Because LEI is at least a two-step process involving laser excitation and collisional ionization steps, many transitions may be preferred for LEI that are not usable by purely optical methods. In other words, low-probability transitions may produce good LEI sensitivity due to proximity of the laser-populated state to the analyte element's ionization potential. The choice of the most sensitive LEI lines is more complicated, but it introduces an important practical advantage for dye laser spectrometry.

The efficiency of an organic dye for laser action over a specific wavelength range is characterized by its gain curve. The visible (fundamental) and ultraviolet (second harmonic) spectral regions are covered by a series of dyes, but the power output is variable with wavelength. Since more analytical lines are available that can provide good sensitivity using LEI spectrometry, the analyst may choose a line in a spectral region where the dye laser performs most efficiently. By choosing the optimum combination of LEI line and dye, superior sensitivity may be achieved. In addition, the high resolution and tunability provided by a dye laser coupled with LEI detection, may permit the sequential determinations of several metals within the gain curve of a single laser dye.

The overlap of molecular spectra with atomic lines, which occurs in optical flame spectrometry, has been less commonly encountered with LEI. Native flame species such as OH and CH are not observed because of their high ionization potentials. Resonantly enhanced multiphoton ionization of molecules such as NO (44, 45) or PO (46) may cause interferences in some cases. The LEI spectra of oxides of lanthanum, scandium, and yttrium have been observed in hydrogen–air flames (47) and may constitute a spectral interference for samples containing these metals. The problem of MPI background has been discussed in Section 3.1.1.

The larger number of analytical wavelengths available with LEI spectrometry is an advantage for avoiding atomic and molecular spectral overlaps. The analyte signal can be identified by scanning the dye laser wavelength across the analytical line. Stepwise excitation may improve the selectivity of LEI spectrometry (4). The probability of both analyte transitions coinciding with two interferent lines is very small. A spectral background may overlap either analytical line, but the LEI signal will be enhanced only by excitation at both wavelengths. Either of the two wavelengths may be scanned while holding the other fixed to distinguish the signal from background. Generally, scanning the visible (fundamental) wavelength is preferred because of the additional complexity of tracking a frequency-doubling crystal while tuning the dye laser. A complete spectral overlap of the analyte line with another atom at one wavelength can also be treated by scanning the second laser and keeping the first laser wavelength fixed (4). An increased baseline signal due to the first step excitation of the interferent is observed, but 100% of the analyte signal can still be recovered.

4. FIGURES OF MERIT

4.1. Detection Limits

The periodic table shown in Fig. 3.7 indicates the elements that have been determined in premixed analytical flames by LEI spectrometry. Limits of detection are given in nanograms of analyte per milliliter of distilled water aspirated into the flame; 1 ng/mL corresponds to an atom density of approximately 10^8/

cm^3 in the flame (48). The table also shows whether the element was determined using a single-wavelength or stepwise excitation scheme. The possible excitation schemes for LEI are identified in Fig. 3.1.

The limits to sensitivity in LEI spectrometry have been discussed recently (49). The ultimate limits of detection obtainable are determined by noise associated with the LEI signal. Several sources of noise have been identified (50), but the limiting noise for optimum conditions is statistical or shot noise. A reasonable ultimate detection limit for LEI may be calculated as 1 part per trillion (1 pg/mL) or 10^5 atoms/cm^3. According to Fig. 3.7, lithium may be detected at this level. Variations in detection limits for other metals may be attributed to incomplete atomization in a given flame and lack of 100% ionization. Of course, when real samples are analyzed, other complications, which have already been discussed, may arise leading to higher detection limits than obtained for pure, aqueous analyte solutions.

4.2. Dynamic Range

Dynamic ranges for LEI have not been studied in detail, but ranges of 4 to 5 orders of magnitude are typical (51).

5. APPLICATIONS

5.1. General

The general application of LEI spectrometry to the determination of trace metals has been somewhat limited, but progress in this area should continue. Alloy analyses are particularly amenable to LEI spectrometry because of the absence of a readily ionizable sample matrix. Indium (303.9 nm) has been determined in a nickel-based high-temperature alloy (24). Atomic absorption spectrometry of this sample requires time-consuming extraction procedures to remove concomitant metals that contribute to spectral interferences. After dissolution of the alloy sample with acids, the concentration of indium was determined to be 35 μg/g by LEI spectrometry. This value agreed within experimental error within a value of 37 μg/g obtained with a graphite furnace. Manganese was also determined in a National Bureau of Standards (NBS) Standard Reference Material (SRM) No. 1261 steel sample to within 0.66% of the certified value (24).

Several NBS low-alloy steels (SRM 362 and 363) and unalloyed copper (SRM 396) have been analyzed by stepwise LEI spectrometry (4). Calibration was performed by bracketing the dissolved SRM samples with standards without matrix matching. The LEI results agreed within experimental error with the certified values in all cases. The stepwise LEI signal was obtained by scanning the dye laser over the second step wavelength with the first step wavelength fixed. The LEI value for tin in the unalloyed copper sample was reported with a factor of 6 less uncertainty than the NBS SRM value. The low detection limit for tin and the absence of any significant spectral background made a high-precision measurement possible by LEI spectrometry.

Rubidium has been determined in pine needles (NBS SRM 1575) to demonstrate the resistance to interferences of the immersed electrode with CW laser excitation (40). The concentration of matrix components such as calcium and potassium were on the order of 40 μg/mL. The measured rubidium concentration of 11.1 μg/g agreed well with the certified value.

5.2. Flame Diagnostics

The application of LEI spectrometry to flame diagnostics has been recently reviewed (52). This is a research area where the unique aspects of the technique can provide some new insights into combustion systems.

Vertical spatial profiles of a hydrogen–air flame have been generated by interposing a small-diameter rod between the anode and cathode plates, as illustrated in Fig. 3.5 (32). An amplitude-modulated cw laser was the excitation source for LEI. In addition to obtaining ion images, this experiment also demonstrated the physical size of the excess ion region produced by laser enhancement and the influence of external voltage, flame velocity, diffusion, and electrostatic expansion on the excess ion region. The rise and decay times of the excess ion current, which were followed at different applied voltages, illustrated the dependence of LEI on ion mobility, electric field, and excited-state ionization rate constant.

Atomic ion mobilities in several flames have been measured by detecting the arrival times of the ions produced by laser enhancement (53). The time from the production to the collection of the ions was measured as a function of the distance from the laser beam to the collecting electrode for a series of applied voltages. An oscilloscope trace for irradiation of an acetylene–air flame containing sodium and uranium with a single laser pulse is shown in Fig. 3.13. At the excitation wavelength, 539.9 nm, there is a discrete two-photon transition and a broadband uranium transition. The arrival of the electron pulse and the ion signals due to sodium and uranium are visible. The mobility measurements for lithium, sodium, potassium, calcium, iron, strontium, barium, indium, thallium, and uranium were compared to the Langevin theory. The model predicts the mobility of some of these ions to within 10%, provided that the dielectric constant of the flame gases is considered. In general, the larger, more polarizable ions are subject to the largest deviation, sometimes as much as 50% larger than predicted. For atomic ions, the Langevin theory is accurate and provides an upper limit for estimating absolute ion mobility.

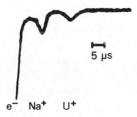

e⁻　Na⁺　U⁺

Fig. 3.13. Oscilloscope trace for irradiation of an acetylene–air flame containing Na and U atoms with a single laser pulse. From (53); reprinted by permission of the Combustion Institute.

Using the same apparatus, the mobility of very small particles at the sooting limit of a premixed flame has been estimated (54). The observed ionization signals are attributed to laser-induced heating and subsequent thermal ionization of the soot particles. An estimate of the particle size was made from the mobility measurements.

The temporal and spatial evolution of the depleted neutral atom density following LEI has been used to characterize the flow velocity of the flame gases in a laminar flow flame (55). A cw dye laser beam was split into two beams and tuned to the 589.0-nm sodium transition. The lower beam was acousto-optically modulated and produced the depleted sodium atom region in the flame. The upper beam (probe) was directed counter to the modulated beam through a Na D-line filter onto a photomultiplier tube. The LEI laser was operated in a pulsed mode, and the output from the photomultiplier which monitored the sodium absorption, was sent to a signal averager. The averaged signal was processed by a small computer that fit the signal to a model for arrival time. This measurement was repeated for two probe beam heights. The difference in the arrival times and the probe beam heights yielded the velocity at the probe's average height above the burner with 2% precision. This technique is limited to reasonably laminar flames because turbulence will dilute the neutral atom depletion region. No flow perturbations were introduced by optical detection of the LEI.

The estimation of flame temperatures by DLI has been proposed (56). The ionic diffusion and mobility coefficients for sodium and lithium in a stoichiometric hydrogen–oxygen–argon flames were determined by DLI and the flame temperature calculated using the Einstein relation. Although the flame velocity was neglected in this treatment, the DLI temperature estimates agree with published temperature measurements by line reversal in similar flames.

Hydrogen (57) and oxygen (58) were detected as atomic ions with biased electrodes in hydrogen–air and hydrogen–oxygen–argon atmospheric pressure flames, respectively. In both cases, crossed laser beams and resonantly enhanced multiphoton ionization schemes were used. Since the first excited states of both hydrogen and oxygen are approximately 10 eV above the ground state, single-photon excitation is difficult. The resonant excited states were populated by two-photon absorption while the ionization was accomplished with a third photon. In the hydrogen experiment, a 266-nm photon (Nd:YAG fourth harmonic) and a 224-nm photon (dye laser second harmonic summed with the YAG fundamental) were resonant with the two-photon transition. One photon from either beam was sufficiently energetic to ionize the excited atom. The two-photon transition in oxygen was excited by two 226-nm photons from a frequency-doubled and mixed dye laser while a third 226-nm photon ionized the excited atom. Parts-per-million sensitivity was estimated for both hydrogen and oxygen. Since relatively few collisions would be capable of supplying the several electron volts necessary for ionization in these experiments, photoionizations should be the predominant mechanism. Atomic hydrogen has also been detected in a hydrogen–oxygen–argon flame by resonance four-photon ionization (59). The sensitivity was comparable to the three-photon determination (57), but the excitation wavelength was 365 nm in the near visible. This wavelength was provided by a dye laser pumped by a XeCl excimer laser. Crossed-beam excitation schemes (57) provide important spatial information, but 365-nm excitation requires less sophisticated, less expensive instrumentation while eliminating problems of laser-induced photolysis and background interferences.

6. THE FUTURE OF LASER-INDUCED IONIZATION METHODS

The distinction among LEI, DLI, and MPI is blurred, particularly when several steps are incorporated into an excitation scheme. Laser-induced ionization techniques have been demonstrated that are highly sensitive and selective methods for analytical determinations in flames. Since an electrical signal is measured directly, detection of laser-induced ionization is insensitive to scattered excitation, flame background, and ambient light. This property accounts for the very low LEI detection limits that have been obtained for metals in a turbulent flow, total consumption flame. These results suggest that LEI spectrometry may be useful for sensitive measurements in adverse sampling environments where conventional optical methods have been unsuccessful. Progress with the detection of nonmetals in flames by direct laser photoionization is also encouraging for flame diagnostics.

In the future, research must be undertaken to demonstrate that the laser-induced ionization techniques are adaptable to a wider variety of samples and analytically useful flames. This will require further consideration of methods to discriminate against or remove low ionization potential interferents. Preliminary results have indicated that the use of an acetylene–nitrous oxide flame for the LEI determination of metals that form refractory oxides exacerbates electrical interferences when samples contain IA elements. The much higher flame temperature produces higher ionization rates whose effects cannot be entirely mitigated by using an immersed electrode.

Although it possesses many advantages in terms of versatility, ease of operation, and low cost, the flame has its limitations for high-sensitivity analytical spectrometry. Further lowering of detection limits will require other sample reservoirs. For example, atomization in a plasma could be coupled to a low-temperature, low-background flame for laser excitation. Electrical interferences might also be eliminated using such a segregated approach.

There is little doubt that laser-induced ionization techniques will continue to occupy prominent positions in analytical methodology. The first International Colloquium on Optogalvanic Spectroscopy and its Applications was held in Aussois, France, in June 1983 under the auspices of C.N.R.S. (62). The sessions devoted to LEI spectrometry (also known as optogalvanic spectroscopy in flames) were indicative of worldwide interest. Because of the simplicity of the detection schemes, laser-induced ionization measurements complement other laser-based techniques in addition to providing primary methods for precise, high-sensitivity spectrochemical determinations.

ACKNOWLEDGMENT

The author wishes to acknowledge the National Science Foundation for partial support of this research. The many people who contributed to the body of work described here are recognized in the references. J. C. Travis, J. F. Hall, and J. D. Messman made many valuable suggestions during the preparation of this chapter. The author would also like to thank P. K. Schenck and G. C. Turk for their help in preparing Fig. 3.7.

REFERENCES

1. R. B. Green, R. A. Keller, P. K. Schenck, J. C. Travis, and G. G. Luther, *J. Am. Chem. Soc.*, **98**, 1517 (1976).
2. J. C. Travis, G. C. Turk, and R. B. Green, *Anal. Chem.*, **54**, 1006A (1982).
3. G. C. Turk, G. W. Mallard, P. K. Schenck, and K. C. Smyth, *Anal. Chem.*, **51**, 2408 (1979).
4. G. C. Turk, J. R. DeVoe, and J. C. Travis, *Anal. Chem.*, **54**, 643 (1982).
5. C. A. van Dijk, F. M. Curran, K. C. Lin, and S. R. Crouch, *Anal. Chem.*, **53**, 1275 (1981).
6. F. M. Curran, K. C. Lin, G. E. Leroi, P. M. Hunt, and S. R. Crouch, *Anal. Chem.*, **55**, 2382 (1983).
7. A. S. Gonchakov, N. B. Zorov, Y. Y. Kuzyakov, and O. I. Matveev, *J. Anal. Chem. USSR*, **34**, 1792 (1980).
8. A. S. Gonchakov, N. B. Zorov, Y. Y. Kuzyakov, and O. I. Matveev, *Anal. Lett.*, **12**, 1037 (1979).

9. N. B. Zorov, Y. Y. Kuzyakov, O. I. Matveev, and V. I. Chaplygin, *J. Anal. Chem.*, **35,** 1108 (1980).

10. J. P. Young, G. S. Hurst, S. D. Kramer, and M. G. Payne, *Anal. Chem.*, **51,** 1050A (1979).

11. G. S. Hurst, M. G. Payne, S. D. Kramer, and J. P. Young, *Rev. Mod. Phys.*, **51,** 767 (1979).

12. G. S. Hurst, *Anal. Chem.*, **53,** 1448A (1983).

13. S. Mayo, T. B. Lucatorto, and G. G. Luther, *Anal. Chem.*, **54,** 553 (1982).

14. T. J. Whitaker and B. A. Bushaw, *Chem. Phys. Lett.*, **79,** 506 (1981).

15. S. D. Kramer, C. E. Bemis, J. P. Young, and G. S. Hurst, *Opt. Lett.*, **3,** 16 (1978).

16. D. Popescu, C. B. Collins, B. W. Johnson, and I. Popescu, *Phys. Rev.*, **A9,** 1182 (1974).

17. P. M. Johnson, M. R. Berman, and D. Zakheim, *J. Chem. Phys.*, **62,** 2500 (1975).

18. J. C. Travis, P. K. Schenck, and G. C. Turk, *Anal. Chem.*, **51,** 1516 (1979).

19. T. J. Hollander, *AIAA J.*, **6,** 385 (1968).

20. K. C. Smyth, P. K. Schenck, and G. W. Mallard, "What Really Does Happen to Electronically Excited Atoms in Flames?" in D. R. Crosley, Ed., *Laser Probes for Combustion Chemistry* (ACS Symp. Ser. 134), American Chemical Society, Washington, D.C., 1980, p. 175.

21. C. A. van Dijk and C. Th. J. Alkemade, *Combust. Flame*, **38,** 37 (1980).

22. G. C. Turk, J. C. Travis, J. D. DeVoe, and T. C. O'Haver, *Anal. Chem.*, **50,** 817 (1978).

23. J. C. Travis, G. C. Turk, and R. B. Green, "Laser-Enhanced Ionization for Trace Metal Analysis in Flames," in G. M. Hieftje, Ed., *New Applications of Lasers to Chemistry* (ACS Symp. Ser., 85), American Chemical Society, Washington, D.C., 1978, p. 91.

24. G. C. Turk, J. C. Travis, J. R. DeVoe, and T. C. O'Haver, *Anal. Chem.*, **51,** 1890 (1979).

25. R. B. Green, G. J. Havrilla, and T. O. Trask, *Appl. Spectrosc.*, **34,** 561 (1980).

26. G. J. Havrilla and R. B. Green, *Anal. Chem.*, **52,** 2376 (1980).

27. G. C. Turk, *Anal. Chem.*, **53,** 1187 (1981).

28. M. A. Nippoldt and R. B. Green, *Anal. Chem.*, **55,** 554 (1983).

29. T. Berthoud, J. Lipinsky, P. Camus, and J. L. Stehle, *Anal. Chem.*, **55,** 959 (1983).

30. G. J. Havrilla, P. K. Schenck, J. C. Travis, and G. C. Turk, *Anal. Chem.*, **56,** 186 (1984).

31. L. Lawton and F. J. Weinberg, *Electrical Aspects of Combustion*, Clarendon Press, Oxford, 1969; (a) p. 167, (b) p. 320, (c) p. 315, (d) p. 91.

32. P. K. Schenck, J. C. Travis, G. C. Turk, and T. C. O'Haver, *J. Phys. Chem.*, **85,** 2547 (1981).

33. F. M. Curran, C. A. van Dijk, and S. R. Crouch, *Appl. Spectrosc.*, **37,** 385 (1983).

34. R. A. Peters and R. B. Green, Abstract, 185th National American Chemical Society Meeting, Seattle, WA, March 24, 1983, No. 209.

35. J. E. Hall and R. B. Green, *Anal. Chem.*, **55,** 1811 (1983).

36. J. E. Hall and R. B. Green, 39th Southwest Regional American Chemical Society Meeting, Tulsa, OK, December 7, 1983, No. 204.

37. V. I. Chaplygin, N. B. Zorov, Y. Y. Kuzyakov, and O. I. Matveev, *Zh. Anal. Khim.*, **38,** 802 (1983).
38. T. O. Trask and R. B. Green, *Spectrochim. Acta*, **38B,** 503 (1983).
39. H. M. Kingston, I. L. Barnes, T. J. Brady, and T. C. Rains, *Anal. Chem.*, **50,** 2064 (1978).
40. G. J. Havrilla, S. J. Weeks, and J. C. Travis, *Anal. Chem.*, **54,** 2566 (1982).
41. G. J. Havrilla and R. B. Green, *Chem. Biomed. Environ. Instrum.*, **11,** 273 (1981).
42. G. H. Vickers, T. O. Trask, J. D. Parli, J. A. Wisman, B. Durham, and R. B. Green, *Chem. Biomed. Environ. Instrum.*, **12,** 289 (1983).
43. T. O. Trask and R. B. Green, *Anal. Chem.*, **53,** 320 (1981).
44. W. G. Mallard, J. H. Miller, and K. C. Smyth, *J. Chem. Phys.*, **76,** 3483 (1982).
45. B. H. Rockney, T. A. Cool, and E. R. Grant, *Chem. Phys. Lett.*, **87,** 141 (1982).
46. K. C. Smyth and G. W. Mallard, *J. Chem. Phys.*, **77,** 1779 (1982).
47. P. K. Schenck, G. W. Mallard, and J. C. Travis, *J. Chem. Phys.*, **69,** 5147 (1978).
48. J. D. Winefordner and T. J. Vickers, *Anal. Chem.*, **36,** 1939 (1964).
49. J. C. Travis, *J. Chem. Educ.*, **59,** 909 (1982).
50. G. C. Turk, Ph.D. dissertation, The University of Maryland, 1978.
51. J. C. Travis, and J. R. DeVoe, "The Optogalvanic Effect," in G. M. Hieftje, J. C. Travis, and F. E. Lytle, Eds., *Lasers in Chemical Analysis*, Humana Press, Clifton, NJ, 1981, p. 93.
52. P. K. Schenck and J. W. Hastie, *Opt. Engineer*, **20,** 522 (1981).
53. W. G. Mallard and K. C. Smyth, *Combust. Flame*, **44,** 61 (1982).
54. K. C. Smyth and G. W. Mallard, *Combust. Sci. Tech.*, **26,** 35 (1981).
55. P. K. Schenck, J. C. Travis, G. C. Turk, and T. C. O'Haver, *Appl. Spectrosc.*, **36,** 168 (1982).
56. K. C. Lin, P. M. Hunt, and S. R. Crouch, *Chem. Phys. Lett.*, **90,** 111 (1982).
57. J. E. M. Goldsmith, *Opt. Lett.*, **7,** 437 (1982).
58. J. E. M. Goldsmith, *J. Chem. Phys.*, **78,** 1610 (1983).
59. P. J. H. Tjossem and T. A. Cool, *Chem. Phys. Lett.*, **100,** 479 (1983).
60. R. B. Green, "Laser-Enhanced Ionization Spectrometry," in F. L. Boschke, Ed., *Analytical Chemistry Progress*, Springer-Verlag, Berlin, 1984.
61. V. I. Chaplygin, Y. Y. Kuzyakov, and N. B. Zorov, *Spectrochim. Acta*, **38B,** Supplement, 386 (1983).
62. Proceedings published in *J. Physique* (Supplement), "Optogalvanic Spectroscopy and its Applications," **44**(C7), 1983.

CHAPTER

4

DETECTION OF SMALL NUMBERS OF ATOMS AND MOLECULES

C. TH. J. ALKEMADE

Physical Laboratory of the State University
Princetonplein 5,
Utrecht, The Netherlands

1. INTRODUCTION AND SCOPE

1.1. Selectivity and Detectability

Free atoms, ions, molecules, or radicals (all of which will usually be referred to as *atoms* for simplicity) of a given chemical species can interact with an optical radiation field at certain wavelengths that are characteristic for that species (Chapter 1). By selecting radiation at a characteristic wavelength, we can thus tag these atoms and distinguish them from other species. The signal produced by the interaction is then a measure for the number of atoms of the selected species. This is the very basis of quantitative analytical spectroscopy. Tunable optical lasers (Chapter 1) are beneficial for improving the selectivity of analysis (because of their small spectral bandwidth) as well as the detectability (because of their high radiant power). Especially in diagnostic applications their directionality is an extra bonus for obtaining high spatial resolution, whereas high temporal resolution can be achieved with pulsed lasers of extremely short duration. In this chapter we shall be mainly interested in the capability of lasers to push the detectability toward the ultimate limit set by the discrete nature of matter. The improvement of the spectral selectivity as such will not be considered, unless detectability is intimately connected with selectivity, as in the detection of rare isotopes in a mixture.

The detectability is characterized by the analytical *limit of detection (LOD)*, which is defined as that concentration, c_m, or quantity of substance, q_m, in the sample, which produces a mean signal equal to k times the root-mean-square (rms) random error (1). A LOD expressed in concentration units (e.g., mol/L, μg/mL) is called a *concentrational limit of detection*, whereas in the other case we shall speak of the *absolute limit of detection* (e.g., in pg). The *statistical confidence factor k* is usually 3. We distinguish here between two categories of random errors. On the one hand, there are errors arising from extrinsic sources of fluctuations, such as fluctuations in the background (produced by concomitants in the sample, by the atomizer, or by scattering of laser radiation), fluctuations in the laser power or in the efficiency of atomization, and detector or amplifier noise. On the other hand, there is an intrinsic error in the signal itself, arising from the discrete nature of matter: First, the number of analyte atoms found in the sample or atomizer will fluctuate around a mean (or expectation) value in a manner described by Poisson statistics. Second, the detection process (e.g., the absorption of a discrete photon and the subsequent release of a photoelectron) is a chance process. The resulting error, which becomes relatively more important for smaller numbers of atoms and detected photons, determines the *intrinsic limit of detection*. This is the ultimate LOD obtainable with a given setup if all extrinsic fluctuations are suppressed. We are, technically, still far from the ideal of complete *atomization* (= conversion of analyte in the sample to free atoms) and complete *probing* (= counting of free atoms). Therefore, we shall define the intrinsic LOD as related to the statistical fluctuation in the number of free atoms being *probed* and *not* in the number of analyte atoms in the *sample*.

1.2. Single-Atom Detection (SAD)

Even under favorable measuring conditions, "conventional" atomic spectrometers (with or without lasers) yield LODs that are set by extrinsic errors. The best absolute LOD, as reported in Table 2.14 and in Omenetto and Winefordner (2) for laser-excited atomic fluorescence (AFS) and atomic absorption spectrometers (AAS) with electrothermal atomizers, are about 0.1 and 1 pg, respectively. If we assume complete atomization, these quantities correspond to a total number of about 10^9–10^{10} free atoms (of median relative atomic mass) in the atomizer. Even if we assume a probe volume, V_p, as small as 1 mm^3 and an (effective) atomizer volume, V_a, as large as 10^4 mm^3, the number of free atoms being probed (counted) at the LOD level is still 10^5 or higher. According to theoretical estimates this number is even expected to be at least 10^7 when conventional AAS is applied with a tube atomizer of 1 mm diameter (3).

The best experimental concentrational LODs found in AFS with flames or inductively coupled plasmas (ICP), using conventional as well as laser sources, are typically 0.1 ng/mL in aqueous solutions [Table 2.13 and (2, 4)]. Assuming V_p to be as small as 1 mm^3 in the flame, this LOD value would still correspond to at least 10^3 free atoms being probed. The best experimental concentrational LODs found in laser-enhanced ionization spectroscopy (LEIS) using flames combined with a premixed or total consumption burner are typically 0.01 ng/mL (Tables 3.1 and 2.13). Assuming $V_p \approx 10^2$ mm^3, the number of free atoms being probed at the LOD level is then estimated to exceed 10^4.

These rough estimates demonstrate clearly that the number of analyte atoms, N_p, probed in conventional laser-based spectrometers at the LOD level is still so high that its relative statistical fluctuation ($1/\sqrt{N_p}$) is irrelevant. The situation is typically different if we apply special laser-based spectrometers that are capable of probing *small numbers* of free atoms. This requires the signal produced by an individual atom or small group of atoms to become detectable above background and detector noise. In the former case we shall speak of *single-atom detection (SAD)* [occasionally also called one-atom detection, OAD (5)]. Although such techniques have been developed mostly for atoms in the strict sense of the word, we shall use the term *SAD* more generally to relate to free ions, molecules, or radicals as well. Similar generalizations will be made as to the meaning of terms like *atomizer* and *atomic spectroscopy*. When we want to refer specifically to free molecules, the term *single-molecule detection (SMD)* will be used.

An SAD technique meets, ideally, the following demands:

1. *Each* (not just *a*!) free analyte atom present in or crossing the probe volume V_p during probing time T is detected and registered as one, and only one count with 100% statistical confidence.

2. The detection is *selective* as to the atomic species sought. Other species in the sample (concomitants) are not detected and do not interfere with the probability of detecting a given free analyte atom. Interferences caused by other species that can be compensated in the calibration procedure are

allowed, such as ionization interference and interference with the atomization of the analyte.

3. *No false signals* (counts) are produced by the background originating from the atomizer (e.g., flame background, thermal radiation from furnace walls, scattering of laser radiation) or by the detector and amplifier (e.g., dark current, spurious electric pulses).

4. The *LOD* and *precision* of the analysis are solely determined by the Poisson statistics of the free analyte atoms present in or crossing V_p during T.

This concept of SAD thus relates to the free atoms appearing *in the probing region*, where they can interact with the laser beam and be observed by the detector. With a given atomizer, the realization of SAD will depend upon the region selected. We should distinguish this restrictive SAD concept from the general SAD concept relating to the detection of a single analyte atom present *in the sample* (usually in a bound state). The latter is the ultimate goal (and absolute limit!) in chemical analysis as far as detectability is concerned. This goal is still so unrealistic that we will confine ourselves to SAD in the restricted sense.

Sophisticated, laser-based techniques have been developed during the last decade to detect smaller and smaller numbers of atoms in the gas phase. Species investigated were most often metallic atoms or ions such as the alkalis, Pb, Yb, Ba, Ba^+, and Mg^+ (with Na as ''topper''; see also Section 3) and occasionally noble-gas atoms (6–8). However, the ideal experimental conditions under which SAD conditions have been (approximately) attained were mostly aimed at special applications, with ''pure'' samples, in physics and chemistry. These conditions are quite different from those encountered in analytical spectroscopy dealing with real-world samples.

Applications in fundamental as well as applied physics and chemistry are connected, for example, with the search for rare or short-lived species (produced, e.g., by nuclear reactions) or for rare events induced by solar neutrinos captured in radiochemical detectors (5, 9–14). SAD techniques have also been proposed (15) for demonstrating the possible existence of Na-quark atoms (having a fractionally charged quark attached to the nucleus) amidst 10^{14} normal Na atoms per cubic centimeter. Another example is the measurement of saturated vapor pressures at low temperatures (15). Other applications concern the ultra high-resolution spectroscopy of rare isotopes (16, 17) and the kinetic or statistical behavior of individual atoms in a gas (5, 18, 19). SAD techniques were also applied in the diagnostics, with high spatial and temporal resolution, of plasmas or reaction systems containing ultratrace concentrations of atoms of given species or in specific excited states (12). SAD was further applied in the experimental verification of the antibunching statistics of the photons that are spontaneously emitted by a single atom excited by a strong radiation field (20–23). In Section 3 a few of these applications will be described in more detail.

1.3. Scope of Chapter

For the analytical spectroscopist who wants to detect smaller and smaller numbers of atoms, the development of these SAD techniques is both a challenge and an opportunity. We shall therefore review in this chapter the main streams of UV-visible laser-based spectroscopies that are most promising to detect small numbers of atoms in the gas phase. Techniques that are (primarily) aimed at ultra high-resolution spectroscopy, isotope separation, or remote sensing will not be considered. The use of high-powered lasers to vaporize, atomize and, possibly, excite or ionize analytical samples will not be discussed either (Chapter 19).

For completeness' sake we note that there exist also nonoptical techniques that can be applied, with special arrangements, to detect, localize, and/or identify individual atoms or ions in the solid or gas phase. Examples are the scanning transmission electron microscope, the field ion microscope ("imaging atom probe") and the accelerator mass spectrometer (which combines a nuclear accelerator with a mass spectrometer).

A good question is, how small is "a small number" in the title of this chapter. No sharp borderline can be drawn. Broadly speaking, we shall focus on those techniques that do allow us to count atoms (not just photons or electrons!) or the results of which are best discussed in terms of atom numbers. However, we shall generously include, for comparison, more conventional laser-based techniques that have successfully been applied in spectrochemistry to yield very low LODs. We shall also bring out some of the best LODs that have been obtained for molecular species with laser-based techniques. This will be done to underline how much farther we still are away from SMD (13, 24–28) than from SAD.

In the following we shall often borrow, by analogy, terms from analytical atomic spectroscopy, when we describe techniques that were not aimed at chemical analysis proper. The meaning of such terms (such as analyte, analysis, atomizer, sample, limit of detection) when applied, for example, to diagnostic techniques, may be evident from the context.

Several laser-based techniques and laser-induced processes are dealt with in other chapters too. We shall occasionally refer to them for a more detailed discussion. LODs are also discussed or tabulated in these other chapters. Our interest, however, is focused on the lower end of the LOD range. In several respects this chapter is thus a cross section through other chapters; repetition could not always be avoided.

With regard to the fast development and great variety of laser-based detection techniques in widespread areas of physics and chemistry, this review can by no means be exhaustive or profound. We shall rather concentrate on simple basic concepts, with only a few, oversimplified formulas, and on general detection schemes (Sections 2 and 3.1). These will be illustrated in Sections 3.2–4 by concrete examples and by experimental data in tabular form.

For further information and more extensive references to the literature, the reader may consult a blend of review papers or book chapters: (2, 27, 29) (mainly fluorescence techniques), (5, 14, 30–33) (mainly ionization techniques), and (3, 13, 26, 34–39) (general).

Terms, definitions, and symbols of the main quantities used are collected in Table 4.1.

2. FROM SAMPLE TO SIGNAL: GENERAL CONSIDERATIONS

2.1. Schematic Layout of Laser-Based Spectrometer

The "atoms" of the species to be detected are produced from a *sample* (Fig. 4.1). The liquid, solid, or gaseous sample may contain these atoms in any chemical state or state of ionization. Their conversion to *free* atoms in the gas phase (*atomization*) takes place in the *atomizer*. The atomizer includes the "atom reservoir" containing the free atoms available for probing.

We distinguish between *continuous-flow atomizers* and *closed atomizers*. Examples of the former class in analytical atomic spectroscopy are the flame and the inductively coupled plasma (ICP), in combination with a pneumatic nebulizer for converting a liquid sample into an aerosol (Fig. 4.2a). Volatilization of the aerosol particles and subsequent dissociation of the analyte compounds occur in the carrier gas heated by combustion or the electric discharge. The

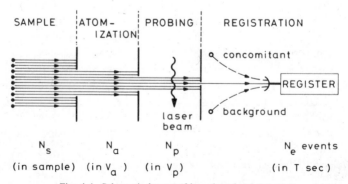

Fig. 4.1. Schematic layout of laser-based spectrometer.

Table 4.1. Symbols, Terms, and Definitions

Symbol	Term and Definition[a]
c_s (c_m)	(Minimum detectable) sample concentration
k	Statistical confidence factor
N_a	Number of free atoms in atomizer
n_a	Number density ($\equiv$ number per unit volume) of free atoms in atomizer
N_e	Number of events at detector during T
N_p	Number of free atoms in V_p
N_s	Number of analyte entities in consumed sample (one free atom derives from one entity)
R	Number of repeated probings
T	Probing time ($\equiv$ duration of a single probing by laser)
t	Time (general)
t_a	Atomization time ($\equiv$ duration of atomization needed for a single probing with a continuous-flow atomizer)
t_m	Total time of measurement for R repeated probings
V_a	Atomizer volume ($\equiv$ volume of closed atomizer or fictitious volume of continuous-flow atomizer; see Section 2.4)
V_p	Probe volume [$\equiv$ (effective) volume irradiated by laser and "observed" by detector]
ϵ_a	Atomization efficiency ($\equiv N_a/N_s$)
ϵ_d[b]	Detection efficiency ($\equiv$ probability that a given free atom appearing in V_p produces an event during T)
ϵ_o	Overall efficiency ($\equiv N_e/N_s$)
ϵ_p	Probing efficiency ($\epsilon_p \cdot N_a \equiv$ number of free atoms present in or passing through V_p during T)
ϕ	Flux ($\equiv$ number per second) of detected photons or ions (electrons) produced by N_p atoms, divided by N_p
τ	Transit or residence time
$\bar{x}$	Mean (or expectation) value of x
$(x)_m$	Value of x at limit of detection
$\tilde{x}$	Standard deviation of x [$\equiv \sqrt{\overline{(\Delta x)^2}}$]

General Abbreviations

LIBDS	Laser-induced beam deflection spectroscopy
LIFS	Laser-induced fluorescence spectroscopy
LIIS	Laser-induced ionization spectroscopy
LOD	Limit of detection
SAD	Single-atom detection
SMD	Single-molecule detection

[a]By generalization, *atom* is the selected chemical species of any kind (atom, ion, molecule, or radical) that interacts with the laser radiation. As a result of this interaction, an *event* may be produced at the detector. *Atomizer* is any device that produces these "atoms" in the gas phase (so-called *free* atoms) from the sample.

[b]In (39) symbol ϵ was used for ϵ_d. This definition of ϵ_d should not be confused with the efficiency of detecting a (fluorescence) *photon*.

open-ended graphite tube and the carbon-rod and carbon-filament atomizers (with pulsed atomization of minute samples), operating with a carrier gas flow, belong to this class, too. A quite different type of continuous-flow atomizer, especially suited for SAD, is the atomic beam (Fig. 4.2b). Here atoms produced, for example, in a furnace containing a metallic sample, effuse through a small hole into a vacuum chamber. Collimation of the atomic beam is achieved by means of a diaphragm placed downstream from the effusive flow.

An example of a closed atomizer is a cuvette filled with a gaseous sample or a cell containing a solid or liquid sample (Fig. 4.2c). The sample may be atomized by thermal or Langmuir evaporation, by atom sputtering in a glow discharge, or by bombardment with a fast-ion beam (in vacuo). In order to simplify our later discussions we shall assume that probing takes place after the sample has been atomized and the atoms have spread more or less uniformly over the atomizer volume. The cuvette or cell is equipped with windows for admitting the laser beam (and for observation in LIFS). The cell may or may not contain a buffer gas. Another example, especially suited for SAD, is the more recently developed *ion* or (neutral) *atom trap* (Fig. 4.2d). Here a few ionized or neutral atoms are trapped in a small confinement in vacuo and stored. This may be realized by means of a rf quadrupole field, an electromagnetic field

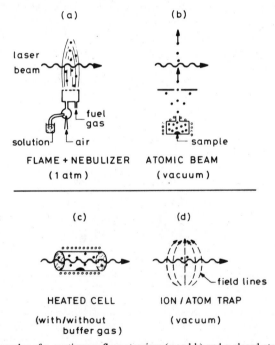

Fig. 4.2. Examples of a continuous-flow atomizer (a and b) and a closed atomizer (c and d).

with appropriate gradients, and/or optical trapping by laser radiation pressure (40–49).

An example of a Ba^+ trap will be described in Section 3.2. The "translational temperature" of the trapped atoms or ions can be reduced to below 10 K by laser cooling. This cooling is brought about by repetitive exchange of linear momentum between the atom and the photons absorbed from a monochromatic laser beam. The atom suffers a net loss of linear momentum after each excitation-emission cycle, if the laser frequency is tuned slightly below the atomic resonance frequency. The resulting reduction of the Doppler broadening of the atomic line and the absence of collision broadening make these traps an attractive atom reservoir for ultra high-resolution spectroscopy and for achieving high spectral selectivity in analysis. A well-collimated atomic beam, if directed perpendicular to the laser beam, offers similar advantages. In this respect, atomic beams and traps are far superior to atom reservoirs filled with a buffer gas at atmospheric pressure at or above room temperature.

It is not always possible to draw a sharp borderline between closed and continuous-flow atomizers. An open-ended tube furnace may be an intermediate case, depending on the chance the atoms have to flow out after being atomized.

The next stage in any laser-based spectrometer is the *probing* of the atoms by one or more laser beam(s) crossing the atomizer and their final *registration* (Fig. 4.1). The advantages and disadvantages of using continuous-wave (cw) or pulsed lasers depend strongly on the detection scheme chosen, the species to be detected, and the need for suppressing interferences from concomitants or background (Section 3).

2.2. Laser Probing of Atoms

The presence of a free atom of a given species is probed by the absorption of one or more photons from the laser beam(s), resulting in the production of a photophysical or photochemical effect.

The *photophysical effect* may be *optical*, *electrical*, or *mechanical* (Fig. 4.3). Here we present a few examples for introduction only. (A more detailed classification and description, especially with regard to SAD, will be given in Section 3.1.) The spontaneous emission of a secondary photon by the atom after laser excitation (fluorescence) is an obvious example of an optical effect (Fig. 4.3a). Any detection technique based on this effect is called here *laser-induced fluorescence spectroscopy* (*LIFS*). (Note that in Chapter 2 it is called laser-excited fluorescence.) The frequency of the fluorescence photons may be the same as that of the laser photons (ν) or different (ν'). We distinguish accordingly between *resonance* and *nonresonance fluorescence* (for a finer distinction, see Fig. 2.1).

The fluorescence photons are emitted in all directions (but not necessarily

A   FLUORESCENCE

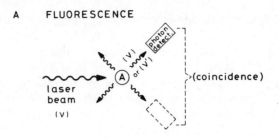

B   IONIZATION

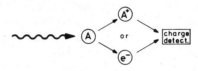

C   ATOM - BEAM DEFLECTION

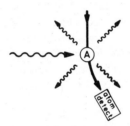

Fig. 4.3. Examples of optical (A), electrical (B), and mechanical (C) laser probing. A is the free atom or molecule probed.

isotropically if a polarized laser beam is used). So the photodetector may be positioned at an arbitrary direction of observation, preferably with a large solid angle of acceptance. Since often a monochromator is not needed (a spectral filter may suffice to select one of the few fluorescence lines or to reject laser scattering if $\nu \neq \nu'$), the angle of acceptance can be made quite large by suitable optical arrangements (Section 3.2). A weak resonance fluorescence signal can be discriminated from laser scattering by application of a photon coincidence technique with the aid of a second photodetector (Fig. 4.3a, Section 3.2). If optical saturation (Chapter 1) is attained, a given atom may produce, on the average, at least one photoelectron in each detector during its (short) passage through the laser beam. If the coincidence interval is matched to the passage time, fluorescence photons, occurring in bunches, will have a greater probability of causing a coincidence signal than the randomly received scattered photons.

In the case of Fig. 4.3b, absorption of one or more photons from one or more laser beams results in the production (directly or indirectly; Section 3.1) of an ion and a free electron. These can be directed by an electric or magnetic field to a charge or particle detector. This type of atom detection technique will here be generally called *laser-induced ionization spectroscopy (LIIS)*.

In the case of Fig. 4.3c, an atom moving in a direction perpendicular to the laser beam acquires transversal linear momentum by absorbing a laser photon. The net recoil effect resulting from multiple absorption–spontaneous emission cycles in the atom deflects its trajectory from the atomic beam axis, in the direction of travel of the laser beam. A particle detector placed off-axis and downstream of the atomic beam counts the atoms that are selectively deflected by the laser. This example of *laser-induced beam deflection spectroscopy (LIBDS)* is thus typically based on a mechanical effect produced by atom–laser interaction; LIBDS has not yet been exploited for SAD, but it may have good potentialities in special applications (13, 34). It has until now been mainly considered as a selective separation technique. We have included LIBDS in this review to stress that SAD might be realized by other than optical or electrical techniques, too.

Neutral atoms of low ionization energy can be detected by surface ionization when they impinge on a hot metal wire. The positive ions leaving the surface are accelerated and directed to an ion multiplier tube, where they produce pulses of secondary electrons.

Photochemical effects have so far but seldom been considered for use in SAD or SMD, in contrast to their exploitation in isotope separation. The feasibility of detecting single alkali–halide molecules by laser-induced photodissociation and LIIS detection of the alkali product has been suggested (5, 50).

As a result of its interaction with laser radiation, the atom undergoes a "process," that is, a change or series of changes in its state or motion. A *general characterization of a laser-induced process* is presented in Table 4.2. A process can be characterized as to (I) the final state of the atom, (II) the type of cyclic process, and (III) the multiplicity of recycling. In LIFS, case Ia may apply, for example, when one of the Na-*D* doublet components is excited by the laser. After excitation the atom returns, directly or indirectly, to the initial (ground) state, thus completing a cycle. An *intraatomic* cycle (case IIa) can take place via different routes: by spontaneous or stimulated emission, and by inelastic collisions leading to "quenching" of the excited state or "population mixing" of the doublet states (51) (Table 2.1). A *regenerative* cycle (case IIb) takes place when, for example, in a flame the laser-excited Na atom becomes first ionized or bound in a molecule before it returns to the initial state by recombination with a free electron or by dissociation. If the probing time (Table 4.1) is long enough, the atom may undergo more than one excitation–deexcitation cycle (case IIIa). We may distinguish between the multiplicities of recycling via different routes.

Case Ib applies, for example, when Na atoms in a beam are excited by a

Table 4.2. General Characterization of a Laser-induced Process[a]

I. As to *final state of the atom*:

a. *Cyclic process*
Atom returns directly or indirectly to initial state in the probe volume within probing time

b. *Consumptive process*
Atom does not return to initial state and/or is removed from probe volume

II. As to *type of cyclic process*:

a. *Intraatomic cycle*
Atom returns to initial state in the probe volume without changing its chemical state or state of ionization

b. *Regenerative cycle*
Atom returns to initial state and to probe volume after a transitory change in chemical state or state of ionization or in position

III. As to *multiplicity of recycling*:

a. *Multiple recycling*
Atom undergoes more than one interaction cycle during probing time

b. *Single cycling*
Atom undergoes no more than one interaction cycle during probing time

[a] We consider here processes that take place in one and the same atom interacting with laser radiation during a single probing. "Initial state" is here defined as the state in which the atom can interact with laser radiation.

narrow-band laser tuned to a particular hyperfine structure (hfs) transition and directed perpendicular to the atomic beam to avoid Doppler broadening. Optical selection rules permitting, the excited Na atom may return to a hyperfine (hf) component of the gound state different from the initial one. As long as it stays there, it cannot be excited again (unless the spectral wing of the associated hf line overlaps with the spectral laser profile). By this "consumptive" process the atom is taken out of circulation. Case Ib is also found when the excited atom becomes trapped by making a spontaneous transition to a metastable level. The atom is then "consumed," unless it is released from the trap by an inelastic collision or by a second excitation step to a higher lying nonmetastable level, using an additional laser. A more general discussion of atom trapping to inaccessible levels is presented in (13, 34, 36, 52). In LIIS, atom consumption naturally occurs, as the ionized atom is removed from the laser interaction region, being captured by an electrode. In principle, it is feasible to release the atom again from the electrode, so that it can return in time to the laser interaction region (case IIb) (34).

When a flame is used as atomizer, this "regeneration" could be realized by immersing the cathode in the flame, near to the laser beam. Regenerative recycling could also be achieved in LIIS when the ionized analyte atom A^+ transfers its charge, by collision, to a different particle M whose ionization energy is lower than that of A. These particles are supposed to have been admixed at a sufficiently large concentration to ensure efficient charge transfer. The M^+ ions and/or free electrons are detected, whereas A is again available for laser ionization. If this recycling process is repeated many times during the probing time (case IIIa), one obtains a considerable amplification of the electric signal produced by one atom A. This is the basis of the resonance ionization spectroscopy with amplification (RISA) proposed in (12); it would be attractive for SAD.

Consumptive interaction processes (case Ib) may have the advantage of preventing a given atom from being counted twice. In the analytical application of laser-enhanced ionization in flames they may have the additional advantage that the resulting depletion of free atoms, such as Li, brings about a shift in the partial equilibrium between atomic Li and LiOH (51, 53). This shift replenishes (partly) the Li atoms consumed in the laser ionization process. This holds at least if the probing time is long enough compared to the equilibration time of the Li $\leftrightarrow$ LiOH reaction. This replenishment might be retarded, however, by the enhanced tendency of laser-excited alkali atoms to form monohydroxides (54–56). (Laser excitation of alkali atoms is an intermediate step in LEIS.)

2.3. Counting of Atoms in the Probe Volume; Detection Efficiency

An atom interacting with the laser beam(s) may produce an *event* in the detector that carries one-bit information about its appearance in the probe volume. In LIFS the event may be the emission of just a single photoelectron or a burst of photoelectrons in one photodetector, or it may be the coincidence between such emissions in two photodetectors (Fig. 4.3a). In LIIS the event may be the capture of an ion (or primary electron) by the collector plate or the ion (electron) multiplier.

After internal and/or external amplification the event is registered as a count or current pulse. The *total number of events*, N_e, occurring in a single probing with probing time T can be found by accumulating the counts, or by integrating the current pulses. If T is long, one can also record the count rate or detector output current as a function of time, smoothed by the recorder time constant. Integration over T then yields a measure for N_e. The statistical error can be improved if one repeats the probing R times (Section 2.6). The total time taken for R repeated probings is here called the *(total) time of measurement* t_m.

With a pulsed laser, the *probing time* T is determined by the pulse duration or the width of the time gate in case a time-gating circuit is applied, whichever is the shorter. Multiple probings are obtained by firing laser shots at *repetition frequency* f_{rep}. By means of a boxcar integrator and a smoothing circuit, the count rate or output current can then be recorded as a function of time and averaged or integrated over t_m. With a cw laser (whether or not intensity modulated) T is determined by the duration of one uninterrupted observation run. When time gating is applied with a cw laser, T is the width of the time gate. With repetitively pulsed lasers the statistics of measuring depend on the *duty cycle* $(T \cdot f_{rep})$, which is usually very small compared to unity. In this respect pulsed lasers are disadvantageous in comparison with cw lasers operated in either the dc or the ac mode, at a given t_m and with the same (peak) power (3). There are, of course, benefits in other respects when working with pulsed lasers (higher peak power, permitting multiphoton excitation or ionization, frequency doubling, and enlargement of the probe volume while maintaining optical saturation (2, 57) (Chapters 2 and 3).

The *probe volume* V_p is determined by the geometry of the (intersecting) laser beam(s) and by the part of the irradiated region that is ''seen'' by the detector. When the radiant energy density in the laser beam or the efficiency of detecting a fluorescence photon or charge carrier varies with position inside V_p, we should rather speak of the *effective* probe volume. The latter depends on whether a one- or multiphoton process is involved and on whether saturation is attained or not. For a given laser beam geometry, the saturation volume may be a function of laser power P_L when a focused beam is used or when the beam irradiance drops gradually with off-axis distance (3, 36). In these cases the saturation curve (fluorescence intensity versus P_L) will not show a proper ''plateau'' in the limit of high P_L (51, 58).

Before establishing the relationship between N_e and the number, N_p, of atoms in V_p, we have to distinguish between two extreme probing conditions: the (*quasi-*)*stationary* and the *nonstationary case* (Fig. 4.4a and b, respectively). Usually V_p is a small part of the atomizer volume V_a and atoms move in and out of V_p freely and independently. Let τ be the (mean) *residence* or *transit time* of an atom in V_p. Following (39) we distinguish between (i) the *flow-controlled case* and (ii) the *diffusion-controlled case*. In case (i), which can occur only with continuous-flow atomizers, τ depends on the *flow velocity*, v_f, and on the linear dimension of V_p in the flow direction. In case (ii), τ is determined by the random motion of the atoms and the (smallest) linear dimension of V_p. If this dimension is less than the mean free path, the thermal velocity distribution of the atoms is decisive; in the opposite case, their diffusional motion should be considered. In a closed atomizer with a uniform atom distribution, only case (ii) can apply. In a continuous-flow atomizer with a carrier gas, transport of atoms takes place by diffusion as well as convection. Case (i) or case (ii) can then apply, depending on whether the one or the other yields the shorter τ value, for given dimensions of V_p.

The (quasi-)stationary case (a) is found if τ is much larger than the probing time T; the nonstationary case (b) occurs if $\tau \ll T$. In the former case, the number, N_p, of atoms in V_p does, virtually, not vary in a time interval of length T, whereas N_p varies stochastically in the latter case. (Of couse, when atoms

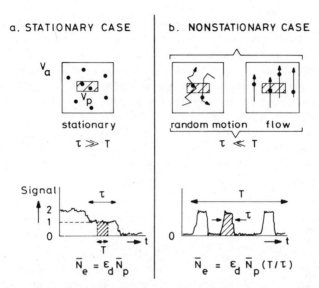

Fig. 4.4. Stationary (a) and nonstationary (b) laser probing. Dots in upper figures denote free atoms. Hatched areas in the bottom figures constitute an "event" produced by an individual atom at the detector. Left upper figure under b represents the diffusion-controlled case (ii) and right upper figure the flow-controlled case (i).

are ionized and captured by a detector as in LIIS, N_p will decrease, by depletion, during probing; but this is not at stake here.) Both with a closed and a continuous-flow atomizer, case (a) can go over to case (b) if T is increased, and conversely.

At the bottom of Fig. 4.4 it is schematically shown how in LIFS the number of events occurring in probing time T could be determined. We suppose that each atom, as long as it stays in V_p and is irradiated by the laser, produces a train of photoelectrons due to multiple recycling. In Fig. 4.4a the signal (photocurrent) is plotted as a function of time t, assuming continuous irradiation. The signal varies by a unit step each time an atom leaves or enters V_p; the mean interval between two consecutive steps is just τ. When time gating is applied with $T \ll \tau$, the signal produced by one atom, integrated over T, counts as *one* event (hatched area in Fig. 4.4a). Alternatively, one could count the photoelectrons released during T. Then the mean number of photoelectrons counted per atom constitutes one event.

At the bottom of Fig. 4.4b the photocurrent is again plotted as a function of t, but now in the nonstationary case. Every time an atom crosses V_p, a pulse of mean duration τ is observed. The mean charge contained in one such pulse constitutes again *one* event (see hatched area). The number of events occurring in T ($\gg \tau$) is then found by dividing the photocurrent, integrated over T, by the pulse charge. Alternatively, one could count the time-resolved pulses during T, each individual pulse corresponding to one event.—The variable magnitude of the events is, as such, irrelevant, as long as we can clearly distinguish them from background and electronic noise. When two events may be recorded simultaneously, as in Fig. 4.4a, we only require that their spread is sufficiently small to discern them from the recording of one or three events, etc.

The distinction between the stationary and nonstationary probing case parallels that made in (59) between the usage of LIF as a density and as a flux detector, respectively.

The *relationship between N_e and N_p* depends on whether the (a) stationary or (b) nonstationary case applies. In the former case we have straightforwardly:

$$\overline{N}_e = \epsilon_d \cdot \overline{N}_p \qquad (4.1a)$$

where ϵ_d is the *detection efficiency* ($0 \leqslant \epsilon_d \leqslant 1$; see Table 4.1). In LIFS ϵ_d depends on the rate of photoexcitation, the solid angle of observation, the quantum efficiency of the photocathode, and so on. The probability that an event remains unobserved because it is obscured by background or electronic noise is *not* included in ϵ_d. ("Detection" thus refers here to the physical process in the photodetector, not to the actual registration of this process.) Similarly, in LIIS ϵ_d depends, for example, on the collection efficiency of the charge carrier(s) and the efficiency of the particle multiplier.

If $\overline{N}_e \ll 1$, it should be interpreted as the *probability* that one event occurs during T. Similarly, if $\overline{N}_p \ll 1$, it represents the probability of finding one atom in V_p at an arbitrary moment. The probability of finding two atoms in V_p is then $(\overline{N}_p)^2$, and so on. For a statistically stationary stochastic system the overbar on N_e or N_p relates to the average over an ensemble of similar systems at any moment as well as to the time average for one system.

In the nonstationary case (b) we have instead of Eq. (4.1a):

$$\overline{N}_e = \epsilon_d \cdot \overline{N}_p \cdot \frac{T}{\tau} \qquad (4.1b)$$

This equation holds because the number of atoms that pass through V_p within T is, on the average, equal to $\overline{N}_p \cdot (T/\tau)$. Even when $\overline{N}_p \ll 1$, its value can be assessed in a *single* probing if (T/τ) is known and sufficiently large to make $\overline{N}_e > 1$.

In the diffusion-controlled case, an atom that has just left V_p and lingers near its boundary has an enhanced chance to reenter V_p and to cause a second event. The time sequence of the pulses in Fig. 4.4b is then not fully random ("bunching" effect) and Eq. (4.1b) ceases to be strictly valid.

2.4. Overall Efficiency of Counting Atoms in a Sample

Fig. 4.1 illustrates, in the form of a flowchart, that usually but a small fraction of the N_s atoms present in the sample "survive" to become detected. There is a waste of atoms at the atomization and probing stages, whereas not all atoms under probing will produce an event. (We leave out of consideration here that, in addition, only a fraction of the events may be actually observed because of background and electronic noises.) We define the *overall efficiency* ϵ_o of a laser spectrometer to count atoms in the sample as the probability that a given atom in the sample produces an event in a single probing or

$$\epsilon_o \equiv \frac{N_e}{N_s} \qquad (4.2)$$

We can decompose ϵ_o into a product of partial efficiencies, namely the *atomization efficiency* ϵ_a, the *probing efficiency* ϵ_p and the *detection efficiency* ϵ_d (for definitions, see Table 4.1):

$$\epsilon_o = \epsilon_a \cdot \epsilon_p \cdot \epsilon_d \qquad (4.3)$$

In order to derive an expression for ϵ_p, we shall first consider the relation between N_a and N_p.

The number density, n_a, of free atoms in a *closed* atomizer with volume V_a containing N_a free atoms is determined by

$$N_a = n_a \cdot V_a \qquad (4.4)$$

With the assumed uniform atom distribution (Section 2.1), we immediately get for the mean number, $\overline{N}_p$, of atoms in the probe volume V_p:

$$\overline{N}_p = n_a \cdot V_p = N_a \frac{V_p}{V_a} \qquad (4.5)$$

In the case of stationary probing (Fig. 4.4a), we obtain from Eq. (4.5) and the definition of ϵ_p:

$$\epsilon_p = \frac{\overline{N}_p}{N_a} = \frac{V_p}{V_a} \qquad (4.6a)$$

In the nonstationary case (Fig. 4.4b), we have instead

$$\epsilon_p = \frac{\overline{N}_p}{N_a} \cdot \frac{T}{\tau} = \frac{V_p}{V_a} \cdot \frac{T}{\tau} \tag{4.6b}$$

as $\overline{N}_p \cdot (T/\tau)$ is the mean number of atoms that have passed through V_p within probing time T [cf. Eq. (4.1b) and annex discussion]. We note that Eq. (4.6b) ceases to hold when $T/\tau \gtrsim V_a/V_p$. For, at $T/\tau \approx V_a/V_p$, a given atom that is initially found anywhere inside V_a, will turn up sometime during T in V_p with a probability close to unity. Then ϵ_p will be close to unity, too.

Expressions (4.6a) and (4.6b) are also applicable for a *continuous-flow* atomizer if we define here N_a as the total number of free atoms produced during *atomization time* t_a (Table 4.1). Accordingly, we introduce formally a (*fictitious*) *atomizer volume* V_a that obeys Eqs. (4.4) and (4.5) where n_a relates to the probe region. In other words, V_a is the volume that would be occupied by all N_a free atoms produced during t_a, if we could keep their density uniform and equal to its value in V_p.

Using Eqs. (4.1a) and (4.6a) in the stationary case and Eqs. (4.1b) and (4.6b) in the nonstationary case, we easily check the validity of expression (4.3) in either case. The idealizations made implicitly in the derivation of these equations should be kept in mind. Nevertheless, in these simplified equations the basic role of the parameters involved stands out more clearly.

In the *stationary* case, where Eq. (4.6a) applies, and with a closed atomizer, ϵ_p would approach unity if we make V_p as large as V_a. For a continuous-flow atomizer, this limit is hard to realize, especially so when working with a flame or ICP. Here n_a will be small and thus the (fictitious) volume V_a large, at given N_a [Eq. (4.4)], in comparison with open electrothermal atomizers. This holds because of the larger carrier gas flow and thermal expansion in the former case (2). With a given continuous-flow atomizer and a given c_s, n_a is fixed. V_a may then be reduced by shortening t_a, resulting in a smaller N_a value, so that fewer sample atoms are wasted before and after probing takes place. Ideally, t_a should match T. A carbon rod or filament atomizer with pulsed atomization of minute samples is attractive in this respect.

In the *nonstationary* case, where Eq. (4.6b) applies, an unfavorable (V_p/V_a) ratio can be compensated by a large (T/τ) ratio. Under flow-controlled measuring conditions, the cross-sectional area (O_a) of the flowing gas stream should match the area (O_p) of V_p, so that each atom passes through V_p. When, for example, atoms are vaporized from a sample deposited on a (hot) filament, this may be obtained by gas-dynamic focusing using a sheathing gas jet (57, 60). In a particular case the gas jet could be constricted to a diameter of about 0.1 mm.

Suppose the gas stream is cylindrical, with a uniform flow velocity v_f. Then the (fictitious) analyzer volume V_a equals $O_a \cdot v_f \cdot t_a$, whereas $V_p = O_p \cdot v_f \cdot \tau$ if τ is flow controlled (Section 2.3). Applying Eq. (4.6b), under nonstationary measuring conditions, we find $\epsilon_p = (O_p/O_a) \cdot (T/t_a)$. Here v_f and τ drop out. At given (O_p/O_a) and T, ϵ_p can then be improved only by reducing t_a.

When we try to maximize ϵ_p by changing the instrumental parameters or measuring conditions, we should be aware that ϵ_a and ϵ_d may be changed, too. A change of T, τ, or V_p (which may also depend on laser power; Section 2.3) could influence ϵ_d because, for example, the number of fluorescence photons emitted per atom in a single probing may depend on these parameters. Also, a change in V_a might affect ϵ_a if the latter depends on the flow velocity or the dimensions of the atomizer. Furthermore, some of these parameters cannot be changed independently of each other. For example, a change of the linear dimension of the probe volume in the flow direction also affects the transit time τ. An increase of τ, in turn, could ensue a changeover from nonstationary to stationary measuring conditions (if T is held constant).

In practice the best way to maximize ϵ_o or ϵ_d will depend strongly on the particular application and detection technique at hand. So we had to content ourselves with a few general guidelines only.

With regard to the distinction made in Section 1.2 between the restricted and the general concept of SAD, we conclude that a necessary (but not sufficient!) condition for the former is $\epsilon_d = 1$, whereas the latter concept implies $\epsilon_o = \epsilon_a \cdot \epsilon_p \cdot \epsilon_d = 1$. Usually we have $\epsilon_a \cdot \epsilon_p \ll 1$, so that a large number of analyte atoms in the sample is wasted for each atom counted in the probe volume.

2.5. Calibration Procedure and Conversion Factors

The *calibration procedure* in (chemical or diagnostic) analysis depends primarily on the kind of quantity sought and on the state of sample to be analyzed. We may, for example, want to measure the concentration or absolute quantity of analyte in a solution, the saturated vapor pressure of an atomic or molecular species, isotope ratios in a solid sample, the absolute rate at which a rare event takes place, or the absolute number of daughter atoms produced by radioactive decay. Calibration would, in general, be straightforward if we would be able to count the events produced by individual atoms, knowing ϵ_o (Sections 2.4 and 4.2). Usually, ϵ_o is hard to measure or to calculate from instrumental parameters, atomic properties, and laser beam characteristics, especially so if atomization of solid or liquid samples is involved. However, ϵ_o drops out when we are only interested in isotope ratios of the same element. In diagnostic applications where one is only interested in the number density, n_a, of atoms in the gas phase, the atomization efficiency plays no role. If the detection efficiency ϵ_d is known, one derives N_p from measured N_e through Eq. (4.1a) under stationary probing conditions. Then n_a follows from N_p through Eq. (4.5), if we know also V_p (the latter might be hard to assess, however; see Section 2.3). In LIFS, ϵ_d can be determined by means of an additional scattering experiment if, for example, a noble gas is present and its pressure and Rayleigh scattering cross section are known (19, 61, 62). In LIIS one can occasionally eliminate one of the factors that determine ϵ_d (such as the ion collection efficiency) by plotting the signal strength versus some suitably chosen instrumental parameter (collector voltage). One works then at the plateau where "saturation," if any, is reached.

In most spectrochemical applications, calibration is obtained in an indirect way by means of a series of reference samples. These samples contain a known amount or concentration of analyte in a chemical environment that closely resembles that of the unknown samples. Interferences by concomitants or matrix effects are thus accounted for. In this way one bypasses the various (unknown) partial efficiencies contained in ϵ_o.

A quite different type of indirect calibration can be applied, for example, when we want to measure absolute vapor pressures at very low temperatures. Calibration can here be obtained by repeating the measurements, under the same experimental conditions, at higher temperatures at which the absolute vapor pressure is already known. This procedure was applied in LIFS, under (near-) SAD conditions, to pure atomic Na and Pb vapors (15, 63–65) (Section 3.2). Interestingly, in one of these experiments with Na this procedure was combined with a direct, absolute calibration using the known vapor cell dimensions and atomic absorption coefficient (15). The latter procedure utilized in an ingeneous way the effect on the fluorescence signal caused by preabsorption of the laser beam before entering the probe volume. This procedure could therefore be followed only at sufficiently high vapor pressures.

The *conversion factor* ψ relates the concentration c_s (or quantity q_s) of analyte in the sample to the number density n_a (or number N_a) of free atoms in the atomizer:

$$\psi \equiv \frac{n_a}{c_s} \quad (\text{or} \equiv N_a/q_s) \tag{4.7}$$

Expressions for ψ have been given, for example, in (2, 35, 51, 64). These expressions involve ϵ_a, which can thus be derived from ψ if we do know the other instrumental factors involved. These factors include the flow rate (mol/s) and thermal expansion of the carrier gas or the velocity and cross section of the carrier gas stream in a continuous-flow atomizer. They can often be readily measured, whereas ψ can be determined by an absolute measurement of n_a at given c_s (51).

We note that ψ differs by a factor $(\beta_v \cdot \beta_a)$ from the "conversion factor" K as defined in (51). β_v accounts for incomplete volatilization of the aerosol particles, whereas β_a accounts for losses of free analyte atoms due to molecule formation or ionization in the gas phase (1). Typical values of ψ (or K) for a flame with pneumatic nebulizer are in the 10^{10}–10^{11} (atoms/cm^3)/(μg/mL) range if $\beta_a \cdot \beta_v = 1$. (The usage of parts per million, etc., as a unit for concentration is discouraged, as it could relate to relative numbers as well as relative masses, in the gas phase or in the sample.) Remarkably, a similar ψ value (3×10^{10}) was found when a Pb solution was evaporated in a graphite-cup continuous-flow atomizer used for analysis by LIFS (64). One would here expect a comparatively much higher value (Section 2.4). This discrepancy might be explained by the low ϵ_a ($\approx 1\%$) and rather high velocity of the carrier gas (≈ 10 cm/s) reported in the same paper.

Although ψ is usually not known accurately enough to calibrate chemical analyses, it may be an interesting quantity for the following reasons:

1. The atomization efficiency can be assessed from ψ (see preceding discussion).

2. ψ enables us to convert LODs reported in the analytical literature to atomic densities in the atomizer, and conversely. This conversion is especially useful when we want to compare analytical LODs with the intrinsic values (Section 2.6).

3. Through ψ we can intercompare different types of atomizers as to their analytical performance for various analytes.

2.6. Statistical Expressions

The intrinsic LOD as defined in Section 1.1 relates to the inherent error in the signal, caused by statistical fluctuations in the number of atoms probed and in the detection process. The minimum detectable signal was defined as k times its intrinsic rms error. When small numbers of atoms are to be counted, it is appropriate to consider the number of events, N_e, counted in a single probing interval T, as the signal. The minimum detectable signal is therefore

$$(\overline{N_e})_m = k \cdot \tilde{N}_e \qquad (4.8)$$

where $\tilde{N}_e \equiv \sqrt{\overline{(\Delta N_e)^2}}$ is the rms intrinsic error in N_e. The relation between $\overline{N}_e$ and $\overline{N}_p$ is given by Eq. (4.1a) or (4.1b) in the stationary or nonstationary case, respectively. Through this relation we can express $(\overline{N}_p)_m$ in $(\overline{N}_e)_m$, whereas we derive $(n_a)_m$ from $(\overline{N}_p)_m$ by using Eq. (4.5). The minimum detectable sample concentraton c_m follows, in turn, from $(n_a)_m$ through the conversion factor ψ [Eq. (4.7)]. Alternatively, we can directly relate $(\overline{N}_e)_m$ to the minimum detectable number of analyte atoms in the sample, $(\overline{N}_s)_m$, by using ϵ_o [Eq. (4.2)].

In the following we present *statistical expressions* for the *mean signal*, the *precision* [$\equiv$ relative standard deviation (1)], the *detection efficiency*, and the *limit of detection* relating to the number (density) of atoms in the probe volume, for single as well as repeated probings. We disregard again extrinsic errors.

Numerous expressions for the precision or LOD involving extrinsic errors have been presented in the spectrochemical literature [see, e.g., (2, 3, 35, 66–68)]. Only rarely was the contribution of the fluctuation in N_p included (3), and then only for the stationary probing case. A detailed study of the effect of particle fluctuations on the scattered light intensity has been made in (69). The possibility of verifying statistical-mechanical expressions for atom number fluctuations by SAD techniques was discussed in (5).

Let us first consider the *stationary case* where $\bar{N}_e = \epsilon_d \cdot \bar{N}_p$ [Eq. (4.1a)]. If $\epsilon_d = 1$, we have $\tilde{N}_e = \tilde{N}_p$ and $\tilde{N}_e = \sqrt{\bar{N}_e}$ if N_p and thus N_e obey Poisson statistics. However, if $\epsilon_d < 1$, N_e will fluctuate even if N_p is fixed, as the occurrence of an event is now a chance process, described by a Bernoulli (or binomial) distribution. (We assume that events produced by different atoms are statistically uncorrelated.) The relative spread of N_e, at fixed N_p, will be the smaller, the closer ϵ_d approaches unity. If N_p obeys Poisson statistics and N_e, at *fixed* N_p, obeys Bernoulli statistics, we obtain exactly for any ϵ_d

$$\tilde{N}_e = \sqrt{\bar{N}_e} \qquad (= \sqrt{\epsilon_d \cdot \bar{N}_p}) \qquad (4.9)$$

In other words, N_e obeys Poisson statistics, too. This outcome is the same as that obtained in the theory of partition shot noise in electronic devices (70).

One should be careful in deriving expressions for the fluctuations in the photocurrent when the number of emitting atoms fluctuates. One must not, in general, add the variance ($\equiv$ squared standard deviation) induced by the atom number fluctuations to that associated with the photocurrent shot noise. This is seen most clearly in the simple case when each atom produces just one photo-electron ($\epsilon_d = 1$). Adding up these variances would then make the standard deviation in the photocurrent $\sqrt{2}$ times too high!

The assumption that the production of an event (in casu: the emission of a photoelectron) by one atom is statistically uncorrelated with the events produced by other atoms implies the condition of *incoherent* radiation detection. This means that the intensities, *not* the amplitudes, of the radiation fields produced by each atom at the photocathode are additive. Then also the probabilities of photoelectron emission—which are proportional to the corresponding radiation intensities—are additive (69). This condition is surely fulfilled in the experimental situations considered here. The fluorescence radiation emitted by the atom is not coherent with the exciting laser field nor with the radiations emitted by other atoms (34).

Equation (4.9) holds also in the *nonstationary case*, if the transient appearance of an atom in V_p during T ($\gg \tau$) is purely random. We have only to replace in the second part of this equation $\epsilon_d \cdot \bar{N}_p$ by the expression given in Eq. (4.1b).

We now derive a statistical expression for ϵ_d, being the probability of an event occurring in a single probing when one atom is present in, or passes through V_p. Let ϕ be the constant probability per second that a photoelectron is generated (in LIFS) or an ion (or primary electron) is detected (in LIIS) as long as there is one atom in V_p. If the atom will generate at most one photo-electron during its whole stay in V_p, each photoelectron generated counts as one event. We then have in the stationary case ($\tau \gg T$)

$$\epsilon_d = \phi \cdot T \qquad (\text{if } \phi \cdot T \ll 1) \qquad (4.10a)$$

and in the nonstationary case ($\tau \ll T$)

$$\epsilon_d = \phi \cdot \tau \qquad (\text{if } \phi \cdot \tau \ll 1) \qquad (4.10b)$$

[The probability, $(\phi \cdot T)^2$ or $(\phi \cdot \tau)^2$, that two photoelectrons are generated per atom is then much less than $\phi \cdot T$ or $\phi \cdot \tau$ and thus negligible, indeed.]

When in LIFS $\phi \cdot T$ or $\phi \cdot \tau$ is arbitrarily large and multiple recycling occurs, we have to consider the *Poisson distribution* for the number, m, of photoelectrons generated per atom during T or τ:

$$p(m) = \frac{1}{m!} \cdot \exp[-\overline{m}] \cdot \overline{m}^m \qquad (4.11)$$

Here we have to substitute $\phi \cdot T$ or $\phi \cdot \tau$ for $\overline{m}$. Since the generation of any number ($m \geq 1$) of photoelectrons by a given atom is, by definition, one event, we arrive at the following general expression for ϵ_d in LIFS:

$$\epsilon_d = \sum_{m=1}^{\infty} p(m) \qquad (4.12)$$

(The special case when a pulse-height discriminator is applied will be considered in Section 3.2.)

When in LIIS or LIFS $\phi \cdot T$ or $\phi \cdot \tau$ are arbitrarily large and no recycling takes place, we generally find for ϵ_d

$$\epsilon_d = 1 - \exp[-\phi \cdot T] \qquad (\text{stationary case}) \qquad (4.13a)$$

$$\epsilon_d = 1 - \exp[-\phi \cdot \tau] \qquad (\text{nonstationary case}) \qquad (4.13b)$$

Here the probing is a consumptive process, so that each atom produces no more than one ion (or primary electron) or one photon. One easily checks that expressions (4.10a) and (4.10b) follow from expressions (4.12) or (4.13a) and (4.13b), respectively, in the limit $\epsilon_d \to 0$.

The Poisson distribution law applies also to the total number of events, N_e, produced by all atoms present in or passing through V_p during probing time T:

$$p(N_e) = \frac{1}{N_e!} \cdot \exp\left[-\bar{N}_e\right] \cdot \bar{N}_e^{N_e} \qquad (4.14)$$

The mean value, $\bar{N}_e$, is again given by Eq. (4.1a) or (4.1b). The validity of Eq. (4.14) was verified in a particular LIIS experiment under nonstationary probing conditions with an Yb and a Na beam (63). In this experiment ϵ_d was close to unity, whereas $\bar{N}_p$ and $\bar{N}_e$ ranged from 0.003 to 0.03 and from 0.35 to 3, respectively. The probing time T was 20 s.

Table 4.3 collects statistical expressions for the mean signal, precision and

Table 4.3. Statistical Expressions[a]

Quantity	Stationary Case (a) ($\tau \gg T$)	Nonstationary Case (b) ($\tau \ll T$)
Mean signal	$\bar{N}_e = \epsilon_d \bar{N}_p$ (4.15)	$\bar{N}_e = \epsilon_d \bar{N}_p (T/\tau)$ (4.16)
	$[\bar{N}_e = \phi T \bar{N}_p \qquad (4.17)]^b$	
Precision		
1 probing	$\tilde{N}_e/\bar{N}_e = 1/\sqrt{\bar{N}_e}$ (4.18a)	
R probings	$\tilde{N}_e/\bar{N}_e = 1/\sqrt{R\,\bar{N}_e}$ (4.18b)	
Intrinsic LOD[c]		
N_e $\begin{cases} \text{1 probing} \\ \\ R \text{ probings} \end{cases}$	$(\bar{N}_e)_m \equiv k\,\tilde{N}_e = k\,\sqrt{(\bar{N}_e)_m}$	
	or: $(\bar{N}_e)_m = k^2$ (4.19a)	
	$(\bar{N}_e)_m = k^2/R$ (4.19b)	
N_p; R probings	$(\bar{N}_p)_m = k^2/\epsilon_d R$ (4.20)	$(\bar{N}_p)_m = (k^2/\epsilon_d R)(\tau/T)$ (4.21)
	$[(\bar{N}_p)_m = k^2/\phi TR \qquad (4.22)]^b$	
n_a; R probings	$(n_a)_m = k^2/\epsilon_d V_p R$ (4.23)	$(n_a)_m = (k^2/\epsilon_d V_p R)(\tau/T)$ (4.24)
	$[(n_a)_m = k^2/\phi TV_p R \qquad (4.25)]^b$	

[a] N_e is number of events *per probing*; for explanation of other symbols, see Table 4.1. For statistical expressions for ϵ_d, see text.

[b] Equations within square brackets refer to the limiting case when the probability of one atom in V_p producing a photoelectron (in LIFS) or an ion (in LIIS) during probing time T (case a) or transit time τ (case b) is small. These equations are placed between the two column headings, because they hold in both cases.

[c] LOD relates here to number (density) of atoms in V_p or number of events at detector (not to sample concentration).

[Reproduced, with slight modifications, from (39) by permission of the Society for Applied Spectroscopy, USA.]

intrinsic LOD in the case of stationary (a) and nonstationary (b) probing. A Poisson distribution is assumed for N_e and N_p, and extrinsic errors are disregarded, as before. Some expressions have already been dealt with; the others follow straightforwardly from them, if we assume that repeated probings are statistically uncorrelated. This assumption need not necessarily hold in case (a), if $f_{rep} > 1/\tau$ *and* $\epsilon_d \approx 1$. If $\epsilon_d \ll 1$, the relative spread in N_e largely exceeds that in N_p, so that consecutive probings are statistically uncorrelated, irrespective of $f_{rep} \cdot \tau$. A few remarks are to be made:

1. At the limiting condition underlying Eq. (4.17) (see footnote *b* of Table 4.3), the mean signal appears to be the same in case (a) as in case (b), although the general expressions (4.15) and (4.16) differ from each other by a factor (T/τ) (> 1). This factor is, however, canceled in Eq. (4.17) because ϵ_d for case (a) exceeds that for case (b) by the same factor [Eqs. (4.10a) and (4.10b)]. This cancellation also occurs in Eqs. (4.22) and (4.25), which hold for the same condition that underlies Eq. (4.17).

2. Equation (4.18a) for the precision holds formally also when $\overline{N}_e \ll 1$. We can still determine such low $\overline{N}_e$ values at any precision by repeating the probings a sufficiently large number of times [Eq. (4.18b)]. The mean total number of events observed in R probings, $\overline{N}_e \cdot R$, should exceed unity for obtaining a statistically significant measurement.

3. Expressions for the intrinsic LOD relating to n_a follow from those relating to N_p by dividing the latter ones by the probe volume V_p. We recall that ϵ_d, τ, or ϕ may depend on the extent and shape of V_p. So an enlargement of V_p need not necessarily lead to an improvement of $(n_a)_m$. Much depends on the particular experimental conditions at hand. The expressions for $(n_a)_m$ are of direct interest in diagnostic studies. They are of indirect interest in chemical analysis because c_m follows from $(n_a)_m/\psi$ (Section 2.5).

4. Remarkably, R appears in the denominator of expressions (4.19b)–(4.25) to the *first* power. This contrasts with the conventional, extrinsic LOD, which varies in direct proportion to $1/\sqrt{R}$ if the background and detector noise spectrum is "white" (51, 71). The dependence of the intrinsic LOD on probing time T is less evident, as ϵ_d occurring in Eqs. (4.20), (4.21), (4.23), and (4.24) may vary with T, too. Only in the particular situation where Eqs. (4.22) and (4.25) apply can we generally state that the intrinsic LOD improves in direct proportion to T. This is again in obvious contrast with the conventional LOD, which improves in direct proportion to $\sqrt{T}$ (2, 51, 71).

5. Expressions that hold under SAD conditions are obtained by taking $\epsilon_d \equiv 1$. Expressions (4.17), (4.22), and (4.25) are then no longer applicable.

6. One should not extrapolate the usual expressions for the analytical (extrinsic) LOD to such low c_s or q_s values that the mean number of atoms probed becomes of the order of unity or less.

3. SPECIFIC DETECTION TECHNIQUES AND EXPERIMENTAL RESULTS

3.1. Classification of Detection Techniques

The broad classification of laser beam detection techniques in Section 2.2 will here be refined, and some examples of experimental realizations and results obtained will be described in the following sections. We shall focus our attention on techniques that enable us, with special arrangements, to detect small numbers of atoms (and molecules) or even single atoms. Figure 4.5 summarizes the basic schemes of those techniques that have proved to be most successful in this respect.—With regard to the existing, wildly flowering, and confusing terminology, we have tried to use a more consistent set of descriptive terms.

Optical Detection Techniques. This category encompasses techniques based on any optical effect—such as fluorescence, absorption, stimulated emission (3,

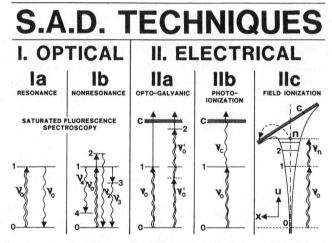

Fig. 4.5. Classification of main types of laser-based single-atom detection (SAD) techniques. Simplified energy level schemes are given with the relevant atomic transitions. A double, upward pointing, wavy arrow denotes a saturated transition induced by a laser tuned to an atomic resonance. Single wavy arrows denote nonsaturated transitions by photon absorption or spontaneous photon emission. Straight lines denote collisional transitions. Energy levels are denoted by 0 (for the ground state), 1, 2, . . . , n; the continuum state is denoted by c. Optical frequencies are indicated by ν. Under IIc the distortion of the potential energy function $U(x)$ of the valence electron, caused by an external dc electric field in the x direction, is schematically shown. Here horizontal bars represent discrete energy levels, whereas the sloping line represents the continuum state. A spontaneous transition from a high-lying bound state n to the continuum state c may occur by quantum mechanical tunneling through a potential barrier. [Reproduced from (39) by permission of the Society for Applied Spectroscopy, USA.]

72), polarization*, Raman scattering (incoherent or coherent) (Chapter 10) (73), birefringence (74), anomalous dispersion (75), and anomalous refraction—that can be induced or detected by laser radiation [for a survey see (27, 35, 76)]. We restrict ourselves to fluorescence and absorption techniques, which are the more promising ones as regards detectability.

Techniques based on *laser-induced fluorescence spectroscopy* (*LIFS*) can be subdivided into *resonance* (*RF*) and *nonresonance fluorescence* (*NF*) techniques (Section 2.2 and Chapter 2). Their basic operation schemes are shown in Fig. 4.5, Ia and Ib. The main advantage of using a laser is that optical saturation (Chapter 1) can be achieved. [The term *saturated optical nonresonance emission spectroscopy* (*SONRES*) has therefore come into use.] The main practical advantage of using NF, instead of RF, is that laser background scattering can be simply rejected. If an optical transition is allowed from the laser-excited level 1 in Fig. 4.5, Ib, to a lower lying excitation level 4, NF can be applied without requiring inelastic collisions in a buffer gas.

Since the emission of a fluorescence photon is preceded by the absorption of a laser photon, the fluorescence excitation spectrum ($\equiv$ fluorescence intensity versus laser detuning) is essentially an absorption spectrum. This holds at least if saturation broadening (51) (Chapter 1) is avoided and if the probability that an excited atom reemits a photon is independent of the detuning (77).

The use of lasers in conventional *atomic* (*AAS*) or *molecular absorption spectroscopy* (*MAS*) may not bring about a substantial gain in LOD as long as the latter is determined by fluctuations in the laser power (3). Then, only the directionality of the laser beam can be an advantage in obtaining a long absorption pathlength l_{abs} by multiple folding of the beam inside the cell. However, the minimum detectable absorption factor α_m ($= k_m \cdot l_{abs}$; $k \equiv$ napierian absorptivity, proportional to number density) can be greatly improved by special laser-based variants of AAS or MAS. These variants include: intracavity absorption spectroscopy (*ICAS*), low frequency and rf laser wavelength modulation spectroscopy, laser Zeeman or Stark modulation spectroscopy (where the absorption line frequency is modulated) and saturated absorption spectroscopy (where the saturation induced by a pump laser is monitored by AS with a relatively weak probe laser). We will discuss here only two of these variants, which have the prospect of obtaining low LODs [for a survey see (26, 27, 35, 76)].

In *ICAS*, where the absorption cell is placed inside the laser cavity (Chapter 12) (26, 27), laser photons pass a large number of times through the atomic vapor before they are coupled out. In this way l_{abs} may be enhanced by a factor of 100–1000 (3). Due to the nonlinear interaction between laser field and atoms,

---

*A LOD of 0.05 ng/mL has recently been attained for sodium in an air-acetylene flame by laser-induced polarization spectroscopy [W. G. Tong and E. S. Yeung, *Anal. Chem.*, **57**, 70 (1985)].

a small ($k \cdot l_{abs}$) value may already drastically change the laser output if the laser operates close above threshold. With multimode (broadband) lasers an extra enhancement of absorption sensitivity occurs, that is proportional to the number of oscillating modes (78). (Mode competition may affect the spectral laser characteristics too.) With long pathlengths, k_m values ranging from 10^{-8} to 10^{-9} cm^{-1} have been attained with standing-wave lasers (35, 79). A value of 10^{-10} cm^{-1} has been attained with a ring traveling-wave laser (79) for an unidentified species in air contained in a cuvette of nearly 150 cm length. For the Na-D doublet lines a value $k \approx 10^{-8}$ cm^{-1} corresponds to $n_a = 4 \times 10^3$ cm^{-3} if the Na atoms are present in a gas at 1 atm pressure and 2000 K and a narrow-band line source is used (51). This illustrates the feasibility of detecting small numbers of atoms by ICAAS in an absorption cell of small internal cross section (≈ 1 mm^2) at reduced gas temperature and pressure. For molecular species the prospects are much less good (27, 35). This is because their ground-state population is distributed over a manifold of rovibrational levels, and the oscillator strength of a rotation line is much smaller than that of the Na-D lines. For similar reasons the detectability of LIFS for molecules is much worse than for atoms.

We note that k_m varies $\sim (l_{abs})^{-1}$, when α_m is fixed for a given detection technique. Since α correlates with the number of atoms per centimeter squared (i.e., $n_a \cdot l_{abs}$), and $N_p = n_a \cdot V_p = n_a \cdot l_{abs} \cdot O_p$ [Eq. (4.5); O_p = laser beam cross section], the minimum detectable *number* of atoms $(N_p)_m$ is $\sim O_p$ but independent of l_{abs}. The minimum detectable *number density* $(n_a)_m$, however, is independent of O_p, but $\sim (l_{abs})^{-1}$.

In contrast with conventional wavelength modulation techniques, *rf laser-wavelength modulation spectroscopy* (usually called frequency modulation spectroscopy) operates at modulation frequencies f_m in the megahertz–gigahertz ranges (80). The wavelength of a single-mode laser is here modulated by an external electro-optical phase modulator. The intensity remains constant but the laser spectrum now contains sidebands mutually separated by f_m hertz. When the laser beam is sent through an (external) absorption cell, while one sideband is tuned at, or near to the absorption line center, the amplitude and relative phase of this sideband will be altered. This results in an *amplitude* modulation of the transmitted laser beam, which can be heterodyne detected. One thus obtains a signal that is a measure of the absorption factor. Apart from being a null method, this technique has the advantage that electronic detection takes place outside of the frequency band of the excess laser power fluctuations. Preliminary experiments with an I_2 vapor cell (80) and a Na-seeded flame (81) did not yield impressive α_m values, as these were determined by absorption background noise. But this noise is not characteristic of the detection capability of this technique as such.

Since measuring stimulated Raman gain is equivalent to measuring *negative* absorption, rf modulation spectroscopy can be applied here, too. This was demonstrated by an experiment with a deuterium gas cell, where the detectability appeared to be shot-noise limited (82).

Electrical Detection Techniques (LIIS). They can be subdivided into *opto-galvanic* (IIa), *photoionization* (IIb) and *field ionization* (IIc) *techniques* (Fig. 4.5). This distinction is based on the different ways in which ionization from the laser-excited level is accomplished. In all cases species selectivity is obtained, as before, by tuning the laser frequency into resonance with one or more successive transitions between discrete atomic levels. (For this reason the term *resonance ionization spectroscopy*, *RIS*, has come into use for technique IIb; this term could, however, also refer to other techniques.)

Following (53), we define the *optogalvanic effect* as a perturbation of the state of ionization in a plasma in response to the absorption of optical radiation, which changes the relative level populations of atomic or molecular constituents of the plasma. Ionization from the laser-excited state is brought about by inelastic collisions with electrons or other particles in the plasma. This may occur in one step or in several steps involving higher-lying excitation levels. Since the collisional ionization and intermediate excitation steps are endoergic (requiring activation energy), a sufficiently high gas or electron temperature is needed. Inelastic collisions may then also produce background ionization in the absence of laser excitation. (In work with flames the term *laser-enhanced ionization spectroscopy*, *LEIS*, has therefore come into use—although background ionization is not essential.) For a more detailed description we refer to Chapter 3 and (32, 33, 53, 83). Two variants based on one- and two-photon excitation, respectively, are shown in Fig. 4.5, IIa.

Since endoergic collisions are involved in the ionization step, the laser-excited level should preferably lie close to the ionization continuum. Levels with excitation energies E_{exc} exceeding the laser photon energy can be populated by choosing a nonresonance absorption line. It is often more efficient to excite (even saturate) a high-lying level from the ground state by a *multistep process* using multiple laser beams tuned at consecutive atomic transitions. In a *multiphoton (MP) process* several photons of the same frequency are absorbed simultaneously. Apart from the energy matching condition ($E_{exc} = 2 \times$ photon energy), the probability of a two-photon process is considerably enhanced when there happens to exist a suitable intermediate level with $E_{exc} \approx$ photon energy. MP processes (including photodissociation and photoionization) are therefore, in general, more probable in molecules than in atoms. Two-photon excitation is to be distinguished from one-photon excitation by a frequency-doubled laser beam because of the different optical selection rules involved.

With the *photoionization technique*, the atoms after being excited by absorption of one or more laser photons, with the same or different frequencies, are transferred to the ionization continuum by nonresonant absorption of another laser photon (Fig. 4.5, IIb). This transfer may, occasionally, also proceed via an unstable autoionizing state excited by resonant photon absorption from a level below the ionization continuum. With this technique collisions are not needed, and the presence of a buffer gas is not essential. In flames processes IIa and IIb may both occur. For a more detailed treatment we refer to (5, 30, 31, 34) (see also Chapters 3 and 18). The potentialities of these techniques for SAD have, in particular, been considered in (5, 34).

With the *field ionization technique* the atoms are first raised by a multistep or a MP process to a Rydberg level lying close under the ionization continuum (Fig. 4.5, IIc). Atoms in Rydberg levels have a long lifetime in the absence of quenching collisions. By applying a pulsed electric dc field shortly after the excitation pulse, the "Rydberg atom" can spontaneously ionize before it has a chance to decay to a lower-lying level. No additional energy is needed here in the ionization step. This autoionization process is brought about by the reduction of the potential energy, U, experienced by the excited valence electron when moving in a direction opposite to the electric field vector (Fig. 4.5, IIc). In this direction the bottom of the continuum state is bent below the Rydberg level energy. Even when there exists a potential barrier for the transition to the continuum state (as depicted in Fig. 4.5, IIc), a spontaneous transition is still possible by quantum mechanical tunneling. The reason that the electric field is switched on *after* the excitation pulse is that otherwise the accompanying Stark effect would blur the Rydberg levels, which are closely packed near the ionization limit. This blurring effect would spoil the selectivity of the (last) excitation step to the Rydberg level. This technique works best with atomic beams in vacuo. For a further discussion, with special emphasis on SAD, the reader is referred to (13, 31, 34, 84).

Mechanical Detection Techniques. This category encompasses the *laser-induced beam deflection technique*, of which one variant based on *radiation pressure* was, exemplarily, described in Section 2.2. Beam deflection can also be brought about in an external electric or magnetic field when, as a result of laser excitation, the *atomic polarizability* or *magnetic moment* is changed (13, 34).

Optoacoustic spectroscopy (OAS) (also called photoacoustic spectroscopy) falls within the same category, as it exploits a mechanical effect, in casu an increase in gas pressure. This increase results from the heating effect of selective (laser) absorption in a closed cell containing the gaseous sample. The resulting (periodic or pulselike) pressure increase is detected by means of a sensitive microphone (Chapter 5) (27, 35). In contrast with the previously discussed techniques, OAS requires conversion of excitation energy into heat through quenching collisions. OAS is usually applied to detect molecules or to study their absorption spectra in the gaseous or condensed phase. It is a complement to reflection and transmission spectroscopy. Detection limits k_m as low as 10^{-10} cm^{-1} have occasionally been reached with lasers in NO pollution analysis. When comparing this achievement with those attainable by optical MAS, it should be realized that in OAS the absorption pathlength is usually restricted to typically 10 cm. However, intracavity arrangements can be applied in OAS too. Analytical detection limits expressed in *relative* particle densities are impressive $(1 : 10^{11})$. But when they are expressed in *absolute* numbers of molecules probed, this technique appears to be much less sensitive than the others. The typically low absorption cross section of molecular spectral lines is mainly to be blamed for that (as we saw earlier in this section).

3.2. Experimental Results by Fluorescence and Absorption Techniques

Some laser-based fluorescence and absorption experiments will be described that illustrate the foregoing, rather schematic, considerations.

Table 4.4 collects data that relate to the experimental conditions and the LODs obtained in experiments using resonance fluorescence (1-4), nonresonance fluorescence (5-10), and intracavity absorption (11) spectroscopy. For comparison, a nonresonance fluorescence experiment (19) dealing with rhodamine 6G molecules in the liquid phase has also been included. The numerical data listed will enable us to judge how close the experimental LODs relating to numbers, $(\overline{N}_p)_m$, or number densities, $(n_a)_m$, of atoms in the *gas* phase approach the intrinsic values or even the SAD limit. Values of V_p are inserted to convert $(\overline{N}_p)_m$ into $(n_a)_m$, or conversely. The ratio τ/T is of interest in helping to select the appropriate formulas in Table 4.3 that hold in the stationary or nonstationary case.

The figures entered are often correct as to order of magnitude only because of insufficient or ambiguous specifications in the literature. In some instances they had to be deduced indirectly from other data reported or to be estimated from analogous situations. (The specifications in experiment 19 are exemplary in their completeness and detail.) It should be admitted that the primary aim of some experiments was not to achieve low detection limits per se. Also, in some analytical applications one was rather interested in LODs pertaining to the sample before atomization. In some of these cases a conversion factor ψ had to be assumed in order to estimate $(n_a)_m$.

The *goals* of application and consequently the atomizer designs used vary largely. Experiments 6-10 and 19 were aimed at lowering the analytical detection limit. It is therefore interesting to compare the result obtained in experiment 9, where a cell with pure, saturated Pb vapor was used to investigate the optimum measuring conditions, with that obtained in experiment 10. In the latter, a real sample (here: a liquid Pb solution) was evaporated in a graphite cup

Table 4.4.

Experiment	Reference	Atomizer[b]	Technique[c]	Species	Laser[d]	Saturation
1	(52)	Be	RF	Na	c	Yes
2	(16, 17)	Be	RF	^{138}Ba	c	Yes
3	(61)	Ce(Ne)[i]	RF[j]	^{20}Na	c	No
4	(19)	Ce(He)	RF[j]	Na	c	Yes
5[l]	(85, 86)	Fl	NF[j]	Na	p	Yes
6	(87)	ETA	NF[j]	Na	c	No
7	(88–90)	Ce(Ar)	NF[j]	Na	c	Yes
8[l]	(88, 90)	Fl	NF[j]	Na	p	Yes
9	(65)	Ce	NF	Pb	p	Yes
10[n]	(64)	ETA	NF	Pb	p	Yes
11	(91)	Ce(Ne)	ICA	Na	c	Yes
12	(92)	Fl	OG	Cs	p	[Yes]
13	(93)	Fl	OG	Li	p	(No)
14	(11, 94)	Ce(Ar)[r]	PI	Cs	p	Yes
15	(28)	Ce(Ar)[r]	PI	Naphthalene	p	(Yes)
16	(84, 95, 96)	Be	FI	Na	p	Yes
17	(9, 84, 97)	Be	FI	^{173}Yb	p	Yes
18	(98, 99)	Fl *Sample cell*	OA	Na	p	Yes
19	(24, 25)	LFC	NF	Rhodamine 6G	c	No

[a] For explanation of symbols see Table 4.1. Numbers in parentheses are highly uncertain. Numbers within brackets were indirectly *derived* from data in the original paper(s) or *assumed*. Many of these numbers are reliable as to order of magnitude only.

[b] Be = atomic beam (usually combined with oven); Ce(G) = closed cell (possibly with buffer gas G; usually in oven); ETA = electrothermal atomizer (as used in AAS); Fl = flame with nebulizer; LFC = liquid flow cytometer.

[c] RF = resonance fluorescence; NF = nonresonance fluorescence; ICA = intracavity absorption; OG = optogalvanic; PI = photoionization; FI = field ionization; OA = optoacoustic.

[d] c = continuous wave; p = pulsed.

[e] Relating or converted to statistical confidence factor $k = 3$.

[f] Taken from (100).

[g] Derived from specified dark count rate and dark count number.

[h] At assumed threshold $m_{thr} = 6$ and 2 for photoelectrons in one burst in case 2 and 1, respectively.

[i] ^{20}Na atoms were produced by reaction of a proton beam with Ne atoms.

[j] Fluorescence of Na-D doublet was observed, while only one of its lines was excited by the laser in cases 3 and 4; in cases 5–8 fluorescence was observed at the other doublet line only.

V_p (cm³)	τ (s)	T (s)	ϵ_d	$(\bar{N}_p)_m$[e]	$(n_a)_m$[e] (cm⁻³)	R
10^{-3}	10^{-5}	$[10]^f$	0.4^h	10^{-4}	0.1	$[1]^f$
5×10^{-4}	1×10^{-6}	$[100]^g$	1×10^{-2h}	$(5 \times 10^{-5})^h$	$(0.1)^h$	1
2×10^{-4}	3×10^{-4}	4×10^{-3}	0.1	0.1	5×10^2	100
$[3 \times 10^{-4}]$	3×10^{-4}	2×10^{-3}	1	2×10^{-3}	10	$[10^3]^k$
10^{-5}	2×10^{-7}	1×10^{-6}	2×10^{-3}	$[10^3]$	$[10^8]$	1
(1)	?	(1)	?	(10^6)	10^6	100
10^{-3}	?	(15)	?	10^{-2}	10	1
10^{-2}	$[10^{-4}]$	5×10^{-7}	$[5 \times 10^{-3}]$	$[50]$	$[5 \times 10^3]$	$[\geq 30]$
0.1	10^{-5}	5×10^{-9}	$[10^{-3}]^m$	30	3×10^2	5×10^3
0.1	$>10^{-5o}$	5×10^{-9}	$[10^{-3}]$	150	1.5×10^3	150
7×10^{-3}	$[10^{-5}]$	60^p	—	(4×10^3)	(6×10^5)	1
(10^{-2})	$[3 \times 10^{-4}]$	1×10^{-8}	?	$[4 \times 10^3]$	$[4 \times 10^5]^q$	10
0.5	$[10^{-3}]$	1×10^{-6}	?	$[5 \times 10^4]$	$[1 \times 10^5]^q$	20
5×10^{-2}	10^{-3}	2×10^{-6}	1	5	100	1
10^{-3}	?	1×10^{-6}	s	(10^4)	(10^7)	t
5×10^{-3}	(3×10^{-6})	1×10^{-8}	0.5	15	3×10^3	1
5×10^{-4}	(10^{-6})	1×10^{-8}	0.3	0.15	3×10^2	120
2	$[10^{-3}]$	1×10^{-6}	—	(2×10^7)	(10^7)	$?^u$
1×10^{-8}	4×10^{-5}	1	7×10^{-2}	1	1×10^8	10

[k] Minimum value required to obtain the reported $(\bar{N}_p)_m$ value under assumed SAD conditions.

[l] See (39) for a discussion of assumed values within brackets.

[m] Assuming a quantum efficiency of 10% of photodetector.

[n] Some data were taken from (65).

[o] The actual τ may exceed the reported value of 10^{-5} s because motion of atoms is limited by diffusion in buffer gas.

[p] Duration of one spectral scan across the laser profile (personal communication by authors).

[q] Derived from reported LOD in sample solution, assuming conversion factor $\psi = 10^{11}$ cm⁻³/(μg/mL) [see also (101) for case 13].

[r] A minor component was admixed to the gas for operation as a gas-proportional counter.

[s] ϵ_d relating only to a molecule in the specific level(s) from which laser excitation takes place, is close to 1; $\epsilon_d \ll 1$ if it relates to a molecule in any level.

[t] Registration occurs by recording the two-photon laser excitation spectrum from the $\nu = 0$ level in the ground state. The number of laser shots required for one registration was not reported.

[u] Only the value $f_{rep} \approx 10$ Hz was reported.

atomizer and a concentrational LOD of 0.05 pg/mL and an absolute LOD of 10^{-15} g were achieved. In both experiments single-step excitation by one frequency-doubled laser beam was applied.

A similar absolute detection limit for Pb was obtained in a graphite rod atomizer by using NF with two-step excitation by two lasers (102). Remarkably, $(n_a)_m$ was several orders of magnitude larger in the latter experiment than in experiment 10. This clearly demonstrates that not only $(n_a)_m$ is decisive for obtaining good analytical LODs but also the atomizer design.

Experiment 5 was designed as a diagnostic tool for measuring temporally and spatially resolved trace concentrations in gas flows of engineering interest. Experiment 1 was especially set up to detect single atoms, whereas experiment 2 demonstrated the possibility of high-resolution spectroscopy in situations where only a small number of atoms (here: Ba isotopes) is available. Experiment 3 was aimed at detecting ^{20}Na isotopes produced by a proton beam reacting with ^{20}Ne atoms, and at studying the kinetic behavior of Na atoms in neon gas. The extension of the saturated Na vapor–pressure curve toward lower temperatures was the goal of experiment 6. Finally, experiment 11 was meant as a study of the effect of ICAAS on the spectral features of a narrow-band laser.

Resonance fluorescence (*RF*) *spectroscopy* has to face the problem of reducing *laser background scattering*. When, for example, one Na-*D* level is excited in an atomic beam experiment in vacuo, the other component does not fluoresce. When the laser is tuned at the first resonance line of Ba ($6^1P_1 \leftarrow 6^1S_0$), the NF line ($6^1P_1 \rightarrow 5^1D_2$), emitted from the laser-excited level, is but weak and lies in the infrared region. One has in these cases no other choice than applying RF. There are, however, various means to suppress laser background scattering (and any other background as well) (86). Under saturation conditions one can apply to this end the *photon-burst technique* (13, 16, 17, 19, 36, 52, 103). If the mean number of photoelectrons, $\phi \cdot \tau$, generated by one atom crossing the laser beam exceeds unity (Section 2.6), the atom has a good chance to generate a burst of at least two photoelectrons during its crossing. By applying a pulse-height discriminator and/or a coincidence setup (Fig. 4.3), one can discriminate the temporally bunched fluorescence photons from the randomly scattered laser photons and dark-current electrons. The gate width should match τ, and the chance of more than one atom being present in the probe volume should be negligible. The detection efficiency ϵ_d is then given by Eqs. (4.11) and (4.12) if we replace therein the lower limit of summation by the threshold value m_{thr} (> 1). For a given mean number, $\overline{m}$, of photoelectrons per burst, ϵ_d will of course become smaller, the higher the threshold. But this decrease will be overcompensated by a still stronger reduction in the randomly released background electrons.

One can make $\overline{m} > 1$ by increasing ϕ or τ. The latter can be increased by expanding the laser beam cross section, but this should not be done at the cost of saturation. ϕ can be increased by improving the photon collection efficiency or by selecting a photodetector with a larger quantum efficiency η. Since we do not need a monochromator in RF detection, the solid angle of observation Ω_F can be made large.

The feasibility of the photon-burst technique may be simply shown for an (effectively) two-level atom with equal statistical weights. Under saturation conditions each atom will spend half of its life in the excited level (Chapter 1). Given a single atom, the probability per second, ϕ, of detecting a photon is then given by $\phi = \frac{1}{2} A \cdot (\Omega_F/4\pi) \cdot \eta$, where A is the Einstein transition probability. Using the relation $\overline{m} = \phi \cdot \tau$ in the nonstationary case and adopting realistic values $A = 3 \times 10^7 \text{ s}^{-1}$, $\Omega_F/4\pi = 0.5$ and $\eta = 0.1$, one finds $\overline{m} = 0.7 \times 10^6 \, \tau$. For $\tau > 1.5 \, \mu\text{s}$, $\overline{m}$ will thus exceed unity. (The possible, slight anisotropy of the spontaneous photon emission has here been disregarded.)

Figure 4.6 illustrates how in atomic beam experiment 2 a solid angle of 0.6 $\times 4\pi$ sr was realized. The result is an $\overline{m}$ value of 1.3 photoelectrons per ^{138}Ba atom, when its first resonance level was saturated. According to Eq. (4.11) the fractional probability of finding 2 and 6 photoelectrons in one burst is then 0.2 and 0.01, respectively. Figure 4.7 shows for different selected m values or multiplicities the hfs excitation spectrum of Ba with natural isotopic composition, obtained by detuning the laser frequency over a range of 200–400 MHz from the line center of the main ^{138}Ba isotope. It is seen that the line-to-back-

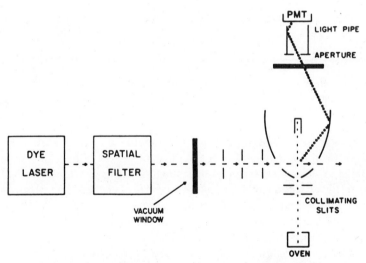

Fig. 4.6. Experimental setup for detection of resonance fluorescence of Ba atoms evaporated in an oven and collimated to form an atomic beam in vacuo. The intersection of this beam and the laser beam is found in the focus of an ellipsoidal mirror, giving a large solid angle of detection. Laser background scattering is suppressed by means of various diaphragms. PMT = photomultiplier tube. [Reproduced from (16) by permission of the American Physical Society.]

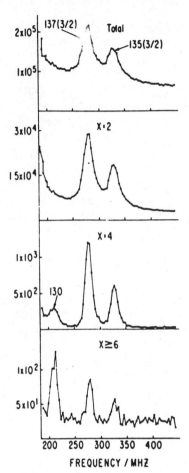

Fig. 4.7. Detail of the hfs excitation spectrum of Ba with natural isotope composition, measured with the setup shown in Fig. 4.6 by application of the photon-burst technique, for various selected multiplicities of x photoelectrons per burst. The top figure was obtained without multiplicity selection. The frequency difference scale refers to the line center of the ^{138}Ba isotope. [Reproduced from (16) by permission of the American Physical Society.]

ground ratio improves considerably when the multiplicity is enhanced. With $m \geq 6$ the residual background was due to the dark current of the photomultiplier tube (which exhibits some temporal bunching, too). It is also seen that the lines become narrower with increasing multiplicity. This beneficial line narrowing in high-resolution spectroscopy (13) arises because $\bar{m}$ is smaller in the line wings than in the line center. As a consequence, ϵ_d drops more sharply with increasing multiplicity at line wing excitation than at line center excitation. Also, the peak height of the ^{130}Ba line grows relative to that of the ^{135}Ba and ^{137}Ba lines. The first peak becomes even dominant at $m_{thr} = 6$, although the corresponding abundance ratios are 0.1, 6.5, and 11%. The line strengths for the 135 and 137 isotopes are comparatively small and saturation was not reached (17).

The photon-burst technique was also applied in Na beam experiment 1, where

$\bar{m}$ was 2 and m_{thr} was chosen equal to 2, yielding $\epsilon_d = 0.4$. The background was dominated by stray light. Pulse-height discrimination was here combined with coincidence measurements with two photodetectors. In experiment 4 the photon-burst technique was realized by measuring the single-clipped autocorrelation function displaying the number of counts in 100 successive time samples. Whenever an atom crossed the laser beam, a bump was seen with a height of about 15 photoelectron counts and a width equal to τ. Since the background (due to stray light and Rayleigh scattering) was relatively low, ϵ_d was, virtually, unity.

The application of the photon-burst technique may be hampered by the occurrence of a consumptive process in the probing (Section 2.2; Table 4.2, case Ib). In beam experiments 1 and 2, where a narrow-band laser was used and collisions did not occur, the Na atom and some of the isotopic Ba species may become trapped in hfs levels of the ground state that are inaccessible to laser excitation. Ba atoms may, moreover, become trapped in the metastable 5^1D_2 level. If the atoms reside long enough in the saturating laser beam, they will all end up in the trap level. The value of $\bar{m}$ as a function of τ then reaches a plateau (16).

Trapping in the $F = 1$ hyperfine (hf) level of the Na ground state can, in principle, be avoided by using a narrow-band laser tuned at the transition from the $F = 2$ level of the ground state to the $F = 3$ level of the $^2P_{3/2}$ state. (Optical selection rules forbid spontaneous transitions from the latter level to the $F = 1$ ground level.) However, saturation broadening may lead to excitation of the nearby $F = 2$ level in the upper state, too. From the latter level spontaneous transitions to the trap level $F = 1$ are allowed. This detrimental effect of saturation broadening was eliminated in experiment 1 by using a circular polarized laser beam. Of course, by using a broadband laser that excites atoms from both hf ground-state levels one could also avoid trapping. But the excitation rate would then also be reduced, at a given laser power.

The loss of atoms due to trapping in a metastable level depends on the branching ratio of the spontaneous transitions to this level and to the ground state. In Ba experiment 2 this branching ratio was small enough to avoid noticeable trapping during $\tau = 1$ μs.

Nonresonance fluorescence (NF) spectroscopy was applied in experiments 5–8 with sodium, where efficient population mixing of the D-doublet levels was brought about by collisions in the buffer or carrier gas. This technique was also applied in experiments 9 and 10 with lead, which emits a strong NF line at 405.8 nm from the saturated 7s $^3P_1^0$ level. The lower level (6p^2 $^3P_2$) of the NF transition is, however, metastable and a noticeable fraction of the Pb atoms may eventually become trapped during the probing time T (= pulse duration) of 5 ns, which was here much less than τ (stationary case). According to (104) the optimum pulse duration would have been 7 ns under saturation conditions. In this case an increase in T would thus have had little effect on the LOD, but an increase in f_{rep} could still have been beneficial. This holds at least as long as f_{rep} does not exceed τ^{-1} or the reciprocal of the metastable lifetime. Since in the experiments f_{rep} was 50 Hz, a considerable gain in LOD, at given t_m, is still feasible. In experiment 9 the extrinsic LOD was determined by fluorescence from the quartz windows at the wavelength of the NF line used.

ICAAS experiments in chemical analysis have not been entered in the table because their LOD performance was rather disappointing (105). For Na solutions nebulized into a slot-burner flame, placed inside the laser cavity, the LOD was no more than one order below the best LODs obtained in conventional AAS (106). Laser instability was mainly responsible for this meager achievement, but this might be considerably remedied.

The listed *limits of detection* $(N_p)_m$ may be compared with the intrinsic values calculated from Eqs. (4.20) and (4.21) in the stationary and nonstationary case, respectively, or from Eq. (4.22) (Table 4.3). This cannot be done, however, for experiments 6 and 7, where ϵ_d is unknown, and for experiment 11, where ϵ_d is not well definable.

With the proviso that the listed data are correct within one order of magnitude, we find that the LODs in experiments 1, 2, 3, 9, and 10 exceed their intrinsic values by no more than one order of magnitude. The LOD found in experiment 4 is intrinsic and determined by atom number fluctuations because $\epsilon_d = 1$. In experiments 5 and 8 $(\overline{N}_p)_m$ comes close to the intrinsic value calculated from Eq. (4.22), but the experimental values were derived indirectly and are thus rather uncertain. Anyway, since here $\epsilon_d \ll 1$, the intrinsic LOD is determined mainly by shot noise in fluorescence detection, not by atom number fluctuations.

Flame experiments 5 and 8 yielded $(n_a)_m$ values that differed by more than 4 orders of magnitude, although both come close to their intrinsic values at similar T and ϵ_d. This large difference is connected with the difference in product $V_p \cdot R$ occurring in Eq. (4.25). This product was very small in experiment 5, which was aimed at flame diagnostics with high spatial (small V_p) and temporal ($R = 1$) resolution. [Note that in Eq. (4.25) τ has dropped out.] The other experiment was aimed at obtaining good analytical LODs (approximately 1 pg/mL for Na).—The detrimental effect of laser-enhanced NaOH formation on the LOD might not have been negligible in these experiments, done with a laser pulse duration of about 1 μs (Section 2.2). Laser background scattering was reported to play no role in both NF experiments, as expected.—Experiment 5 was erroneously listed in (39) as RF.

Single-atom detection exists only if $\epsilon_d = 1$ *and* the intrinsic LOD is attained. Only experiment 4 meets these conditions, whereas experiment 1 approaches SAD capability within one order of magnitude.

SAD has also been realized by applying RF in combination with a saturating cw laser beam and an ion trap, which allows observation of single ions over a protracted period of time (47, 48). Figure 4.8 illustrates the application of a rf

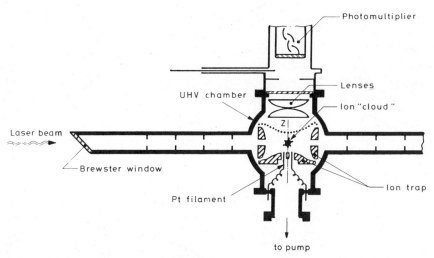

Fig. 4.8. Experimental setup for detection of resonance fluorescence of Ba$^+$ ions trapped in a rf quadrupole field. The Ba$^+$ ions were produced by evaporation from a metallic Ba sample deposited on a hot Pt filament. [Reproduced, with slight modifications, from (47) by permission of Springer Verlag, Berlin.]

quadrupole trap for Ba$^+$ ions that were evaporated, after surface ionization, from a Ba sample deposited on a hot platinum filament (47). Figure 4.9 displays the photon count rate, Z, as a function of t. The clearly distinguishable steplike structures in the latter figure correspond to consecutive escapes of single ions from the trap (compare Fig. 4.4a). The mean residence time, τ, of an ion in the trap appears to be about 2 min. The probe volume, V_p, is about 5% of the trap volume, V_a. But each ion oscillates rapidly inside V_a, thereby crossing frequently V_p. The count rate Z measured was actually an average over many repeated crossings. Therefore, we can effectively equalize V_p and V_a, which makes $\epsilon_p = 1$ [Eq. (4.6a)]. Since the time T (≈ 1 s) taken for a single probing in Fig. 4.9 is obviously less than τ, we should apply Eq. (4.20), holding in the stationary case, to calculate the intrinsic LOD. Because ϵ_d is obviously 1, we get $(\overline{N}_p)_m = 3^2/R$ if repeated probings are separated in time by at least τ seconds. [Equation (4.20) holds only for statistically *un*correlated probings.) Another condition is that the mean number of atoms, $\overline{N}_a$, should not decay (as in Fig. 4.9) because of uncompensated escape losses during t_m. Upon closer inspection of the scatter in the individual points in Fig. 4.9, we conclude that the rms error due to background counts is small relative to the mean step height. So extrinsic measurement errors were negligible indeed.—Since V_p was not specified in (47) we could not calculate $(n_a)_m$. An absolute LOD in the sample of 10^5 Ba atoms has been attained, which was restricted by the ionization and trapping efficiencies. This result implies an overall efficiency $\epsilon_0 \approx 10^{-5}$. (Personal communication by Dr. W. Ruster, 1985.)

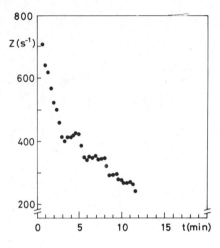

Fig. 4.9. Photon count rate Z of stored Ba^+ ions, measured as a function of t with the setup shown in Fig. 4.8. [Reproduced from (47) by permission of Springer Verlag, Berlin.]

In contrast with the photon-burst technique combined with pulse-height discrimination, the described ion trap technique does not require that at most one atom is present at a time in the probe volume.

There exists also in Ba^+ a metastable doublet level ($5d^2$ ^2D) between the laser-excited ($6p^2$ $^2P_{3/2}$) level and the ground state ($6s^2$ ^2S). In the reported experiment, accumulation of Ba^+ ions in the metastable level was prevented by adding a low-pressure buffer gas (H_2) to quench this level.

In experiment 19 with a rhodamine 6G solution the intrinsic LOD as calculated from Eq. (4.21) is about 4 orders of magnitude lower than the experimental value. Raman scattering in the solvent was mainly responsible for that. The detection efficiency ϵ_d is remarkably high (≈ 0.1). The probing efficiency ϵ_p may also be high, as every molecule was forced to pass through V_p. This was achieved by hydrodynamic focusing, that is, by constricting the capillary diameter to the waist of the laser beam. Nevertheless, SMD has by far not been reached here.

3.3. Experimental Results by Ionization and Optoacoustic Techniques

Table 4.4 lists the experimental conditions and results obtained for a few optogalvanic (OG), photoionization (PI), and field ionization (FI) experiments with lasers. Data are also listed for a single optoacoustic (OA) experiment with Na in a flame.

The *goal* of OG and PI experiments 12, 13, and 15 was the improvement of analytical LODs. PI experiment 14 demonstrated the utility of a gas-proportional counter as an electron detector for realizing SAD of rare gaseous species. By applying a pulsed laser the statistical and kinetic behavior of a small ensemble of atoms was studied here, too. FI experiments 16 and 17 demonstrated the potentialities of this technique in SAD and in high-resolution spectroscopy of rare or short-lived species.

The best analytical LOD in *optogalvanic spectroscopy* up to now has been obtained in experiment 13 with Li excited to its first resonance level by a flashlamp-pumped dye laser in an acetylene–air flame. The reported analytical LOD of 1 pg/mL holds for $k = 3$ and was converted to $(n_a)_m$ by using the reported conversion factor ψ. This result is more remarkable as Li has a pronounced tendency to form LiOH in flames (Section 2.2). It is relevant to note here the use of a relatively long laser pulse (see ibidem). LODs of the same order of magnitude were realized for Co, Cs (experiment 12), In, Tl, and Mg (Fig. 3.7) (53, 107). In the case of Co, having a relatively high ionization energy, two-step laser excitation was applied. Cs atoms were directly excited to their second resonance level. Most LODs were set by shot noise in the current induced by flame background ionization. The use of hydrocarbon flames having a high natural ionization level (51) is thus unfortunate in this respect.

The problem of atom trapping in inaccessible levels does not arise in flames at 1 atm pressure because of the generally high rates of inelastic collisions (51).

A classic example of the *photoionization technique* is experiment 14. The experimental setup used in the detection of single Cs atoms by two-photon ionization via the 7P state is shown in Fig. 4.10. Cs vapor is produced by evaporation from a pure Cs sample contained in the side arm of a gas-proportional counter with quartz windows, filled with Ar at a typical pressure of 100 torr and a small admixture (10%) of methane. The secondary electrons were

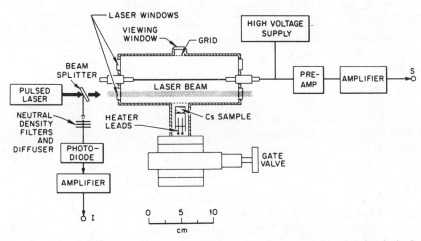

Fig. 4.10. Experimental setup used in the detection of single Cs atoms by the photoionization technique with the help of a gas-proportional electron counter. [Reproduced from (94) by permission of Springer Verlag, Berlin.]

collected on a central wire at a high positive voltage. Background ionization was suppressed by time gating. The authors reported that nearly every atom appearing in the laser beam produced an observed event, in this case a burst of secondary electrons. The intermediate 7P state was saturated so that the depletion of this state by photoionization was instantaneously supplemented by atoms from the ground state, whose population was in quasiequilibrium with the 7P population. The photoionization rate was large enough to attain "saturation" in the production of ion–electron pairs, too. Trapping of Cs atoms in a lower, metastable state, populated by radiative decay from the 7P state, turned out not to be a limiting factor, as the photoionization rate from this metastable state was also large enough.

A similar setup was used in experiment 15 for the detection of naphthalene molecules in the vapor phase. The $(n_a)_m$ value listed corresponds to the detection of about 1 molecule per 10^{12} buffer gas atoms. Similarly good results have been obtained by this technique for aniline (108).

The PI technique with various excitation schemes can, in principle, be applied to most elements of the periodic table, including, for example, ^{85}Kr (5, 109). Resonantly enhanced multiphoton ionization (MPI) is also coming more and more into use in chemical analysis and gas diagnostics for the detection of simple as well as complex molecules or radicals. This technique is then often combined with other methods such as gas chromatography and mass spectrometry (109). On the other hand, MPI of molecular constituents of the buffer gas or concomitants in the sample could contribute to background ionization in intense laser fields (5). PI preceded by laser-induced photodissociation might be used to detect, for example, CsI at SMD level (Section 2.2).

Selectivity is a greater problem with molecules, which have more closely spaced absorption lines, than with atoms. However, the selectivity (and detectability) for molecules can be improved by using supersonic jet expansion, which effectively "cools" the internal degrees of freedom (Chapter 7) and reduces Doppler broadening (just as in the case of atomic beams; Section 2.2).

The usage of a gas-proportional counter could be a problem when real samples are to be introduced or atomized in the proportional counter. In experiment 15 with a sample of pure naphthalene vapor this problem did not arise. The usage of ion or electron particle multipliers poses another problem, as they require vacuum conditions. It is also possible to collect the primary ions or electrons, without internal amplification, on two parallel-plane electrodes or on an electric probe. Interestingly, using a graphite furnace atomizer provided with a tungsten rod electrode, an absolute analytical LOD of 10^{-15} g has been obtained for Na by a multistep PI technique (110) (Chapter 3).

The application of the *field ionization technique* is exemplified by experiments 16 and 17. In experiment 16, a Na beam was produced from an oven containing the metallic element. Na atoms were raised to their 13D Rydberg level via the

3P level in a two-step photoexcitation process with two pulsed dye lasers. Saturation of the Rydberg level was attained so that about $\frac{2}{3}$ of all available atoms were brought into this level, according to the ratio of the statistical weight factors involved. In experiment 17, Yb isotopes were produced by irradiating a hot tantalum target by an intense beam of energetic protons. The ionized Yb isotopes were spatially separated by a mass separator and implanted, after acceleration, into different regions of a tantalum foil. Heating of this foil and proper collimation of the released atoms by means of a narrow cylindrical tube produced the atomic beam required for high-resolution spectroscopy. Because of the high ionization energy, a three-step excitation process was required to saturate the high-lying 17^3P_2 level, in which $\frac{5}{12}$ of the selected isotopic species was found.

The experimental setup used to detect the resonantly excited Yb atoms by field ionization is schematically shown in Fig. 4.11. The atomic beam, which is crossed at right angles by the triple laser beams, passes in between two parallel-plane electrodes to which a high-voltage pulse with a rectangular shape is applied about 20–30 ns after termination of the laser pulses. "Saturation" of the ion current was obtained with an electric field strength of about 12 kV/cm. The Yb^+ ions passing through a slit in the cathode were registered by means of a secondary-emission multiplier with an efficiency close to unity. A reference

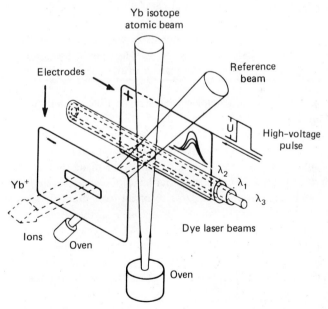

Fig. 4.11. Arrangement of coaxial dye laser beams, particle beams, and electrodes used in the detection of single Yb atoms by the field ionization technique. [Reproduced from (9) by permission of Springer Verlag, Berlin.]

beam effusing from an oven containing metallic Yb with natural abundance ratios was used for tuning the laser frequencies to the atomic transitions.

In experiments 16 and 17 ϵ_d was limited only by the fractional quasiequilibrium population of the saturated Rydberg level and was, therefore, on the order of unity (Table 4.4). Since collisions were absent, the lifetime of this level was sufficiently long to ensure that its population did not markedly change during the delay time of the high-voltage pulse.

In (96) it was claimed that the FI technique is more universally applicable to other elements than the LIFS techniques, as Rydberg levels exist for every element. Also, the laser power required is much less than in the PI technique because of the relatively low photoionization cross sections. On the other hand, the simultaneous tuning of two or more laser beams to consecutive atomic transitions may be a disadvantage. Since the laser pulse duration is short (less than 10 ns), the application of stationary saturation formulas might be questioned. The applicability of the FI technique is, however, restricted to atomizers where inelastic collisions are absent and no discharge occurs as a consequence of the applied electric field.

The *detection limits* listed in Table 4.4 are extrinsic in OG experiments 12 and 13, as they were determined by noise in the flame background signal or in the electronic circuit. In PI experiment 15 the reported LOD was determined by (unspecified) background noise and should thus be classified as extrinsic, too. In PI and FI experiments 14, 16, and 17 the intrinsic LOD, calculated from Eq. (4.20) or (4.23), was attained, whereas ϵ_d was on the order of unity.* In these cases the ideal of SAD was therefore closely approached. We note that in the stationary case, the value of τ drops out in the calculation of the intrinsic LOD.

As to LIIS techniques in general we can state that they usually require pulsed lasers for efficient ionization (Chapter 3) (5, 13, 53). Pulsed lasers are indispensable anyway, here as in LIFS, when a MP process or frequency doubling or mixing is involved. Since cross sections for photoionization from an excited level are comparatively small, highest demands are made on the irradiance of the ionizing laser beam in PI. On the other hand, the spectral bandwidth of this beam is not as critical as that used for resonant excitation of discrete atomic levels. The laser power requirements depend also on the competition between photoexcitation and collisional quenching. However, once full optical saturation has been reached, neither quenching nor the laser irradiance play a role anymore (Chapter 1).

When multiple, pulsed laser beams are to be used, their temporal and spatial

---

*In a recent PI experiment [A. T. Tursunov and N. B. Eshkabilov, *Sov. Phys. Tech. Phys.*, **29**, 93 (1984)] the intrinsic LOD was also attained for gallium atoms in an atomic beam with $\epsilon_d = 0.3$, $V_p = 6 \times 10^{-4}$ cm^3, laser pulse duration $T = 8 \times 10^{-9}$ s and $\tau \gg T$.

overlap in the probe volume as well as their simultaneous tuning need careful attention. This holds especially so when focused laser beams (diameter ≤ 0.1 mm) and/or short pulses (≤ 1 ns) are applied (Chapter 3). Multistep excitation by multiple, resonantly tuned laser beams gives the gratification of enhancing the overall selectivity (13).

The degree of focusing, which anyway is limited by diffraction and possibly also by the laser mode structure, should not be pushed too far because of saturation broadening. This broadening of the excitation line profile may cause the spectral selectivity to deteriorate. As a general rule, one should rather try to enlarge the saturated probe volume, at a given laser power, unless high spatial resolution is wanted, as in diagnostic applications.

The LOD, at given measuring time t_m, may be improved by increasing f_{rep} and thus the duty cycle. The existence of an optimum value for f_{rep} should be kept in mind, however (Section 3.2).—These general considerations are equally valid in LIFS.

The potentiality of PI and FI techniques to detect small numbers of atoms (Table 4.4) implies (nearly) unit ionization efficiency in V_p as well as efficient collection of the charge carriers. Unit collection efficiency is easy to attain (13). Such favorable measuring conditions may also be found in OGS if the effective ionization rate constant times the duration of the saturating laser pulse significantly exceeds unity and the electrode voltage surpasses its "saturation" value (53) (Chapter 3). However, background ionization in the flame is likely to prevent the attainment of intrinsic LODs. This is probably the cost to be paid for the usage of this versatile atomizer in chemical analysis. The atomizers applicable in PI and FI spectroscopy are less suited for handling real-world samples. But here the effect of residual background ionization may be suppressed by applying a coincidence technique in the detection of electron–ion pairs (13).

Optoacoustic detection has only incidentally been applied to free atoms (experiment 18). Efficient conversion of absorbed photon energy into heat by quenching collisions is expected in the acetylene–air flame used (51). By applying a pulsed laser, acoustic pulses were generated that were detected by a microphone placed close to the flame. The reported LOD value is still large when compared to the best results obtained by fluorescence and optogalvanic techniques with flames. But this preliminary LOD value could probably be improved.

3.4. Combined Detection Techniques

In recent years efforts have been made not so much in inventing new laser-based detection techniques as in combining existing techniques with each other or with other methods of analysis such as mass spectrometry and chromatography. This hybridization can lead to improved sensitivity, selectivity, and background suppression in chemical analysis. It can also enlarge the coverage of atomic and molecular analyte species and the kinds of samples amenable to analysis. Many of these combinations are still in a stage of development or were ventilated merely as possibilities. Although some combinations are good prospects for the detection of small numbers of atoms and molecules or even for SAD/SMD, hard facts and experimental data are still lacking. We mention here only a few combinations [see also (38, 109)].

The combination of the PI technique with a mass spectrometer is a most obvious one, as it bypasses the need of a separate ionization chamber (Chapter 18). This combination was first applied in fundamental and diagnostic studies. In chemical analysis the technique called resonance ionization mass spectrometry (RIMS) is now reaching maturity (57, 111–115). The versatility of this technique depends on atomizing the analyte from a solid sample under vacuum conditions. Briefly, RIMS provides selectivity as to mass number A (by mass analysis) as well as to atom number Z (by resonance ionization). It is therefore especially useful in eliminating isobaric interferences in mass spectrometry when isotope ratios are to be determined. (Note that this elimination can also be achieved by accelerator–mass spectrometry, if a nuclear accelerator happens to be available; Section 1.3.) RIMS has worked in the intracavity mode too (57). The addition of a (simple) mass spectrometer has been considered for rejection of background ions in photoionization spectroscopy (6). The prospects of this combination for SAD have been theoretically considered in (37).

Whereas in the preceding combination PI comes first, it comes last when used as an element-specific detector in LIFS or ICAAS (116, 117). This detector is an auxiliary cell (or flame) that contains free atoms of the same species as the analyte at sufficiently high concentration. Resonance line radiation from the analytical cell, absorbed in the detector, acts as the first step in a multistep laser-assisted photoionization process (or OG process in the flame). The ion current produced is then a measure for the incident radiation intensity. The quantum efficiency of the detector is improved by making the ionization and ion collection efficiencies as large as possible (Section 3.3.). Theoretically this combinational technique might have SAD capability (116).

Combinations of laser-induced beam deflection with LIFS or LIIS have been suggested for improving the overall detection selectivity and for suppressing background (13, 34). The latter goal might be achieved by measuring the deflected beam signal and fluorescence signal in coincidence.

An interesting combination of magnetic-state selection by beam deflection and mass spectrometry has been described for the study of optical resonances in short-lived Na isotopes produced by nuclear spallation of Al by 150 MeV protons (118). This technique is based on the orientation of the electron spin in the Na($^2S_{1/2}$) ground state brought about by optical pumping of one of its Zeeman sublevels $m_J = \pm\frac{1}{2}$. Optical pumping is accomplished by tuning a saturating laser beam at one of the hfs lines of the D doublet. A magnetic filter focuses atoms in the pumped sublevel onto a hot rhenium surface and defocuses atoms in the other sublevel. The ions produced by surface ionization on the hot rhenium are finally analyzed according to their mass. This technique has been developed for high-resolution spectroscopy, but could be useful also in ultrasensitive detection of short-lived species.

4. CONCLUSIONS AND RECOMMENDATIONS

4.1. Conclusions

Selective detection of small numbers of atoms in the gas phase has been experimentally proven to be possible for a large variety of elements by applying laser-based fluorescence, photoionization (PI), and field ionization (FI) techniques. Results have been obtained in atomic beams, vapor cells, and electromagnetic ion traps, working under ideal conditions and with sophisticated measuring arrangements. In a few experiments the SAD limit has even been attained. With more conventional analytical atomizers such as the flame and electrothermal atomizer, free atom numbers as low as 100 have been detected by the use of nonresonance fluorescence (NF) techniques.—Good detectability and high selectivity often appear to go hand in hand.

Most of these experiments were not aimed at chemical analysis. Low limits of detection (LOD) relating to free atoms in the atomizer do not imply, per se, low absolute or concentrational LODs in the analytical sample. The efficiencies of atomization (ϵ_a) and probing (ϵ_p) play an important role here, too.

Apart from the special case of the alkali halides (Section 2.2), SMD lies beyond the horizon of present experimental possibilities. The reason why molecules are much more difficult to detect by laser probing than atoms is theoretically clear. In trace analysis of gas mixtures, however, it is often the *relative* concentration that matters. Relative LODs as low as $1:10^{11}$ are attainable for molecular species by laser-based detection techniques.

Intrinsic LODs have been obtained not only in the SAD experiments proper but also in some others. However, in most of the latter experiments ϵ_d was small compared to unity. This implies that shot noise in the detection process and not atom number fluctuations determined the intrinsic signal fluctuations.

The minimum number, $(\overline{N}_p)_m$, of free atoms detectable in probe volume V_p can be much lower than just 1 atom (Section 2.3). Of course, a statistically meaningful assessment of $(\overline{N}_p)_m$ ($\ll 1$) is possible only if during the total measurement time (t_m) at least some atoms have actually been observed. Under stationary probing conditions (Fig. 4.4) this requires that statistically uncorrelated probings be repeated a sufficiently large number of times (R). Under nonstationary probing conditions a single probing of duration T may suffice, if enough atoms pass through V_p during T ($\gg$ transit time τ) (Section 2.3).

Apart from the widely varying goals pursued in the experiments entered in Table 4.4, the listed LOD values should not be used, as such, for merit rating. These values were obtained under largely different measuring conditions (R, T, V_p, t_m). Some normalization of these parameters would be appropriate, while allowing for the different dependencies of the intrinsic and the extrinsic LODs on R and T (Section 2.6).

One handicap to a more quantitative evaluation of the experimental results is the often insufficient specification of the experimental conditions in the literature. Another handicap is that the theoretical expressions presented hold only for strongly idealized situations. No account was taken, for example, of the nonuniform spatial distribution of the atoms in the atomizer and the nonuniform spatial, spectral, or temporal distributions of the (pulsed) laser irradiance. Also, statistical correlations between repeated probings or between successive atom crossings through V_p were disregarded.

4.2. Recommendations for Spectrochemical Analysis

The spectrochemist who wants to use gas-phase detection for smaller and smaller numbers of analyte atoms or molecules in condensed-phase samples is faced with the following problems:

(i)   Improvement of the detection efficiency ϵ_d for atoms or molecules in the gas phase,

(ii)   Suppression of background noise,

(iii) Improvement of the atomization efficiency ϵ_a,

(iv)   Improvement of the probing efficiency ϵ_p.

In solving problems (i) and (ii) the spectrochemist should take notice of the special laser-based techniques described, possibly in combination with other methods of analysis or separation existing in analytical chemistry. Some of these laser-based techniques operate under vacuum conditions. The analytical atomic spectroscopist should therefore overcome any possible *horror vacui* in designing atomizers that produce free analyte atoms from real-world samples in an evacuated compartment. The "atomizers" used in the described FI and ion trap experiments with Yb and Ba^+ are especially instructive.

The PI technique is certainly a good candidate for chemical analysis at SAD levels. But here we are confronted with the extra problem of achieving efficient and versatile sample atomization without disturbing, by contamination, the performance of the ion or electron detector present inside of the atomizer confinement.

Atomizers handling real-world samples are expected to produce higher background levels. Ample attention should therefore be given to the special methods of improving the signal-to-background ratio that were occasionally mentioned in Section 3.

Problem (iii) is, of course, not new in analytical atomic spectroscopy. But special care should be taken that the atomization does not affect the laser–atom interaction, and to the avoidance of analyte contamination. There is a great variety of means to produce free atoms or ions from a (minute) solid sample deposited or collected on a filament or foil, and from a solid sample surface under vacuum or low-pressure conditions (Section 3). In ultratrace analysis, especially when measuring isotope ratios, one may consider the feasibility of multistage atomization. Here the analyte is preselected by a laser-assisted separation technique and/or mass separator and accumulated on a foil—in one or more cycles—before final atomization for laser probing takes place (36). In this way concomitant interferences and matrix effects may be suppressed, while preconcentration of the analyte species is achieved.

Problem (iv) is specific for laser-based techniques. The possible expansion of the laser beam by optical means is limited by the high irradiance required for efficient probing and detection. The atomizer design should thus be matched to the optimal shape and extent of the probe volume, which depend also on the photon or ion collection system used. Under stationary probing conditions and in a closed atomizer this means, ideally, that V_a should be made equal to V_p [Eq. (4.6a)], or that the probings should be repeated so often that each free atom present in V_a appears in V_p during at least one of the probings. Under nonstationary probing conditions, it is not necessary that all free atoms are confined inside V_p during probing. It suffices that they all pass through V_p during T at a transit time τ ($< T$) long enough to ensure their efficient detection. If in a closed atomizer $V_p/V_a \ll \tau/T$ [Eq. (4.6b)], the optimum number of repeated probings is $R \approx (V_a/V_p) \cdot (\tau/T)$. With continuous-flow atomizers the atomization time t_a should match the total measurement time t_m, in order to reduce analyte losses when working with microsamples. When pulsed lasers are to be used, due consideration should be given to analyte losses resulting from a bad duty cycle. The laser repetition frequency should then be raised close to its optimum value (Section 3.2) or pulsed atomization, synchronized with the laser pulses, may be attempted.

Some of these matching problems are discussed in Chapter 3 in relation to optogalvanic spectroscopy (OGS) with flames. Flames seem, in general, not to be suited for realizing SAD because of their high background levels, limited atomization efficiencies, unfavorable conversion factors, analyte contamination in the combustion gases and difficulties with handling ultramicrosamples. But OGS combined with a quadrupole ion mass spectrometer might improve the LODs in flames that are limited by flame background ions (39). Ion counting techniques would then be applicable, too.

Space–charge amplification, already applied in research on Rydberg atoms in combination with PI and OG detection (119), might be considered for chemical analysis by LIIS (39). It can easily lead to a 100-fold enhancement of the electric pulse generated by an individual analyte ion. But it requires the presence of a hot cathode filament, acting as an electron emitter, close to the (narrow) laser beam. This may put special demands on the atomizer and the filament material. For example, the use of an oxygen-lean flame would be mandatory.

In general, the spectrochemist pursuing extremely low LODs, should have an open mind to tricks and techniques developed in other areas of physical and chemical research or chemical analysis. We mention, for example, the advanced ion trap technology developed for detection in gas chromatography. The technique of accumulating analyte species by trapping them in the lower part of a flame on a cooled silica tube may be borrowed from flame AAS (120) to enhance the sensitivity.

For advancing the detectability of *molecules* other than alkali halides toward SMD levels, resonantly enhanced multiphoton ionization in combination with supersonic jet expansion and a conventional mass spectrometer is promising (121) (Section 3.3). Nevertheless, there is still a long way to go before molecular LODs will be realized that are comparable to the best ones obtained for atoms. There is ample opportunity here for surprising innovations. The utilization of laser-induced selective photochemical effects might be considered for probing complex molecules. Also, the suggestions made in (25) for improving the LOD of molecules probed in the liquid phase by LIFS should be taken to heart in the analysis of gaseous molecules, too.

Calibration by reference samples could pose an extra problem when working at SAD levels, because of analyte contamination. Linear extrapolation of the calibration curve toward the lower end of the concentration scale is one possible solution. It requires, however, that contamination levels are reproducible and not too high. If SAD conditions are realized in the gas phase, atom counting combined with knowledge of ϵ_a and ϵ_p may provide absolute calibration. The value of ϵ_p may be calculated using expressions given in Section 2.4 or more refined expressions, depending on the probing conditions at hand. The value of ϵ_a can be determined by measuring the conversion factor (Section 2.5) using higher concentration reference samples. The ϵ_a value thus measured allows also for interferences by concomitants and matrix effects, if these are constant in the relevant concentration range.

Since effort and effect are, unfortunately, always balanced, a high price tag is attached to the application of any of the described ultrasensitive detection techniques to chemical analysis. Besides, the handling and maintenance of lasers working at the top of their performance requires an expert's skill (and patience!). The same may apply to the use of atomizers especially designed for work at SAD levels under vacuum conditions. Most of the techniques reviewed here are therefore not expected to bring about a breakthrough in routine analysis.—As to the cost problem of laser-based analytical techniques, we refer to the enlightening paper (38).

"Small is beautiful" but also difficile, when one tries to detect a small number of atoms or even a single atom in a whole sample. It will remain a thrilling challenge in the years ahead to pursue this noble goal.

REFERENCES

1. "Nomenclature, Symbols, Units and Their Usage in Spectrochemical Analysis— II. Data Interpretation," *Pure Appl. Chem.*, **45**, 99 (1976); "III. Analytical Flame Spectroscopy and Associated Non-flame Procedures," *Pure Appl. Chem.*, **45**, 105 (1976).

2. N. Omenetto and J. D. Winefordner, *Prog. Analyt. Atom. Spectrosc.*, **2**, 1 (1979).

3. H. Falk, *Prog. Analyt. Atom. Spectrosc.*, **3**, 181 (1980).

4. M. L. Parsons, S. Major, and A. R. Forster, *Appl. Spectrosc.*, **37**, 411 (1983).

5. G. S. Hurst, M. G. Payne, S. D. Kramer, and J. P. Young, *Rev. Mod. Phys.*, **51**, 767 (1979).

6. C. H. Chen, G. S. Hurst, and M. G. Payne, *Chem. Phys. Lett.*, **75**, 473 (1980).

7. G. S. Hurst, in *Symposium on Laser-based Ultrasensitive Analysis*, Amer. Chemical Soc., Abstracts of papers, 1984.

8. T. J. Whitaker and B. D. Cannon, in *Symposium on Laser-based Ultrasensitive Analysis*, Amer. Chemical Soc., Abstracts of papers, 1984.

9. G. I. Bekov, E. P. Vidolova-Angelova, V. S. Letokhov, and V. I. Mishin, in H. Walther and K. W. Rothe, Eds., *Laser Spectroscopy IV*, Springer, Berlin, 1979.

10. H. Walther and K. W. Rothe, Eds., *Laser Spectroscopy IV*, Springer, Berlin, 1979.

11. G. S. Hurst, M. H. Nayfeh, and J. P. Young, *Appl. Phys. Lett.*, **30**, 229 (1977); *Phys. Rev.*, **A15**, 2283 (1977).

12. G. S. Hurst, M. G. Payne, S. D. Kramer, and J. P. Young, *Chem. Phys. Lett.*, **63**, 1 (1979).

13. V. S. Letokhov, "Laser Selective Detection of Single Atoms," in C. Bradley-Moore, Ed., *Chemical and Biochemical Applications of Lasers*, Academic Press, New York, 1980.

14. G. S. Hurst, M. G. Payne, S. D. Kramer, and C. H. Chen, *Physics Today*, **33** (9), 24 (1980).

15. W. M. Fairbank, T. W. Hänsch, and A. L. Schawlow, *J. Opt. Soc. Amer.*, **65**, 199 (1975).

16. D. A. Lewis, J. F. Tonn, S. L. Kaufman, and G. W. Greenlees, *Phys. Rev.*, **A19**, 1580 (1979).

17. G. W. Greenlees, D. L. Clark, S. L. Kaufman, D. A. Lewis, J. F. Tonn, and J. H. Broadhurst, *Opt. Commun.*, **23**, 236 (1977).

18. C. Y. She, W. M. Fairbank, and K. W. Billman, *Opt. Lett.*, **2**, 30 (1978).

19. C. L. Pan, J. V. Prodan, W. M. Fairbank, and C. Y. She, *Opt. Lett.*, **5**, 459 (1980).
20. E. Jakeman, E. R. Pike, P. N. Pusey, and J. M. Vaughan, *J. Phys.*, **A10**, L257 (1977).
21. M. Dagenais and L. Mandel, *Phys. Rev.*, **A18**, 2217 (1978).
22. H. Paul, *Rev. Mod. Phys.*, **54**, 1061 (1982).
23. R. Short and L. Mandel, *Phys. Rev. Lett.*, **51**, 384 (1983).
24. M. Trkula, N. J. Dovichi, J. C. Martin, J. H. Jett, and R. A. Keller, "Prospects for Single-Molecule Detection in Liquids by Laser-Induced Fluorescence," in W. S. Lyon, Ed., *Analytical Spectroscopy*, Elsevier Science, Amsterdam, 1984.
25. N. J. Dovichi, J. C. Martin, J. H. Jett, M. Trkula, and R. A. Keller, *Anal. Chem.*, **56**, 348 (1984).
26. R. A. Keller and J. C. Travis, "Recent Advances in Analytical Laser Spectroscopy," in N. Omenetto, Ed., *Analytical Laser Spectroscopy*, Wiley, New York, 1979. QD 71.C4 v.50 Chem
27. W. Demtröder, "Molecular Absorption and Fluorescence Spectroscopy with Lasers," in N. Omenetto, Ed., *Analytical Laser Spectroscopy*, Wiley, New York, 1979.
28. R. Frueholz, J. Wessel, and E. Wheatley, *Anal. Chem.*, **52**, 281 (1980).
29. N. Omenetto and J. D. Winefordner, "Atomic Fluorescence Spectroscopy with Laser Excitation," in N. Omenetto, Ed., *Analytical Laser Spectroscopy*, Wiley, New York, 1979.
30. M. G. Payne, C. H. Chen, G. S. Hurst, and G. W. Foltz, "Applications of Resonance Ionization Spectroscopy in Atomic and Molecular Physics," in D. R. Bates and B. Bederson, Eds., *Advances in Atomic and Molecular Physics*, Vol. 17, Academic Press, New York, 1981.
31. V. S. Letokhov, V. I. Mishin, and A. A. Puretzky, *Prog. Quant. Electr.*, **5**, 139 (1977).
32. J. C. Travis, G. C. Turk, and R. B. Green, *Anal. Chem.*, **54**, 1006A (1982).
33. G. C. Turk, J. C. Travis, and J. R. DeVoe, *J. de Physique*, **44**, Supplém. No. 11, Colloq. C7, 301 (1983).
34. V. I. Balykin, G. I. Bekov, V. S. Letokhov, and V. I. Mishin, *Usp. Fiz. Nauk*, **132**, 293 (1980) [*Sov. Phys. Usp.*, **23**, 651 (1980)].
35. N. Omenetto and J. D. Winefordner, *CRC Critical Reviews in Analytical Chemistry*, (December 1981).
36. R. A. Keller, D. S. Bomse, and D. A. Cremers, *Laser Focus*, p. 75 (October 1981).
37. O. I. Matveev, N. B. Zorov, and Yu. Ya. Kuzyakov, *Izv. Moskwa Univers.*, *Ser. Khim.*, **19**, 537 (1978).
38. A. L. Robinson, *Science*, **199**, 1191 (1978).
39. C. Th. J. Alkemade, *Appl. Spectrosc.*, **35**, 1 (1981).
40. R. E. Drullinger and D. J. Wineland, in H. Walther and K. W. Rothe, Eds., *Laser Spectroscopy IV*, Springer, Berlin, 1979.
41. W. Neuhauser, M. Hohenstatt, and P. E. Toschek, in H. Walther and K. W. Rothe, Eds., *Laser Spectroscopy IV*, Springer, Berlin, 1979.

42. H. A. Schuessler, "Stored Ion Spectroscopy," in W. Hanle and H. Kleinpoppen, Eds., *Progress in Atomic Spectroscopy, Part B,* Plenum, New York, 1979.

43. J. E. Bjorkholm, R. R. Freeman, A. Ashkin, and D. B. Pearson, in H. Walther and K. W. Rothe, Eds., *Laser Spectroscopy IV,* Springer, Berlin, 1979.

44. W. Neuhauser, M. Hohenstatt, P. E. Toschek, and H. Dehmelt, *Phys. Rev.,* **A22,** 1137 (1980).

45. W. D. Phillips, Ed., *Laser-cooled and Trapped Atoms,* Ntl. Bureau of Standards, Spec. Publ. 653, Gaithersburg MD, 1983.

46. A. Ashkin, *Phys. Rev. Lett.,* **4,** 729 (1978).

47. W. Ruster, J. Bonn, P. Peuser, and N. Trautmann, *Appl. Phys.,* **B30,** 83 (1983).

48. D. J. Wineland and W. M. Itano, *Phys. Lett.,* **82A,** 75 (1981).

49. D. J. Wineland, W. M. Itano, J. J. Bollinger, J. C. Bergquist, and H. Hemmati, in R. A. Keller, Ed., *Laser-based Ultrasensitive Spectroscopy and Detection V,* Proc. SPIE, Vol. 426, Bellingham, WA, 1983. QC 454. L3 2334 Eng.

50. L. W. Grossman, G. S. Hurst, M. G. Payne, and S. L. Allman, *Chem. Phys. Lett.,* **50,** 70 (1977).

51. C. Th. J. Alkemade, Tj. Hollander, W. Snelleman, and P. J. Th. Zeegers, *Metal Vapours in Flames,* Pergamon, Oxford, 1982.

52. V. I. Balykin, V. S. Letokhov, and V. I. Mishin, *Zh. Eksp. Teor. Fiz.,* **77,** 2221 (1979) [*Sov. Phys. JETP,* **50,** 1066 (1979)].

53. J. C. Travis, G. C. Turk, J. R. DeVoe, P. K. Schenck, and C. A. van Dijk, *Prog. Analyt. Atom. Spectrosc.,* **7,** 199 (1984).

54. C. H. Muller, K. Schofield, and M. Steinberg, in J. W. Hastie, Ed., *Characterization of High Temperature Vapors and Gases,* Ntl. Bureau of Standards, Spec. Publ. 561, Washington, D.C., 1979.

55. C. H. Muller, K. Schofield, and M. Steinberg, *Chem. Phys. Lett.,* **57,** 364 (1978).

56. C. A. van den Wijngaart, H. A. Dijkerman, Tj. Hollander, and C. Th. J. Alkemade, *Combust. Flame,* **59,** 135 (1985).

57. N. S. Nogar, S. W. Downey, R. A. Keller, and C. M. Miller, "Resonance Ionization Mass Spectrometry at Los Alamos National Laboratory," in W. S. Lyon, Ed., *Analytical Spectroscopy,* Elsevier Science, Amsterdam, 1984.

58. R. A. van Calcar, M. J. M. van de Ven, B. K. van Uitert, K. J. Biewenga, Tj. Hollander, and C. Th. J. Alkemade, *J. Quant. Spectrosc. Radiat. Transfer,* **21,** 11 (1979).

59. R. Altkorn and R. N. Zare, *Ann. Rev. Phys. Chem.,* **35,** 265 (1984).

60. R. A. Keller and N. S. Nogar, *Appl. Opt.,* **23,** 2146 (1984).

61. F. C. M. Coolen and P. Menger, in J. G. A. Hölscher and D. C. Schram, Eds., *Proceedings of the 12th Intern. Conference on Phenomena in Ionized Gases, Part 1,* North-Holland, Amsterdam, 1975.

62. V. I. Balykin, V. S. Letokhov, V. I. Mishin, and V. A. Semchishen, *Pis'ma Zh. Eksp. Teor. Fiz.,* **24,** 475 (1976) [*Sov. Phys. JETP Lett.,* **24,** 436 (1976)].

63. F. C. M. Coolen, L. C. J. Baghuis, H. L. Hagedoorn, and J. A. van der Heide, *J. Opt. Soc. Am.,* **64,** 482 (1974).

64. M. A. Bolshov, A. V. Zybin, V. G. Koloshnikov, and M. V. Vasnetsov, *Spectrochim. Acta,* **36B,** 345 (1981).

65. M. A. Bolshov, A. V. Zybin, and V. G. Koloshnikov, *Kvantovaya Elektron. (Moscow)*, **7**, 1808 (1980) [*Sov. J. Quantum Electron.*, **10**, 1042 (1980)].
66. N. Omenetto, L. M. Fraser, and J. D. Winefordner, *Appl. Spectrosc. Revs.*, **7**, 147 (1973).
67. P. W. J. M. Boumans, R. J. McKenna, and M. Bosveld, *Spectrochim. Acta*, **36B**, 1031 (1981).
68. E. D. Prudnikov, *Spectrochim. Acta*, **36B**, 385 (1981).
69. E. O. Schulz-DuBois, H. Koppe, and R. Brummer, *Appl. Phys.*, **21**, 369 (1980).
70. A. van der Ziel, *Noise in Measurements*, Wiley, New York, 1976.
71. C. Th. J. Alkemade, W. Snelleman, G. D. Boutilier, B. D. Pollard, J. D. Winefordner, T. L. Chester, and N. Omenetto, *Spectrochim. Acta*, **33B**, 3831 (1978).
72. P. K. Wittman and J. D. Winefordner, *Appl. Spectrosc.*, **37**, 208 (1983).
73. M. Lapp and C. M. Penney, Eds., *Laser Raman Gas Diagnostics*, Plenum Press, New York, 1974.
74. H. Debus, W. Hanle, A. Scharmann, and P. Wirz, *Spectrochim. Acta*, **36B**, 1015 (1981).
75. M. C. E. Huber, in D. S. Dosanjh, Ed., *Modern Optical Methods in Gas Dynamic Research*, Plenum Press, New York, 1971.
76. C. Th. J. Alkemade, *Spectrochim. Acta*, **38B**, 1395 (1983).
77. M. J. Jongerius, Tj. Hollander, and C. Th. J. Alkemade, *J. Quant. Spectrosc. Radiat. Transfer*, **26**, 285 (1981).
78. T. W. Hänsch, A. L. Schawlow, and P. E. Toschek, *IEEE J. Quantum Electron.*, **QE-8**, 802 (1972).
79. A. A. Kachanov and T. V. Plakhotnik, *Opt. Commun.*, **47**, 257 (1983).
80. G. C. Bjorklund, *Opt. Lett.*, **5**, 15 (1980).
81. E. A. Whittaker, P. Pokrowsky, W. Zapka, K. Roche, and G. C. Bjorklund, *J. Quant. Spectrosc. Radiat. Transfer*, **30**, 289 (1983).
82. M. D. Levenson, W. E. Moerner, and D. E. Horne, *Opt. Lett.*, **8**, 108 (1983).
83. O. Axner, T. Berglind, J. L. Heully, I. Lindgren, and H. Rubinsztein-Dunlop, *J. Appl. Phys.*, **55**, 3215 (1984).
84. G. I. Bekov, V. S. Letokhov, O. I. Matveev, and V. I. Mishin, *Zh. Eksp. Teor. Fiz.*, **75**, 2092 (1978) [*Sov. Phys. JETP*, **48**, 1062 (1978)].
85. J. W. Daily and C. Chan, *Combust. Flame*, **33**, 47 (1978).
86. J. W. Daily, *Appl. Opt.*, **17**, 1610 (1978).
87. S. Mayo, R. A. Keller, J. C. Travis, and R. B. Green, *J. Appl. Phys.*, **47**, 4012 (1976).
88. J. A. Gelbwachs, C. F. Klein, and J. E. Wessel, *IEEE J. Quantum Electron.*, **QE-13**, No. 9, 11D (1977).
89. J. A. Gelbwachs, C. F. Klein, and J. E. Wessel, *IEEE J. Quantum Electron.*, **QE-14**, No. 2, 121 (1978).
90. J. A. Gelbwachs, C. F. Klein, and J. E. Wessel, *Appl. Phys. Lett.*, **30**, 489 (1977).
91. A. Dönszelmann, J. Neijzen, and H. Benschop, *Physica*, **83C**, 389 (1976).
92. V. I. Chaplygin, N. B. Zorov, and Yu. Ya. Kuzyakov, *Talanta*, **30**, 505 (1983).

93. G. C. Turk, J. C. Travis, J. R. DeVoe, and T. C. O'Haver, *Anal. Chem.*, **51**, 1890 (1979).

94. G. S. Hurst, M. H. Nayfeh, J. P. Young, M. G. Payne, and L. W. Grossman, in J. L. Hall and J. L. Carlsten, Eds., *Laser Spectroscopy III*, Springer, Berlin, 1977.

95. G. I. Bekov, V. S. Letokhov, and V. I. Mishin, *Zh. Eksp. Teor. Fiz.*, **73**, 157 (1977) [*Sov. Phys. JETP*, **46**, 81 (1977)].

96. G. I. Bekov, V. S. Letokhov, and V. I. Mishin, *Pis'ma Zh. Eksp. Teor. Fiz.*, **27**, 52 (1978) [*Sov. Phys. JETP Lett.*, **27**, 47 (1978)].

97. G. I. Bekov, V. S. Letokhov, O. I. Matveev, and V. I. Mishin, *Opt. Lett.*, **3**, 159 (1978).

98. J. E. Allen, W. R. Anderson, and D. R. Crosley, *Opt. Lett.*, **1**, 118 (1977).

99. W. R. Anderson, J. E. Allen, T. D. Fansler, and D. R. Crosley, in J. W. Hastie, Ed., *Characterization of High Temperature Vapors and Gases*, Ntl. Bureau of Standards, Spec. Publ. 561, Washington, D.C., 1979.

100. V. I. Balykin, V. S. Letokhov, V. I. Mishin, and V. A. Semchishen, *Pis'ma Zh. Eksp. Teor. Fiz.*, **26**, 492 (1977) [*Sov. Phys. JETP Lett.*, **26**, 357 (1977)].

101. J. C. Travis, P. K. Schenck, G. C. Turk, and W. G. Mallard, *Anal. Chem.*, **51**, 1516 (1979).

102. A. W. Miziolek and R. J. Willis, *Opt. Lett.*, **6**, 528 (1981).

103. G. Schatz, in H. Walther and K. W. Rothe, Eds., *Laser Spectroscopy IV*, Springer, Berlin, 1979.

104. M. A. Bolshov, A. V. Zybin, V. G. Koloshnikov, and K. N. Koshelev, *Spectrochim. Acta*, **32B**, 279 (1977).

105. E. H. Piepmeier, "Atomic Absorption Spectroscopy with Laser Primary Sources," in N. Omenetto, Ed., *Analytical Laser Spectroscopy*, Wiley, New York, 1979.

106. M. Maeda, F. Ishitsuka, and Y. Miyazoe, *Opt. Commun.*, **13**, 314 (1975).

107. N. Omenetto, T. Berthoud, P. Cavalli, and G. Rossi, *Anal. Chem.*, **57**, 1256 (1985).

108. J. H. Brophy and C. T. Rettner, *Opt. Lett.*, **4**, 337 (1979).

109. *Symposium on Laser-based Ultrasensitive Analysis*, Amer. Chemical Soc., Abstracts of papers, 1984.

110. A. S. Gonchakov, N. B. Zorov, Yu. Ya. Kuzyakov, and O. I. Matveev, *Zh. Analyt. Khim.*, **34**, 2312 (1979) [*J. Anal. Chem. USSR*, **34**, 1792 (1980)].

111. J. D. Fassett, J. C. Travis, L. J. Moore, and F. E. Lytle, *Anal. Chem.*, **55**, 765 (1983).

112. J. D. Fassett, L. J. Moore, J. C. Travis, and F. E. Lytle, *Int. J. Mass Spectrom. Ion Processes*, **54**, 201 (1983).

113. J. C. Travis, J. D. Fassett, and L. J. Moore, *Proceed. of 2nd Int. Symposium on Resonance Ionization Spectroscopy and its Applications*, Knoxville (Tenn.), 1984, The Institute of Physics, Conference Series No. 71, New York, 1984.

114. N. S. Nogar, S. W. Downey, and C. M. Miller, *Proceed. of 2nd Int. Symposium on Resonance Ionization Spectroscopy and its Applications*, Knoxville (Tenn.), 1984, The Institute of Physics, Conference Series No. 71, New York, 1984.

115. R. A. Keller, Ed., *Laser-based Ultrasensitive Spectroscopy and Detection V*, Proc. SPIE, Vol. 426, Bellingham, WA, 1983, various papers, pp. 1–32.

116. O. I. Matveev, *Zh. Analyt. Khim.*, **38**, 736 (1983) [*J. Anal. Chem. USSR*, **38**, 561 (1983)].

117. O. I. Matveev, N. B. Zorov, and Yu. Ya. Kuzyakov, *Talanta*, **27**, 907 (1980).

118. G. Huber, C. Thibault, R. Klapisch, H. T. Duong, J. L. Vialle, J. Pinard, P. Juncar, and P. Jacquinot, *Phys. Rev. Lett.*, **34**, 1209 (1975).

119. P. Camus, *J. de Physique*, **44**, Supplém. No. 11, Colloq. C7, p. 87 (1983).

120. J. Khalighie, A. M. Ure, and T. S. West, *Anal. Chim. Acta*, **131**, 27 (1981).

121. M. V. Johnston, in *Symposium on Laser-based Ultrasensitive Analysis*, Amer. Chemical Soc., Abstracts of papers, 1984.

CHAPTER

5

OPTOACOUSTIC SPECTROSCOPY

ANDREW C. TAM

IBM Almaden Research Center
San Jose, California

1. INTRODUCTION

The optoacoustic (OA) or photoacoustic (PA) effect may be described as the "splashing" of photons in matter, that is, the generation of acoustic waves by electromagnetic waves or other types of radiation incident on a sample. This effect was discovered by A. G. Bell in 1880, who observed that audible sound is produced when chopped sunlight is absorbed at a surface. Although the OA effect has been known for over 100 years, there has been a recent resurgence of interest in the phenomenon, both in theoretical and experimental studies. This renewed interest stems from several possible reasons. Intense light sources, such as lasers and arc lamps, are more readily available. High-sensitivity detection tools for measuring acoustic waves have been developed. OA techniques have been shown to be capable of detecting weak absorption features (e.g., due to trace contaminants) in gases as well as in condensed matter. Furthermore, OA methods are finding many unique applications, such as in spectroscopic studies of opaque or powdered materials, studies of energy conversion processes, and nondestructive evaluation or imaging of invisible subsurface defects in solids.

The essential feature of OA spectroscopy is that the heat deposited in the sample by the *absorption* of a modulated light beam is detected. Thus, OA spectroscopy has the following advantages: (a) It is a zero-background method, unlike conventional extinction measurements; (b) it can be used for strongly light-scattering materials; and (c) the OA signal is dependent on the excited-state decay pathways resulting in acoustic generation. These features distinguish OA spectroscopy from other spectroscopic methods. In this chapter we shall examine OA generation processes, discuss the acoustic detection methods, and review the spectroscopic applications in gases and in condensed matter. This chapter is meant to cover the important points relevant to analytical application of lasers, but not to be an in-depth review. Detailed reviews of spectroscopic and nonspectroscopic applications of OA techniques have been given, for example, by Pao (1), Rosencwaig (2), Coufal et al. (3), Patel and Tam (4), and Tam (5).

2. OPTOACOUSTIC GENERATION

OA generation can be due to diversified processes. Some of the possible OA generation mechanisms are shown in Fig. 5.1, where the OA generation efficiency η (i.e., acoustic energy generated/light energy absorbed) generally increases downward for the mechanisms listed. For electrostriction or thermal expansion mechanisms, η is small, for example on the order of 10^{-12}–10^{-8}, while for breakdown mechanisms, η can be as large as 0.3, as reported by Teslenko (6). However, generally it is only the thermal expansion mechanism that is useful for the purpose of spectroscopy, and further discussions are limited to this so-called thermal-elastic OA generation, which can be grouped into two general cases: direct and indirect. In the direct thermal OA generation, treated in detail by numerous workers including White (7), Gourney (8), Hu (9), Liu (10), Lai and Young (11), Heritier (12), and Sullivan and Tam (13), the modulated excitation beam produces a time-dependent heating of the sample, and

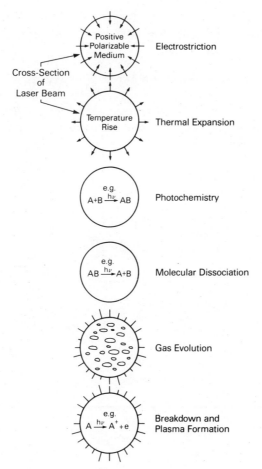

Fig. 5.1. Some common mechanisms of OA generation. The technique of OA spectroscopy almost always uses the thermal expansion mechanism.

thus an acoustic wave in it. In the indirect OA generation, treated by Rosencwaig and Gersho (14), Adamodt and Murphy (15), Wetsel and McDonald (16), Tam and Wong (17), and others, the excitation beam produces a modulated temperature at the surface of a solid or liquid sample that is in contact with a transparent coupling fluid (typically a gas); thus, time-dependent expansion of the coupling fluid is produced, and this pressure wave can be sensed by a microphone. Both direct and indirect OA generation are widely used for spectroscopic purposes. Indirect OA generation was frequently called "photoacoustic" (PA) generation in the earlier literature; both "OA" and "PA" are nowadays commonly used with the same meaning. For clarity, we shall use the name PA here to indicate indirect OA generation involving heat coupling to a transparent fluid adjacent to a sample.

2.1. Simple Theory for Direct OA Generation

The detailed general theories of OA generation tend to be rather involved; to show the important parameters, we consider a simple case of direct and of indirect OA generation.

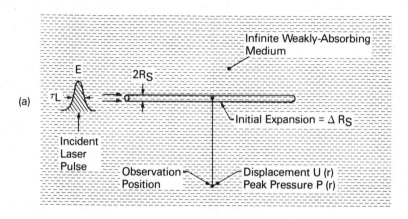

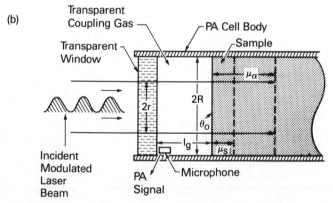

Fig. 5.2. Schematic explanation of the two cases of thermal OA generation, direct (a) and indirect (b). For clarity, we simply use OA to refer to direct acoustic generation as exemplified in (a) and use photoacoustics (PA) to refer to indirect acoustic generation utilizing a transparent coupling fluid as exemplified in (b).

The simplest case of direct thermal OA generation is indicated in Fig. 5.2a, for the case of an infinite weakly absorbing medium excited by a narrow pulsed laser beam producing an OA source of radius R_s. We assume that the laser pulse width τ_L is sufficiently short so that thermal diffusion effects can be neglected. The initial expansion ΔR_s of the source radius R_s immediately after the laser pulse is given by

$$\pi(R_s + \Delta R_s)^2 l - \pi R_s^2 l = \beta V \Delta T \qquad (5.1)$$

with the initial temperature rise

$$\Delta T = \frac{E\alpha l}{\rho V C_p} \qquad (5.2)$$

where l is the length of the OA source (assumed long), β is the expansion coefficient, $V = \pi R_s^2 l$ is the source volume, E is the laser pulse energy, α is the absorption length (with $\alpha l \ll 1$), ρ is the density, and C_p is the specific heat at constant pressure. Combining Eqs. (5.1) and (5.2), and assuming $\Delta R_s \ll R_s$ (true in all cases we are considering), we have

$$\Delta R_s = \frac{\beta E\alpha}{2\pi R_s \rho C_p} \qquad (5.3)$$

which has been given, for example, by Patel and Tam (4). The peak displacement $U(r)$ at the observation point at distance r from the OA source (for $r \ll l$) varies as $r^{1/2}$ because of conservation of acoustic energy, as described by Landau and Lifshitz (18) for a cylindrical acoustic wave:

$$U(r) = \Delta R_s \left(\frac{R_s}{r}\right)^{1/2} = \frac{\beta E\alpha}{2\pi R_s^{1/2} \rho C_p r^{1/2}} \qquad (5.4)$$

The peak acoustic pressure $P(r)$ at position r is related to the acoustic displacement $U(r)$ and sound velocity c by

$$P(r) \approx \frac{c\rho U(r)}{\tau_L} \qquad (5.5)$$

Substituting Eq. (5.4) into (5.5), we obtain the peak OA pressure observed at r for small source radius as

$$P(r) \approx \frac{\beta c E\alpha}{2\pi R_s^{1/2} C_p \tau_L r^{1/2}} \qquad (5.6)$$

Equation (5.6) shows the basis of OA spectroscopy based on direct OA generation. It indicates that the normalized OA signal, defined as the detected acoustic pressure amplitude P divided by the laser pulse energy E, is proportional to the absorption coefficient α, with a proportionality constant K that depends on geometry and thermo-elastic properties. Thus, if the laser beam is tunable, the normalized OA spectrum provides an uncalibrated absorption spectrum if K is unknown. Absolute calibration is possible by evaluating $K = \beta c / (2\pi R_s^{1/2} C_p \tau_L r^{1/2})$, or more practically, by measuring the normalized OA signal for a known absorber at one wavelength and thus empirically finding K. Equation (5.6) also indicates clearly the advantages and features of OA spectroscopy of weak absorption by direct OA generation. This method is zero background, since $P \rightarrow 0$ if $\alpha \rightarrow 0$. For detecting small α, the signal magnitude P is increased by using intense laser pulses. Equation (5.6) indicates that the OA generation efficiency η defined earlier is proportional to the laser intensity; this does not violate energy conservation since small absorption approximation is assumed here. Although the treatment here is semiquantitative for the simplest case, more detailed theories [e.g., done by Patel and Tam (4)] give essentially the same conclusions.

2.2. Simple Theory for (Indirect) PA Generation

A simple case of (indirect) PA generation is indicated in Fig. 5.2b. In general, PA generation does not provide as high a sensitivity as the direct OA generation for detecting weak absorptions, basically because it is only a thin layer of the thickness of a thermal diffusion length at the surface of the solid sample that is thermally coupled to the gas, producing the detected acoustic wave. However, PA generation is very valuable for the case opposite to weak absorption, that is, when the optical absorption is so strong that no light passes through the sample and hence conventional transmission monitoring fails. This case is illustrated in Fig. 5.2b. The mathematics can be semiquantitatively described as follows, and more detailed mathematical treatment has been given by Rosencwaig and Gersho (14).

Let the incident laser beam of radius r and modulated at frequency f be incident on the sample of thickness l_s in a cylindrical cell of radius R and coupling gas thickness l_g. Let the sample optical attenuation coefficient be α at the excitation wavelength, and the optical absorption length be $\mu_\alpha = 1/\alpha$. The modulated component of the laser-induced heating is distributed over a diffusion length μ_s given by

$$\mu_s = \left(\frac{D_s}{\pi f}\right)^{1/2} \tag{5.7}$$

where D_s is the thermal diffusivity of the sample. We assume that the optical wavelength and the modulation frequency are chosen so that the lengths l_s, μ_α, and μ_s are in decreasing magnitudes. This represents one of the most interesting cases for PA spectroscopy. Let the modulated laser beam intensity absorbed by the sample be represented by

$$I(t) = \tfrac{1}{2} I_0(1 + \sin 2\pi ft) \qquad (5.8)$$

The modulated heat produced within the diffusion length μ_s is only a fraction μ_s/μ_α of the power input I_0 that is absorbed over a depth μ_α. The heat conduction in the geometry of Fig. 5.2b can be described as follows:

Thermal conductivity $\times$ thermal gradient

$\qquad\qquad$ = thermal power within diffusion length $\quad$ (5.9)

which means

$$k_s \frac{\theta_0}{\mu_s} \approx I_0 \frac{\mu_s}{\mu_\alpha} \qquad (5.10)$$

where k_s is the sample conductivity and θ_0 is the amplitude of the temperature variation on the sample surface, which is thermally coupled to an active volume V_{act} of the gas, given by

$$V_{act} \approx \pi r^2 \mu_g \qquad (\text{for } l_g > \mu_g) \qquad (5.11)$$

where μ_g is the gas thermal diffusion length. Using the ideal-gas law, the amplitude δV of the volume change of V_{act} is

$$\delta V = \frac{V_{act}\theta_0}{T} \qquad (5.12)$$

where T is the absolute temperature. Now the volume fluctuation δV causes a pressure fluctuation δP at the microphone. Assuming the adiabatic pressure–volume relation, we have

$$\delta P = \frac{\gamma P \, \delta V}{V} \qquad (5.13)$$

where γ is the ratio of the specific heats and V is the total cell volume, given by

$$V = \pi R^2 l_g \tag{5.14}$$

Combining Eqs. (5.10)–(5.14), we obtain the PA amplitude δP as

$$\delta P \approx \frac{\gamma P \mu_g \mu_s^2 I_0 r^2}{\mu_\alpha k_s l_g T R^2} \tag{5.15}$$

Equation (5.15), which agrees with the more detailed work of Rosencwaig and Gersho (14), indicates that the PA magnitude is proportional to the sample absorption coefficient $\alpha = 1/\mu_\alpha$, and the normalized PA signal $\delta P/I_0$ measured for a range of excitation wavelength λ can provide the absorption spectrum $\alpha(\lambda)$ as in the direct OA generation case. The unusual advantage here is that spectra of totally opaque or highly light-scattering materials can now be measured.

3. DETECTION

3.1. Microphones

The sound transducer used in a gas-phase OA cell (whether it is a gaseous sample that is being studied or a condensed matter sample studied via the PA generation process monitored in a coupling gas) is usually a commercial microphone, which is typically a capacitance sensor that senses the deflection of a diaphragm in contact with the gases. Electrets and piezoelectric microphones are also sometimes used by researchers. Many commercial brands are available, and the choice should depend on the best compromise among sensitivity desired, bandwidth, size, and noise. Sensitivity as high as ~ 100 mV/Pa is available in commercial microphones (e.g., produced by Bruel and Kjaer), and bandwidth ~ 100 kHz is also available; however, high sensitivity is usually available only as a trade-off for lower bandwidths and larger microphone size. Exotic "microphones" are also available. For example, Miles et al. (19) have described a thin-filament device that "follows" the gas flows and can be used for measuring the small pressure fluctuations in OA spectroscopy. Also, Choi and Diebold (20) have developed a "laser Schlieren microphone" that relies on the use of a "probe laser" to monitor the deflection of a diaphragm due to the pressure modulation in an OA cell.

3.2 Piezoelectric Transducers

For detecting the direct OA signal produced in condensed matter, microphones are typically not suitable because of the serious acoustic impedance mismatch at the sample–gas interface, which means that typically less than 10^{-4} of the acoustic pressure amplitude is transmitted from a solid sample into a coupling gas. Piezoelectric ceramic materials are more suitable for detecting the direct OA signal in condensed matter. Many types of piezoelectric ceramics or crystals are commercially available, for example, lead zirconate titanate (PZT), lead metaniobate, lithium niobate, crystalline quartz, and so on, and reviews on these transducers are given by Mason and Thurston (21). For OA detection, the piezoelectric element with metallized electrodes should be mounted in a suitable manner. One way of mounting is described by Patel and Tam (4). A PZT cylinder (PZT 5A from Vernitron, Ohio) of 4 mm diameter and 4 mm height is pressed against a front stainless steel diaphragm polished on both sides. The PZT element is enclosed in the stainless steel casing so that the following noise sources are minimized: electromagnetic pickup, possible corrosion due to contact with reactive samples, and absorption of stray light that is scattered toward the transducer. The sensitivity of such a PZT transducer is typically ~ 3 V/atm. This is much smaller than that of a sensitive microphone (e.g., B&K model 4166) with a sensitivity $\sim 5 \times 10^3$ V/atm. However, PZT transducers are preferred for OA detection in condensed matter because of the much faster rise times and better acoustic impedance matching for PZT compared to microphones.

Thin polymeric films that are made piezoelectric are also frequently used for OA monitoring in condensed matter. These are highly insulating polymeric films that can be poled in a strong electric field at elevated temperatures or can be subjected to charged-beam bombardment so that they become polarized and exhibit piezoelectric character. Such films include polyvinylidene difluoride (PVF_2), Teflon, Mylar, and so on, with PVF_2 being the most commonly used. There is strong interest in the use of PVF_2 film as transducers for acoustic imaging in the human body because of the nonringing characteristic of the film (Q is much lower than PZT), fast rise time, flexibility, and good acoustic impedance matching to liquids like water. A disadvantage of PVF_2 is that its sensitivity is typically much lower than PZT. Many ways to mount a PVF_2 film are possible, and a way is described by Bui et al. (22). Tam and Coufal (23) have shown that such a PVF_2 transducer is capable of ringing-free detection of OA pulses with widths ~ 10 ns excited by a pulsed N_2 laser.

3.3. Other Types of Transducers

Other types of transducers for detecting pulsed acoustic signal have been described in the literature. For example, Dewhurst and co-workers (24, 25) have used capacitance transducers for detection of OA pulses generated in metallic samples by pulsed Nd:YAG lasers. Amer and co-workers (26) have used a continuous probe laser beam directed at the sample surface to detect the surface distortions due to the thermal or acoustic effects excited by a pulsed laser. Tam et al. (27, 28) have used a continuous probe laser beam to detect the transient refractive index profile of the OA pulse in gases as well as in liquids; such an acoustic refractive index profile causes the probe beam to undergo a transient deflection proportional to the spatial derivative of the refractive index.

The methods of optical probing of OA surface distortions or refractive index changes represent a class of noncontact transducers. There are many cases where noncontact or remote OA monitoring are necessary, as in the case of an inaccessible sample (e.g., in a vacuum chamber), or samples in hostile environments, or samples that cannot be contaminated. A good example is the OA spectroscopy of free radicals like OH in flames, as performed by Gupta and coworkers (29). In their work, they used a tunable pulsed ultraviolet laser beam to excite OH-free radicals in a methane–oxygen flame and used a continuous HeNe probe beam to detect the thermal as well as acoustic refractive index gradient caused by a laser pulse of 1-μs duration. Their work clearly indicates that totally noncontact OA spectroscopy can be performed in a localized region in a highly hostile environment. Besides spectroscopy, flame temperature profiles can also be obtained by monitoring the acoustic speeds, as shown earlier by Zapka et al. (30).

3.4. Absorption Modulation and the OA Cell

OA generation requires a modulated absorption that can be obtained from a modulated light beam or modulated absorption characteristics of the sample. Modulation methods for the light source include Q switching, mode locking, flashlamp pulsing, wavelength switching, and the use of mechanical modulators (choppers), electro-optic modulators, acousto-optic modulators, wavelength-modulating devices, and so on. Modulation of the absorption characteristics of the sample is possible, for example, by applying modulated magnetic or electric fields to the sample; thus, Kavaya et al. (31) have demonstrated that Stark modulation of a gas sample by a modulated electric field is excellent for OA detection because the background signal in the Stark modulation mode is about 500 times smaller than that in the same OA cell in the conventional chopped light beam mode. This is because the background absorption (e.g., due to the cell windows or due to the buffer gases) has little dependence on the electric field. In a sense, the Stark modulation method is equivalent to a wavelength modulation method, as is done by Lahmann et al. (32) for OA studies of liquids. Castleden et al. (33) have described a simple wavelength-modulated PA spectrometer that generates differentiated PA spectra: This provides an apparent enhancement in the resolution and results in an increased precision of locating absorption features in the sample.

The OA cell is a container for the sample and the microphone or transducer such that the incident excitation beam can be absorbed by the sample to produce an acoustic signal. At low enough modulation frequencies or long enough pulse durations of the excitation beam, the exact geometry of the OA cell is important since the acoustic wave is reflected from the cell walls, and acoustic interference and resonances can occur. Indeed, many researchers [e.g., Hess (34)] have exploited the effect of acoustic resonance to enhance the OA detection sensitivity. However, at a very high modulation frequency or for very short pulse duration of the excitation beam, effects of reflections from cell walls are unimportant, and so is the geometry of the OA cell; indeed, "leaky" OA cells can be used in such cases.

The above design considerations for an OA cell are aimed at two factors: minimizing noise (i.e., spurious signals) and maximizing the signal. Further useful design considerations include the use of Brewster windows for minimizing light scattering, multipassing for increased sensitivity, a recessed microphone to avoid scattering light onto it, acoustic baffles and shields for reducing effects of unwanted acoustic sources, and locating the excitation beam entrance and exit positions at acoustic nodes in a resonant cell.

4. OA SPECTROSCOPY IN GASES

Modern interest in OA spectroscopic detection in gases begins with the work of Kreuzer (35) in 1971, who reported a detection limit of 10 ppb of methane in air, using a 0.015-W infrared laser for excitation; Kreuzer also indicated that a more powerful infrared laser source may make possible 0.1 part per trillion impurity detection. This work simulates many later experiments for achieving high OA detection sensitivity in various gases. Laser OA in gases has now advanced to the ability to detect an absorption coefficient of $\sim 10^{-10}$ cm^{-1} with a cell length of ~ 10 cm [see Patel and Kerl (36)]. This high sensitivity cannot be matched by other conventional absorption techniques such as extinction measurement, which includes absorption plus all scattering losses, and cannot readily be used to monitor absorption coefficients less than $\sim 10^{-3}$ cm^{-1}. There are a few other ultrasensitive techniques for spectroscopic detection that may have sensitivities better than the laser OA technique, notably the single-atom detection method of Hurst et al. (37) using multistep laser excitation and ionization, and also luminescence monitoring with pulsed laser excitation. Although the laser OA method lags behind the multiphoton ionization spectroscopy or the luminescence monitoring spectroscopy method in sensitivity, it does offer other advantages over these methods such as simplicity and suitability for atmospheric conditions. The simplicity of laser OA detection means that only a sound transducer is required. The suitability for atmospheric conditions means that nonradiative thermal relaxation in a high-density gas system occurs very generally (may be only partial in certain highly fluorescent or chemically active systems), while ion attachment, diffusion, recombinations, and quenching phenomena may pose serious limits to the other detection techniques.

4.1. Weak One-Photon Absorptions

The high detection sensitivity of a laser OA method has the obvious use for detecting very weak absorptions, such as those that are forbidden by dipole selection rules or by spin conservation. An example is an overtone vibrational transition. If the vibrational potential of the molecule can be represented by a simple harmonic oscillator potential, only the fundamental vibrational excitation is allowed for optical absorption, and all higher harmonics have zero dipole matrix elements. However, in a real molecule, higher harmonic transitions are possible because the potential well is at least slightly anharmonic. The use of dye laser OA spectroscopy to examine overtone absorption in gases was demonstrated by Stella et al. (38). In their experiment, the 619.0-nm overtone absorption band of CH_4 and the 645.0-nm overtone absorption band of NH_3 were measured by placing the OA cell in the cavity of a dye laser and scanning the laser across the absorption bands. They were able to resolve rotational features. Subsequently, numerous researchers have performed intracavity laser OA spectroscopy for measuring weak overtone vibrational absorptions in molecules, for example, Bray and Berry (39), Smith and Gelfand (40), and Fang and Swofford (41).

4.2. Trace Detection

The high sensitivity of laser OA detection also permits measurements of weak absorptions due to a strong absorption line of a trace constituent. The idea of trace detection by "spectrophone" detection in conjunction with laser excitations was first demonstrated by Kerr and Atwood (42), and the later work of Kreuzer (35) greatly enhanced the interest. Patel (43) reported some pioneering work of laser OA monitoring of pollutants. His light source is a spin-flip Raman laser (SFRL) with InSb as the active medium that is pumped by a CO or CO_2 laser. The IR output from the SFRL is tunable by a magnetic field, which is typically a superconducting magnet situated in the same liquid helium Dewar as the InSb crystal. The detection technique turns out to be extremely useful for in situ pollutant detection, both in terrestrial stations or in the upper atmosphere. Many other workers have recently demonstrated outstanding successes in detecting trace pollutants in nitrogen or in atmospheric air. Angus et al. (44) have reported OA detection of 10 ppb of NO_2 using a modulated continuous-wave (cw) dye laser excitation. Clapsy et al. (45) reported a study of laser OA detection of explosive vapors, including nitroglycerine, ethylene glycol dinitrate, and dinitrotoluene. A CO_2 laser source tuned to suitable lines in the 9–11-μm spectral range is used to minimize background absorption due to other normal constituents in air. Koch and Lahmann (46) have used a cw frequency-doubled dye laser of 1 W power to detect SO_2 concentrations as low as 0.1 ppb. Vansteenkiste et al. (47) have reported the use of a $PbS_{1-x}Se_x$ diode laser of 96 μW power at 4.8 μm to detect the absorption of CO in a small nonresonant OA cell. A concentration of 50 ppm of CO in N_2 can be detected by operating the cell in a double-pass mode. Gerlach and Amer (48) earlier reported a detection of 0.15 ppm of CO by using a higher-power laser.

4.3. Detection of Excited States

Since trace amounts of excited states can be produced in a gaseous system by suitable excitation mechanisms (optical, discharge, chemical, etc.), it is logical to expect OA detection methods to be well suited for measuring excited-state spectra and collision dynamics. This was first demonstrated by Patel et al. (49) for a gaseous system of NO that was stepwise excited by two infrared lasers. Also, chemically reactive gases may produce transient intermediate chemical species, and the high sensitivity available with OA detection may be useful to identify some of the intermediate products, and hence provide important understanding of the reaction channel. In continuous photolysis, concurrent spectroscopic identification of photolysis products is generally very difficult since the steady-state concentration of intermediates is usually too small for spectroscopic measurements without using a matrix isolation procedure. Colles et al. (50) have shown that in a gas-phase continuous photolysis experiment, OA detection of intermediates is possible using a tunable dye laser.

4.4. Nonlinear Optical Absorptions

The high-sensitivity OA method is also ideally suited for measuring nonlinear absorption or nonlinear optical scattering effects since some degree of heat deposition (often small amounts) in the gas usually occurs due to these nonlinear optical interactions, for example, Raman gain spectroscopy, Doppler-free saturation spectroscopy, and multiphoton absorption spectroscopy.

Photoacoustic Raman-gain spectroscoy (PARS) was first suggested by Nechaev and Ponomarev (51), and was first observed by Barrett and Berry (52) in gases. In their experiments, the pump beam was a continuous Ar^+ laser beam chopped at 573 Hz, and the signal beam was a continuous dye laser beam tuned near the 605.4-nm line of CH_4. With microphone detection, the symmetric stretch vibrational mode of methane near 2900 cm^{-1} was detected as a PARS signal when the photon energy difference of the two laser beams was resonant with this Raman mode. In the more recent experiments on PARS, West and Barrett (53) reported that greatly enhanced sensitivity is obtained using pulsed lasers for the pump and probe beams.

The OA spectroscopy experiments performed so far are mostly of resolution at best equal to the Doppler width. In principle, the high sensitivity of detection implies that the technique of saturation spectroscopy (already shown for fluorescence monitoring, absorption monitoring or optogalvanic monitoring) can be used with OA monitoring to allow a Doppler-free linewidth, that is, with the spectra linewidth being limited by the molecular lifetimes. This is demonstrated by the experiment of Marinero and Stuke (54). In their work, the P(193) line of 11–0 band of the B ← X transition of I_2 is measured using a single-longitudinal-mode cw dye laser beam that is split into two opposite beams chopped at $f_1 = 757$ Hz and $f_2 = 454$ Hz. The OA signal is observed to contain a sum modulation frequency component (i.e., at $f_1 + f_2$) when the dye laser is within the natural linewidth of an absorption line.

5. OA SPECTROSCOPY OF CONDENSED MATTER

5.1. (Indirect) PA Generation

The earlier work in condensed matter generally involved (indirect) PA generation (referred to here as PA spectroscopy). For example, Harshbarger and Robin (55) and Rosencwaig (56) simply extended the gas-phase OA spectroscopy technique of Kreuzer (35) to solids; namely, they used a gas-phase microphone to sense the heating and cooling of a thin gas layer in contact with the sample illuminated by a chopped light beam. This gas-phase-microphone PA technique for condensed matter relies on the inefficient thermal diffusion in the sample and in the coupling gas: hence, this technique lacks sensitivity and is typically useful only for optical absorptions larger than 1%. However, an essential advantage of PA spectroscopy is that many types of "difficult" samples can be measured without any "sample preparation," as would be required by conventional transmission-monitoring methods. In the review by Rosencwaig (2), it is well documented that PA spectra can be obtained for highly light scattering and/ or opaque materials like powders, pigments, polymers, catalysts, hemoproteins, plant matter, bacteria, in vivo cells, and so on. However, to obtain a meaningful PA spectrum, two precautions are required: avoid "PA saturation" and minimize light scattering onto any absorbing surface of the PA cell and especially of the microphone.

The phenomenon of PA saturation is well recognized and is explained in detail by Rosencwaig and Gersho (14) and others. The intuitive explanation of PA saturation [already implied in Eq. (5.10)] for a totally opaque sample with wavelength-dependent absorption length μ_α is as follows. Suppose the laser modulation frequency f is high enough so that the sample thermal diffusion length μ_s [see Eq. (5.7)] is shorter than μ_α throughout the wavelength range of interest. In this case, at each wavelength, λ, the effective heat $H(\lambda)$ is $I_0 \mu_s/\mu_\alpha$ since the incident light amplitude $I_0(\lambda)$ is absorbed in a thickness μ_α, but only a thinner layer of thickness μ_s can communicate with the coupling gas and hence contribute to the PA signal. Hence, we see that $H(\lambda)$ is proportional to the absorption coefficient $\alpha(\lambda)$, and hence the PA signal magnitude is linear to $\alpha(\lambda)$. This is nonsaturation. However, if the modulation frequency f is deceased so that μ_s is equal to μ_α at a certain wavelength λ_1, the heat $H(\lambda_1)$ is now equal to $I_0(\lambda_1)$ since all the heat generated in the depth μ_α at this wavelength can communicate with the coupling gas. This indicates that $H(\lambda_1)$ has reached the maximum value $I_0(\lambda_1)$, independent of $\alpha(\lambda_1)$. Thus, PA saturation has occurred. A way to avoid PA saturation for opaque samples is to make sure that the modulation frequency is large enough.

The second precaution mentioned earlier is the minimization of light scattering causing spurious signal generation at cell walls or microphone surfaces. One way to minimize this is to use a PA cell with a narrow passage connecting the sample and the microphone, as done by McClelland and Kniseley (57), Monahan and Nolle (58), Aamodt and Murphy (15), Bechthold et al. (59), and others. The passage can serve several purposes: (a) Light scattering from the sample, sample holder, and window onto the microphone is reduced. (b) By varying the volume of the sample or the microphone chamber, Helmholtz resonance can be obtained, thus enhancing the PA signal. (c) By using a long enough connection passage, the sample chamber can be kept very cold or very hot to perform PA spectroscopy at these temperatures with the microphone being kept at room temperature.

The optical absorptions by aerosols, colloids, powders, and all other forms of particulates provide good examples of the applicability of PA methods because conventional methods work poorly at best. Absorptions due to fine particles is very important to measure, both in basic science and in applied technologies like the motor industry, smog control, pigment manufacturing, coal conversions, and so on. The PA absorption measurements of powders using gas-coupling methods have been reported by several authors; examples of studies of inorganic powders (e.g., Ho_2O_3, metal powders, etc.) can be found in Rosencwaig (2). Very practical applications of PA spectroscopy can be found in the investigations of diesel smoke particles by Roessler (60) and Bruce and Richardson (61).

5.2. Direct OA Generations

Although the possibility of directly generating acoustic waves in condensed matter by a modulated laser beam was pointed out by White (7) in 1963, it was not until the work of Hordvik and Schlossberg reported in 1977 (62) that the direct-coupling OA technique was first demonstrated as useful for detecting weak absorptions in condensed matter. In their experiment, a cw laser (CO_2, CO, Ar^+, etc.) was used; the chopped beam is incident on a highly transparent solid sample. A piezoelectric transducer is attached on one surface of the sample with epoxy, while the other similar transducer is positioned close to the opposite surface of the sample without contact. The purpose of the latter transducer is to measure the effect of laser light scattered onto the transducers. Using laser powers of several hundred milliwatts, and chopping frequencies of 0.15–3 kHz, Hordvik and Schlossberg (62) achieved an absorption sensitivity of 10^{-4}–10^{-5} cm^{-1}, depending on the amount of light scattering in the solid. In a latter experiment with a similar apparatus, Hordvik and Skolnik (63) reported that surface and bulk absorptions in solids can be separately identified because surface absorption produces a hemispherical wave, while bulk absorption produces a cylindrical wave, distinguishable by spatial and temporal characteristics.

Besides the above direct-coupling OA spectroscopy of solids, similar experiments for liquids have been performed. For example, Lahmann et al. (32) have demonstrated the detection of 0.012 ppb of β-carotene in chloroform. Here, an Ar ion laser producing simultaneously the 488.0-nm and 514.5-nm lines is used. Modulation is performed with a dual chopper so that light of periodically alternating wavelengths is incident onto the liquid. This wavelength alternation permits a considerable reduction of the background signal due to light scattering and window absorptions. They thus achieved a detection limit of 10^{-5} cm^{-1} absorption by the solution.

All the above direct-coupling OA techniques for solids or liquids rely on the use of a chopped light beam, typically at chopping frequencies below a few kilohertz. Window absorptions and effects of light scattering onto the transducer are usually the origins of detection limitations, although other factors such as heating of the liquid by the strong cw laser beam ~ 1 W power (causing convection currents, self-defocusing, etc.), mechanical noise at low frequencies, electrical pickup noise at multiples of line frequencies, and so on, also pose limitations. These limitations are minimized in the pulsed OA technique developed by Patel and Tam (64) using pulsed laser beams of low duty cycle (e.g., 1 μs, 1-mJ pulses at 10 Hz). Time gating the desired OA signal (which travels ballistically from the illuminated region of the sample to the transducer) is used to discriminate against window absorption (which usually arrives later than the desired OA signal) and against light scattering (which occurs almost instantaneously, i.e., earlier than the desired OA signal). Noise effects due to steady heating are minimized by using a low average laser power of ~ 10 mW. Mechanical noise and electrical pickup noises are large at frequencies $\lesssim 10$ kHz and are eliminated by using high-frequency bandpass filters (e.g., passing between 0.1–0.5 MHz); this is possible because the pulsed OA signal produced by short laser pulses ($\lesssim 1$ μs) is of high acoustic frequencies ($\gtrsim 0.1$ MHz), in contrast to the low-frequency OA signals produced by chopped cw excitation beams considered above.

5.2.1. Weak One-Photon Absorption

Similar to the gas-phase OA case, the high sensitivity available with the use of piezoelectric transducers in direct contact with the condensed sample has been utilized to measure weak absorptions. The gas-microphone PA method is seldom used for measuring weak absorptions because of the comparatively lower sensitivity, although McDavid et al. (65) have used it in conjunction with high intensity CO_2 laser sources to measure weak absorptions of alkali halide samples at 10.6 μm. The investigation of Tam et al. (66) indicates that pulsed OA spectroscopy can be used to measure weak absorption coefficients $\sim 10^{-6}$ cm^{-1} in liquids with pathlengths of a few centimeters by using laser pulses of energy 1 mJ and duration 1 μs, and using PZT transducers that are in direct contact with the sample. An important example of such a method of pulsed spectroscopy in liquids is the measurement of the visible absorption spectrum of water by Tam and Patel (67). Water is, of course, the most important liquid in many respects (environmental, biological, technological, geological, etc.), and hence many workers have previously measured its absorption spectrum, mostly by using long pathlength absorption techniques. Accuracies of the previous measurements were severely limited by scattering losses at windows and in the liquid, by refractive index effects causing small changes in the collimation of the light beam through the liquid onto the detector, by contamination of the liquid due to the containing vessel, and so on. Thus, disagreement of previously available data is as large as a factor of 10 at the "green minimum." However, the pulsed OA spectroscopy method of Tam and Patel (67) avoids the above-mentioned limitations, and provides what is believed to be the most reliable absorption spectra in the visible range for pure ordinary water as well as for heavy water at room temperature.

5.2.2. Trace Detection

The use of direct OA spectroscopy for trace detection in liquids for example, by Lahmann et al. (32), was mentioned earlier. Obviously, for routine measurement of trace constituents in a liquid, a cell that is not vulnerable to contamination and corrosion is desired, and such a cell is described by Tam and Patel (68). Another design for trace analysis is a flow-through cell described by Sawada et al. (69) who have used it to detect carcinogenic dyes in solutions. Such flow-through OA cells, in conjunction with pulsed or suitably modulated cw lasers, should be quite valuable for continuous real-time sampling of liquids.

A new flow cell, suitable for OA detection as well as for other modes of detection simultaneously, is described by Voigtman and Windfordner (70).

5.2.3. Thin-Film Absorptions

The monitoring of surface absorption or thin-film absorption is one of the most powerful applications of OA spectroscopy. Important examples include absorptions by glass surfaces, laser mirror coating, and thin-film chromatography plates. Some of the earlier important work can be found in the series of papers by Nordal and Kanstad (71, 72). In their work, various substances and complexes on metal surfaces are investigated. This may have important applications for surface chemistry and surface catalysis on metal surfaces. Such surface reactivity studies are best studied by infrared (IR) photoacoustic spectroscopy (to probe vibrational transitions) as shown by Low and Parodi (73). Pulsed OA detection has been used to measure absorptions due to a thin film of powdered sample [Tam and Patel (74)] or due to a liquid film of several micron thickness trapped between substrates [Patel and Tam (75)]. In more recent developments, Coufal and co-workers (76, 77) demonstrated OA spectroscopy of adsorbed molecules by using an infrared-modulated laser beam and a PZT transducer onto which a silver substrate layer and an adsorbing gas (e.g., NH_3) are deposited under ultrahigh vacuum conditions. An OA spectrum of a sub-monolayer surface molecule layer was detected. Coufal et al. (77) also show that by using polarization modulation instead of the previously employed intensity modulation, the background signal originating from the absorption of light by the substrate material can be dramatically reduced. Intensity fluctuations of the exciting light source are canceled, thus improving the detection sensitivity for the adsorbate considerably. In the case of ammonia on silver films, a few thousandths of a molecular monolayer can be detected without intensity stabilization of the light source.

5.2.4. Nonlinear Optical Absorptions

In nonlinear spectroscopy, the absorption is usually detectable only when high-intensity light sources (like pulsed lasers) are used. Hence, the pulsed OA technique developed by Patel and Tam (64) is ideally suitable for measuring weak nonlinear optical absorption/emission effects; for example, Raman scattering or multiphoton absorption. The first demonstration of OA Raman-gain spectroscopy (OARS) for liquids was reported by Patel and Tam (78). Two synchronized flashlamp-pumped dye lasers are used to provide the pump and the signal pulses, and gated OA measurement with a boxcar integrator is used to detect the energy deposited into the liquid due to the stimulated Raman scattering. The measurement of weak two-photon absorption cross sections by an OA method was also demonstrated by Tam and Patel (79). Previously, sound generation in liquids due to two-photon absorption was known, but the effect had never been applied for quantitative measurements of two-photon cross sections. Another important example of OA monitoring of two-photon absorption in insulators or semiconductors is the work of Van Stryland and Woodall (80) who have reported measurements of two-photon absorption in CdTe and CdSe using pulsed dye lasers and piezolectric transducers coupled to the sample with a suitable liquid.

Other developments in nonlinear OA spectroscopy in condensed matter include the use of very short laser pulses to study excited-state lifetimes. The sample is excited by two short laser pulses that are temporarily separated. The first laser pulse produces certain excited states, and the second laser pulse that can be controllably delayed from the earlier one by time t_d causes further excitation of the excited states. Thus, the OA signal depends on t_d, and this dependence provides the excited-state lifetimes. This stepwise delayed excitation is first done by Rockley and Devlin (81) using nanosecond laser pulses, and is much improved in time resolution by Bernstein et al. (82) and by Heritier and Siegman (83) with the use of picosecond laser pulses.

6. DISCUSSIONS AND SUMMARY

The technique of OA spectroscopy relies on the production and detection of acoustic waves due to a modulated excitation light beam. As the wavelength of the excitation beam is scanned, the magnitude of the acoustic wave or pulse changes, and thus an excitation spectrum commonly called an OA or PA spectrum is obtained. The phase or shape of the acoustic wave or pulse may provide additional information on the thermal properties or excitation lifetimes of the sample. Important advantages of OA spectroscopy are zero background, high sensitivity, applicability to difficult samples, depth-profiling capability, simplicity, and the detector being "ultrapanchromatic" (i.e., useful for any excitation wavelength).

The meaning of the term *optoacoustic spectroscopy*, as generally understood and as used in this chapter is a technique using light-induced sound generation for *optical* spectroscopy. Since the optoacoustic effect involves optical and acoustical waves, we may also expect an "optoacoustic spectroscopy of the second kind," (OAS II), which is a technique using light-induced sound generation for *acoustical* spectroscopy. This has recently been demonstrated by Tam and Leung (84) in gases. Their technique relies on the use of a short-duration laser pulse to generate reliably a narrow acoustic pulse containing a broad Fourier frequency spectrum; as this pulse propagates, the various Fourier components are absorbed differently, resulting in a pulse distortion that is probed by a focused cw laser beam. Fast Fourier transform of the transient probe deflection signal provides the acoustic absorption spectrum; this is much faster than conventional acoustical spectroscopy, which involves transducers for generation and detection, and point-wise frequency measurement. This new technique of OAS II, which is all-optical and noncontact, represents a new analytical application of lasers that generates and probes acoustic pulses in a medium to obtain important properties affecting acoustic propagations like compositions, temperatures, and inelastic collisions.

The concept of OA spectroscopy in its generalized form (i.e., including other types of acoustic generation mechanisms and other types of incident energetic beams) is reviewed in Fig. 5.3. The incident beam can be electromagnetic radiation in the visible on near-visible range (as is the case in most OA studies) or in other spectral regions from RF to X-rays; it can also be a particle beam of electrons, protons, muons, neutrinos, and so on. Several investigations describing acoustic generation by nonoptical beams have been reported in the

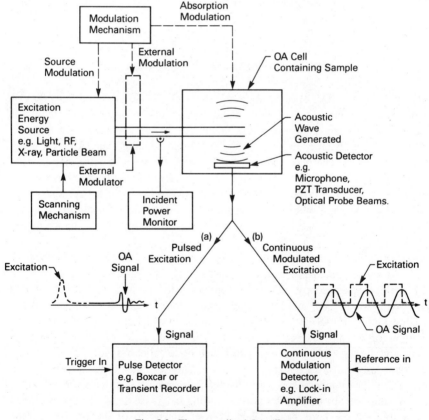

Fig. 5.3. The generalized OA effect.

literature; for example, Melcher (85) has reported the acoustic detection of RF absorption in electron paramagnetic resonance, and Learned (86) has discussed the acoustic detection of energetic particles (cosmic ray muons or neutrinos) in deep oceans. To cause acoustic generation, the incident beam can be pulsed or modulated at close to 50% duty cycle; as shown in Fig. 5.3, the corresponding OA signal is a transient acoustic signal with a well-defined delay time or a modulated acoustic signal with a well-defined phase shift. Typically, the magnitude of the OA signal provides a measure of the energy or intensity of the incident beam or its absorption, while the delay time or phase shift provide information on thermal diffusion, acoustic propagation, or deexcitation times in the sample. It is now obvious that the acoustic detection techniques discussed are not only useful for optical spectroscopy but can also be advantageously used to detect many types of energetic beams and study their interactions with matter.

In this chapter, we mainly have been concerned with the technique of OA spectroscopy based on the thermal expansion OA generation mechanism. We have already indicated that other OA generation mechanisms are also possible. This means that many other interesting effects can be studied by OA monitoring; for example, chain reactions in photochemical reactions can cause strong acoustic amplifications [Diebold and Hayden (87)]; weak laser pulses of 1-mJ energy can generate an acoustic shock wave (via breakdown in a vapor) that is observable many centimeters away from the breakdown region [Tam et al. (27)]; and micro-mechanical motions in the form of droplet ejection from a nozzle can be controlled by pulsed OA generation (probably involving boiling) in liquids [Tam and Gill (88)]. Furthermore, ultrashort acoustic pulses of ≤ 1-nsec duration can be generated for new ultrasonic testing of materials [Tam (89)]. The propagation of such short acoustic pulses can also be detected by optical deflection methods or by interferometry [Sontag and Tam (90, 91)]. The study and applications of these various OA generation mechanisms in different systems will be an area of fruitful research, and are further discussed elsewhere [Tam (92)].

ACKNOWLEDGMENTS

This work is supported in part by the Office of Naval Research. I would like to thank Hans Coufal for very helpful discussions.

REFERENCES

1. Y.-H. Pao, *Opto-Acoustic Spectroscopy and Detection*, Academic, New York, 1977.
2. A. Rosencwaig, *Photoacoustics and Photoacoustic Spectroscopy*, Wiley, New York, 1980.

3. H. Coufal, P. Korpiun, E. Lüscher, S. Schneider, and R. Tilgner, *Photoacoustics— Principles and Applications*, Vieweg Verlag, Braunschweig, West Germany, 1983.
4. C. K. N. Patel and A. C. Tam, *Rev. Mod. Phys.*, **53**, 517 (1981).
5. A. C. Tam, "Photo-acoustics: Spectroscopy and Other Applications," in D. Kliger, Ed., *Ultrasensitive Laser Spectroscopy*, Academic, New York, 1983, p. 1.
6. V. S. Teslenko, *Sov. J. Quant. Elect.*, **7**, 981 (1977).
7. R. M. White, *J. Appl. Phys.*, **34**, 3559 (1963).
8. L. S. Gourney, *J. Acoust. Soc. Am.*, **40**, 1322 (1966).
9. C. L. Hu, *J. Acoust. Soc. Am.*, **46**, 728 (1979).
10. G. Liu, *Appl. Optics*, **21**, 955 (1982).
11. H. M. Lai and K. Young, *J. Acoust. Soc. Am.*, **72**, 2000 (1982).
12. J. M. Heritier, *Opt. Comm.*, **44**, 267 (1983).
13. B. Sullivan and A. C. Tam, *J. Acoust. Soc. Am.*, **75**, 437 (1984).
14. A. Rosencwaig and A. Gersho, *J. Appl. Phys.*, **47**, 64 (1976).
15. L. C. Aamodt and J. C. Murphy, *J. Appl. Phys.*, **48**, 3502 (1977).
16. G. C. Wetsel, Jr., and F. A. McDonald, *Appl. Phys. Lett.*, **30**, 252 (1977).
17. A. C. Tam and Y. H. Wong, *Appl. Phys. Lett.*, **36**, 471 (1980).
18. L. D. Landau and E. M. Lifshitz, *Fluid Mechanics*, Pergamon, New York, 1959.
19. R. B. Miles, J. Gelfand, and E. Wilczek, *J. Appl. Phys.*, **51**, 4543 (1980).
20. J. G. Choi and G. J. Diebold, *Appl. Opt.*, **21**, 4087 (1982).
21. W. P. Mason and R. N. Thurston, Ed., *Physical Accounts*, Vol. XIV, Academic, New York, 1979.
22. L. Bui, H. J. Shaw, and L. T. Zitelli, *Electronics Lett.*, **12**, 393 (1976).
23. A. C. Tam and H. Coufal, *Appl. Phys. Lett.*, **42**, 33 (1983).
24. D. A. Hutchins, R. J. Dewhurst, S. B. Palmer, and C. B. Scruby, *Appl. Phys. Lett.*, **38**, 677 (1981).
25. A. M. Aindow, R. J. Dewhurst, D. A. Hutchins, and S. B. Palmer, *J. Acoust. Soc. Am.*, **69**, 449 (1981).
26. M. Olmstead, N. M. Amer, D. Fournier, and A. C. Boccara, *Appl. Phys. A*, **32**, 141 (1983).
27. A. C. Tam, W. Zapka, K. Chiang, and W. Imaino, *Appl. Opt.*, **21**, 69 (1982).
28. W. Zapka and A. C. Tam, *Appl. Phys. Lett.*, **40**, 310 (1982).
29. A. Rose, G. J. Salamo, and R. Gupta, *Appl. Opt.*, **23**, 781 (1984).
30. W. Zapka, P. Pokrowsky, and A. C. Tam, *Opt. Lett.*, **7**, 477 (1981).
31. M. J. Kavaya, J. S. Margolis, and M. S. Schumate, *Appl. Opt.*, **18**, 2602 (1979).
32. W. Lahmann, H. J. Ludewig, and H. Welling, *Anal. Chem.*, **49**, 549 (1977).
33. S. L. Castleden, G. F. Kirkbright, and D. E. M. Spillane, *Anal. Chem.*, **53**, 2228 (1981).
34. P. Hess, "Resonant Photoacoustic Spectroscopy," in F. L. Boschke, Ed., *Topics in Current Chemistry*, Vol. 111, Springer, Berlin, 1983.
35. L. B. Kruezer, *J. Appl. Phys.*, **42**, 2934 (1971).
36. C. K. N. Patel and R. J. Kerl, *Appl. Phys. Lett.*, **30**, 578 (1977).
37. G. S. Hurst, M. G. Payne, S. D. Kramer, and J. P. Young, *Rev. Mod. Phys.*, **51**, 767 (1979).
38. G. Stella, J. Gelfand, and W. H. Smith, *Chem. Phys. Lett.*, **39**, 146 (1976).
39. R. G. Bray and M. J. Berry, *J. Chem. Phys.*, **71**, 4909 (1979).
40. W. H. Smith and J. Gelfand, *J. Quant. Spectrosc. Radiat. Transfer*, **24**, 15 (1980).

41. H. L. Fang and R. L. Swofford, *Appl. Opt.*, **21**, 55 (1982).
42. E. L. Kerr and J. G. Atwood, *Appl. Opt.*, **7**, 915 (1968).
43. C. K. N. Patel, *Science*, **202**, 157 (1978).
44. A. M. Angus, E. E. Marinero, and M. J. Colles, *Opt. Commun.*, **16**, 470 (1980).
45. P. C. Claspy, Y. H. Pao, S. Kwong, and E. Nodov, *Appl. Opt.*, **15**, 1506 (1976).
46. K. P. Koch and W. Lahmann, *Appl. Phys. Lett.*, **32**, 289 (1978).
47. T. H. Vansteenkiste, F. R. Faxvog, and D. M. Roessler, *Appl. Spectrosc.*, **35**, 194 (1981).
48. R. Gerlach and N. M. Amer, *Appl. Phys. Lett.*, **32**, 228 (1978).
49. C. K. N. Patel, R. J. Kerl, and E. G. Burkhardt, *Phys. Rev. Lett.*, **38**, 1204 (1977).
50. M. J. Colles, A. M. Angus, and E. E. Marinero, *Nature (London)*, **262**, 681 (1976).
51. S. Ya Nechaev and N. Yu Ponomarev, *Sov. J. Quant. Elect. (Engl. Transl)*, **5**, 752 (1975).
52. J. J. Barrett and M. J. Berry, *Appl. Phys. Lett.*, **34**, 144 (1979).
53. G. A. West and J. J. Barrett, *Opt. Lett.*, **4**, 395 (1979).
54. E. E. Marinero, and M. Stuke, *Opt. Commun.*, **30**, 349 (1979).
55. W. R. Harshbarger and M. B. Robin, *Acc. Chem. Res.*, **6**, 329 (1973).
56. A. Rosencwaig, *Opt. Commun.*, **7**, 305 (1973).
57. J. F. McClelland and R. N. Kniseley, *Appl. Opt.*, **15**, 2967 (1976).
58. E. M. Monahan and A. W. Nolle, *J. Appl. Phys.*, **48**, 3519 (1977).
59. P. A. Bechthold, M. Campagna, and J. Chatzipetros, *Opt. Commun.*, **36**, 309 (1981).
60. D. M. Roessler, *Appl. Opt.*, **21**, 4077 (1982).
61. C. W. Bruce and N. M. Richardson, *Appl. Opt.*, **23**, 13 (1984).
62. A. Hordvik and H. Schlossberg, *Appl. Opt.*, **16**, 101 (1977).
63. A. Hordvik and L. Skolnik, *Appl. Opt.*, **16**, 2919 (1977).
64. C. K. N. Patel and A. C. Tam, *Appl. Phys. Lett.*, **34**, 467 (1979).
65. J. M. McDavid, K. L. Lee, S. S. Yee, and M. A. Afromowitz, *J. Appl. Phys.*, **49**, 6112 (1978).
66. A. C. Tam, C. K. N. Patel, and R. J. Kerl, *Opt. Lett.*, **4**, 81 (1979).
67. A. C. Tam and C. K. N. Patel, *Appl. Opt.*, **18**, 3348 (1979).
68. A. C. Tam and C. K. N. Patel, *Opt. Lett.*, **5**, 27 (1980).
69. T. Sawada, H. Shimizu, and S. Oda, *Jpn. J. Appl. Phys.*, **20**, L25 (1981).
70. E. Voightman and J. Winefordner, *Anal. Chem.*, **54**, 1834 (1982).
71. P. E. Nordal and S. O. Kanstad, *Opt. Commun.*, **22**, 185 (1977).
72. P. E. Nordal and S. O. Kanstad, *Opt. Commun.*, **24**, 95 (1978).
73. M. J. D. Low and G. A. Parodi, *Appl. Spectrosc.*, **34**, 76 (1980).
74. A. C. Tam and C. K. N. Patel, *Appl. Phys. Lett.*, **35**, 843 (1979).
75. C. K. N. Patel and A. C. Tam, *Appl. Phys. Lett.*, **36**, 7 (1980).
76. F. Träger, H. Coufal, and T. J. Chuang, *Phys. Rev. Lett.*, **49**, 1720 (1982).
77. H. Coufal, F. Träger, T. J. Chuang, and A. C. Tam, *Surf. Sci*, **145**, L504 (1984).
78. C. K. N. Patel and A. C. Tam, *Appl. Phys. Lett.*, **34**, 760 (1979).
79. A. C. Tam and C. K. N. Patel, *Nature (London)*, **280**, 304 (1979).
80. E. W. Van Stryland and M. A. Woodall, in *Laser Damage Conference*, Boulder, Colorado, 1980.
81. M. G. Rockley and J. P. Devlin, *Appl. Phys. Lett.*, **31**, 24 (1977).

82. M. Bernstein, L. J. Rothberg, and K. S. Peters, *Chem. Phys. Lett.*, **91,** 315 (1982).
83. J.-M. Heritier and A. E. Siegman, *IEEE J. Quant. Elect.*, **QE-19,** 1551 (1983).
84. A. C. Tam and W. P. Leung, *Phys. Rev. Lett.*, **53,** 560 (1984).
85. R. L. Melcher, *Appl. Phys. Lett.*, **37,** 895 (1980).
86. J. G. Learned, *Phys. Rev.*, **D19,** 3293 (1979).
87. G. J. Diebold, and J. S. Hayden, *Chem. Phys.*, **49,** 429 (1980).
88. A. C. Tam and W. D. Gill, *Appl. Opt.*, **21,** 1891 (1982).
89. A. C. Tam, *Appl. Phys. Lett.*, **45,** 510 (1984).
90. H. Sontag and A. C. Tam, *Appl. Phys. Lett.*, **46,** 725 (1985).
91. H. Sontag and A. C. Tam, *IEEE Trans. Ultrasonics, Ferroelectrics, Freq. Contr.*, (to be published).
92. A. C. Tam, *Rev. Mod. Phys.*, (Apr, 1986).

CHAPTER

6

INFRARED ABSORPTION SPECTROSCOPY

EDWARD S. YEUNG

Department of Chemistry and Ames Laboratory
Iowa State University
Ames, Iowa

1. INTRODUCTION

The ability to provide structural information from characteristic vibrational frequencies has firmly established infrared (IR) spectroscopy as an important analytical tool. The development of the laser has made a significant impact on the
way many infrared absorption measurements are performed today. Unlike the
complementary technique of Raman spectroscopy (Chapter 10), infrared was
already a popular analytical method prior to the invention of the laser. With the
laser, the impact is not so much due to the introduction, for example, of nonlinear techniques but rather to the replacement of the generally poor conventional

infrared light sources. Also, during the past two decades, when laser technology has made significant advances, there is the parallel development of Fourier transform infrared (FTIR) methods. Substantial gain in spectral resolution and in sensitivity has been achieved using FTIR. So, other than serving as a reference for the location of the mirror in FTIR spectrometers, the laser still has not seen much use in routine IR instruments.

In this chapter we shall examine the various characteristics of infrared laser sources, the types of measurements that can take advantage of these characteristics, and the potentials and limitations of these new techniques. Since optoacoustic spectroscopy has been separately discussed in Chapter 5, its infrared versions will be omitted from our discussion. It must be recognized, however, that optoacoustic spectroscopy is an important application of lasers in the IR region and is often the technique of choice to monitor absorption.

2. TYPES OF INFRARED LASERS

Table 6.1 shows the various types of IR lasers. The operation and special features of each type are discussed below.

2.1. Optical Parametric Oscillators (OPO) and Wave Mixing

When two light sources at frequencies ω_1 and ω_2 are mixed in a nonlinear crystal, the difference frequency at $\omega_1 - \omega_2$ can be generated. The efficiency of this

Table 6.1. Properties of Tunable Infrared Lasers

	Wavelength Region			Typical Power (W)	
Tunable Source	Overall Range (μm)	Continuous Range (cm^{-1})	Highest Resolution Obtained (cm^{-1})	CW	Pulsed
OPO	0.55–3.5 (LiNbO$_3$) 1.2–8.5 (Ag$_3$AsS$_3$) 8–12 (CdSe)	3000	1×10^{-3}	10^{-2}	10^5
Diode laser	1–34	2	3×10^{-6}	10^{-3}	10
SFR laser	3 (HF)	15		1	10^3
	5–6 (CO)	50	1×10^{-6}		
	9–14 (CO$_2$)	100	3×10^{-5}		
LPG laser	9–11 (CO$_2$)	Discrete lines	1×10^{-6}	40	10^3
HPG laser	9–11 (CO$_2$)	10	3×10^{-2}	1	10^6
Color-center laser	0.88–3.3	500	10^{-5}	10^{-2}	

wave-mixing process is dependent on laser power, so that typically high-power pulsed lasers are used. For example, a ruby laser and a dye laser can be mixed in a LiNbO$_3$ crystal to generate light in the 3–4 μm range (1). Since pulsed lasers generally do not have very narrow spectral outputs, the corresponding IR radiation is also spectrally broad. To produce light suitable for high-resolution work, two continuous-wave (cw) lasers can be mixed (2). The schematic diagram of such a difference frequency spectrometer is shown in Fig. 6.1. Phase matching is required for high efficiency, so the nonlinear crystal is used at 90° from the c axis and placed in an oven, the temperature of which is scanned together with the dye laser. With about 100 mW in the argon laser and 10 mW in the dye laser, a few microwatts of IR radiation is obtained. Most importantly, this power level is available in a bandwidth of a few megahertz (10^{-4} cm^{-1}) and is far superior to those from conventional IR sources.

The OPO is a variation of wave mixing in which a single visible or near IR photon can be made to produce two IR photons whose total energy equals that of the incident photon. The nonlinear crystal is placed in a resonant cavity for one of the IR frequencies. The other frequency is then generated following energy conservation. The schematic diagram of an OPO is shown in Fig. 6.2. Coarse tuning is achieved by changing the phase-matching condition, for example, by controlling the temperature of the oven. Etalons are required to further reduce the output bandwidth. Typically, a bandwidth of 30 MHz can be achieved, with powers in the milliwatt range.

2.2. Diode Lasers

If current is applied to the *pn* junction of a semiconductor, the recombination of electrons and holes across the junction can result in the emission of a photon with a wavelength corresponding to approximately the energy bandgap. If the active region is contained within the polished faces of the crystal, oscillations can then occur to produce laser action. A typical diode laser is shown in Fig. 6.3. The linewidth can be as narrow as 54 kHz (3), but normally a few longi-

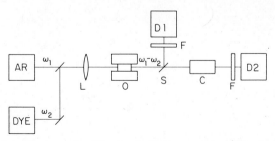

Fig. 6.1. Experimental scheme for difference frequency generation. AR, single-mode argon ion laser; DYE, single-mode dye laser; L, lens; O, oven with nonlinear crystal; S, beam splitter; F, filter; C, sample cell; D1, reference photodetector; D2, transmittance photodetector.

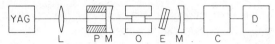

Fig. 6.2. An optical parameter oscillator. YAG, pulse Nd:YAG laser; P, piezoelectric driver; M, cavity mirror; O, oven with nonlinear crystal; E, tunable etalon; C, sample cell; D, photodetector.

tudinal modes oscillate together to give distinct output spaced about 1 cm^{-1} apart. The most useful IR wavelengths are produced by the lead–salt semiconductors (4). To reduce thermal population of the lasing level, cryogenic cooling is normally required for lasers operating in this spectral region. The spatial divergence of these lasers are quite large because of the short cavities, and proper collimating optics must be used. Also, the power output (cw) is quite low (milliwatt range), and sensitive detectors are needed. So, all these combine to make the operation nontrivial, although commercial systems are available.

To change the frequency of the laser, one can change the composition of the semiconductor, for example, $Pb_{1-x}Sn_xSe$ (5). Each class of semiconductors can be used over several hundreds of reciprocal centimeters. For a given diode, the frequency can be tuned by changing the temperature (6), pressure (7), magnetic field (8), and the current applied (9). The last method is essentially the same as temperature tuning, which changes both the bandgap and the effective cavity length of the diode. None of these tuning methods by themselves produces a spectrally continuous output. Mode hopping occurs after about 1 cm^{-1}. Furthermore, the tuning rate is nonlinear, and independent wavelength calibration is necessary. Lastly, temperature cycling to cryogenic temperatures plus interdiffusion effects can alter the bandgap over repeated usage. So, much attention is needed to take full advantage of the high-resolution capabilities of these lasers.

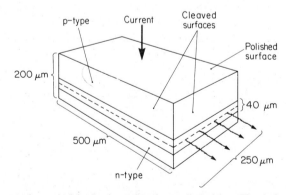

Fig. 6.3. Diagram of a typical lead–salt semiconductor diode laser. The dashed line shows the position of the pn junction. The radiation is emitted from an area $40 \times 250 \ \mu m$.

2.3. Spin-Flip Raman (SFR) Laser

The need to change semiconductors in diode lasers over even a moderate spectral region and the low powers available are incentives for finding alternatives to make use of the semiconductor bandgap. When a coherent pump laser irradiates the lowest spin states of the electrons in the conduction band of a semiconductor in a uniform external magnetic field, excitation involving a change of spin states results in Raman gain, and laser action is produced at the Stokes frequency. Conversely, deexcitation of spin states leads to laser action at the anti-Stokes frequency. No population inversion is required in SFR. The laser frequency is tunable because the lowest spin states are separated by an energy that varies approximately linearly with the external magnetic field. A possible semiconductor is InSb (10). The relationships among the various fields in SFR is shown in Fig. 6.4.

The InSb crystal is typically $2 \times 2 \times 5$ mm^3 with the end faces polished. The carrier concentration is between 1×10^{14} to 2×10^{16} cm^{-3}. Again, the crystal is cooled cryogenically. The pump laser is normally a CO_2 or a CO laser, pulsed or cw (11). The laser output can be as narrow as 100 kHz if an external cavity is used with an antireflection coated semiconductor crystal, so that the cavity length can be stabilized (12). Power levels in the 100-mW range can be obtained due to the high conversion efficiency of the SFR process. The experimental complication is the magnet since a minimum field is necessary for operation. This can range on the order of kilogauss to tens of kilogauss. The utility of SFR has been limited by the available spectral range, which has so far been 1820–1900 cm^{-1}, even though truly continuous tuning is possible.

2.4. Gas Lasers

Many gases when excited with an electrical discharge will produce population inversion between vibrational–rotational levels, and can be used as lasers. The most common of these are the CO_2 and the CO lasers. When a low gas pressure (LPG) is used the lines are on the order of several megahertz wide, and tens of watts are available on each line. The disadvantage is that the locations of the lines are dictated by the molecule used, and cannot be changed except if an

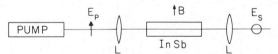

Fig. 6.4. A spin-flip Raman laser. E_p, polarization direction of pump laser; E_s, polarization direction of Stokes emission; L, lens; InSb, single-semiconductor crystal; B, direction of applied magnetic field.

external field is applied. Spectroscopy will have to be based on accidental co-incidences between such laser lines and the analyte species. The CO continuous-wave LPG laser has the further complication of requiring cryogenic cooling. To provide a choice among the available discrete lines for a given gas, the laser cavity usually incorporates a diffraction grating. Frequency stabilization within the gain curve of a laser line can be achieved by piezoelectric control of the cavity length. By operating at atmospheric or higher pressures (HPG), the gas transition is pressure broadened. This allows some tunability over the broadened gain curve of the transition (13). More interesting is the ability to produce pulses with very high peak powers that are important in certain analytical applications. Another method to extend the tunability of gas lasers is to introduce an infrared waveguide modulator that is driven at microwave frequencies (14). This way, sidebands up to 12 mW can be generated on either side of the discrete gas laser line. Since the microwave frequency (around 10 GHz) is well defined, the resulting sideband output has a frequency that is automatically calibrated relative to the gas laser line.

2.5. Color Center Lasers

The lasing medium in these lasers is the color center of alkali or alkali-doped halide crystals (15). The change in ionic configuration from the vacancy con-figuration to the double-well configuration is responsible for the laser transition. In operation, they are similar to cw dye lasers. Optical excitation is provided by an argon ion or krypton ion laser, depending on the absorption band. Up to 240 mW cw can be generated with a linewidth about 0.1 MHz. The tuning range for each type of crystal can be the order of 500 cm^{-1}. The techniques for frequency selection and stabilization is normally based on etalons and piezo-electric control of the cavity length. The entire region between 0.8 and 3.3 μm can be covered. To produce laser action, these crystals have to be cooled to liquid nitrogen temperatures. But, unlike diode lasers, these crystals seem to be stable over extended usage. The main drawback for color center lasers is the inability to cover the mid-infrared spectral region.

2.6. Raman Laser

A Raman laser is essentially stimulated Raman scattering adapted to produce long-wavelength output from other laser sources. Under intense excitation, it is possible to generate high-order Raman emission, that is, $\omega_L \pm n\omega_R$, where ω_L is the excitation source, ω_R is the energy of the Raman transition in the medium (usually a vibrational-rotational line in a molecule), and n is an integer. Using a tunable visible dye laser and a waveguide Raman cell of compressed hydrogen gas, output in the 0.7–7.0 μm region can be continuously covered (16). The peak power can reach 80 kW for $n = 3$. Unfortunately, the inherent linewidth of the Raman medium limits the available spectral resolution to a few reciprocal centimeters.

3. EXPERIMENTAL METHODS

Conventional IR absorption measurements are limited by both the light source and the detector. In principle, the higher radiant energies available in IR lasers can improve detectability in absorption measurements by extending the shot-noise limit. In practice, however, the slow response time of simple IR detectors, the presence of a large background at room temperature, and the absorption by interfering species make shot-noise-limited detection highly improbable in standard instruments. One must then try to develop special optical and electronic arrangements to take advantage of the special properties of the laser, that is, collimation, monochromaticity, power, temporal properties, and coherence. Some examples are listed below. Since IR spectroscopy of trace components in condensed phases is generally limited by absorption of the major component (solvent), the use of a laser does not lead to substantial improvements (e.g., compared to FTIR). The following discussion will therefore only be concerned with gas-phase IR methods.

3.1. Long-Path Absorption

The low number density in the gas phase requires a long absorption path for reasonable signal-to-noise ratios. The highly collimated nature of the laser beam makes it readily adaptable for traveling large distances. Long-path gas cells are available to provide an effective pathlength of 40 m in a physical length of 1.5 m. To discriminate against the background radiation, mechanical modulation is often used (17). A typical arrangement is shown in Fig. 6.5. The detectability depends on the individual species and the particular molecular transition. Taking a reasonable absorption strength (18) of 0.1 cm^{-1} torr^{-1} at the line center for gas samples at 1 atm pressure, one can estimate that a 0.1 ppm concentration of a trace component will give an absorbance of 0.04 for a 40-m pathlength. This is well within the capabilities of the absorption detector. This, of course, assumes that no interfering absorption lines are present. In fact, it is possible to construct multiple reflection cells with an effective pathlength up to 1.5 km (19). Actual detectabilities range from 300 ppb for N_2O to 0.05 ppb for NH_3 mainly due to absorption strength differences (20). In the free atmosphere, a retroreflector is used to return the laser light so that it can be monitored near the laser source. The main problem is scattering and atmospheric turbulence. However, if the laser frequency is scanned, the background can be effectively

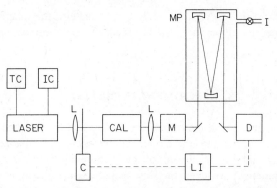

Fig. 6.5. Semiconductor laser used for absorption in point sampling. TC, temperature control; IC, current control; L, lens; C, mechanical chopper; CAL, gas cell for calibration; M, monochromator; MP, multiple-pass gas cell; I, gas inlet; D, detector; LI, lock-in amplifier.

suppressed. For atmospheric studies, a complete spin-flip Raman laser has been put on a balloon to a height of 28 km (21).

Instead of tuning the laser, the differential absorption at two discrete wavelengths of LPG lasers can be used. The advantage is the much higher power levels available in gas lasers for background suppression. However, these measurements are extremely sensitive to the actual environment since temperature, pressure, and interfering species can greatly affect the absorption coefficients at the laser lines. For example, even though the CO_2 laser operates in a relatively transparent region of the atmosphere, some of its lines overlap with H_2O absorption lines. The high (and variable) concentration of H_2O makes its interference effect highly unpredictable. The importance of parameters such as line position, line strength, linewidth, and line shape of common atmospheric species, including isotopic variants, to reliable long-path absorption measurement has been summarized previously (22). It may be possible to use more than just one pair of lines to monitor each atmospheric species, so that unusual interference at one of the lines can be recognized and properly accounted for. Using a combination of CO_2, CO, HF, and DF lasers, the types of species that can be monitored in the atmosphere include H_2O, HDO, D_2O, CO_2, various hydrocarbons and halogenated hydrocarbons, N_2O, NH_3, NO, O_3, SO_2, HF, HCl, and ethyl mercaptan. A sophisticated instrument (23) is based on a CO_2 laser operating at 100 Hz and 1 J/pulse. The IR radiation is collected by a 62-cm telescope and detected by a Ge–Hg detector at 10 K. Fourier transform is used to further improve the spatial resolution, which was limited by the 0.3-μs pulse. The potential for these differential absorption measurements is much enhanced if the OPO is used due to its wavelength tunability. For example, even the weakly absorbing overtone transition of methane can be used (6008 cm^{-1}) over a path of 5.4 km to determine concentrations in the 4 ppm range (24).

3.2. Heterodyne Detection

The weak IR signals can be detected to the quantum noise limit by heterodyne detection. This technique has been described in detail (25). Briefly, when two light waves are mixed together at a detector, the beat frequency between them, that is, the difference frequency, will appear as an amplitude-modulated signal. If the detector has high enough frequency response, the signal can be extracted free from other contributions in the background. A typical arrangement for heterodyne detection is shown in Fig. 6.6 (26). A blackbody source is used for the absorption measurement. A local oscillator of well-defined, stable frequency is then combined with the absorbed beam at a beam splitter, and the two are focused onto a fast detector. A wide-bandwidth, low-noise amplifier is used to treat the signal, which is then sent to a narrow-bandwidth detector. To further amplify the signal, a lock-in amplifier in the kilohertz range is synchronized with the chopped blackbody radiation. This way, one monitors the radiation intensity at a spectral resolution dictated by the narrow-bandwidth detector, and at a spectral location displaced from the local oscillator by exactly its central frequency.

It has been shown that the best detectability is achieved when the power of the local oscillator is larger than $2kT_nh\nu/e^2\eta R$, where T_n is the temperature equivalent of the amplifier noise, ν is the frequency of the light, η is the efficiency of mixing, and R is the load resistance of the detector. This translates to several milliwatts of power, which is compatible with most IR lasers. The mixer is a high-speed infrared detector, for example, HgCdTe, which allows a bandwidth of 2 GHz. This implies that the laser must be within this frequency range from the desired spectral location. For tunable diode lasers, it can be seen that the effective tuning range is not extended by a substantial amount. The advantage

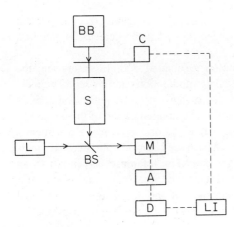

Fig. 6.6. Experimental scheme for heterodyne detection. L, laser for local oscillator; BB, blackbody radiator; C, mechanical chopper; S, sample cell; BS, beam splitter; M, mixer; A, amplifier; D, crystal detector; and LI, lock-in amplifier.

is that an accurate relative frequency calibration is possible by operating the diode at a fixed frequency and scanning the detector frequency. The disadvantage is that diode lasers have low powers that result in degradation of the signal-to-noise ratio, poor beam qualities for mixing, and additional noise due to competition among the many modes in the laser. For gas lasers (e.g. CO_2), the quantum noise limit is easily achieved. Since the gas laser lines are about 50 GHz apart, continuous tuning is still not possible. The widest tuning range of several GHz is obtained if the heterodyne technique is combined with a high-pressure gas laser. Since CO_2 lasers have short-term stabilities in the 10–100-kHz range, excellent spectral resolution is achieved. The blackbody source is typically a natural emitter like the sun, which can be used conveniently to monitor the gaseous species in the earth's atmosphere (27). An interesting application in astronomy is the study of NH_3 gas emission from Jupiter (28). The spectral resolution of the heterodyne technique is also dependent on the stability of the detector. Usually a crystal detector is used and stability is much better than that of the laser source. However, it is also possible to use a series of intermediate frequency channels, each with adjacent frequency ranges, to span the entire frequency spectrum allowed by the mixer (29). This way, one gains the multiplex advantage.

3.3. Modulation Spectroscopy

Since background radiation and detector noise are much more serious in the IR than the UV-visible spectral regions, various modulation techniques have been employed to extract the weak signal. Amplitude modulation of the source by conventional light choppers is the simplest modulation technique. This, however, does not take advantage of any of the special properties of the laser. A straightforward refinement of amplitude modulation is to rely on the short pulses that can be derived from infrared lasers. For example, CO_2 lasers with 0.1-μs pulses operated at 100 Hz can improve discrimination against the background by a factor of 5×10^4 at the same average power, if the detector is gated on only during the pulse.

Because of the high spectral resolution associated with IR lasers and the narrow absorption lines of gases, wavelength modulation can be used. The most direct method is to rely on the dependence of the output frequency on applied current in diode lasers. A small ac current can be applied on top of the dc current used for scanning to produce wavelength modulation. A typical experimental arrangement is shown in Fig. 6.7 (30). The laser beam is split into two parts to a reference gas cell and a sample gas cell. The reference cell produces an absorption signal from the species of interest at low pressures, and is synchronously detected at the fundamental frequency of the ac current. The normal absorption line (without the ac modulation) causes a change in transmitted in-

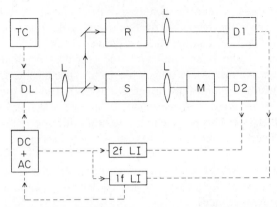

Fig. 6.7. Experimental scheme for derivative spectroscopy. DL, diode laser; TC, temperature control; DC, current for wavelength tuning; AC, modulation current; L, lens; R, reference gas cell to control laser frequency; S, sample gas cell; D1, reference photodetector; D2, sample photodetector; M, monochromator; LI, lock-in amplifier; 1f, fundamental frequency; 2f, second-harmonic frequency.

tensity as shown in Fig. 6.8a. The lock-in amplifier at the fundamental frequency thus produces an output as shown in Fig. 6.8b, which is the first derivative of the absorption line. Now, the zero-crossing point can be used to stabilize the dc level of the diode laser, so that its frequency is locked onto the line center. The sample gas cell is generally also at reduced pressure to minimize the absorption linewidth. As one reduces the gas pressure, the number density of the species of interest decreases, but the absorption coefficient at line center increases due to reduced pressure broadening. Laser light transmitted by the sample cell must be isolated into the individual modes by a monochromator before reaching the detector. The second photodetector is monitored at the second harmonic of the ac frequency, so that the second derivative signal, Fig. 6.8c,

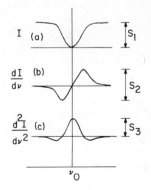

Fig. 6.8. Experimental shapes of signals as a function of frequency for (a) transmitted intensity at D2 in Fig. 6.7, (b) output at 1f lock-in amplifier, and (c) output at 2f lock-in amplifier. ν_0, frequency at line center.

is produced. The signal level S_3 is then used to determine the concentration. It is also possible to use the second derivative signal to lock the laser onto the side of the absorption line, and use the first derivative signal, S_2, to determine the concentrations. There is an optimum choice for the magnitude of the ac modulation current. If the modulation is too large, spectral interference from additional lines nearby becomes important. If the modulation is too small, one loses sensitivity unnecessarily. A reasonable value is 1.5–2 times the spectral linewidth (FWHM) (31). This procedure is also useful for discriminating against absorption from interfering species that have absorption lines in the same general spectral region. The normal absorption signal, S_1 in Fig. 6.8a, will show contributions from the line wings of the interfering species. Because these are much more slowly varying with frequency, the second derivative signal is influenced less by these overlapping lines.

It is possible to incorporate wavelength modulation when using discrete-line gas lasers. The idea is to alternate between two adjacent laser lines, which are typically one to several reciprocal centimeters apart. Since most gas lasers have a diffraction grating as a tuning element, one can use a piezoelectric device to modulate the grating between two fixed positions. An adaptation of the CO_2 laser is shown in Fig. 6.9 (32). Since the laser was equipped with a 80-lines/mm grating used in the Littrow arrangement, the angle change to go from 10.513 to 10.532 μm is about 8×10^{-4} rad. This implies a 113-μm linear movement for the mount with an axis 13.5 cm long. Commercial piezoelectric pushers conveniently provide a 40-μm movement. This is actually sufficient if one can sacrifice a little power so that the modulation is not from center to center on the gain curves. The modulation frequency is in the 100-Hz range or lower. The benefits of wavelength modulation can be fully utilized only if the amplitude of the laser is not modulated simultaneously. A sensitive way to obtain a stable intensity during modulation is to introduce a photoacoustic gas cell filled with a nonabsorbing gas such as N_2. Since window absorption is spectrally broad, the acoustic signal produced in this reference cell is only due to amplitude modulation. Minimizing this acoustic signal by aligning the grating assures that

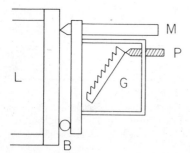

Fig. 6.9. Modifications of CO_2 gas laser for wavelength modulation. L, laser cavity; B, ball bearing pivot; M, coarse tuning micrometer; P, piezoelectric pusher for modulation; G, diffraction grating.

amplitude modulation is minimized, regardless of the laser line pair used (32). With about 59% wavelength modulation, we have been able to improve the detectability of ethylene in nitrogen by more than two orders of magnitude compared to that for 100% amplitude modulation.

Stark and Zeeman effects can be used for modulation, either based on the laser medium or the species of interest. To modulate the laser, the gas medium must have large Stark or Zeeman constants to allow the use of a moderate field over the entire length of the laser tube. The Stark effect is generally not very useful because the gas discharge allows only limited perturbations in the electric field. Zeeman modulation has been demonstrated (33), but is not generally applicable to all IR gas lasers. A Stark cell for modulating the absorption line of the species of interest consists of two highly polished, optically flat stainless steel electrodes, 40 cm × 5 cm × 2.5 cm each, separated by 1-mm quartz spacers (34). The sample pressure is in the few millitorr range to minimize the linewidth, and to be able to avoid discharge at fields as high as 60 kV/cm. In operation, a dc high-voltage bias is used at one electrode to scan across the absorption line for excitation at a fixed CO_2 laser line. At the same time, ac modulation is produced by a high-voltage operational amplifier to the other electrode, and a lock-in amplifier is used to record the signal. The spectrum thus appears as the first derivative of the absorption profile. Detectabilities of vinyl chloride and acrylonitrile are found to be in the sub-parts-per-million range (34). The sample can also be modulated using the Zeeman effect. This is accomplished by placing the sample cell inside a solenoid (35) that is driven by a modulated current. The sensitivity there was increased so that even absorption from excited vibrational levels can be monitored.

Another concept for modulation is based on the velocity of the species (36). For ions in an electric field, there is a drift velocity that is comparable to normal gas velocities. If the electric field is reversed on alternate half-cycles, the velocities of these ions will be modulated at the same rate. The corresponding blue and red Doppler shift of the ion absorption relative to a laser beam along the electric field will cause a modulation in the transmitted intensity, which can be detected with good sensitivity with a lock-in amplifier. The neutral species in the cell, which are normally at much higher concentrations, will not be modulated, and can be selectively rejected. A typical arrangement for velocity modulation is shown in Fig. 6.10. The chief limitation of this technique is that it is only applicable to charged species.

3.4. Wavelength Calibration

To take advantage of the very narrow linewidths of the IR lasers, it is necessary to provide wavelength calibration to the same spectral resolution. This is particularly important when the tuning curves are not linear, such as in the diode

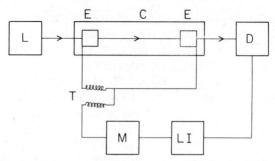

Fig. 6.10. Velocity modulation spectroscopy of ions. L, tunable laser; E, electrode; C, discharge tube; D, photodetector; T, high-voltage transformer; M, modulation driver; LI, lock-in amplifier.

lasers. The most reliable is naturally a direct frequency measurement that can be traced to the frequency of the Cs atomic clock (36). Briefly, this is based on wave-mixing techniques using nonlinear material, so that the high harmonics of a lower-frequency standard can be mixed with a higher-frequency standard. The beat frequency of the mixing can be measured accurately by the appropriate counters to determine the actual frequency of the higher-frequency standard. For example, the twelfth harmonic of the HCN laser at 337 μm can be mixed with the H_2O laser at 28 μm to determine the frequency of the latter. This way, spectral regions adjacent to discrete gas laser lines, that is, within the bandwidth of the heterodyne detector (see Section 2), can all be calibrated relative to the gas laser lines. Absolute wavelength calibration can also be provided by a wavelength meter (37). This is essentially a traveling interferometer that is used to compare the number of interference fringes produced by the laser and that produced by a standard laser (e.g., a single-mode HeNe laser) over the length of travel of one of the mirrors. A 0.5-m interferometer results in 0.001-cm^{-1} accuracy in the 10-μm region.

Usually, a relative frequency calibration is satisfactory, whenever there are well-known absorption lines from other gaseous species in the same spectral region. The relative tuning rate, for example, of a diode laser can be monitored with an etalon, such as a 2.5-cm solid Ge etalon or a 33-cm air-spaced etalon (38). Relative line positions can be established to within 2×10^{-4} cm^{-1}. In fact, commercial diode laser systems are normally equipped with these etalons to deal with the nonlinear tuning rates of diodes. Another relative calibration scheme is based on the well-defined Doppler profiles of gases at low pressures (39). The familiar expression for the Doppler width (FWHM) is

$$\Delta\omega_D = 7.162 \times 10^{-7} \omega_0 \sqrt{T/M} \qquad (6.1)$$

where ω_0 is the frequency at line center, T is the absolute temperature, and M

is the molecular weight. Even at cell pressures of a tenth of a torr, self-broadening cannot be neglected. However, it is possible to fit the observed spectral shape with a Voigt profile to uniquely determine the Doppler contribution specified by Eq. (6.1). Since all parameters in Eq. (6.1) can be determined to an accuracy of at least 1 part in 10^4, the scan rate of the laser at the absorption line can be determined to 1×10^{-4} the Doppler width, or better than 10^{-6} cm^{-1}, which is the width of the laser emission. Since many gaseous absorption lines are available in any given spectral region, tuning-rate calibration can be performed at each line. Some polynomial fit can then be used to interpolate in between lines. Using this procedure (39), we have determined several line positions of SO_2 to ± 3 MHz, and have measured the self-broadening coefficient of NH_3 per torr to ± 0.4 MHz.

3.5. Interferometry

When absorption occurs in the infrared region, essentially all absorbed energy eventually becomes heat. This induces a temperature change in the gas, and thus a change in the refractive index in the sample. Interferometry is perhaps the most sensitive way to detect small changes in refractive index (RI), and can therefore be used to monitor IR absorption indirectly. The advantage is that one is measuring a small signal from an essentially null background (when no absorption occurs), as opposed to the conventional transmission measurement, which depends on the small difference between two large signals.

One arrangement of indirect absorption measurement is based on the Mach–Zender interferometer (40). The optical scheme is shown in Fig. 6.11. The sample cell is placed in one of the two arms of the interferometer. A pulsed CO_2 laser passes into the cell through one of the cavity mirrors and is absorbed by the species of interest. The interferometer utilizes a single-frequency HeNe laser, and is servo-stabilized by a piezoelectric transducer on one of its mirrors. Without absorption, the interferometer is set to the point of the interference fringe that has the maximum slope, that is, midway between constructive and

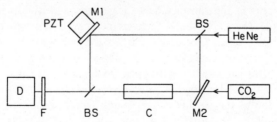

Fig. 6.11. Mach–Zender interferometer for absorption measurements. HeNe, single-mode laser; CO_2, modulated laser; M1, M2 mirrors for interferometer; C, gas cell; PZT, piezoelectric pusher for servo control; F, filter; D, photodetector.

destructive interference. When absorption-induced heating occurs, the refractive index of the sample changes, causing a shift in the interference pattern and thus a change in the transmitted intensity. A detectability of 10 ppb of SF_6 in air was reported (40). The pulsed measurement suffers from radio-frequency interference and does not allow narrow-band detection of the signal. An improved version is based on a continuous-wave CO_2 laser (41). Modulation is provided by either a mechanical chopper or an ac Stark field on the species of interest. To further isolate the HeNe laser and the CO_2 laser, a six-mirror arrangement is used to allow the use of Ge flats to couple the CO_2 laser. The entire interferometer is put in an acoustic enclosure, since pressure waves also change the refractive index. Using a 20-cm cell, absorptions as low as 10^{-8} cm^{-1} can be measured. Similar to optoacoustic spectroscopy (Chapter 5), window absorption also causes heating and thus a background signal. So, the incorporation of Stark modulation was necessary to allow the detection of 5 ppb NH_3 in air. A further advantage of Stark modulation is that absorption from interfering species can be suppressed if their Stark coefficients are sufficiently small compared to the species of interest. When the cell is replaced by a flow tube from a gas chromatograph, this can serve as an absorption detector (42). A detectability of 0.08 ppb of injected SF_6, or a total injected quantity of 3 pg (S/N = 3), was feasible. The sensitivity of this apparatus also allows it to be used to measure small absorptions in aerosols in the submicron size range (43). Sulfate levels as low as 100 $\mu g/m^3$ can be detected. The limit there was due to background absorption from water vapor.

The sensitivity of interferometry in detecting refractive index changes (and hence absorptions) is dependent on the slope on the interference fringe, that is, the intensity change for a given small change in mode position. So, one can expect the sensitivity of Fabry–Perot interferometers to be superior to two-beam interferometers such as the Mach–Zender arrangement. The main criterion is the finesse of the system, which is two for two-beam arrangements but can be the order of 200 for multiple-beam arrangements. We have previously demonstrated a refractive index detectability of 4×10^{-9} units (S/N = 3) and an absorbance detectability of 2.6×10^{-6} units (S/N = 3) (45) for a liquid inside the interferometer (44). The arrangement for gas-phase measurements is shown in Fig. 6.12 (46). The Fabry–Perot interferometer is 15 cm long with mirrors coated for the HeNe laser. One of the mirrors is controlled by a piezoelectric driver from a voltage provided by a high-voltage operational amplifier. Light reaching the detector is compared with a reference level, normally equal to one-half the intensity at total constructive interference, so that the interferometer can be servo-controlled to stay at this optical length with a response time of about 1 s. A wavelength-modulated CO_2 cw laser (see Section 3.3.) is coupled into the inteferometer cavity by a sodium chloride window at Brewster's angle rel-

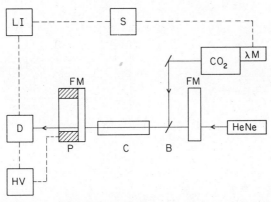

Fig. 6.12. Fabry–Perot interferometer for absorption measurements. FM, interferometer mirror; C, gas cell; B, beam splitter; P, piezoelectric driver; HeNe, single-mode laser; CO_2, wavelength-modulated laser; λM, modulator as in Fig. 6.9; D, photodetector; HV, high-voltage operational amplifier; S, signal generator; LI, lock-in amplifier.

ative to the HeNe laser, so that about 10% of the 2-W CO_2 laser light is used. The signal modulated synchronously with the laser grating is detected by a lock-in amplifier. A parallel experiment using amplitude modulation on the laser showed no advantage over normal absorption measurements because window absorption leads to a dominating background signal. Using wavelength modulation, we have been able to detect ethylene at a concentration of 0.1 ppm flowing in the gas cell at 200-mW laser power. This is a better detectability compared to the Mach–Zender arrangement for the same laser power, confirming that the Fabry–Perot arrangement is preferable. In a study using visible laser absorption (47) similar detectabilities are found. Since interfering absorption in the visible region is much less serious, amplitude modulation is sufficient for those measurements.

A different form of interferometry is based on the Doppler effect (48). The optical scheme is shown in Fig. 6.13. A HeNe laser produces two Zeeman-split frequencies f_1 and f_2 at 632.8 nm, approximately separated by 1.8 MHz. A small amount of this output is sent to detector D1 so that the difference frequency can be used as a local oscillator for the mixer. The major part of the beam is sent through two paths by a dichroic beam splitter, each with a corner cube retro-reflector. The beams are then combined onto detector D2. When a CO_2 laser irradiates the sample, heating produces an optical path change, which in turn results in a Doppler frequency shift for f_2 in the instrument. D2 thus produces a frequency at $f_1 - f_2 \pm df$. In the mixer, the value of df can be determined with good precision. For low absorbance measurements, this has not yet been shown to be superior to the other interferometric techniques.

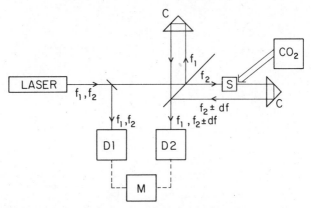

Fig. 6.13. Doppler interferometer for absorption measurements. CO_2 cw laser; C, corner-cube reflectors; S, sample; D1, D2, difference frequency detectors; M, mixer.

3.6. Thermal Lensing and Photodeflection

While interferometry probes the bulk refractive index change via absorption-induced heating, thermal lensing and photodeflection probe the refractive index gradient that is produced. The axially symmetric refractive index gradient is essentially a lens that diverges the laser beam. A change in intensity is then observed if the detector at far field is limited by an aperture. This is the basis for thermal lens spectroscopy, which is discussed in more detail in Chapter 13. An application to the measurement of IR absorption in solutions has been demonstrated using the 3.39-μm line of the HeNe laser (49). Using about 6 mW of laser light, a detectability of 5×10^{-4} absorbance units was found. The detectability is expected to improve with laser power, to the limit of absorption in the solvent. However, unlike visible lasers, other IR lasers normally do not exhibit good spatial qualities (e.g., a truly Gaussian beam), and the interpretation of the thermal lensing signal is much more difficult.

Photodeflection spectroscopy is based on probing the RI gradient at the edge of the absorption region. The entire probe beam, which is much smaller than the excitation beam, is deflected to a different part of a position-sensitive detector (50). The experimental arrangement is shown in Fig. 6.14. The advantage of this over thermal lensing is that the excitation beam need not possess such good beam quality. The sensitivity is in principle also higher because the region of maximum change is monitored. Using a CO_2 laser at 1 W and chopped at 16 Hz, the deflection of a HeNe laser at 0.5 mW is sufficient to detect the order of 5 ppb of ethylene in air (10^{-7} cm^{-1} absorbance). The limiting factors are the beam pointing stability of the laser and the intensity noise in the laser. The RI gradient can be probed either collinearly as in Fig. 6.14, or transversely.

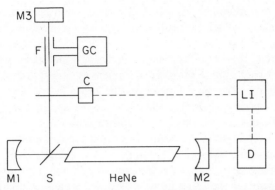

Fig. 6.14. Experimental scheme for intracavity absorption based on mode competition. M1, M2, cavity mirrors for HeNe laser; S, beam splitter; C, mechanical chopper; F, flow tube for gas sample; GC, gas chromatograph; M3, extra feedback mirror; and LI, lock-in amplifier.

The former allows a longer interaction length (thus increased sensitivity), but the latter allows better spatial resolution and easier separation of the pump and probe beams. The details of the signal dependence on the experimental conditions have been discussed previously (51).

3.7. IR Optogalvanic Spectroscopy

The principles of optogalvanic spectroscopy are discussed in more detail in Chapter 3. Briefly, the absorption of radiation by molecules in the IR region can also enhance the formation of ions in a discharge. If a tunable IR laser is used to scan across the absorption line, Doppler-limited spectral resolution can be obtained by monitoring the impedance of the discharge. A successful application has been reported for low-pressure samples of NH_3 and NO_2 (52). About 1 mW of radiation from diode lasers is sufficient for spectroscopy in a cell 5 cm long and at 0.5 torr of gas. Both positive and negative impedance changes are observed. Although the mechanism is still not well understood, such optogalvanic signals can be conveniently used to lock the frequencies of tunable IR lasers or for frequency calibration. Sensitivity is limited by the stability of the discharge, so that the ultimate analytical utility may not be as good as optogalvanic spectroscopy in the visible or UV regions.

3.8. Intracavity Absorption

Inside the laser cavity, there are two mechanisms for enhancing the detectability for absorption. The laser cavity allows multiple reflection of the light beam, hence resulting in a multipass geometry. The absorption leads to a decrease in the gain of the laser, which normally translates to exponential changes in the output intensity. An extreme case is to operate the laser right at its threshold. Any absorption can then stop laser action entirely, which is essentially an absorption detector with infinite sensitivity. Intracavity absorption is discussed in detail in Chapter 12. The main problem is the lack of control over the lasing process and thus poor quantitative accuracy, unless special arrangements are provided to make the gain constant in the laser cavity (53).

In the infrared region, one can make use of some special conditions for mode competition to indirectly determine absorption (54). A HeNe laser can be made to operate simultaneously at two wavelengths, 0.633 and 3.39 μm. These two transitions compete for the same group of excited states in the electrical discharge. Now if one of these wavelengths falls in the absorption band of a molecular species inside the laser cavity, the particular transition will show less gain. The competition for excited molecules is then decreased, and the other transition can then achieve a higher power level. So, absorption at one of the HeNe laser wavelengths is reflected in the increase in output intensity at the other wavelength. A working system is shown in Fig. 6.14 (54). About 1.5 mW of output is obtained at each wavelength without any absorbing species. When 10 μL of methane is injected into the gas chromatograph, the absorption produced at the 3.39-μm laser line is enhanced by a factor of 1.2 for the 5-cm flow cell inside the cavity compared to the absorption outside the cavity. In contrast, the change in the laser intensity at 0.633 μm due to the presence of the eluted methane is a factor of 46.7. This can be explained since a small change in the upper state population induced by absorption at the high-gain infrared line results in a large change in the output of the low-gain visible line. The linear dynamic range of this detector is about 8 orders of magnitude in concentration, with a detectability of 9×10^{-11} g/cm^3. The fixed wavelength of 3.39 μm limits the utility to the various hydrocarbons, however.

4. SUMMARY

We are obviously just beginning to take advantage of the many special properties of the laser for IR absorption spectroscopy. The most obvious applications in chemical analysis are those associated with gas lasers since these are very rugged, reliable instruments. However, semiconductor devices and F-center lasers have been gaining in spectral coverage and in output power. Their wavelength tunability is a necessity in many cases. Technological advances in these laser systems will be important to their success as routine analytical tools.

ACKNOWLEDGMENTS

The author thanks the many co-workers in his laboratory that have contributed to the projects described in this work, particularly R. N. Morris, T. Y. Chang, and B. C. Yip. This work is supported by the U.S. Department of Energy, Office of Basic Energy Sciences, Division of Chemical Sciences, through contract No. W-7405-eng-82.

REFERENCES

1. C. F. Dewey, Jr., and L. O. Hocker, *Appl. Phys. Lett.*, **18**, 58 (1971).
2. A. S. Pine, *J. Opt. Soc. Am.*, **64**, 1683 (1974).
3. E. D. Hinkley and C. Freed, *Phys. Rev. Lett.*, **23**, 277 (1969).
4. J. F. Butler, A. R. Calawa, R. J. Phelan, Jr., A. J. Strauss, and R. H. Rediker, *Solid State Comm.*, **2**, 301 (1964).
5. H. Preier, *Appl. Phys.*, **20**, 189 (1979).
6. R. W. Ralston, I. Melngailis, A. R. Calawa, and W. T. Lindley, *IEEE J. Quant. Electron.*, **QE-9**, 350 (1979).
7. J. M. Besson, W. Paul, and A. R. Calawa, *Phys. Rev.*, **173**, 699 (1968).
8. F. Allario, C. H. Bair, and J. F. Butler, *IEEE J. Quant. Electron.*, **QE-11**, 205 (1975).
9. E. D. Hinkley, A. R. Calawa, P. C. Kelley, and S. A. Clough, *J. Appl. Phys.*, **43**, 3222 (1972).
10. C. K. N. Patel and E. D. Shaw, *Phys. Rev. Lett.*, **24**, 451 (1970).
11. A. Mooradian, S. R. J. Brueck, and F. A. Blum, *Appl. Phys. Lett.*, **17**, 481 (1970).
12. M. H. Mozolowski, R. B. Dennis, and H. A. Mackenzie, *Appl. Phys.*, **19**, 205 (1979).
13. D. E. Evans, S. L. Prunty, and M. C. Sexton, *Infrared Phys.*, **20**, 21 (1980).
14. P. K. Cheo and R. Wagner, *IEEE J. Quant. Electron.*, **QE-13**, 159 (1977).
15. L. F. Mollenauer and D. H. Olson, *Appl. Phys. Lett.* **24**, 386 (1974).
16. W. Hartig and W. Schmidt, *Appl. Phys.*, **18**, 235 (1979).
17. R. S. Eng and A. W. Mantz, *J. Mol. Spectrosc.*, **74**, 388 (1979).
18. H. J. Gerritsen, *Phys. of Quant. Electron.*, P. L. Kelley et al., Eds., McGraw-Hill, New York, 1966.
19. K. C. Kim, E. Griggs, and W. B. Person, *Appl. Opt.*, **17**, 2511 (1978).
20. J. Reid, J. Shewchun, B. K. Garside, and E. A. Ballik, *Appl. Opt.*, **17**, 300 (1978).
21. C. K. N. Patel, E. G. Burkhardt, and C. A. Lambert, *Science*, **184**, 1173 (1974).
22. B. M. Golden and E. S. Yeung, *Anal. Chem.*, **47**, 2132 (1975).
23. W. Baumer, K. W. Rothe, and H. Walther, *Eur. Spectrosc. News.*, **22**, 41 (1979).
24. R. A. Baumgartner and R. L. Byer, *Appl. Opt.*, **17**, 3555 (1979).
25. R. T. Menzies, *Laser Monitoring of the Atmosphere*, E. D. Hinkley, Ed., Springer-Verlag, New York, 1976, Chapter 7.

26. R. T. Ku and D. L. Spears, *Optics Lett.*, **1,** 84 (1977).
27. R. T. Menzies, *Geophys. Research Lett.*, **6,** 151 (1979).
28. M. J. Mumma, T. Kostiuk, and D. Buhl, *Opt. Eng.*, **1,** 50 (1978).
29. M. M. Abbas, G. L. Shapiro, F. Allario, and J. M. Alvarez, *Laser Spectroscopy for Sensitive Detection*, J. A. Gelbwachs, Ed., *SPIE Proceedings*, **286,** 73 (1981).
30. J. A. Mucha, *Appl. Spectrosc.*, **36,** 393 (1982).
31. G. V. H. Wilson, *J. Appl. Phys.*, **34,** 3276 (1963).
32. B. C. Yip and E. S. Yeung, *Anal. Chem.*, **55,** 978 (1983).
33. M. Takami and K. Shimoda, *Japan J. Appl. Phys.*, **11,** 1648 (1972).
34. D. M. Sweger and J. C. Travis, *Appl. Spectrosc.*, **33,** 46 (1979).
35. C. K. N. Patel, R. J. Kerl and E. G. Burkhardt, *Phys. Rev. Lett.*, **38,** 1024 (1977).
36. K. M. Evenson, D. A. Jennings, F. R. Peterson, and J. S. Wells, *Laser Spectrosc. III*, J. L. Hall and J. L. Carlsten, Eds., Springer-Verlag, New York, 1977, p. 56.
37. K. Nagai, K. Kawaguchi, C. Yamoda, K. Hagakawa, Y. Takagi, and E. Hirota, *J. Mol. Spectrosc.*, **84,** 197 (1980).
38. S. P. Reddy, W. Ivanic, V. Malathy Devi, A. Baldacci, K. N. Rao, A. W. Mantz, and R. E. Eng, *Appl. Opt.*, **18,** 1350 (1979).
39. T. Y. Chang, R. N. Morris, and E. S. Yeung, *Appl. Spectrosc.*, **35,** 587 (1981).
40. C. C. Davis, *Appl. Phys. Lett.*, **36,** 515 (1980).
41. A. J. Campillo, H. B. Lin, C. J. Dodge, and C. C. Davis, *Optics Lett.*, **5,** 424 (1980).
42. H. B. Lin, J. S. Gaffney, and A. J. Campillo, *J. Chromatogr.*, **206,** 205 (1981).
43. A. J. Campillo and H. B. Lin, *Laser Spectrosc. for Sensitive Detection*, J. A. Gelwachs, Ed., *Proceedings SPIE*, **286,** 24 (1981).
44. S. D. Woodruff and E. S. Yeung, *Anal. Chem.*, **54,** 2124 (1982).
45. S. D. Woodruff and E. S. Yeung, *Anal. Chem.*, **54,** 1174 (1982).
46. B. C. Yip and E. S. Yeung, *Anal. Chim. Acta,* **169,** 385 (1985).
47. A. J. Campillo, S. J. Petuchowski, C. C. Davis, and H. B. Lin, *Appl. Phys. Lett.*, **41,** 327 (1982).
48. L. H. Skolnik, A. Hordvik, and A. Kahan, *Appl. Phys. Lett.*, **23,** 477 (1973).
49. C. A. Carter, J. M. Brady, and J. M. Harris, *Appl. Spectrosc.*, **36,** 309 (1982).
50. D. Fournier, A. C. Boccara, N. M. Amer, and R. Gerlach, *Appl. Phys. Lett.*, **37,** 519 (1980).
51. W. B. Jackson, N. M. Amer, A. C. Boccara, and D. Fournier, *Appl. Opt.*, **20,** 1333 (1981).
52. C. R. Webster and R. T. Menzies, *J. Chem. Phys.*, **78,** 2121 (1983).
53. J. S. Shirk, T. D. Harris, and J. W. Mitchell, *Anal. Chem.*, **52,** 1701 (1980).
54. J. D. Paril, D. W. Paul, and R. B. Green, *Anal. Chem.*, **54,** 1969 (1982).

PART

III

METHODS WITH IMPROVED SPECTRAL RESOLUTION

CHAPTER

7

CRYOGENIC MOLECULAR
FLUORESCENCE SPECTROMETRY

E. L. WEHRY

Department of Chemistry
University of Tennessee
Knoxville, Tennessee

1. INTRODUCTION

The use of lasers as excitation sources in molecular fluorescence and phosphorescence measurements continues to expand, as the advantages of laser excitation become more evident and commercial tunable dye laser systems appear that exhibit the wavelength tuning range, amplitude stability, and reliability required

211

for analytical applications. Molecular luminescence in cryogenic media is a technique especially suited to laser excitation because of dramatic reductions in the bandwidths of molecular electronic absorption spectra often observed under low-temperature measurement conditions.

The principal motivation for performing analytical luminescence measurements in cryogenic media is *enhanced selectivity*. Under the proper conditions, molecular electronic absorption spectral features in low-temperature samples may exhibit bandwidths of 1 nm or less (perhaps much less), as compared with bandwidths of 20 nm or more typically observed for molecular luminophores in room temperature liquid or gaseous samples. If the extent of absorption band narrowing for both the analyte and potential interferents in a complex sample is sufficiently large, selective excitation of fluorescence (or phosphorescence) from the analyte can be achieved, free from interference by luminescence from other sample constituents. It is then possible to identify and quantify the analyte without prior separation from other luminescent sample constituents. Numerous applications, some of them quite spectacular, of such "selective excitation" procedures have been reported; selected examples are described later in this chapter. The high spectral purity of radiation obtainable from lasers is an important element in the success of such analytical procedures.

A closely related motivation for using cryogenic media in fluorometric analysis is the possibility of *improved limits of detection*, especially for very complex samples. Most molecular luminescence measurements are "blank limited," meaning that the limit of detection for a particular analyte ultimately depends on the ability of the instrumentation to distinguish photons emitted by the analyte from photons produced by other (undesired) processes. The most important "blank" signals in fluorescence spectrometry are Rayleigh and Raman scattering; reflections from optical surfaces (e.g., cell windows); and luminescence of the sample cell, the solvent, or other luminescent constituents of the sample itself (1, 2). Often the latter of these contributions is most significant, especially for very complex samples. Although significant discrimination of these undesired signals (particularly those arising from scattering or reflection) from the analyte fluorescence signal often can be accomplished by time resolution (3–6), resolution in the spectral domain often is required to reduce sufficiently the blank arising from interfering luminescence.

Consequently, for analytical situations in which background luminescence is important, improvements in spectral selectivity may translate directly into improved limits of detection. In some cases, this statement will hold true even if the technique used to improve the selectivity of the measurement adversely affects its sensitivity.[†] In any fluorescence measurement, one detects only a small

---

[†]Here we mean "calibration sensitivity" as defined by Ingle (7): the slope of the analytical calibration curve (assuming a linear calibration curve, as is generally the case in low-temperature fluorescence spectrometry).

fraction of the photons emitted by the analyte. A procedure that reduces this already small fraction may actually pay off in an improved limit of detection for the analyte if that procedure improves the extent to which undesired optical signals are discriminated against.

In many cryogenic media, quenching of luminescence of the analyte by other sample constituents is diminished or totally suppressed. This is particularly important whenever the emitting excited state is long lived. Classically, low-temperature matrices were viewed as essential for the observation of phosphorescence from organic molecules, due to the relatively long mean lifetimes of emitting triplet states that render them very susceptible to quenching (8). The universal use of cryogenic media for phosphorescence measurements has ceased as alternative procedures using solid surfaces (9, 10) or surfactant media (11) have been demonstrated capable of detecting phosphorescence from many molecules at room temperature. However, low-temperature sample preparation techniques remain useful for phosphorescence measurement whenever high spectral selectivity is desired.

Because the use of low-temperature techniques results in high spectral resolution and freedom from quenching, limits of detection for specific analytes tend to be much less strongly dependent on the nature of the sample than is the case when conventional solution fluorometry is used for their determination. These factors also combine to extend the linear dynamic range and to cause it to be relatively invariant (for any particular analyte) irrespective of the nature of the sample in which that analyte occurs. Low-temperature fluorescence signals linear in analyte concentration over 6–8 decades are not uncommon.

The principal disadvantages of cryogenic luminescence spectrometry arise from the need to use special sample containers. The nature of this apparatus depends on the specific type of cryogenic medium to be used, as discussed in subsequent sections. Although the cost of such apparatus is not negligible, it is usually much less than the cost of a laser. Likewise, in a laser cryogenic fluorometer, the laser (and not the cryogenic apparatus) is likely to be the least reliable component. Both laser and cryogenic technologies are improving steadily, and the experimental difficulties attributed to laser low-temperature luminescence measurements often are severely exaggerated.

There are two fundamentally different approaches to the preparation of samples for low-temperature fluorometric analysis. In *solid-state* cryogenic spectrometry, the sample molecules are embedded in a rigid solid, often at a temperature of 20 K or less. The various solid-state techniques (Section 2) differ from each other in the type of solid matrix and the manner in which the guest (sample) molecules are incorporated into the host (solid matrix). A fundamentally different approach to cryogenic spectrometry is use of a *supersonic expansion*, wherein a gas-phase sample (usually diluted by a noble gas) is expanded through an orifice into a vacuum. The spectral behavior of the "cooled" molecules in the gaseous expansion is then interrogated (Section 3).

2. SOLID-STATE TECHNIQUES

2.1. General Considerations

Solid-state techniques for low-temperature spectroscopy have been studied thoroughly and have been used for analytical purposes for nearly 30 years. The two general techniques presently in widespread use for preparing low-temperature solids for spectroscopic examination are:

1. *Frozen-solution* methods, in which a liquid solution of the sample constituents in a suitable liquid solvent is frozen into an optically transparent solid (12).

2. *Matrix isolation* methods, in which the sample constituents (if liquid or solid at room temperature) are vaporized, diluted with a large excess of a diluent ("matrix") gas, and deposited on a cold surface as a solid (13).

A thorough review of experimental techniques in solid-state low-temperature fluorometry has appeared (14).

It is important to realize that the viscosity of low-temperature solids is much greater than that of fluid media; thus, diffusional quenching phenomena are much less probable in a cryogenic solid than in a liquid solution. Some low-temperature solid matrices are sufficiently rigid that the "solvent cage" surrounding a luminescent guest molecule is virtually fixed on the fluorescence time scale. The solvent cage relaxation that occurs after electronic excitation (but prior to emission) of a solute molecule in liquid solution therefore is largely "frozen out" in many cryogenic solid matrices (15). In such cases, the "Franck–Condon" and "relaxed" excited states of the guest molecule (16) are characterized by similar "solvent" environments. The averaging of molecular microenvironments over time that occurs in fluid media often is largely suppressed in low-temperature solid hosts. Thus, the immediate environment "seen" by a particular guest molecule in a low-temperature host may remain virtually unchanged throughout a fairly lengthy spectroscopic experiment.[†] In such cases, different molecules of a particular analyte that happen to be situated in different environments within the host may be spectroscopically distinguishable; their absorption and emission spectra may differ substantially, and in some cases even their fluorescence decay times are significantly different. This fact can represent either a nuisance or a boon, depending on the specific circumstances of a particular experiment, and it is a fact that we shall encounter several times in subsequent discussion.

[†]Important exceptions to this statement are discussed in Section 2.3.4 ("photophysical hole burning").

Irrespective of whether frozen-solution or matrix isolation techniques are used to prepare the sample, the basic criteria for choice of host are the anticipated spectral bandwidths of the analyte and potential interferents in that matrix. Incorporation of a compound in a cryogenic host does *not* automatically bring about a dramatic decrease in its absorption or emission spectral bandwidth. The spectral bandwidth observed for a particular compound in a specific low-temperature matrix usually is dominated by the extent of *inhomogeneous broadening* in that host. Inhomogeneous broadening arises from the fact that different molecules of a particular compound, if located in slightly different stable environments, absorb and emit at different energies. The extent of those differences depends on the environmental differences perceived by the different molecules, and upon the sensitivity of the spectra to those differences.

Real fluorescence measurements presently involve large ensembles of molecules, though that situation is subject to change (17). The observed spectrum of an ensemble of molecules affected by inhomogeneous broadening is significantly broader than the spectrum that would be observed for any single molecule of that compound (14, 18). Thus, to achieve high selectivity in low-temperature solid-state fluorometry (which is the basic purpose of the experiment), we must minimize the extent to which inhomogeneous broadening occurs. If we cannot satisfactorily decrease the extent of inhomogeneous broadening, we must find some way to ameliorate its deleterious effects on spectral selectivity. Either of the basic approaches hinted at in the preceding two sentences can be implemented in low-temperature solids.

Conceptually, the simpler approach is to minimize inhomogeneous broadening. The most obvious requirement for so doing is to choose a host in which all, or at least most, molecules of the analyte are constrained to occupy virtually identical environments, or "sites." From this standpoint, the ideal host is a perfect crystal into which the guest can be incorporated substitutionally without appreciable distortion of the host crystal lattice. Crystalline materials have indeed been used with considerable success in the laser-induced luminescence determination of certain metal ions (19).

Since the perfect crystal approach is seldom feasible, especially for organic analytes, we will usually be faced with the fact that different guest molecules may occupy different sites in the host. We are then very interested in ensuring that such microenvironmental heterogeneity exerts the smallest possible spectroscopic cost (i.e., spectral bandwidth). Thus, we usually wish to choose a host that interacts very weakly with both ground and excited electronic states of the guest, so that the sensitivity of the absorption and emission spectra of the guest to molecular environment is minimal. Unfortunately, this means that the optimal "spectroscopic" host for a particular guest often is a very poor "chemical" host for that guest. This fact can have important ramifications in frozen-solution spectroscopy, because it means that the guest may not be soluble in the best spectroscopic liquid solvent host materials.

2.2. Shpol'skii Spectrometry

In organic luminescence analysis, the choice of host to minimize inhomogeneous broadening is best exemplified by the "Shpol'skii effect" (20), in which the host is usually an *n*-alkane and the guest either a polycyclic aromatic hydrocarbon or a porphyrin species. Cryogenic samples of guest molecules in Shpol'skii hosts can be prepared either by frozen-solution (12) or matrix isolation (21) techniques; their relative merits are considered later. In a "classical" Shpol'skii experiment, the guest is a planar aromatic hydrocarbon having one molecular dimension that is rather closely matched to the length of the carbon chain of a host *n*-alkane (22–24).

2.2.1. Fundamental Principles

In the most favorable cases, spectacular spectral resolution is observed for luminescent molecules in Shpol'skii matrices. See Fig. 7.1 for a particularly favorable example (the fluorescence spectrum of the polycyclic aromatic hydrocarbon benzo[*a*]pyrene in solid *n*-octane). Such spectra, which can be produced by both frozen-solution and matrix isolation sample preparation methods, often

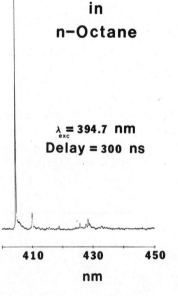

BaP

in

n–Octane

$\lambda_{exc} = 394.7$ nm

Delay = 300 ns

410   430   450

nm

Fig. 7.1. Fluorescence spectrum of benzo[*a*]pyrene matrix isolated in *n*-octane at 15 K. From Conrad et al. (63); reproduced by permission of ASTM.

are referred to in the literature as "quasilinear" emission spectra. What is particularly important from the analytical standpoint is that similarly high resolution is observed in the *absorption* spectra of suitable guest molecules in Shpol'skii hosts. Thus, selective excitation (see Section 1) is readily achieved if a narrow-bandwidth source (laser) is used.

The spectrum shown in Fig. 7.1 is somewhat atypical for Shpol'skii fluorometry. A much more common situation is that depicted in Fig. 7.2 (the fluorescence of benzo[*a*]pyrene in solid *n*-heptane), wherein both the absorption and emission spectra of the guest consist of several closely spaced sharp multiplets. Although it can be something of a nuisance in quantitative Shpol'skii fluorometry, the multiplet structure often observed in Shpol'skii fluorescence spectra is extremely useful for identification of individual constituents of complex samples. Clearly, the fluorescence spectrum of benzo[*a*]pyrene shown in Fig. 7.2 is much more of a "fingerprint" for the compound than that in Fig. 7.1. It is rather surprising (but highly fortunate) that the relative intensities of the individual multiplet components in a spectrum such as that shown in Fig. 7.2 can be made highly reproducible in both frozen-solution (12) and matrix isolation (25) Shpol'skii fluorometry. Moreover, it is possible to excite selec-

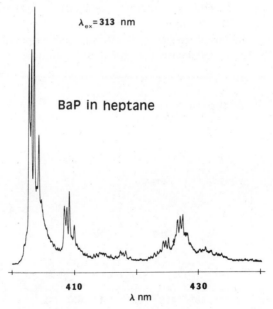

Fig. 7.2. Fluorescence spectrum of benzo[*a*]pyrene matrix isolated in *n*-heptane at 15 K. Note the multiplet structure associated with each feature, which is generally attributed to four distinct trapping sites in the Shpol'skii matrix. From Conrad et al. (63); reproduced by permission of ASTM.

tively fluorescence of each of the individual multiplets (compare Figs. 7.2 and 7.3).

Highly resolved phosphorescence spectra also can be observed for solute molecules in Shpol'skii matrices via excitation in the singlet excited-state manifold of the guest, and laser excitation of selected individual phosphorescence multiplets can be achieved by proper choice of excitation wavelength in the singlet–singlet absorption spectral region (26). In Shpol'skii fluorescence spectra exhibiting multiplet structure, the measured fluorescence decay times (27, 28) and degrees of fluorescence polarization (29) for the various multiplet components may be significantly different.

The fact that any particular guest exhibits quasilinear spectral behavior in only a restricted (usually very small) number of solid hosts implies that steric compatibility of host and guest molecules is a crucial requirement for the effect to be observed (14, 24). Frozen n-alkanes are polycrystalline solids in which the orientation of the individual alkane molecules is as depicted in Fig. 7.4. The guest is thought to be incorporated into the host by substitutional displacement of several alkane molecules, as also shown in Fig. 7.4. It is generally presumed that multiplets (e.g., Fig. 7.2) correspond to individual ensembles of guest molecules situated in different types of discrete sites within the host, with each ensemble being perturbed by a slightly different crystal field (30–34). In some cases, multiplet spectral structure may result from insertion of guest molecules subjected to different ground- and/or excited-state structural distortions into geometrically similar lattice sites (35–37). Still another possibility (for conformationally flexible molecules) is that different conformers of the guest may

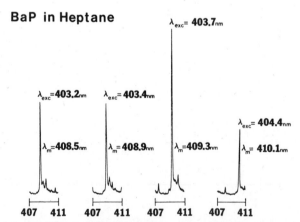

Fig. 7.3. Selective excitation of each of the multiplets in the fluorescence spectrum shown in Fig. 7.2. The excitation and emission wavelength for optimal site-selective excitation of each multiplet is shown. From Conrad et al. (63); reproduced by permission of ASTM.

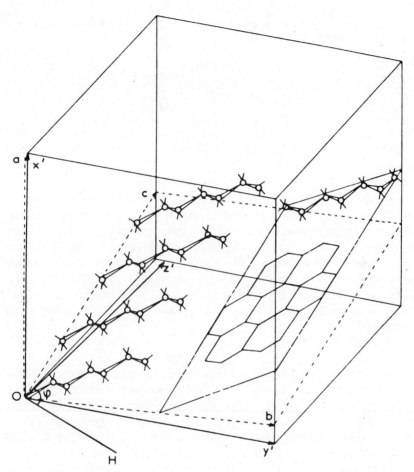

Fig. 7.4. Representation of the orientation of a guest molecule (coronene) in a single-crystal Shpol'skii matrix (*n*-heptane). Note the replacement of three heptane molecules by one coronene molecule. From Merle et al. (30); reprinted by permission of the North-Holland Publishing Company.

be trapped preferentially in different types of "sites" (27). Comparing Figs. 7.1 and 7.2, we would regard heptane as a "four-site," and octane as a "one-site," matrix for benzo[*a*]pyrene.

In view of the foregoing picture, it is not surprising that the Shpol'skii effect is much more readily observed for flat, planar, conformationally rigid molecules than for molecules having bulky substituent groups and/or great conformational flexibility (38). Thus, the effect is not universal; some fluorescent analytes of interest do not exhibit the effect. However, with ingenuity (and perhaps a bit of luck), Shpol'skii spectra can be observed in some seemingly unlikely cases. For example, 3-methyllumiflavin

is not an especially good candidate for Shpol'skii fluorometry; the molecule is not flat and it possesses numerous substituents including several polar groups that render it virtually insoluble in the classical (alkane) Shpol'skii solvents. However, the molecule can be derivatized by incorporation of a long alkyl chain at position *10* ($C_{11}H_{23}$); when that is done, the derivatized compound exhibits adequate solubility in *n*-decane, and surprisingly sharp excitation and emission spectra are observed *if* site-selective laser excitation is used (Fig. 7.5) (39). Whether or not the "derivatization" approach to Shpol'skii luminescence spectrometry is anaytically practical is not presently clear, but the idea is worth consideration.

The manner in which guest molecules are incorporated into Shpol'skii hosts can be complex (14, 24, 28, 32, 34, 40–42). The simplest situation arises when molecules of guest and host are virtually identical in size and shape, such as benzene in a cyclohexane matrix (43). Then, isomorphous substitution of one guest for one host molecule can occur without appreciable disruption of the host lattice, and exceedingly sharp spectra for the guest can be obtained. More commonly, one dimension (not necessarily the longest) of the guest is virtually equal to the longest dimension of the host, but the remaining dimension(s) of the guest are much too large to permit one-for-one isomorphous substitution (24). In the most favorable cases, the relative dimensions of the guest and host are such that one guest can be accommodated by displacement of several host molecules, in such a way that the host lattice is not severely disrupted and the guest can be accommodated in substitutional host lattice vacancies in a small number of well-defined types of orientations or sites. If the latter condition is not met, extremely complex multiplet patterns are obtained in both absorption and emission spectra. Examples in which the behavior outlined above appears to be substantiated include triphenylene (31) and coronene (30) (Fig. 7.4) in heptane hosts, wherein the guests displace two and three host molecules, respectively, and free base porphin in octane (44), in which each guest replaces two adjacent octane molecules.

In Shpol'skii systems, the quasilines may be accompanied by a broad background (usually of low intensity), attributable to guest aggregates or "preag-

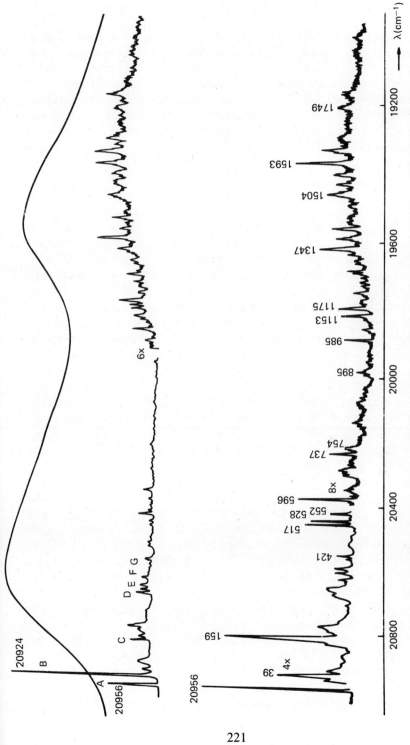

Fig. 7.5. Fluorescence spectra of N_3-undecyllumiflavin in *n*-decane. Top: room temperature solution. Center: single-crystal frozen decane matrix at 4.2 K using broadband excitation; the various incorporation site multiplets are indicated by letters. Bottom: selective excitation of site A at 4.2 K using laser radiation at 21,110 cm^{-1}. From Platenkamp et al. (39); reproduced by permission of the North-Holland Publishing Company.

221

gregates'' (40). The optimal conditions for maximizing the fraction of guest molecules exhibiting quasilinear spectral behavior are often not identical with those that give rise to the smallest spectral bandwidths in the quasilinear portion of the guest emission spectrum (42). Although our understanding of the fundamentals of Shpol'skii spectrometry is increasing, there are still characteristics of the phenomenon that are not fully understood (42), and there remains a certain amount of art in Shpol'skii fluorometry irrespective of whether frozen-solution or matrix isolation techniques are used.

2.2.2. Experimental Techniques

The Shpol'skii effect was first reported over 30 years ago, and its analytical potential became evident immediately. Traditionally, continuum sources (usually xenon or mercury–xenon lamps) have been used to excite fluorescence, and the frozen-solution sample preparation methods of classical low-temperature phosphorometry have been employed (8). Liquid nitrogen Dewars usually were used, producing a minimum sample temperature of 77 K. Both spectral bandwidths and relative intensities of multiplets in Shpol'skii fluorescence spectra are sensitive to the rate at which the initial liquid solution sample is cooled (45–48). Thus, when a liquid nitrogen Dewar is used, care must be devoted to controlling the cooling rate. Several detailed descriptions of the design and use of lamp source, nitrogen Dewar instruments for Shpol'skii fluorometry have appeared (45, 49, 50).

Recent modifications in instrumentation for Shpol'skii frozen-solution fluorometry have stressed the use of closed-cycle refrigerators, rather than liquid nitrogen Dewars. A closed-cycle refrigerator (51) offers two major advantages over a nitrogen Dewar. First, a closed-cycle refrigerator achieves a minimum temperature of 10–15 K; the lower temperature may bring about improved spectral resolution. Second, and more important, the rate of cooling of a sample can be controlled precisely by use of a closed-cycle refrigerator, and the temperature of the sample can be varied from the low-temperature limit of the cryostat to the melting point of the Shpol'skii host. Shpol'skii emission spectra often are temperature sensitive. By making fluorescence measurements at several different temperatures, additional information useful for ''fingerprinting'' specific sample constituents can be acquired (52). This is particularly useful when the excitation is broadband and selective excitation of individual fluorophores is not possible. Techniques for use of closed-cycle cryostats in frozen-solution Shpol'skii luminescence spectrometry have been described (12, 53–55).

Two disadvantages of closed-cycle refrigerators also should be noted. First, the rate of cooling may be lower in a closed-cycle cryostat than in a nitrogen Dewar. Aside from whatever effects this may have on the quality of Shpol'skii luminescence spectra (46–48), it adds to the total time required for an analysis. Second, a closed-cycle cryostat is more expensive than a nitrogen Dewar.

While the option of using a nitrogen Dewar or a closed-cycle refrigerator is available in frozen-solution Shpol'skii luminescence, sample preparation by matrix isolation (MI) requires use of a closed-cycle cryostat. Although MI is conventionally regarded as entailing use of matrix materials that are gases at room temperature (56, 57), virtually all Shpol'skii solvents exhibit sufficient vapor pressures to be used as matrix "gases" in MI. An apparatus for MI consists of a Knudsen cell for vaporizing the sample constituents, a vacuum line and delivery system for the matrix gas, an area in which the sample and matrix vapors are thoroughly mixed, and a cold substrate for deposition of the diluted sample as a solid. The latter two components usually are located inside the cold head of a closed-cycle cryostat. Sapphire and gold-plated copper are the most commonly used deposition surfaces. Figure 7.6 is a schematic diagram of a typical apparatus for use in MI Shpol'skii fluorometry. Designs of Knudsen cells for vaporization of samples are somewhat dependent on the nature of the samples to be analyzed; detailed design discussions have been presented elsewhere (58–60).

A solid deposit prepared by MI deposition of a vaporized sample diluted by a vapor-phase Shpol'skii "solvent" does not exhibit quasilinear fluorescence spectra unless the matrix is "annealed" (heated) for a brief period following

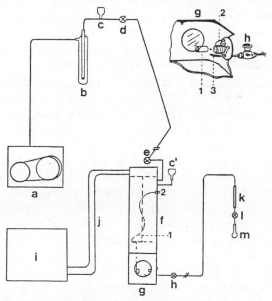

Fig. 7.6. Apparatus for preparing samples for matrix isolation spectroscopy. Components: a, vacuum pump; b, trap; c, pressure gauge; d and e, valves; f, shroud of closed-cycle refrigerator; g, cryostat head (cutaway view at upper right: 1, sample cell; 2, Knudsen oven to volatilize sample constituents; 3, line for matrix gas); h, valve; i, cryostat compressor; k, flowmeter; l, valve; m, reservoir for liquid matrix (e.g., Shpol'skii solvent). From Perry (60).

deposition (21). For example, in a typical MI Shpol'skii fluorometric procedure, the sample constituents (diluted with n-octane vapor) are deposited at the minimum operating temperature of a closed-cycle refrigerator (15 K). After deposition has concluded, the deposit is annealed at 155 K for 5 min and then cycled back to 15 K for measurement of fluorescence (61). In the absence of annealing, badly broadened excitation and emission spectra are obtained, presumably because the initial deposit is amorphous. The annealing temperature must be sufficiently high to permit reorientation of the alkane chains comprising the host, but not sufficiently high to allow diffusion of guest molecules and resulting aggregation phenomena (21). Fortunately, for the n-alkanes most widely used as Shpol'skii hosts, the available temperature range for annealing is quite broad. Relative multiplet intensities (such as those in Fig. 7.2) are sensitive to the annealing temperature, so the temperature (but not the duration) of annealing should be reproduced carefully from sample to sample. The precision with which the temperature of a deposit can be controlled when a commercial closed-cycle cryostat is utilized is extremely high. Thus, multiplet intensity patterns can be reproduced without difficulty in MI Shpol'skii fluorometry.

The experimental procedures of matrix isolation spectroscopy have been described in great detail in the books by Hallam (56) and Meyer (57).

Two principal advantages are offered by use of lasers as excitation sources in Shpol'skii fluorometry. First, the high monochromaticity and high optical power per unit wavelength interval of a laser is of immense importance if selective excitation of individual analytes in mixtures is to be achieved. This point has already been stated and need not be further belabored.

A second key advantage of laser excitation is the possibility of achieving time-resolved measurement of fluorescence under analytically realistic conditions. Occasionally, time resolution can be used to distinguish fluorophores whose absorption and fluorescence spectra overlap, if their fluorescence decay times differ sufficiently (62). More frequently, the principal use of time resolution in Shpol'skii fluorometry is to reduce background contributions arising from scattering, reflections, and broad background luminescence. It must be recalled that not all fluorescent compounds exhibit the Shpol'skii effect in any single host. Thus, even under laser excitation, the Shpol'skii fluorescence spectrum of an analyte in a very complex sample may be submerged in broad (non-Shpol'skii) fluorescence of other sample constituents. Fortunately, such broad background fluorescence very often exhibits a shorter decay time than that of the Shpol'skii guest species of interest, and thus can be partially or completely time resolved from the spectrum of the analyte (63). Of course, background contributions arising from Rayleigh or Raman scattering or reflections from window or cell surfaces can be virtually eliminated by proper use of time resolution (1, 64). Eliminating these background contributions contributes to improved detection limits even though many time-resolved fluorescence measurement schemes are inherently insensitive (because they sample only a small portion of the fluorescence decay curve of the analyte and thus reject a large fraction of the potentially available signal).

The most important characteristic that a laser must exhibit to be useful for Shpol'skii fluorometry is a broad tuning range; it is also very helpful if the reproducibility of wavelength tuning is high. Dye lasers pumped by nitrogen, excimer, or Nd:YAG lasers all are suitable. Though the requirements for output laser bandwidth are not particularly stringent (intracavity etalons are rarely required), significant levels of amplified spontaneous emission in the laser output can wreak havoc whenever a selective excitation experiment is performed. Thus, the level of stray dye fluorescence in the laser output is a more important specification than the bandwidth per se. In order to exploit the advantages of time resolution, it is desired that the laser exhibit reasonably short pulse durations (preferably no greater than 2 ns).

The Shpol'skii effect was discovered long before dye lasers became practical as laboratory sources, and many impressive applications [reviewed elsewhere (14, 65)] of lamp-induced Shpol'skii fluorescence have been reported. The advent of laser instrumentation does not automatically relegate lamp excitation to the sidelines (50). Indeed, even if the ultimate objective of an analysis is selective excitation by a laser, a preliminary survey of a Shpol'skii sample using broadband illumination can be extremely useful. The optimum wavelength for excitation and measurement of fluorescence of an analyte in a Shpol'skii host may depend on the nature of spectral interferences that are likely to be encountered, which can vary dramatically from one sample to another (61, 66). The disadvantage of lasers as sources (50) include cost, complexity of instrumentation for processing signals produced by pulsed lasers, low-duty factors (ca. 10^{-7}) for many pulsed lasers, radiofrequency interference produced by pulsed lasers (67), and the current inaccessibility of fluorometrically important wavelengths (240–320 nm) without recourse to second-harmonic generation or other nonlinear processes. Also, synchronous fluorometry (68), which can enhance further the already extremely high selectivity of Shpol'skii spectrometry (69), is more easily performed with lamp instrumentation.

2.2.3. Analytical Applications

To date, virtually all reported applications of *laser-excited* Shpol'skii fluorometry to chemical analysis have emanated from Fassel, D'Silva, and co-workers (in frozen-solution media) and Wehry, Mamantov, and co-workers (using matrix isolation). The work of both groups has been reviewed in considerable detail elsewhere (13, 14, 24, 63, 70), and only a brief selection of the most important results will be considered here.

Applications of Shpol'skii fluorometry to energy-related samples have been particularly prominent, in part because polycyclic aromatic hydrocarbons (which are particularly good Shpol'skii guests) are ubiquitous constituents of fossil-energy-derived materials. Many of these samples are exceedingly complex and represent extraordinarily challenging tests of any analytical procedure. A good example is a shale oil issued as a Standard Reference Material (SRM-1580) by the National Bureau of Standards (71), which contains many polycyclic aromatic hydrocarbons (PAHs) at 3–400 ppm levels in the presence of substantial excesses of phenols and nitrogen heterocycles. The ability of Shpol'skii laser fluorometry to deal with samples of this complexity is exemplified by Fig. 7.7–7.9, showing, respectively, selectively excited fluorescence spectra of perylene (Fig. 7.7) and benzo[a]pyrene (Fig. 7.8) in frozen octane matrices (72) and benzo[a]pyrene in vapor-deposited heptane and octane matrices (60) (Fig. 7.9; compare with Fig. 7.1 and 7.2). A particularly noteworthy aspect of Fig. 7.7 and 7.8 is the fact that the fluorescence spectra of the perdeutero analogs of the PAHs are sufficiently different from those of the parent compounds to be resolved easily. Thus, the perdeutero analog can be used as an internal standard for quantification of the corresponding PAH (72, 73). In each of these studies (60, 72), the shale oil

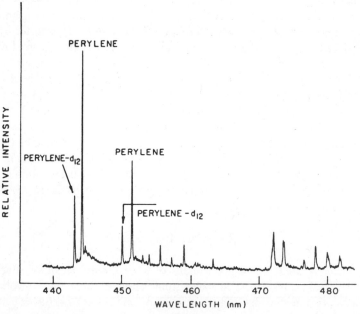

Fig. 7.7. Laser-induced fluorescence spectrum of perylene in a Wilmington crude oil sample with 160 ppb perdeuteroperylene added as an internal standard. The spectrum was obtained in a n-octane frozen-solution matrix at 15 K. Note the shift of the spectrum produced by deuteration. From Yang, D'Silva, and Fassel (72); reproduced by permission of the American Chemical Society.

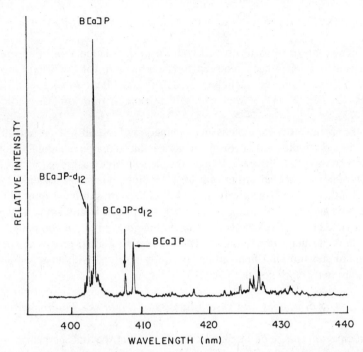

Fig. 7.8. Laser-induced fluorescence spectrum of benzo[*a*]pyrene in a shale oil sample with 10 ppb perdeuterobenzo[*a*]pyrene as internal standard, in a *n*-octane frozen-solution matrix at 15 K. From Yang, D'Silva, and Fassel (72); reproduced by permission of the American Chemical Society.

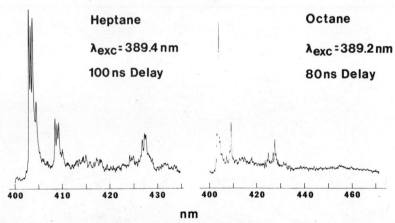

Fig. 7.9. Laser-induced fluorescence spectra of benzo[*a*]pyrene in a solid solvent-refined coal sample in vapor-deposited heptane (left) and octane (right) matrices at 15 K. Compare with the "pure compound" spectra in Figs. 7.1 and 7.2. From Perry, Wehry, and Mamantov (61); reproduced by permission of the American Chemical Society.

227

sample was analyzed "as received" with no prior fractionation or separation. Similar selectivity has been reported for determination of the PAHs in other liquid fossil-fuel samples, including crude oil and liquid solvent-refined coal samples (12, 24, 72, 74), as well as in solvent extracts from airborne particulate matter (73).

Another noteworthy example is an examination of a solid solvent-refined coal (again with no sample pretreatement) by laser-induced matrix isolation Shpol'skii fluorescence (60, 61). Figure 7.10 exhibits the selectivity achieved in this analysis. The fluorescence spectrum of pure 7,10-dimethylbenzo[a]pyrene is compared with that excited selectively in the PAHs vacuum sublimed from the solid coal-derived sample. This sample contained the parent hydrocarbon as well as other alkylated derivatives thereof. As noted previously in pure-compound studies (12, 66), isomeric derivatives of PAHs containing alkyl groups at different positions of substitution often can be spectrally resolved without separation by laser fluorometry in Shpol'skii matrices.

2.2.4. Advantages and Disadvantages of Shpol'skii Fluorometry

The principal *advantage* of the Shpol'skii effect is that spectral sharpening occurs both in absorption and emission. Thus, line narrowing and selective excitation

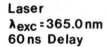

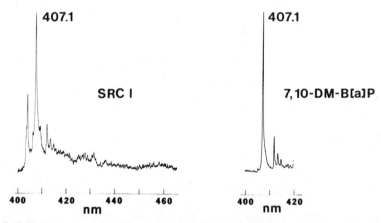

Fig. 7.10 Fluorescence spectra of pure 7,10-dimethylbenzo[a]pyrene (right) and the same compound in a solid solvent-refined coal sample (left), both obtained by matrix isolation in octane at 15 K. From Perry, Wehry, and Mamantov (61); reproduced by permission of the American Chemical Society.

in fluorescence is achieved without an automatic sacrifice in sensitivity (compare with energy selection as described in Section 2.3). Although complexities are introduced by the multiplet structure seen in many Shpol'skii spectra, these usually do not cause significant analytical difficulty if sufficient care is exercised in Shpol'skii sample preparation. Thus, at present the Shpol'skii effect is usually regarded as the high-resolution fluorometric technique of choice for any compound for which Shpol'skii spectra can be observed without undue difficulty.

The most obvious disadvantage of the Shpol'skii effect is its lack of universality; some compounds of great analytical importance do not exhibit the effect. In addition to studies of PAHs and porphyrin derivatives, which represent the vast majority of reported Shpol'skii fluorescence observations, Shpol'skii luminescence spectra also have been obtained for such other classes of compounds as nitrogen (26, 75–77) and sulfur (78) heterocycles, halogenated PAHs (79), and aromatic aldehydes (80) and ketones (81–84). Although many of the highly substituted or polar compounds produce very complex site multiplet patterns, it is possible to simplify the spectra by site-selective excitation (26, 79, 82). It also is worth recalling the possible derivatization of fluorescent analytes to render them sterically compatible with Shpol'skii hosts (39). Likewise, though n-alkanes have been used almost universally as Shpol'skii hosts, there are other possibilities [e.g., tetrahydrofuran (76) and nitrobenzene (27)]. There is considerable room for expansion of analytical applicability of the Shpol'skii effect beyond the determination of polycyclic aromatic hydrocarbons.

The fact that a limited number of solvents can serve as Shpol'skii hosts for any particular analyte is a source of difficulty in frozen-solution spectroscopy if the analyte is insoluable in those solvents. Such problems sometimes can be circumvented by adding a co-solute to the solvent which solubilizes the analyte. For example n-alkanes are Shpol'skii hosts for many phthalocyanines, but the solubility of such compounds in alkanes usually is extremely low. However, phthalocyanines are reasonably soluble in alkane–1-chloronaphthalene mixtures (85). Whether this approach can have analytical applicability is presently unclear. This problem, of course, does not arise in matrix isolation Shpol'skii fluorometry, wherein the guests are "dissolved" in the Shpol'skii solvent in the vapor phase.

Guest molecules in Shpol'skii matrices may undergo photochemical decomposition especially under intense laser illumination. It has been our experience that PAHs are considerably more susceptible to photolysis in Shpol'skii hosts than in rare-gas or nitrogen matrices (25).[†] There is evidence for chemical participation of the alkane matrix in these phenomena (86, 87). Although the quantum yields tend to be very small (86, 87), appreciable photolysis can be

---

[†]These photochemical processes are to be distinguished from "photochemical hole-burning" phenomena (Section 2.3.4), which can be reversible.

observed upon relatively brief laser excitation (25). Photolysis of the analyte obviously is undesirable in an analytical fluorescence measurement. On the brighter side, the use of the Shpol'skii phenomenon as a diagnostic tool in mechanistic photochemistry has interesting possibilities. Photolysis of molecules incorporated in Shpol'skii matrices has been observed to afford the opportunity to obtain electronic spectra of unstable photochemical intermediates, such as radicals, under conditions of very high spectral resolution (88, 89).

2.2.5. Organic Fluorometry in Crystalline Matrices

All applications of Shpol'skii fluorometry that we have described thus far use host materials that are liquids at room temperature. The Shpol'skii effect acts to diminish the extent of spectral inhomogeneous broadening for the guest by incorporating the guest in a limited number of discrete sites in a polycrystalline solid. An extension of this principle to the use of hosts that are crystalline solids at room temperature recently has been introduced by Thornberg and Maple (90). For example, methylnaphthalenes do not exhibit analytically useful spectral line narrowing in any conventional Shpol'skii matrix. However, they can be incorporated into the crystal lattice of durene (1,2,4,5-tetramethylbenzene; melting point 82°C) in such a way that "Shpol'skii quality" spectra can be obtained. For example, Fig. 7.11 compares the laser-induced fluorescence spectrum of pure 2,3-dimethylnaphthalene in durene at 10 K with that observed for the pure compound in an unfractionated shale oil (NBS SRM-1580; cf. Section 2.2.3). Analogous selective excitation experiments for alkylnaphthalenes are impossible in classical Shpol'skii matrices.

Since the host is a solid at room temperature, conventional frozen-solution sample preparation techniques are inapplicable. Two general approaches have been used (91). In the first, a mixture of the pure (solid) host and sample (which may be liquid or solid) is placed in a glass capillary tube which is then inserted into a zone-melting oven at sufficiently high temperature to melt the matrix. Crystals of the host, within which are incorporated sample molecules, are then formed from the melt by slowly removing the capillary from the oven ("normal freezing"). An alternative method is to vaporize both the sample constituents and the matrix and then form the crystalline matrix by vapor deposition (i.e., matrix isolation).

The selectivity that can be achieved by this technique is particularly noteworthy for derivatives (particular those having polar substituents) of aromatic hydrocarbons, many of which exhibit very broad absorption and luminescence spectra in classical Shpol'skii hosts. The range of host materials available for use is enormous, and it is possible to "tailor" a host to be suitable for a particular analyte of interest. For example, substituted derivatives of a PAH can be incorporated into the parent hydrocarbon used as host (91). The idea is that de-

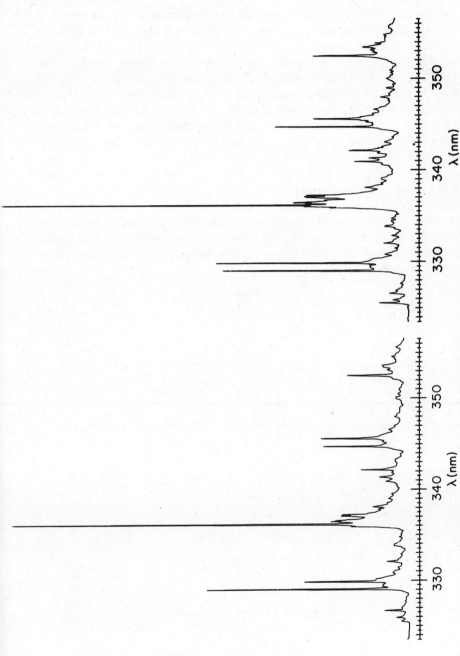

Fig. 7.11. Fluorescence spectrum of pure 2,3-dimethylnaphthalene (left) and the same compound in a shale oil sample (right), both in durene crystal matrices at 10 K. The excitation wavelength was 321.1 nm in both cases. From Maple (91).

231

rivatives of a PAH often will incorporate into the crystal lattice of that PAH in such a way that the extent of spectral inhomogeneous broadening will be small. This approach also exploits the fact that the lowest singlet–singlet absorption and fluorescence transition is usually at lower energy for a PAH derivative than for the parent. Thus, the derivatives can be excited without absorption by the host (or guest–host energy transfer) representing a major interference. An example of the spectral resolution that can be achieved in this way is shown in Fig. 7.12 (a fluorescence spectrum of 2-chloronaphthalene in a naphthalene matrix). If, for a particular analyte, the parent PAH is not a satisfactory host, other derivatives of that PAH are likely to be suitable. In this way, it is possible in principle to observe quasilinear absorption and luminescence spectra for a wide variety of aromatic compounds, though obviously different hosts will be optimal for different analytes.

The possibilities of this crystalline matrix approach have barely begun to be explored. It can be regarded as a logical extension both of the Shpol'skii effect and the crystalline–host techniques for inorganic analytes developed by Wright and co-workers (19).

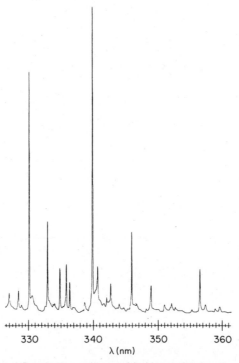

Fig. 7.12. Laser-induced fluorescence spectrum of 2-chloronaphthalene in crystalline naphthalene host at 10 K. The excitation wavelength was 324.6 nm. From Maple (91).

2.3. Transition Energy Selection Spectrometry

2.3.1. Principles

All techniques described in Section 2.2 seek to reduce inhomogeneous broadening contributions to a luminescence spectrum by doing so in the absorption spectrum. The principal disadvantages of this approach are that different host matrices inevitably must be used for different guests (unless the guests are structurally similar), and that some analytes will not exhibit fluorescence (or absorption) line narrowing in any reasonable host. It would be useful to have an alternative solid-state high-resolution fluorometric technique not dependent on host–guest steric similarity.

To achieve this goal, we must accept the existence of considerable inhomogeneous broadening in the absorption spectrum of the analyte, yet alleviate its effects on the fluorescence spectrum. The basic technique for doing so appears in the literature under various names, including "optical site selection," "fluorescence line narrowing," and "transition energy selection." The last of these names is most accurate and descriptive of the way in which the molecules are selected for observation, and it therefore will be used in this chapter.

In transition energy selection (or "energy selection" for short), an excitation source is used that has a spectral bandwidth much smaller than the width of the inhomogeneously broadened absorption spectrum of the analyte. Under certain specific circumstances, described below, the resulting fluorescence spectrum can be much sharper than the absorption spectrum. In the most favorable case, the bandwidth of the fluorescence spectrum will equal that of the incident exciting light. An example of an energy selection spectrum (compared with that obtained in the same system by broadband excitation) is shown in Fig. 7.13.

It is immediately apparent that the energy selection technique interrogates only a fraction (perhaps a very small fraction) of the molecules of a particular analyte in a particular solid host. Thus, it is reasonable to expect energy selection spectrometry to be less sensitive[†] than the Shpol'skii effect (which can interrogate all, or at least a large fraction, of the guest molecules in a sample). This does not mean that its limits of detection automatically are catastrophically inferior to those of Shpol'skii fluorometry. The blank-limited nature of fluorescence measurements and the intrinsically high resolution of energy selection techniques also must be considered. Nevertheless, the possibility that an energy selection technique for a particular analyte may exhibit an undesirably high limit of detection cannot be ignored.

In order for an energy selection experiment actually to achieve results such as those shown in Fig. 7.13, several important requirements must be satisfied

[†]"Calibration sensitivity" as defined by Ingle (7) as the slope of the analytical calibration curve.

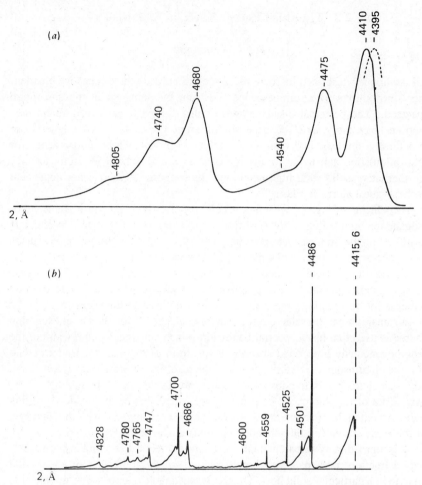

Fig. 7.13 Comparison of broadband-excited (top) and energy selection (bottom) fluorescence spectra of perylene in ethanol frozen-solution matrices at 4.2 K. For the energy selection spectrum, the excitation wavelength was 443.0 nm and the excitation bandwidth was 0.0125 nm. From Personov, Al'shits, and Bykovskaya (140); reproduced by permission of the North-Holland Publishing Company.

[see Kohler (18), Wehry and Mamantov (14), and Personov (92) for greater detail]:

1. For the experiment to work, those analyte molecules having the same absorption frequency (i.e., those molecules "selected" by the incident laser beam) must exhibit virtually the same emission frequency. This is usually observed to be true only if the excited electronic state populated by absorption is

the state whose radiative decay is to be detected (93). Thus, for example, phosphorescence excited via energy selection usually exhibits the desired degree of line narrowing only if the emitting triplet state is populated by direct absorption from the ground singlet state (93–100), which is impractical for analytical purposes due to the tiny absorption coefficients for singlet-to-triplet transitions. Presumably if the "selected" analyte molecules occupied nearly identical sites, singlet–singlet excitation would produce line narrowing in phosphorescence, as well as fluorescence, spectra. Since this is not generally the case, *energy selection* (93) appears to be more accurate than *site selection* as a name for the technique.

2. For the selected molecules to exhibit fluorescence at virtually the same wavelength, the molecular environments they "see" must remain virtually unaltered during the lifetime of the excited state. Narrow-line excitation of a fluorophore in liquid solution seldom achieves detectable fluorescence band narrowing, due to the solvent cage reorientation that occurs after absorption but prior to fluorescence. Thus, energy selection experiments must be performed in rigid solids at low temperature.

3. Excitation either into the 0–0 band or a resolved vibronic band of the S_0–S_1^* absorption transition is required. Otherwise, molecules with widely varying emission energies will be selected and the experiment will fail. For example, if one molecule of the analyte absorbs into the 0–1 band while another absorbs into the 0–0 band at the same wavelength, their fluorescence wavelengths are bound to differ because of the vibrational relaxation that will occur for the molecule promoted into a vibrationally excited level of S_1^*. Moreover, if a vibrationally excited S_1^* molecule is formed by absorption, it is conceivable that energy released to the matrix via vibrational relaxation of the excited guest molecule could cause changes in the environment of the guest molecule, violating rule 2 (*above*) (94). Hence, optimal energy selection generally occurs only if excitation is within ca. 1500 cm^{-1} of the 0–0 absorption band (94, 101). For some fluorescent molecules, the 0–0 component of $S_0 \rightarrow S_1^*$ absorption is relatively weak, exacerbating potential limitations in sensitivity, and excitation into a vibronic band then is unavoidable.

4. One wishes to obtain fluorescence spectra free, or virtually free, of phonon sidebands. Here a complication (that we have previously ignored) in solid-state spectroscopy must be confronted. Any electronic transition of a guest molecule in a solid host may be accompanied by changes in the vibrational state of the host lattice. Quanta of lattice vibrational energy are called "phonons." The host therefore provides a large number of additional vibrational degrees of freedom coupled to the electronic transitions of the guest. If this "electron–phonon coupling" is weak, its manifestations are virtually absent in the electronic absorption and emission spectra of the guest. This is the usual situation in Shpol'skii matrices; thus, we were able to avoid dealing with the matter in Section 2.2.

If, on the other hand, the extent of electron–phonon coupling is fairly great, phonon sidebands are observed both in the emission and absorption spectra of the guest, as shown in Fig. 7.14. The phonon sidebands typically are much broader than the "zero-phonon" line of the guest, which is what we are interested in observing in the energy selection fluorescence spectrum of the guest. In exteme cases, the phonon sidebands dominate the spectra to the extent that the zero-phonon lines are submerged in them and are invisible in both absorption and emission. If this is the case, then the line-narrowing objective of an energy selection experiment cannot be expected to succeed.

To minimize phonon sideband contributions, it is necessary to choose a host whose lattice vibrational modes do not couple strongly with the electronic transitions of the guest. Thus, energy selection spectrometry does not allow arbitrary choices of host–guest combinations, although the restrictions usually are less severe than the molecular geometric constraints that must be satisfied in Shpol'skii fluorometry. Unfortunately, hosts that satisfy this requirement are likely to be substances that do not interact strongly with the ground electronic state of the guest (102, 103); thus, a good host for energy selection fluorometry of a particular guest is likely to be a poor chemical solvent for that guest. Energy selection is most easily achieved for guest molecules that are relatively rigid and do not undergo drastic changes in shape upon excitation (100, 102); such molecules usually exhibit relatively small Stokes shifts in room temperature

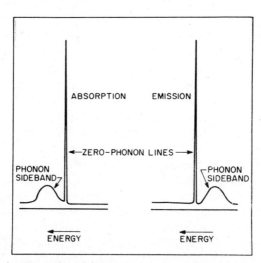

Fig. 7.14. Idealized representation of the appearance of zero-phonon lines and phonon sidebands in the absorption (left) and emission (right) spectra of a guest molecule in a low-temperature solid host. Note that the phonon sideband occurs at higher energy than the zero-phonon line in absorption, but at lower energy in emission.

spectra. Molecules of a particular guest situated in recognizably different sites in the host may exhibit different degrees of electron–phonon coupling (104).

The fraction of integrated intensity of the 0–0 absorption band of a particular guest that appears in the zero-phonon line (rather than the phonon sideband) is called the "Debye–Waller factor." Its magnitude usually decreases rapidly with increasing temperatures (18, 93). Thus, observation of a relatively intense emission zero-phonon line in the absence of a major phonon sideband contribution may be possible only at very low temperatures. In some cases, temperatures achievable by closed-cycle refrigerators (Section 2.2.2) may not be low enough, and it then becomes necessary to resort to liquid helium cryostats.

The presence of phonon sidebands in molecular absorption spectra measured in solid matrices reinforces the constraints already mentioned upon choice of excitation wavelength for energy selection fluorometry. As indicated in Fig. 7.14, the phonon sideband is located at higher energy (lower wavelength) than the zero-phonon line in absorption. Thus, as the excitation energy is increased within the 0–0 absorption band of the guest, the probability increases that some of the molecules selected for absorption will absorb in the phonon sideband, rather than the zero-phonon line. Because the phonon sideband is much broader than the zero-phonon line, excitation into the phonon sideband partially vitiates the energy selectivity of the experiment; not all molecules excited at a particular frequency into their phonon sidebands exhibit emission zero-phonon lines at the same frequency. Thus, excitation as close to the low-frequency onset of absorption as feasible is preferred to achieve maximal line narrowing. Of course, this problem is minimized by choosing a host (and a temperature) in which the absorption spectrum of the guest is relatively free of phonon sidebands (i.e., for which the Debye–Waller factor is large).

5. Finally, the guest molecules observed in emission should be the same ones that were selected in the initial absorption process. Thus, electronic energy transfer from the selected molecules to others of the same guest should not occur with appreciable efficiency. When guest–guest energy transfer occurs, the observed fluorescence bandwidth increases as the time interval between pulsed excitation and fluorescence detection in a time-resolved fluorescence experiment is increased (105). Fortunately, energy transfer usually does not pose a significant problem if the guest is diluted sufficiently by the host and aggregation or microcrystallite formation is avoided.

The above conditions may appear rather restrictive, and it seems clear that there are fluorescent molecules for which energy selection luminescence (at sufficiently high resolution to be analytically useful) cannot be observed in any reasonable host. However, the effect has been observed for many molecules in glassy frozen-solution hosts; included among these are a number of very polar (97, 98, 102, 103, 106–113) and even ionic (114) compounds. Energy selection also has been observed for guest molecules in thin polymer films (114, 115) and in vapor-deposited (matrix isolated) samples using rare-gas and fluorocarbon matrices (63, 109, 116–119).

In glassy frozen solutions, absorption spectra of guest molecules usually are broad and featureless. The guest molecules can be thought of as experiencing a continuum of microenvironments; there are no discrete types of sites. However, vapor-deposited rare-gas matrices (Ne, Ar, Kr, Xe) exhibit significant crystallinity (57) and absorption or broadband excited fluorescence spectra of guests in such matrices often exhibit well-defined fine structure usually attributed to distinct ensembles of guest molecules occupying different, rather well-defined, lattice sites (118, 119). For example, the broadband excited fluorescence of dibenz[a, j] acridine (Fig. 7.15, left) exhibits well-defined multiplet structure; each of the fluorescence multiplets can be excited selectively by use of the appropriate excitation wavelength (e.g., Fig. 7.5, right). These matrices therefore may be thought of as intermediate between Shpol'skii or organic crystal hosts at the one extreme and vitreous frozen solutions at the other extreme. The presence of "site structure" in the inhomogeneously broadened absorption spectrum of a guest compound in a particular host is a virtual guarantee that line narrowing for the guest can be achieved by site-selective excitation in the host. Whether the "site distributions" for the guest in that host are sufficiently reproducible from sample to sample for adequate quantitative precision to be

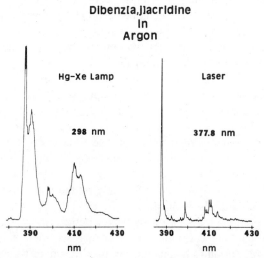

Fig. 7.15. Broadband (left) and site selection (right) fluorescence spectra of dibenz[a, j]acridine matrix isolated in argon at 15 K. Note the site structure in the lamp-excited spectrum, including the closely spaced doublet at ca. 390 nm. In the "site selection" spectrum, the excitation wavelength was 377.8 nm and the higher-wavelength feature of the aforementioned doublet is selectively excited. From Maple and Wehry (116); reproduced by permission of the American Chemical Society.

achieved may be doubtful; some evidence bearing on this point is discussed in Section 2.3.3.

It is obvious from the nature of the phenomenon that energy selection fluorescence spectra are exceedingly sensitive to changes in excitation wavelength. In glassy hosts, the positions of the fluorescence spectral maxima shift in the same direction (and by the same amount in cm^{-1}) as the frequency of the exciting light is changed (see Fig. 7.16 for an example), and this fact can be used as an empirical test for achievement of energy selection for a guest in a glassy host (102). That the emission spectral peaks shift with changing excitation frequency means that wavelength (or more properly wave number) modulation can be used to distinguish energy selection fluorescence of an analyte from broad interfering background fluorescence, which does not shift as the excitation wave

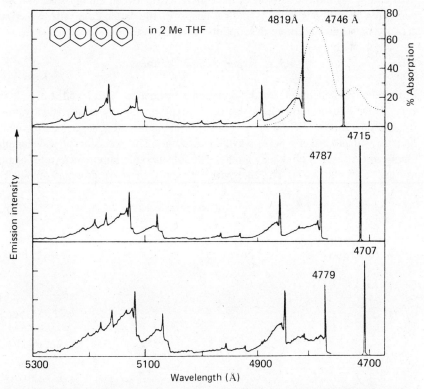

Fig. 7.16. Energy selection fluorescence spectra of tetracene in a 2-methyltetrahydrofuran glassy frozen solution at 4.2 K as a function of excitation wavelength. The excitation wavelengths are: 4746 Å (top), 4715 Å (center), and 4707 Å (bottom); note the corresponding shifts in the fluorescence spectra. From McColgin, Marchetti, and Eberly (102); reproduced by permission of the American Chemical Society.

number is modulated (120). This is potentially a very powerful technique for enhancing both the selectivity and signal-to-noise ratio of energy selection fluorometry in glassy hosts, though (surprisingly) no use of the method in analytical energy selection fluorometry has yet been reported.

The behavior of energy selection fluorescence spectra in rare-gas matrices as a function of excitation wavelength is somewhat different. Fig. 7.17 shows the broadband induced fluorescence spectrum of benzo[a]pyrene matrix isolated in argon at 15 K. Note that the major feature at ca. 400 nm contains site structure. Fig. 7.18 shows the manner in which the appearance of that major feature is altered by changing the laser wavelength. In a case of this type, the site-selected fluorescence does not necessarily shift along with the laser line as the latter is scanned, though obviously the fluorescence spectrum is exceedingly sensitive to slight variations in the wavelength of excitation. This example (compare Figs. 7.15 and 7.17) indicates some of the differences between energy selection experiments in vitreous and crystalline hosts; in the latter case, site selection actually is a rather accurate description of the experiment.

2.3.2. Experimental Techniques

Little need be added here to the comments in Sections 2.2.2 and 2.3.1. It is very difficult to perform energy selection experiments under analytically reasonable conditions unless a laser is used for excitation. The laser must be continuously tunable over a wide wavelength range. The precision of wavelength tuning of the laser is obviously a matter of considerable importance (refer again to Figs. 7.15 and 7.17). The energy selectivity of these experiments obviously

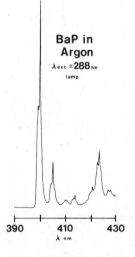

Fig. 7.17. Broadband-excited fluorescence spectrum of benzo[a]-pyrene matrix isolated in argon at 15 K. Note the presence of apparent site structure in the major feature at ca. 400 nm. From Conrad et al. (63); reproduced by permission of ASTM.

BaP in Argon

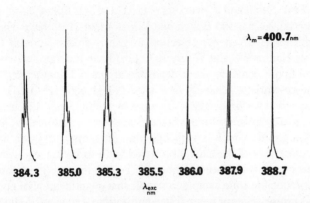

$\lambda_m = 400.7_{nm}$

384.3   385.0   385.3   385.5   386.0   387.9   388.7

λ_{exc}
nm

Fig. 7.18. Variation in the major fluorescence feature of benzo[a]pyrene (matrix isolated in argon at 15 K) as the function of excitation wavelength (numbers below individual fluorescence spectra) using a dye laser (bandwidth: 0.02 nm) source. From Conrad et al. (63); reproduced by permission of ASTM.

depends strongly on the spectral purity of the laser radiation. Though many energy selection experiments have been performed using dye lasers fitted with intracavity etalons, the cost in output power thus incurred is likely to be unacceptable for most analytical applications of the phenomenon. Stray dye fluorescence in the laser output is even less tolerable in energy selection fluorometry than in Shpol'skii spectrometry. Because the emission zero-phonon line usually falls quite close to the excitation frequency, the use of time resolution to reduce scatter background is nearly always advantageous. Thus, the duration and shape of pulses produced by a given laser also is an inportant parameter in energy selection experiments.

Optimal results are achieved in energy selection spectrometry when the sample temperature is as close to 0 K as possible (17). For some guest–host combinations, use of a closed-cycle refrigerator ($T \geq 10$ K) produces quite satisfactory results; for others, lower temperatures (attainable via liquid helium cryostats) must be used. Commercial liquid helium continuous-flow cryostats actually are less expensive than closed-cycle refrigerators (though the cost over time of the liquid helium is not negligible), but liquid helium refrigerators are less convenient to use than closed-cycle refrigerators, and the time required to obtain a spectrum usually is significantly extended if use of a liquid helium cryostat is required. Further detail regarding the design and use of liquid helium cryostats is available elsewhere (51, 57).

2.3.3. Analytical Applications

Thus far, energy selection has received much less analytical use than the Shpol'skii effect. Small and co-workers (111, 121–124); Bykovskaya, Personov, and Romanovskii (125); and Bolton and Winefordner (126) have described analytical applications of energy selection in glassy frozen-solution hosts; and Maple, Gore, Hammons, and Wehry (63, 116) have reported analytical use of site selection fluorescence in vapor-deposited argon and fluorocarbon matrices.

Small and co-workers generally have used 1 : 1 glycerol–H_2O or 1 : 1 : 1 glycerol–H_2O–dimethyl sulfoxide glasses as hosts in a liquid helium Dewar. A major advantage of glassy frozen solutions is their excellent optical quality, minimizing scattering background which is very important in energy selection fluorometry (due to the closeness of the excitation wavelength to the energy-selected fluorescence spectrum). It has been claimed (121–123) that the glass-forming process is so reproducible from sample to sample that quantification of guest species does not require use either of internal standardization or standard addition, though doubts as to the accuracy of this assertion have been voiced by other workers (12), and internal standard or standard addition calibrations were employed in similar measurements by Bykovskaya et al. (125).

Another significant advantage of organic solvent glasses is that their composition can be modified (within limits) as required for specific analytes. For example, polar compounds, such as amino derivatives of polycyclic aromatic hydrocarbons, do not generally exhibit line narrowing under energy selection conditions in aqueous glycerol glasses, presumably because the extent of electron–phonon coupling associated with the intramolecular charge transfer character of S_1^* in these compounds is large. However, acidification of the aqueous glycerol solvent prior to freezing protonates the amino group, significantly decreasing the electron–phonon interaction and enabling energy selection to be achieved (111, 124). Figs. 7.19 and 7.20 illustrate two examples. The spectrum of 1-aminopyrene (Fig. 7.19) is a very favorable case; the spectral features are extremely sharp. That of 1-aminofluorene (Fig. 7.20) is something of an opposite extreme; even under narrow-band excitation in an acidified glass, the phonon sideband intensity is relatively great.

Complex real samples have been analyzed for polycyclic aromatic hydrocarbons (PAHs) by energy selection in frozen glassy hosts (122, 123, 125). These samples have included liquid solvent-refined coal, gasoline, solid paraffin, and solvent extracts of cigarette smoke condensate and soot from combustion of vacuum grease. An illustration of the selectivity that can be achieved for PAHs in unfractionated solvent-refined coal is shown in Fig. 7.21.

The limits of detection that can be achieved by energy selection fluorometry are of interest, due to the lower sensitivity of the technique as compared with the Shpol'skii effect (see Section 2.3.1). Brown, Duncanson, and Small have

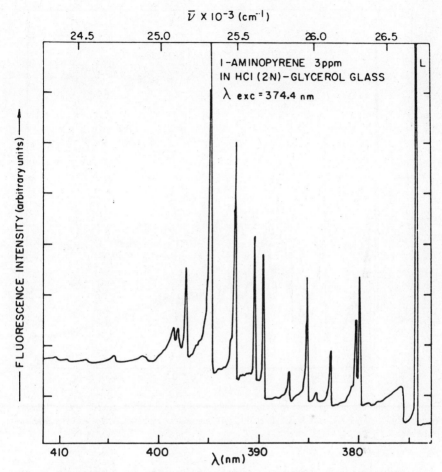

Fig. 7.19. Fluorescence spectrum of 1-aminopyrene in an acidified glycerol–water glassy frozen solution at 4.2 K. From Chiang, Hayes, and Small (111); reproduced by permission of the American Chemical Society.

reported detection limits for PAHs as low as 20 parts per trillion (122), with a linear dynamic range of at least 3 decades in PAH concentration. The best limit of detection reported by Bykovskaya, et al. (125) was 10^{-11} g/mL (10 parts per trillion). Each of these detection limits was reported for perylene, a very intensely fluorescent PAH. Bolton and Winefordner (126) reported detection limits for six PAHs, including perylene; their results were much less encouraging, indicating detection limits of 10^{-7} g/mL or greater (i.e., ≥ 0.1 ppm) and a linear dynamic range typically less than 3 decades. However, these studies were not performed using optimized fluorometric instrumentation and may accord-

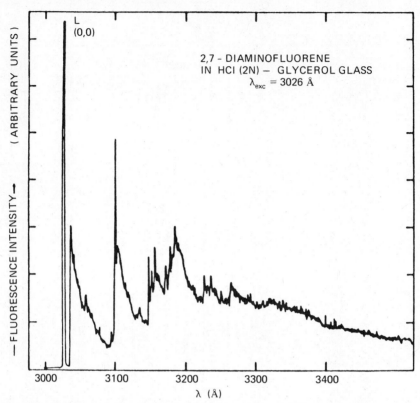

Fig. 7.20. Fluorescence spectrum of 2,7-diaminofluorene in an acidified aqueous glycerol glassy frozen solution at 4.2 K. From McGlade et al. (124); reproduced by permission of Elsevier Science Publisers.

ingly paint an overly pessimistic picture. At present, the ultimate limits of detection for energy selection fluorometry in glassy matrices that can realistically be achieved for complex samples have not been established with certainty.

Maple and Wehry (116) have described an alternative energy selection fluorometric method, using vapor-deposited rare-gas or fluorocarbon hosts rather than glassy frozen solutions. As discussed previously (Section 2.3.1 and Fig. 7.15), rare-gas matrices are crystalline in nature and fluorescence spectra of guest species in such matrices usually are stuctured. This also appears to be generally true for fluorocarbon matrices, such as perfluoro-*n*-hexane; see Fig. 7.22 for examples of broadband and energy selection fluorescence spectra of hydroxynaphthalenes in a fluorocarbon host. These spectra were obtained at 15 K using a closed-cycle cryostat. Fluorocarbons are especially useful as hosts for fluorometric analysis of polar analytes because they interact less strongly with

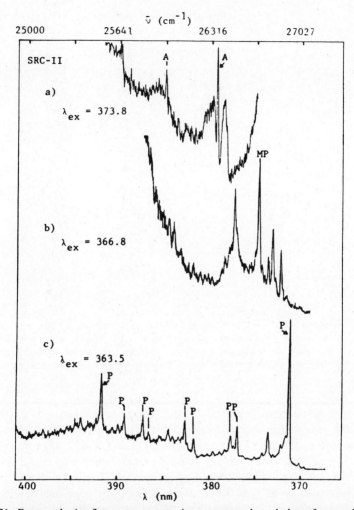

Fig. 7.21. Energy selection fluorescence spectra in an aqueous glycerol glassy frozen solution at 4.2 K for anthracene (top), 1-methylpyrene (center), and pyrene (bottom) in an untreated liquid solvent-refined coal sample. From Brown, Duncanson, and Small (122); reproduced by permission of the American Chemical Society.

the ground and electronically excited states of polar guests than any other common matrix materials, with the possible exception of neon. Not surprisingly, therefore, polar aromatics tend to be insoluble in liquid fluorocarbons. Therefore, they are not amenable to frozen-solution fluorometry and matrix isolation methods must generally be used.

Limits of detection (ca. 5 ng) for polar aromatics by matrix isolation energy

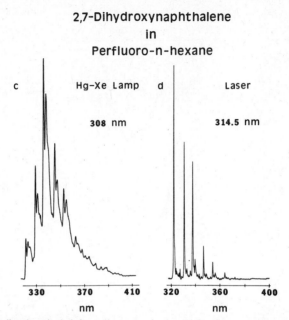

Fig. 7.22. Broadband-excited (left) and energy selection (right) fluorescence spectra of 2,7-dihydroxynaphthalene matrix isolated in perfluoro-*n*-hexane at 15 K. The excitation wavelengths are indicated on the figure; the bandpasses were 7 nm for lamp, and 0.02 nm for laser, excitation. From Maple and Wehry (116); reproduced by permission of the American Chemical Society.

selection fluorometry in fluorocarbon or rare-gas matrices were 1–3 orders of magnitude greater than those observed for PAHs in Shpol'skii matrices using identical instrumentation (116). Again, therefore, the "sensitivity versus selectivity trade-off" inherent in energy selection spectrometry may cause difficulty when extremely low limits of detection are needed. The linear dynamic range was observed to be at least 3 decades in quantity of polar analyte. Relative standard deviations for determinations of hydroxynaphthalenes in argon and perfluorohexane were in the 4–7% range (116); this result indicates that the site distributions implicit in the structured broadband-excited fluorescence spectra can be reproduced from sample to sample with surprisingly high reliability via matrix isolation.

It is almost certain that the greatest analytical impact of energy selection fluorometry (either via matrix isolation or frozen-solution spectrometry) will be in the characterization of very polar, hydrogen-bonding compounds that yield badly broadened absorption spectra in virtually all hosts, even at very low temperature. Small and co-workers recently have shown that energy selection spectrometry in glycerol–water–ethanol glasses at 4.2 K can yield sharp fluorescence spectra of PAH metabolites and adducts thereof with DNA (124, 127). For example, Fig. 7.23 shows an energy selection fluorescence spectrum of an adduct of DNA with a diol epoxide of benzo[a]pyrene. The possibility that the technique can be used to identify DNA adducts obtained after in vivo exposure to mixtures of PAHs or their metabolites, without the very complex separations usually required in analyses of such difficult samples, has been suggested (124, 127). No other fluorometric technique has yet been shown to exhibit such high selectivity for individual fluorophores in mixtures of polar compounds.

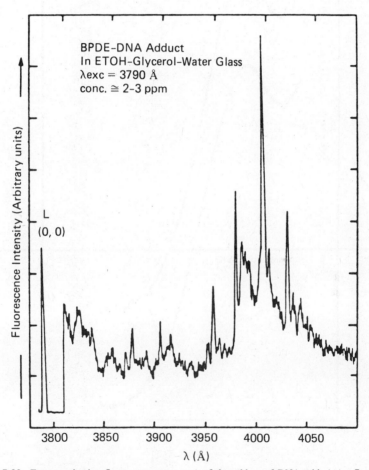

Fig. 7.23. Energy selection fluorescence spectrum of the adduct of DNA with ($\pm$)-r-7,t-8-dihydroxy-t-9,10-epoxy-7,8,9,10-tetrahydrobenzo[a]pyrene in an ethanol–glycerol–water glassy frozen solution at 4.2 K. L denotes the wavelength of excitation. From McGlade et al. (124); reproduced by permission of Elsevier Scientific Publishers.

2.3.4. Hole Burning

Whenever intense laser radiation is used to induce fluorescence by excitation into an inhomogeneously broadened absorption band, the possibility of "hole burning" can arise. An example of the phenomenon is shown in Fig. 7.24,

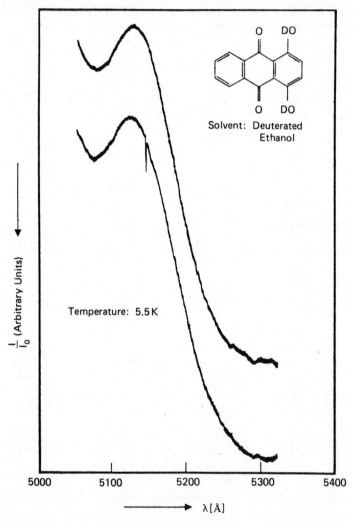

Fig. 7.24. Absorption spectra of 1,4-dihydroxyanthraquinone in a perdeuteroethanol glassy frozen solution at 5.5 K before (top) and after (bottom) illumination of the sample for 3 min at 514.5 nm (Ar[+] laser, 7 mW/cm[2]). Note the "hole" burned in the absorption spectrum at the laser wavelength. From Friedrich, Wolfrum, and Haarer (132); reproduced by permission of the American Institute of Physics.

comparing the absorption spectra of 1,4-dihydroxyanthraquinone in a glassy frozen-solution host at 5.5 K before and after 3-min illumination of the sample at 514.5 nm. Note the narrow but pronounced dip in the absorption spectrum at 514.5 nm after illumination. Because the laser excites only those guest molecules absorbing at the laser wavelength, the result of hole burning is a time-dependent decrease in the measured energy-selected fluorescence intensity for the guest.

Hole burning can occur by both photochemical and nonphotochemical processes. In "photochemical hole burning," some of the excited guest molecules undergo transformation to product species that absorb at different wavelengths than the precursor (128–133). In some cases, the photochemical reaction is reversible [e.g., intramolecular proton transfers (134)]. In such a system, illumination of the "hole-burned" sample with white light may restore the original inhomogeneously broadened absorption profile of the guest, or at least achieve a more or less stable equilibrium between photochemical precursor and product(s) (134). In fact, one analytical study of energy selection fluorometry has taken this precaution in an attempt to minimize quantitative errors caused by reversible photochemical hole-burning phenomena (125). If the photochemical reaction is irreversible, then this procedure obviously is of little value. In such cases, it is desirable to obtain the energy selection fluorescence spectrum of the analyte as rapidly as possible in order to complete the measurement prior to the occurrence of appreciable hole burning. The temptation is great to compensate for the intrinsically low sensitivity[†] of energy selection fluorometry by using high laser power in excitation, thus increasing the probability that photochemical hole burning will occur. For intermolecular photochemical reactions, use of a chemically inert host (135) is a way around the problem, but changing the host material has little effect if the photochemical reaction is intramolecular. Photochemical hole burning has been observed for guest molecules in glassy frozen solutions (132, 135), Shpol'skii frozen solutions (134), aromatic crystal hosts (136), and polymer films (137–139).

Hole burning also can occur for photochemically stable guests. In this second case ("nonphotochemical" or "photophysical" hole burning), the absorption wavelength of an affected molecule changes without any chemical transformation of the molecule. The hole in the absorption spectrum of the guest usually can be "refilled" by annealing the sample or illuminating the sample with intense white light (141, 142). The nonphotochemical hole-burning phenomenon is thought to arise as a result of slight alterations in the host environments "seen" by the affected molecules, with a small (often <2 cm^{-1}) shift in their absorption frequencies (143). The phenomenon has been treated, in terms of a double-well potential for the interaction between guest and host molecules that

---

[†]"Calibration sensitivity" as defined by Ingle (7) as the slope of the analytical calibration curve.

is large enough to prevent "site" interconversion for guest molecules in their ground electronic states but low enough for excited guest molecules to undergo "site" interconversion by thermally activated or phonon-assisted tunneling mechanisms, by Small and co-workers (143–146)[†]. In the absence of "hole refilling," holes produced by this nonphotochemical process can persist for extended periods of time (141–143).

A second nonphotochemical hole-burning mechanism, the "triplet bottleneck" process, occurs when a guest molecule exhibits a large intersystem crossing quantum yield and has a relatively long-lived triplet state (130, 147). In this case, when intense laser radiation is used, the fraction of guest molecules converted to triplets may be large, and the concentration of ground-state guest molecules is depleted correspondingly. The hole so produced in the absorption spectrum persists (upon cessation of laser illumination) only for a time period corresponding to the triplet lifetime, unless the triplet state is photochemically reactive. This problem can be significant whenever an attempt is made to measure an energy selection fluorescence spectrum for a molecule having a fluorescence quantum yield much smaller than that for intersystem crossing, because the obvious temptation in such a difficult case is to use very intense illumination of the sample.

From the standpoint of analytical chemistry, hole burning generally is a nuisance and one wishes to avoid it. However, both the photochemical and photophysical hole-burning processes are very important and useful in fundamental spectroscopy, enabling detailed studies of such important matters as measuring homogeneous spectral line widths for guest molecules in low-temperature solids, establishing mechanisms for photochemical reactions, acquiring accurate vibrational frequencies for molecular excited states, and probing structural characteristics of the host (129, 132, 134, 136, 139, 148, 149).

2.4. Choice of Method

The analytical literature pertaining to solid-state low-temperature fluorometry is fairly large and contains numerous claims and counterclaims as to which of these techniques is "best." Perhaps the best approach for our purposes is to quote verbatim the recommendations of Harris and Lytle, knowledgable but (presumably) disinterested observers (150):

For molecular line narrowing in low-temperature solvents three possible approaches exist. A suggested set of differentiating criteria might be the following. (i) For a first

---

[†]We place "site" in quotation marks because this phenomenon is usually associated with molecules in glassy hosts for which discrete "site structure" is not observed in the inhomogeneously broadened absorption spectra.

attempt start with the vitreous solvents. A large comparative data set exists in the literature and the technique experimentally resembles low-temperature phosphorescence so far as quantification, etc., are concerned. These solvents have been demonstrated to handle a wide range of structural types and the resultant line narrowing has been successfully employed in mixture analyses. (ii) If the sample contains nearly equal-sized polycyclic aromatic hydrocarbons and if vitreous solvents do not provide sufficient spectral resolution, then try to use a Shpol'skii solvent. The main aggravation will be learning the experimental nuances necessary to control the site distribution and utilize internal standards. (iii) If polar compounds are to be studied or a broad range of problems are to be tackled, use matrix isolation. This suggestion is made because the instrument is extremely flexible and can be used to research extensively the analytical problem before more routine methodology is developed. (iv) If phosphorescence spectra are to be utilized (nonfluorescing compounds), then a Shpol'skii solvent should be considered. Dinse and Winscomb (26) have reported such line narrowing where triplet population was achieved by pumping the $S_0 \rightarrow S_1^*$ transition. This is in direct contrast with vitreous solvents, where direct $S_0 \rightarrow T_1^*$ excitation is required.

Two notes of caution have to be made concerning these last three line-narrowing techniques.[†] First, there is a real sensitivity/selectivity trade. This is because the homogeneous population is much, much lower than the inhomogeneous one. Thus, as the laser line is made narrower and narrower, the potentially fluorescing population decreases. Second, the selectivity of the techniques has to be evaluated very carefully. One cannot simply add up resolution elements and square the result (excitation/emission intensity matrix). This is because the entire fluorescence spectrum will slide with the excitation frequency over the inhomogeneous $0 \rightarrow 0$ transition linewidth. Also, excitation spectra will begin to wash out about 1500 cm^{-1} above this level. Here both matrix isolation and Shpol'skii solvents have a slight edge over vitreous solvents.

It remains only to note that all of the methods described in this section are still subjects of active research, and the preceding recommendations may be buttressed (or vitiated) by results as yet unpublished.

3. FLUORESCENCE SPECTROMETRY IN SUPERSONIC EXPANSIONS

3.1. Principles

Spectra of guest molecules in solids inevitably are perturbed, in one way or another, by interactions with the host matrix. It would be very nice if the matrix could be eliminated, and the spectra of isolated ultracold fluorescent molecules could be acquired. Subject to certain limitations noted below, this objective can be achieved by use of a gas-phase technique wherein the sample (diluted by a

---

[†]This paragraph refers to energy selection fluorometry in glassy frozen solutions.

suitable carrier gas) is expanded through an orifice into a vacuum and the resulting expansion is interrogated by a laser beam. The fundamental principles of optical spectroscopy in supersonic expansions have been reviewed by Levy, Wharton, and Smalley (151–153); Hayes and Small (154); and Miller (155); while analytical studies have been reviewed by Hayes and Small (154) and Johnston (156). The literature in this area is growing with astonishing speed, and only those studies that seem of greatest analytical relevance are discussed here.

A schematic diagram of a supersonic expansion is shown in Fig. 7.25 (156). The sample gas, diluted with one of the noble gases (most commonly helium), is confined in a chamber at a "stagnation pressure" ranging anywhere from 0.2 to 100 atm. On one wall of this chamber is an orifice of < 0.1 mm diameter; gas molecules can pass through this orifice into another chamber that is held at low pressure. The diameter of the orifice is much larger than the mean free path of the gas molecules in the stagnation reservoir. As depicted in Fig. 7.25, the motion of the gas molecules in the stagnation chamber is random, and the direction of motion of any individual molecule is subject to sudden change as it collides with other molecules. Of the molecules near the orifice, only those moving in the proper direction can escape through the orifice. Thus, the motion of the stream of gas molecules on the other side of the orifice is directed, rather than random.

As the gas that has passed through the orifice expands, its density eventually becomes so low that collisions between molecules virtually cease to occur. However, in the early stages of the expansion (i.e., in the vicinity of the orifice), many collisions occur. Their effect is to "monochromatize" the molecular velocities (155). Because the sample is diluted extensively by the carrier gas (if

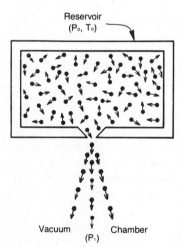

Fig. 7.25. Schematic diagram of a supersonic expansion. The circles represent gas molecules and the arrows their direction of motion at a particular time. The stagnation (or reservoir) pressure P_0 is much greater than the pressure in the expansion chamber (P_1). From Johnston (156); reproduced by permission of Elsevier Scientific Publishers.

the experiment is performed properly), the vast majority of collisions involving sample molecules are with those of the carrier gas. The carrier gas atoms are much less massive than the sample molecules and therefore travel at higher velocities in the early stages of the expansion. Collisions between sample and carrier molecules thus serve to increase the velocity of sample molecules to that of the carrier. Once that has happened, the highly directed nature of the gas flow and the great similarity in the velocities of individual molecules mean that any additional collisions are of relatively low energy.

Thus, the key feature of the expansion is production of a highly directional molecular beam having a very narrow distribution of molecular velocities. In terms of the kinetic theory of gases, this is equivalent to reducing the translational temperature of the gas, in favorable cases to less than 0.5 K.[†] This itself is of little spectroscopic significance. What is important is the fact that the internal degrees of freedom of the sample molecules (i.e., rotations and vibrations) also are cooled as they equilibrate with the cold translational bath. (Of course, the atomic carrier gas has no internal degrees of freedom.) The molecular beam is shielded from collisions with hot background molecules by a shock wave that forms around the expansion.

Because the expansion reaches such low gas densities that molecular collisions cease, the extent to which the internal degrees of freedom of sample molecules are cooled ultimately depends on kinetics. The rate of equilibration between translational and rotational degrees of freedom is usually large; hence, the rotational temperature of a supersonic expansion becomes nearly as low as the translational temperature. Vibrational-translational equilibration generally is much slower, so the extent of vibrational cooling is smaller, though usually significant. For this reason, the spectroscopic technique based on laser-induced fluorescence of analytes seeded into supersonic expansions has been named "rotationally cooled laser-induced fluorescence spectroscopy" (157).

Despite the very low effective temperatures achieved in a supersonic expansion, the gas does not immediately condense (if the experiment is properly performed) because the three-body collisions needed for nucleation are rare at the low pressures characteristic of the rapidly expanding gas (151). Thus, the sample molecules that enter the collision-free region of the expansion are translationally and rotationally cold and vibrationally chilly, yet they still exist as isolated gas-phase molecules. Thus, one is able to obtain molecular spectra at low temperature that are essentially free of the matrix perturbations characteristic of solid samples.

The absence of "complete" vibrational cooling is not as detrimental as one might initially suppose, because the extreme rotational cooling means that ro-

---

[†]The expansion is "supersonic" not because the gas velocity is so enormous but because the velocity of sound is small at the low effective temperature of the expansion (152).

tational structure is absent in the vibrational spectrum of a sample molecule. Moreover, enough vibrational cooling usually occurs that the occupation of excited vibrational levels in the absence of a radiation field is very small. Therefore, both the electronic absorption and emission spectra of large molecules in supersonic expansions are exceedingly sharp; see Fig. 7.26 for an example excitation spectrum. Selective excitation of fluorescence from individual constituents of complex samples therefore is easy in principle, and the high selectivity is not accompanied by any intrinsic trade-off of sensitivity. Moreover (and this is especially important for anlytical applications), the "isolated" nature of the cold molecules means that polar, hydrogen-bonding molecules (which often cause problems in solid hosts) should not necessarily pose overwhelming difficulties in a supersonic expansion.

That molecular collisions occur in early stages of the expansion can cause problems. If a solute is not diluted sufficiently by the carrier gas, dimer formation may occur; the monomer and dimer spectra are different (159–162), and the resulting spectral complication is undesirable. Of greater concern is the formation of complexes between the analyte and other sample constituents, or with the carrier gas itself. For example, Fig. 7.27 shows excitation spectra of the aromatic hydrocarbon tetracene and its complexes with several small molecules (163), and Fig. 7.28 shows excitation spectra of tetracene in an argon expansion

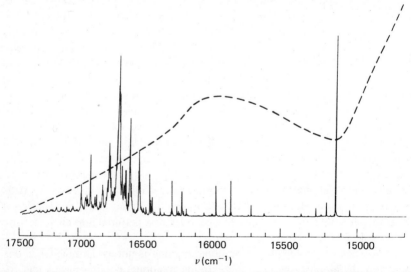

$\nu\,(\text{cm}^{-1})$

Fig. 7.26. Fluorescence excitation spectrum of free base phthalocyanine in a helium supersonic expansion (solid curve) compared with the absorption spectrum of the same compound obtained in a static gas cell. From Fitch, Haynam, and Levy (158); reproduced by permission of the American Institute of Physics.

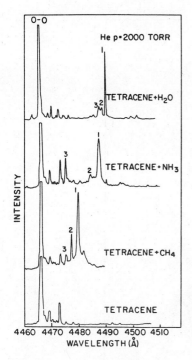

Fig. 7.27. Fluorescence excitation spectra of tetracene by itself (bottom) and in the presence of (from bottom) 20 torr CH_4, 6 torr NH_3, and 1 torr H_2O. The carrier gas was He (stagnation pressure 2000 torr); the vapor pressure of tetracene was ca. 0.1 torr. From Even and Jortner (163); reproduced by permission of the American Institute of Physics.

as it varies with the stagnation pressure of argon (164). In extreme cases (very high reservoir pressures), the spectra lose their characteristic sharp structure and are of limited analytical use.

Such spectral changes indicate formation of weak complexes that are not observed at higher temperatures. Tetracene is not an especially reactive molecule (nor, obviously, is argon!). The aromatic hydrocarbon–rare gas complexes presumably are bound by van der Waals forces; obviously, such weakly bound species can be observed only at exceedingly low temperature. One of the most important fundamental applications of molecular spectroscopy in supersonic expansions is to probe such phenomena as van der Waals complex formation (163–166), hydrogen bonding (167, 168), and solvation processes (169–171). In most analytical studies, it is desired to avoid such complications. For "matrix perturbations" truly to be absent in fluorescence spectra measured in supersonic expansions, proper precautions must be taken, as discussed in the following section.

3.2. Experimental Techniques

Apparatus for spectroscopy in supersonic expansions includes a stagnation reservoir, a means to introduce samples into the expansion, the orifice, a vacuum

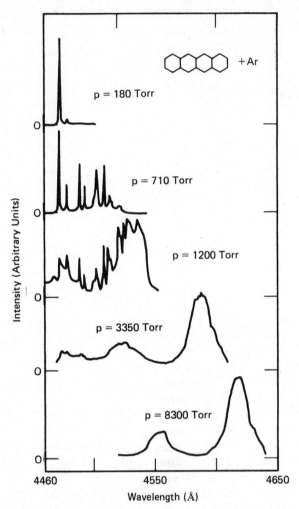

Fig. 7.28. Fluorescence excitation spectra of tetracene in argon supersonic expansions as a function of the stagnation pressure of argon. The vapor pressure of tetracene was ca. 0.1 torr. From Amirav, Even, and Jortner (164); reproduced by permission of the North-Holland Publishing Company.

system, an excitation source, and optics for detecting fluorescence. Because most dye lasers suitable for fluorometry are pulsed rather than continuous wave (cw), it is desirable to pulse the diluted sample through the orifice rather than admitting it continuously into the vacuum chamber. One thereby avoids wasting sample and also decreases the required capacity and speed of the vacuum pump(s) (172). At one time, automobile fuel injectors were popular as valves for producing pulsed expansions (172). More recently, several pulsed valves designed specifically for the purpose have been described (173–179) and suitable pulsed beam valves now are commercially available. Such valves emit gas pulses much longer than the typical 10-ns pulses of a dye laser, but they operate at the low repetition rates typical of pulsed dye lasers and thus reduce wastage of sample. For relatively involatile analytes, the orifice can be heated to prevent condensation, which has the extra benefit of decreasing the extent of complex formation in the expansion (158). The usual orifice is a pinhole in a metal substrate. If the substrate is heated, the possibility of metal-catalyzed thermal decomposition of the analyte may arise, in which case use of a nozzle fabricated from quartz or glass is preferable (158).

Because gases are less dense than solids, the ultimate sensitivity of supersonic jet spectrometry depends on the number of analyte molecules that can be illuminated by a laser beam directed into the expansion. This in turn depends on the path length of sample through which the beam passes and the concentration of analyte molecules in the illuminated volume. For typical supersonic expansions using circular orifices, the path length is of the order of 0.3 cm (180), and the total molecular density can be as great as 10^{14} molecules cm^{-3} (180–183). Hence, an analyte seeded into the expansion at a mole fraction of 10^{-6} (i.e., 1 ppm on a mole basis; >1 ppm on a weight basis) is present at a concentration of ca. 10^8 molecules cm^{-3}. In principle, gas-phase concentrations much smaller than this can be detected by fluorescence if proper care is taken to minimize background contributions from scattering and reflection. Moreover, use of a rectangular ("slit") nozzle reportedly produces a planar expansion characterized by molecular densities 100–1000 times larger and approximately a tenfold increase in optical path length (184–186), as compared with those attainable via a circular orifice.

Introduction of samples (especially those that are not gases at room temperature) into a supersonic expansion in such a way as to achieve precise fluorescence intensity data must be considered. One approach to the problem is to use a gas chromatograph as a sample introduction device to achieve quantitative transfer of analyte through the orifice (157). A schematic diagram of this experimental arrangement is shown in Fig. 7.29. A continuous expansion was used in these studies, though a pulsed valve synchronized with the firing of a pulsed laser presumably could be employed to improve the efficiency with which sample molecules are utilized in the experiment. Note that the gas chromatograph serves as a sample introduction device, not as a preseparation tool, in this instrument. Obviously, any separation that the column produces is an additional benefit, suggesting the use of spectroscopy in a supersonic expansion as a detection technique for gas chromatography (187).

The nature of the vacuum system required to achieve effective cooling in the expansion must next be considered. Pumping plants described in the literature for this purpose range from very high speed [e.g., 20,000 L s^{-1} (173)] systems

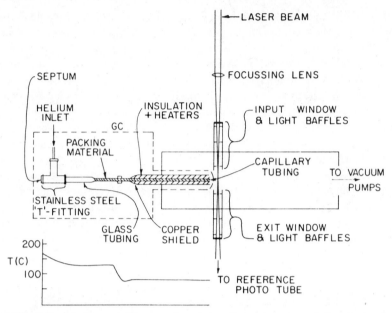

Fig. 7.29. Schematic diagram of the use of a gas chromatograph as a sample introduction device for fluorescence spectrometry in a supersonic expansion. Everything inside the dashed line is part of the gas chromatograph. From Hayes and Small (157); reproduced by permission of the American Chemical Society.

of rather colossal size and cost to "poor-man's" pumping systems consisting of a single low-speed (200–400 L s^{-1}) diffusion pump (186, 188). The latter alternative obviously is preferable. Pulsing the expansion decreases the required pumping capacity. Another key is to use the lowest reservoir pressure compatible with efficient cooling in the expansion. It is helpful to use heavy rare gases (Kr or Xe) rather than He as the carrier gas (151, 188) for large analyte molecules. For example, the gas flow rate needed to achieve a specified degree of rotational cooling for the four-ring aromatic hydrocarbon tetracene was a factor of 400 smaller for Xe than for He as carrier gas (188). Under these conditions, stagnation pressures < 1 atm can be used, greatly decreasing the demands on the pumping plant. The trade-off is an increased probability of van der Waals complex formation with the solute because the heavy rare gases are much more polarizable than He. This problem is considered further in Section 3.3.

A number of optical systems for illuminating, collecting the fluorescence, and detecting the signal from samples seeded into supersonic expansions have been described in detail (154, 161, 173, 181, 183, 189–194). Most of these spectrometers were designed to obtain excitation spectra of pure fluorescent compounds and therefore do not include an emission monochromator; if an emission monochromator was used, it was rather low in resolution. Addition of a high-resolution emission monochromator to such an instrument should be straightforward.

Two reasons for relying upon excitation, rather than emission, spectra whenever possible can be cited. First, in a fluorescence excitation spectrum, the observed bandwidth should ultimately be limited by the spectral purity of the laser radiation rather than the bandpass of a monochromator. Second, this approach increases the number of emitted photons striking the detector per unit time and thus should improve the signal/noise (S/N) ratio. There are also disadvantages in relying upon excitation spectra for analytical purposes. Foremost among these is the limited wavelength scan range of a dye laser as compared with that of a monochromator. For example, scanning the absorption spectrum of nickelocene by multiple-photon ionization over the wavelength range 372.5–540 nm required nine laser dye changes (199). Also, in using relatively broadband fluorescence detection and depending on differences in excitation wavelengths for different fluorophores to achieve selectivity, one sacrifices the principal "selectivity advantage" of fluorescence over absorption spectroscopy—the availability of two wavelengths to use in distinguishing one sample constituent from another. Of course, this latter consideration is irrelevant if the required selectivity can in fact be achieved solely by selective excitation. As yet, there is insufficient analytical literature for this question to be dealt with conclusively.

It is very important to collect as large a fraction of the emitted photons as possible because the emitting molecules in a supersonic expansion move rapidly in and out of the interrogation region. Obviously, it also is vital to minimize the collection of scattered or reflected photons. Thus, proper attention to lensing, baffling, and suitable blackening of the illumination chamber is essential. Unlike solid-state cryogenic fluorometry, wherein the illumination and collection arrangements are now reasonably standardized, there are rather wide variations in optical arrangements for supersonic expansion fluorometry from one laboratory to another. Many of the experimental systems cited above were not designed with trace analytical measurements in mind, and therefore may be rather wasteful of sample, not amenable to use with corrosive or reactive compounds, and/or not optimized for maximal S/N ratio performance.

The important characteristics of lasers for fluorometry in supersonic expansions are virtually identical with those noted previously (Sections 2.2.2 and 2.3.2). In an ideal supersonic expansion, the excitation spectra of all sample constituents are exceedingly sharp. Thus, the reproducibility of the wavelength-tuning mechanism in a dye laser is very important. Also, for scanning excitation spectra, it is highly desirable that the smallest wavelength step size (assuming a stepper motor scan mechanism) be smaller than the bandwidths of the spectral features to be measured. This is a very demanding condition that few, if any, present commercial dye lasers can satisfy (192).

3.3. Analytical Applications

As of this writing (July 1984), analytical applications of laser-induced fluorescence of molecules seeded into supersonic expansions have been reported by Small and co-workers (123, 154, 157, 195); Amirav, Even, and Jortner (186); Imasaka, et al. (196); and D'Silva et al. (197). Practically all reported work thus far has concerned polycyclic aromatic hydrocarbons (PAHs), although (in principle) the selectivity advantages of supersonic expansions over the solid-state techniques should be most pronounced for polar fluorophores. Examples of the spectral resolution that can be obtained for multicomponent samples are shown in Figs. 7.30 and 7.31. Figure 7.30 is an excitation spectrum of a mixture of naphthalene and its 1- and 2-methyl derivatives (195), measured under relatively low-resolution emission conditions such that fluorescence of all three compounds was detected. The possibilities for selective excitation of fluorescence from each of the three sample constituents are apparent. Figure 7.31 compares the fluorescence excitation spectrum of pure fluorene with that of an impurity present in a sample of biphenyl (186). Both the positions of the spectral features and the fluorescence decay time of the impurity were used to assign the impurity fluorescence to fluorene.

An additional illustration of the selectivity of fluorometry in supersonic expansions is a measurement of the chlorine isotopic abundance in 9,10-dichloroanthracene, exploiting the face that the different isotopically substituted com-

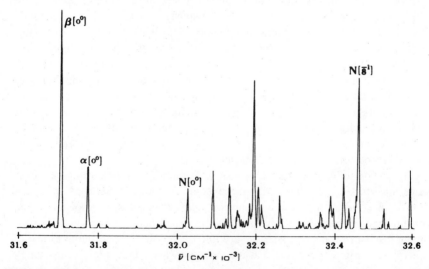

Fig. 7.30. Fluorescence excitation spectrum of a mixture of naphthalene (N) and 1- and 2-methylnaphthalene (α and β, respectively) in a helium supersonic expansion. From Warren et al. (195); reproduced by permission of the American Chemical Society.

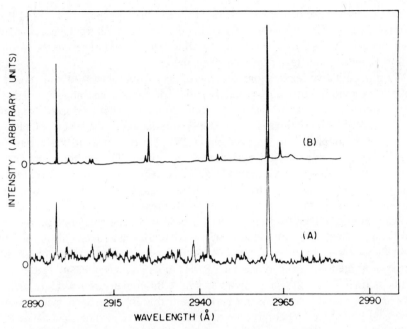

Fig. 7.31. Fluorescence excitation spectra of (top) pure fluorene and (bottom) an impurity in a sample of biphenyl, both in argon supersonic expansions. From Amirav et al. (186); reproduced by permission of the American Chemical Society.

pounds are spectrally distinguishable (186). Thus, use of isotopically substituted analytes as internal standards (Section 2.2.3 and Figs. 7.7 and 7.8) should be useful in supersonic expansions (197), just as in Shpol'skii solid matrices.

Very little information currently is available regarding the quantitative analytical capabilities of supersonic expansion fluorometry. Use of a gas chromatograph for introducing samples into the expansion should be a general technique, though interfacing of a gas chromatograph to a pulsed supersonic expansion has not yet been reported. Using a continuous expansion and a pulsed Nd:YAG-pumped dye laser (10-Hz pulse repetition rate), Hayes and Small achieved limits of detection for naphthalene and the methylnaphthalenes of ca. 10 ng with a linear dynamic range of 4 decades (157). Also, using a continuous expansion and a pulsed nitrogen-pumped dye laser (repetition rate not specified), Imasaka, et al. attained a detection limit for perylene of 100 ng with a linear dynamic range of slightly greater than 1 decade (196). These detection limits are several orders of magnitude inferior to those reported in the past for solid-state frozen-solution and matrix isolation fluorometric methods, particularly in Shpol'skii matrices. However, the instrumentation employed in these studies was not optimized, and the limits of detection should be capable of substantial improvement via use of a pulsed nozzle (157) and planar expansions (184–186). Estimates of the improvements in detectivity that should result from these changes have been made (157), but as of this writing (July 1984) experimental verification of these estimates has not been reported.

Two quantitative applications of supersonic jet fluorometry to real samples have been reported to date. A determination of naphthalene and the methylnaphthalenes in a crude oil sample (at 200–500 ppm levels) produced results in excellent agreement with an independent determination of these compounds by gas chromatography–mass spectrometry (157). No details as to the nature of calibration or standardization procedures were given. The fluorene content in the aforementioned biphenyl sample (Fig. 7.31) was estimated by Amirav, Even, and Jortner to be ca. 1 ppm, though that figure was only an order-of-magnitude estimate and no quantitative calibration procedures were used (186).

An extremely interesting preliminary study of a very challenging problem has been reported by D'Silva et al. (197), who interfaced a supersonic expansion laser fluorescence spectrometer to a small fluidized-bed coal combustor for the purpose of identifying and quantifying PAHs in its gaseous effluent. Traditional methods (including solid-state low-temperature fluorometry) for tackling such a problem proceed by trapping the PAHs from the effluent. Such methods are slow and of dubious reliability due to difficulties in achieving quantitative recoveries of the PAHs. Supersonic expansion fluorometry appears well suited to this problem because the fluorescence measurement can be performed for the PAHs in the vapor phase without trapping. In this apparatus, the combuster effluent was passed through a quartz fiber filter to remove particles, and then through a circular orifice in a heated pyrex substrate. A 250-L s^{-1} Roots blower served as pumping plant. It was concluded that spectra of suitably high resolution could be obtained without diluting the combustor effluent in a carrier gas, which has obvious implications for sensitivity. Spectra of the effluent reported to date indicate the presence of five-ring PAHs, and a tentative identification of benzo[a]pyrene as a specific constituent has been suggested. While this study is preliminary and has not appeared in the journal literature, it represents the most demanding analytical test of the analytical capabilities of supersonic expansion fluorescence spectrometry reported thus far, and the ultimate outcome of this study should greatly help in assessing the analytical performance of the technique.

The question of van der Waals complex formation and its possible adverse effects in analytical applications of supersonic jet spectrometry (Section 3.1) has not yet been addressed, in part because the problem should not be significant for PAHs (186) and the more polar analytes for which difficulties may arise have not yet been studied under analtyically realistic conditions. To quote Amirav et al. (188): "For each diluent [carrier gas], there is a narrow range of stagnation pressure where the [solute] molecule is rotationally-vibrationally cooled, while the contribution of van der Waals complexes is negligible." How narrow that pressure range is, and the way in which it might vary from one analyte to another in a mixture, remain to be determined for polar fluorophores. Use of helium as carrier gas greatly diminishes the problem, but practical considerations of pumping speed (Section 3.2) may dictate use of krypton or xenon in analytical instruments. Complex formation of sample molecules with each other can be alleviated by diluting the sample sufficiently with carrier gas (198) and modifying the nozzle design (200). Until these matters are examined for mixtures of such classes of compounds as phenols and nitrogen heterocycles, statements that fluorescence spectra of organic molecules in supersonic expansions are totally free of matrix perturbations appear premature.

3.4. Advantages and Disadvantages of Supersonic Expansions

Lengthy consideration of this topic is not warranted by the limited analytical data base that presently exists, but a few general statements appear unassailable. First, in the absence of complex formation, it is clear that higher spectral resolution can be achieved for many compounds in supersonic expansions than in any solid host, and without the need for hosts that are "tailored" for specific analytes. Second, the variety of useful analytical measurements that can be performed using a supersonic expansion exceeds that for low-temperature solids. One obvious example is mass spectrometry. The use of laser-induced multiple-photon ionization coupled with mass spectrometric identification of parent and fragment ions (201–206) promises to be a technique of very high selectivity (Chapter 18). The possibility of performing both fluorescence and ionization experiments for the same sample in a supersonic expansion offers the promise of acquiring an enormous amount of analytical information. Although techniques for obtaining mass spectra of cryogenic solids (e.g., those formed by matrix isolation) have been developed (207), neither the resolution nor sensitivity thus far reported matches those obtainable in supersonic expansions.

The most obvious disadvantage of supersonic expansions (relative to solid-state samples) is the rapidly flowing sample; a solid of course is stationary. Analyte molecules in an expansion rapidly enter and leave the interrogation zone, whereas they remain in the illumination region of a solid for as long as the experimenter wishes to prolong the measurement. This has obvious implications for the use of signal averaging to enhance S/N ratios without excessive sample consumption. This fact, coupled with the smaller number of analyte molecules per unit volume in a supersonic expansion than in a solid, raises questions as to whether limits of detection in supersonic expansion fluorometry will ever match those attainable in Shpol'skii solids. So little information regarding limits of detection by fluorometry in supersonic expansions is presently available that no firm conclusions are possible.

Whether the experimental apparatus required for spectroscopy in supersonic expansions is more or less complex than that used in cryogenic solid fluorometry probably is a matter of opinion and prior experience in gas-phase versus condensed-phase spectroscopy. In either case, the laser is the most expensive capital item.

Supersonic expansions and matrix isolation share an inapplicability to compounds that cannot be volatilized without decomposition; frozen-solution techniques are immune to this concern. In practice, any compound that can be forced through a gas chromatograph or mass spectrometer is amenable to study by either technique. Though they have not yet been applied to optical spectroscopy, the various new techniques for dealing with involatile compounds in mass spectrometry (208, 209) may also be applicable to supersonic expansion and/or matrix isolation optical spectroscopy. The intriguing suggestion also has been made that involatile analytes may be introduced into a supersonic expansion after being dissolved in a supercritical fluid (156). Thus, the "analyte volatility" requirements of supersonic expansion spectrometry should not be overstated.

Supersonic expansions have attracted enormous interest as media for fundamental spectroscopic studies, and the few analytical investigations reported thus far indicate considerable potential for the technique. As for all analytical measurements, the ultimate value of fluorescence spectrometry in supersonic expansions, relative to that of competing techniques, must be established by experiment.

4. CONCLUSION

Of the laser-based molecular fluorescence techniques considered in this chapter, only Shpol'skii spectrometry can be regarded as approaching maturity. Transition energy selection in vapor-deposited matrices or glassy frozen solutions and fluorometry in supersonic expansions have intriguing possibilities, but as yet their sensitivity, quantitative reliability, applicability to analytes other than aromatic hydrocarbons, and broad utility to analysis of real samples have received insufficient study. It is reasonable to suspect that these techniques will ultimately emerge as complementary rather than competitive.

The extent to which any of these procedures ultimately will become established in organic analysis will depend in large part on improvements in laser tunability, reliability, and cost, No "turnkey" commercial instrumentation for any of these techniques is available or appears to be on the immediate horizon. "Laserphobia" is rather widespread, and at least partially justified, among users of molecular fluorescence spectrometry. It is hoped that this chapter demonstrates the intriguing analytical possibilities, as well as potential pitfalls, of these extraordinarily selective measurements.

REFERENCES

1. F. E. Lytle, *J. Chem. Educ.*, **59**, 915 (1982).
2. T. G. Matthews and F. E. Lytle, *Anal. Chem.*, **51**, 583 (1979).
3. F. J. Knorr and J. M. Harris, *Anal. Chem.*, **53**, 272 (1981).
4. G. R. Haugen and F. E. Lytle, *Anal. Chem.*, **53**, 1554 (1981).
5. N. Strojny and J. A. F. de Silva, *Anal. Chem.*, **52**, 1554 (1980).
6. S. Yamada, K. Kano, and T. Ogawa, *Anal. Chim. Acta.*, **143**, 21 (1982).
7. J. D. Ingle, Jr., *J. Chem. Educ.*, **51**, 100 (1974).
8. J. D. Winefordner, W. J. McCarthy, and P. A. St. John, *Methods Biochem. Anal.*, **15**, 367 (1967).
9. R. J. Hurtubise, *Solid Surface Luminescence Analysis: Theory Instrumentation, and Applications*, Dekker, New York, 1981.
10. T. Vo-Dinh, *Room Temperature Phosphorimetry for Chemical Analysis*, Wiley, New York, 1984.
11. L. J. Cline Love, J. G. Habarta, and M. Skrilec, *Anal. Chem.*, **53**, 437 (1981).
12. Y. Yang, A. P. D'Silva, and V. A. Fassel, *Anal. Chem.*, **53**, 894 (1981).
13. E. L. Wehry and G. Mamantov, *Anal. Chem.*, **51**, 643A (1979).
14. E. L. Wehry and G. Mamantov, in E. L. Wehry, Ed., *Modern Fluorescence Spectroscopy*, Vol. 4, Plenum, New York, 1981.
15. D. M. Hercules and L. B. Rogers, *J. Phys. Chem.*, **64**, 397 (1960).
16. C. A. Parker, *Photoluminescence of Solutions*, Elsevier, Amsterdam, 1968.
17. N. J. Dovichi, J. C. Martin, J. H. Jett, M. Trkula, and R. A. Keller, *Anal. Chem.*, **56**, 348 (1984).
18. B. E. Kohler, in C. B. Moore, Ed., *Chemical and Biochemical Applications of Lasers*, Vol. 4, Academic Press, New York, 1979.
19. J. C. Wright, in E. L. Wehry, Ed., *Modern Fluorescence Spectroscopy*, Vol. 4, Plenum, New York, 1981.
20. E. V. Shpol'skii and T. N. Bolotnikova, *Pure Appl. Chem.*, **37**, 183 (1974).
21. P. Tokousbalides, E. L. Wehry, and G. Mamantov, *J. Phys. Chem.*, **81**, 1769 (1977).
22. E. V. Shpol'skii, *Soviet Phys. Usp.*, **6**, 411 (1963).
23. J. L. Richards and S. A. Rice, *J. Chem. Phys.*, **54**, 2014 (1971).
24. A. P. D'Silva and V. A. Fassel, *Anal. Chem.*, **56**, 985A (1984).
25. J. R. Maple, E. L. Wehry, and G. Mamantov, *Anal. Chem.*, **52**, 920 (1980).
26. K. P. Dinse and C. L. Winscom, *J. Luminescence*, **18–19**, 500 (1979).
27. R. Tamkivi, I. Renge, and R. Avarmaa, *Chem. Phys. Lett.*, **103**, 103 (1983).
28. F. A. Ermalitski, S. M. Gorbachev, I. E. Zalesskii, and V. V. Nizhnikov, *Opt. Spectrosc.*, **55**, 172 (1983).
29. S. M. Gorbachev, I. E. Zalesskii, and V. V. Nizhnikov, *Opt. Spectrosc.*, **49**, 37 (1980).
30. A. M. Merle, M. Lamotte, S. Risemberg, C. Hauw, J. Gaultier, and J. P. Grivet, *Chem. Phys.*, **22**, 207 (1977).
31. A. M. Merle, M. F. Nicol, and M. A. El-Sayed, *Chem. Phys. Lett.*, **59**, 386 (1978).

32. G. Jansen, M. Noort, N. van Dijk, and J. H. van der Waals, *Mol. Phys.*, **39**, 865 (1980).

33. R. J. Platenkamp and M. Noort, *Mol. Phys.*, **45**, 97 (1982).

34. T.-H. Huang, K. E. Rieckhoff, and E. M. Voigt, *J. Chem. Phys.*, **77**, 3424 (1982).

35. W. M. Pitts, A.-M. Merle, and M. A. El-Sayed, *Chem. Phys.*, **36**, 437 (1979).

36. M. Lamotte, S. Risemberg, A.-M. Merle, and J. Joussot-Dubien, *J. Chem. Phys.*, **69**, 3639 (1978).

37. C. Braeuchle, H. Kabza, J. Voitländer, and E. Clar, *Chem. Phys.*, **32**, 63 (1978).

38. K. A. Muszkat and T. Wismontski-Knittel, *Chem. Phys. Lett.*, **83**, 87 (1981).

39. R. J. Platenkamp, H. D. van Osnabrugge, and A. J. W. G. Visser, *Chem. Phys. Lett.*, **72**, 104 (1980).

40. J. Rima, M. Lamotte, and A. M. Merle, *Nouv. J. Chim.*, **5**, 605 (1981).

41. C. Pfister, *Chem. Phys.*, **2**, 171 (1973).

42. L. A. Nakhimovsky, in *Proceedings of the 1983 Battelle Symposium on Polynuclear Aromatic Hydrocarbons*, Battelle Press, Columbus, OH, 1985.

43. J. D. Spangler and N. G. Kilmer, *J. Chem. Phys.*, **48**, 698 (1968).

44. T. R. Koehler, *Mol. Cryst. Liq. Cryst.*, **50**, 93 (1980).

45. B. S. Causey, G. F. Kirkbright, and C. G. de Lima, *Analyst*, **101**, 367 (1976).

46. G. L. LeBel and J. D. Laposa, *J. Mol. Spectrosc.*, **41**, 249 (1972).

47. A. Colmsjö and U. Stenberg, *Chem. Scr.*, **9**, 227 (1976).

48. D. M. Grebenschikov, N. A. Kovizhnykh, and S. A. Kozlov, *Opt. Spectrosc.*, **37**, 155 (1974).

49. A. Colmsjö and U. Stenberg, *Anal. Chem.*, **51**, 145 (1979).

50. E. P. Lai, E. L. Inman, Jr., and J. D. Winefordner, *Talanta*, **29**, 601 (1982).

51. H. E. Hallam and G. F. Scrimshaw, in H. E. Hallam, Ed., *Vibrational Spectroscopy of Trapped Species*, Wiley, London, 1973.

52. A. Colmsjö and U. Stenberg, *Chem. Sci.*, **11**, 220 (1977).

53. P. Garrigues, M. Lamotte, M. Ewald, and J. Joussot-Dubien, *C. R. Acad. Sci. Paris*, **293**, 567 (1981).

54. T. P. Carter and G. D. Gillispie, *Rev. Sci. Instrum.*, **53**, 1783 (1982).

55. L. Paturel, J. Jarosz, C. Fachinger, and J. Suptil, *Anal. Chim. Acta*, **147**, 293 (1983).

56. H. E. Hallam, *Vibrational Spectroscopy of Trapped Species*, Wiley, London, 1973.

57. B. Meyer, *Low Temperature Spectroscopy*, American Elsevier, New York, 1971.

58. D. M. Hembree, E. R. Hinton, Jr., R. R. Kemmerer, G. Mamantov, and E. L. Wehry, *Appl. Spectrosc.*, **33**, 477 (1979).

59. R. C. Stroupe, P. Tokousbalides, R. B. Dickinson, Jr., E. L. Wehry, and G. Mamantov, *Anal. Chem.*, **49**, 701 (1979).

60. M. B. Perry, Ph.D. Dissertation, University of Tennessee, Knoxville, Tennessee, 1983.

61. M. B. Perry, E. L. Wehry, and G. Mamantov, *Anal. Chem.*, **55**, 1893 (1983).

62. R. B. Dickinson, Jr., and E. L. Wehry, *Anal. Chem.*, **51**, 778 (1979).

63. V. B. Conrad, R. R. Gore, J. L. Hammons, J. R. Maple, M. B. Perry, and E.

L. Wehry, in D. Eastwood, Ed., *New Directions in Molecular Luminescence*, ASTM, Philadelphia, 1983.

64. T. D. Harris and F. E. Lytle, in D. S. Kliger, Ed., *Ultrasensitive Laser Spectroscopy*, Academic Press, New York, 1983.

65. G. F. Kirkbright and C. G. de Lima, *Analyst*, **99**, 338 (1974).

66. V. B. Conrad and E. L. Wehry, *Appl. Spectrosc.*, **37**, 46 (1983).

67. J. N. Demas, *Excited State Lifetime Measurements*, Academic Press, New York, 1983.

68. T. Vo-Dinh, in E. L. Wehry, Ed., *Modern Fluorescence Spectroscopy*, Vol. 4, Plenum, New York, 1981.

69. E. L. Inman, Jr., M. J. Kerkhoff, and J. D. Winefordner, *Spectrochim. Acta*, **39A**, 245 (1983).

70. E. L. Wehry, V. B. Conrad, J. L. Hammons, J. R. Maple, and M. B. Perry, *Opt. Eng.*, **22**, 558 (1983).

71. H. S. Hertz, J. M. Brown, S. N. Chesler, F. R. Guenther, L. R. Hilpert, W. E. May, R. M. Parris, and S. A. Wise, *Anal. Chem.*, **52**, 1650 (1980).

72. Y. Yang, A. P. D'Silva, and V. A. Fassel, *Anal. Chem.*, **53**, 2107 (1981).

73. G. D. Renkes, S. N. Walters, C. S. Woo, M. K. Iles, A. P. D'Silva, and V. A. Fassel, *Anal. Chem.*, **55**, 2229 (1983).

74. Y. Yang, A. P. D'Silva, V. A. Fassel, and M. Iles, *Anal. Chem.*, **52**, 1350 (1980).

75. P. Garrigues, R. De Vazelhes, M. Ewald, J. Joussot-Dubien, J.-M. Schmitter, and G. Guiochon, *Anal. Chem.*, **55**, 138 (1983).

76. G. F. Kirkbright and C. G. de Lima, *Chem. Phys. Lett.*, **37**, 165 (1976).

77. H. J. Haink and J. R. Huber, *J. Mol. Spectrosc.*, **60**, 31 (1976).

78. A. Colmsjö, Y. U. Zebühr, and C. E. Östman, *Anal. Chem.*, **54**, 1673 (1982).

79. T. P. Carter and G. D. Gillispie, *J. Phys. Chem.*, **86**, 2691 (1982).

80. I.-F. Hung, D. F. Williams, and R. Yip, *J. Luminescence*, **15**, 231 (1977).

81. O. S. Khalil and L. Goodman, *J. Phys. Chem.*, **80**, 2170 (1976).

82. R. N. Capps and M. Vala, *Photochem. Photobiol.*, **33**, 673 (1981).

83. H. J. Griesser and R. Bramley, *Chem. Phys. Lett.*, **88**, 27 (1982).

84. T. P. Carter, M. H. Van Benthem, and G. D. Gillispie, *J. Phys. Chem.*, **87**, 1891 (1983).

85. T.-H. Huang, K. E. Rieckhoff, and E. M. Voigt, *Chem. Phys.*, **36**, 423 (1979).

86. J. Joussot-Dubien, M. Lamotte, and J. Pereyre, *J. Photochem.*, **17**, 347 (1981).

87. M. Lamotte, *J. Phys. Chem.*, **85**, 2632 (1981).

88. W. P. Cofino, G. P. Hoornweg, C. Gooijer, C. MacLean, and N. H. Velthorst, *Spectrochim. Acta*, **39A**, 283 (1983).

89. V. Lejeune, A. Despres, and E. Migirdicyan, *J. Phys. Chem.*, **88**, 2719 (1984).

90. S. M. Thornberg and J. R. Maple, *Anal. Chem.*, **56**, 1542 (1984).

91. J. R. Maple, University of New Mexico, personal communication, May 1984.

92. R. I. Personov, *Spectrochim. Acta*, **38B**, 1533 (1983).

93. H. J. Griesser and U. P. Wild, *J. Chem. Phys.*, **73**, 4715 (1980).

94. K. Cunningham, J. M. Morris, J. Fünfschilling, and D. F. Williams, *Chem. Phys. Lett.*, **32**, 581 (1975).

95. E. I. Al'shits, R. I. Personov, and B. M. Kharlamov, *Chem. Phys. Lett.*, **40**, 116 (1976).

96. E. I. Al'shits, R. I. Personov, and B. M. Kharlamov, *Opt. Spectrosc.*, **41**, 474 (1976).

97. G. W. Suter and U. P. Wild, *J. Luminescence*, **24–25**, 497 (1981).

98. R. L. Williamson and A. L. Kwiram, *J. Phys. Chem.*, **83**, 3393 (1979).

99. J. P. Lemaistre and A. H. Zewail, *Chem. Phys. Lett.*, **68**, 302 (1979).

100. B. E. Kohler and R. T. Loda, *J. Chem. Phys.*, **74**, 5424 (1981).

101. R. I. Personov and B. M. Khablamovanov, *Opt. Commun.*, **7**, 417 (1973).

102. W. C. McColgin, A. P. Marchetti, and J. H. Eberly, *J. Am. Chem. Soc.*, **100**, 5622 (1978).

103. P. J. Angiolillo, J. S. Leigh, Jr., and J. M. Vanderkooi, *Photochem. Photobiol.*, **36**, 133 (1982).

104. G. Flatscher and J. Friedrich, *Chem. Phys. Lett.*, **50**, 32 (1977).

105. P. Avouris, A. Campion, and M. A. El-Sayed, *Chem. Phys. Lett.*, **50**, 9 (1977).

106. L. A. Bykovskaya, R. I. Personov, and B. M. Kharlamov, *Chem. Phys. Lett.*, **27**, 80 (1974).

107. R. I. Personov and E. I. Al'shits, *Chem. Phys. Lett.*, **33**, 85 (1975).

108. J. Fünfschilling and D. F. Williams, *Photochem. Photobiol.*, **26**, 109 (1977).

109. R. Rossetti, R. C. Haddon, and L. E. Brus, *J. Am. Chem. Soc.*, **102**, 6913 (1980).

110. K. Brenner, Z. Ruziewicz, G. Suter, and U. P. Wild, *Chem. Phys.*, **59**, 157 (1981).

111. I. Chiang, J. M. Hayes, and G. J. Small, *Anal. Chem.*, **54**, 315 (1982).

112. F. A. Burkhalter and U. P. Wild, *Chem. Phys.*, **66**, 327 (1982).

113. M. N. Sapozhnikov and V. I. Alekseev, *Chem. Phys. Lett.*, **97**, 331 (1983).

114. A. P. Marchetti, M. Scozzafava, and R. H. Young, *Chem. Phys. Lett.*, **51**, 424 (1977).

115. W. P. Cofino, J. W. Hofstraat, G. P. Hoornweg, C. Gooijer, C. MacLean, and N. H. Velthorst, *Chem. Phys. Lett.*, **89**, 17 (1982).

116. J. R. Maple and E. L. Wehry, *Anal. Chem.*, **53**, 266 (1981).

117. H. J. Dewey and J. Michl, *J. Luminescence*, **24–25**, 527 (1981).

118. J. Najbar, A. M. Turek, and T. D. S. Hamilton, *J. Luminescence*, **26**, 281 (1982).

119. J. Najbar and A. M. Turek, *Chem. Phys. Lett.*, **73**, 536 (1980).

120. J. Fünfschilling and D. F. Williams, *Appl. Spectrosc.*, **30**, 443 (1976).

121. J. C. Brown, M. C. Edelson, and G. J. Small, *Anal. Chem.*, **50**, 1394 (1978).

122. J. C. Brown, J. A. Duncanson, Jr., and G. J. Small, *Anal. Chem.*, **52**, 1711 (1980).

123. J. C. Brown, J. M. Hayes, J. A. Warren, and G. J. Small, in G. M. Hieftje, J. C. Travis, and F. E. Lytle, Eds., *Lasers in Chemical Analysis*, Humana Press, Clifton, NJ, 1981.

124. M. J. McGlade, J. M. Hayes, G. J. Small, V. Heisig, and A. M. Jeffrey, in W. S. Lyon, Ed., *Analytical Spectroscopy*, Elsevier, Amsterdam, 1984, p. 31.

125. L. A. Bykovskaya, R. I. Personov, and Y. V. Romanovskii, *Anal. Chim., Acta*, **125**, 1 (1981).

126. D. Bolton and J. D. Winefordner, *Talanta*, **30**, 713 (1983).

127. V. Heisig, A. M. Jeffrey, M. J. McGlade, and G. J. Small, *Science*, **223**, 289 (1984).
128. A. A. Gorokhovskii, R. K. Kaarli, and L. A. Rebane, *JETP Lett. (Eng. tr.)*, **20**, 216 (1974).
129. L. A. Rebane, A. A. Gorokhovskii, and J. V. Kikas, *Appl. Phys.*, **B29**, 235 (1982).
130. H. de Vries and D. A. Wiersma, *J. Chem. Phys.*, **72**, 1851 (1980).
131. S. Voelker, R. M. Macfarlane, A. Z. Genack, H. P. Trommsdorff, and J. H. van der Waals, *J. Chem. Phys.*, **67**, 1759 (1977).
132. J. Friedrich, H. Wolfrum, and D. Haarer, *J. Chem. Phys.*, **77**, 2309 (1982).
133. B. Jackson and R. Silbey, *Chem. Phys. Lett.*, **99**, 331 (1983).
134. S. Völker and R. M. Macfarlane, *J. Chem. Phys.*, **73**, 4476 (1980).
135. J. Friedrich and D. Haarer, *J. Chem. Phys.*, **76**, 61 (1982).
136. R. W. Olson, H. W. H. Lee, F. G. Patterson, M. D. Fayer, R. M. Shelby, D. P. Burum, and R. M. Macfarlane, *J. Chem. Phys.*, **77**, 2283 (1982).
137. E. Cuellar and G. Castro, *Chem. Phys.*, **54**, 217 (1981).
138. F. A. Burkhalter, G. W. Suter, U. P. Wild, V. D. Samoilenko, N. V. Rasumova, and R. I. Personov, *Chem. Phys. Lett.*, **94**, 483 (1983).
139. H. P. H. Thijssen, R. van den Berg, and S. Völker, *Chem. Phys. Lett.*, **97**, 295 (1983).
140. R. I. Personov, E. I. Al'shits, and L. A. Bykovskaya, *Opt. Commun.*, **6**, 169 (1972).
141. B. M. Kharlamov, R. I. Personov, and L. A. Bykovskaya, *Opt. Commun.*, **12**, 191 (1974).
142. B. M. Kharlamov, R. I. Personov, and L. A. Bykovskaya, *Opt. Commun.*, **39**, 137 (1975).
143. J. M. Hayes and G. J. Small, *Chem. Phys.*, **27**, 151 (1978).
144. J. M. Hayes and G. J. Small, *Chem. Phys. Lett.*, **54**, 435 (1978).
145. J. M. Hayes and G. J. Small, *J. Luminescence*, **18–19**, 219 (1979).
146. J. M. Hayes, R. P. Stout, and G. J. Small, *J. Chem. Phys.*, **74**, 4266 (1981).
147. R. M. Shelby and R. M. Macfarlane, *Chem. Phys. Lett.*, **64**, 545 (1979).
148. B. L. Feary, T. P. Carter, and G. J. Small, *J. Phys. Chem.*, **87**, 3590 (1983).
149. W. Breinl, J. Friedrich, and D. Haarer, *Chem. Phys. Lett.*, **106**, 487 (1984).
150. T. D. Harris and F. E. Lytle, in D. S. Kliger, Ed., *Ultrasensitive Laser Spectroscopy*, Academic Press, New York, 1983.
151. D. H. Levy, L. Wharton, and R. E. Smalley, in C. B. Moore, Ed., *Chemical and Biochemical Applications of Lasers*, Vol. 2, Academic Press, New York, 1977.
152. R. E. Smalley, L. Wharton, and D. H. Levy, *Acc. Chem. Res.*, **10**, 139 (1977).
153. D. H. Levy, *Ann. Rev. Phys. Chem.*, **31**, 187 (1980).
154. J. M. Hayes and G. J. Small, *Anal. Chem.*, **55**, 565A (1983).
155. T. A. Miller, *Science*, **223**, 545 (1984).
156. M. V. Johnston, *Trends Anal. Chem.*, **3**, 58 (1984).
157. J. M. Hayes and G. J. Small, *Anal. Chem.*, **54**, 1202 (1982).
158. P. S. H. Fitch, C. A. Haynam, and D. H. Levy, *J. Chem. Phys.*, **73**, 1064 (1980).

159. C. A. Haynam, D. V. Brumbaugh, and D. H. Levy, *J. Chem. Phys.*, **79**, 1581 (1983).
160. D. H. Levy, C. A. Haynam, and D. V. Brumbaugh, *Faraday Discuss. Chem. Soc.*, **73**, 137 (1982).
161. D. E. Poeltl and J. K. McVey, *J. Chem. Phys.*, **78**, 4349 (1983).
162. K. H. Fung, H. L. Selzle, and E. W. Schlag, *J. Phys. Chem.*, **87**, 5113 (1983).
163. U. Even and J. Jortner, *J. Chem. Phys.*, **78**, 3445 (1983).
164. A. Amirav, U. Even, and J. Jortner, *Chem. Phys. Lett.*, **72**, 16 (1980).
165. A. Amirav, U. Even, and J. Jortner, *J. Chem. Phys.*, **75**, 2489 (1981).
166. C. A. Haynam, D. V. Brumbaugh, and D. H. Levy, *J. Chem. Phys.*, **80**, 2256 (1984).
167. Y. Nibu, H. Abe, N. Mikami, and M. Ito, *J. Phys. Chem.*, **87**, 3898 (1983).
168. H. Abe, N. Mikami, M. Ito, and Y. Udagawa, *J. Phys. Chem.*, **86**, 2567 (1982).
169. P. M. Felker and A. H. Zewail, *Chem. Phys. Lett.*, **94**, 454 (1983).
170. A. Oikawa, H. Abe, N. Mikami, and M. Ito, *J. Phys. Chem.*, **87**, 5083 (1983).
171. N. Gonohe, H. Abe, N. Mikami, and M. Ito, *J. Phys. Chem.*, **87**, 4406 (1983).
172. F. M. Behlen, N. Mikami, and S. A. Rice, *Chem. Phys. Lett.*, **60**, 364 (1979).
173. M. G. Liverman, S. M. Beck, D. L. Monts, and R. E. Smalley, *J. Chem. Phys.*, **70**, 192 (1979).
174. C. E. Otis and P. M. Johnson, *Rev. Sci. Instrum.*, **51**, 1128 (1980).
175. A. Auerbach and R. McDiarmid, *Rev. Sci. Instrum.*, **51**, 1273 (1980).
176. T. E. Adams, B. H. Rockney, R. J. S. Morrison, and E. R. Grant, *Rev. Sci. Instrum.*, **52**, 1469 (1981).
177. M. R. Adriaens, W. Allison, and B. Feuerbacher, *J. Phys. E*, **14**, 1375 (1981).
178. R. L. Byer and M. D. Duncan, *J. Chem. Phys.*, **74**, 2174 (1981).
179. J. B. Cross and J. J. Valentine, *Rev. Sci., Instrum.*, **53**, 38 (1982).
180. B. M. DeKoven, D. H. Levy, H. H. Harris, B. R. Zegarski, and T. A. Miller, *J. Chem. Phys.*, **74**, 5659 (1981).
181. A. Amirav, U. Even, and J. Jortner, *J. Chem. Phys.*, **75**, 3770 (1981).
182. D. M. Lubman, *J. Phys. Chem.*, **85**, 3752 (1981).
183. M. Heaven, T. Sears, V. E. Bondybey, and T. A. Miller, *J. Chem. Phys.*, **75**, 5271 (1981).
184. A. Amirav, U. Even, and J. Jortner, *Chem. Phys. Lett.*, **83**, 1 (1981).
185. A. Amirav, U. Even, and J. Jortner, *Chem. Phys.*, **67**, 1 (1982).
186. A. Amirav, U. Even, and J. Jortner, *Anal. Chem.*, **54**, 1666 (1982).
187. T. Imasaka, T. Shigezumi, and N. Ishibashi, *Analyst*, **109**, 277 (1984).
188. A. Amirav, U. Even, and J. Jortner, *Chem. Phys.*, **51**, 31 (1980).
189. A. M. Griffiths and P. A. Freedman, *J. Chem. Soc. Faraday Trans. 2*, **78**, 391 (1982).
190. N. Mikami, A. Hiraya, I. Fujiwara, and M. Ito, *Chem. Phys. Lett.*, **74**, 531 (1980).
191. R. E. Smalley, D. H. Levy, and L. Wharton, *J. Chem. Phys.*, **64**, 3266 (1976).
192. F. M. Behlen and S. A. Rice, *J. Chem. Phys.*, **75**, 5672 (1981).
193. C. H. Chen, S. D. Kramer, D. W. Clark, and M. G. Payne, *Chem. Phys. Lett.*, **65**, 419 (1979).

194. G. M. McClelland, K. L. Saengar, J. J. Valentine, and D. R. Herschbach, *J. Phys. Chem.*, **83,** 947 (1979).

195. J. A. Warren, J. M. Hayes, and G. J. Small, *Anal. Chem.*, **54,** 138 (1982).

196. T. Imasaka, H. Fukuoka, T. Hayashi, and N. Ishibashi, *Anal. Chim. Acta*, **156,** 111 (1984).

197. A. P. D'Silva, M. Iles, G. Rice, and V. A. Fassel, Technical Report IS-4556 UC-90E to U.S. Department of Energy (1984); A. P. D'Silva, Iowa State University, personal communication (June 1984).

198. T. G. Dietz, M. A. Duncan, A. C. Pulu, and R. E. Smalley, *J. Phys. Chem.*, **86,** 4026 (1982).

199. S. Leutwyler, U. Even, and J. Jortner, *Chem. Phys.*, **58,** 409 (1981).

200. Lasertechnics, Inc., Albuquerque, NM, Model LPV pulsed molecular beam valve specifications (1982).

201. R. E. Smalley, *J. Chem. Educ.*, **59,** 934 (1982).

202. R. B. Bernstein, *J. Phys. Chem.*, **86,** 1178 (1982).

203. V. S. Letokhov, *Nonlinear Laser Chemistry*, Springer-Verlag, Berlin, 1983.

204. S. Leutwyler and U. Even, *Chem. Phys. Lett.*, **81,** 578 (1981).

205. D. M. Lubman and M. N. Kronick, *Anal. Chem.*, **54,** 660 (1982).

206. H. Shinohara, *J. Chem. Phys.*, **79,** 1732 (1983).

207. R. G. Orth, H. T. Jonkman, and J. Michl, *Int. J. Mass. Spectrom. Ion Phys.*, **43,** 41 (1982).

208. R. G. Cooks, K. L. Busch, and G. L. Glish, *Science*, **222,** 273 (1983).

209. M. L. Vestal, *Mass Spectrom. Rev.*, **2,** 447 (1983).

CHAPTER

8

LINEAR AND NONLINEAR
SITE-SELECTIVE
LASER SPECTROSCOPY

J. C. WRIGHT, D. C. NGUYEN, J. K. STEEHLER,
M. A. VALENTINI, and R. J. HASKELL

Department of Chemistry
University of Wisconsin
Madison, Wisconsin

1. INTRODUCTION

Laser-excited fluorescence spectroscopy is one of the most sensitive analytical techniques available (1). Modern tunable laser sources are able to deliver 10^{15}–10^{18} photons/second to a sample within its absorption linewidth. Modern spectroscopic instrumentation can measure light levels of 1 photon/second from a sample. This tremendous range allows even small sample absorption to be easily observed when fluorescence is used as the monitor. The measurement of particular components in a sample is typically limited by background arising from the wealth of materials that are present at very low concentrations (2). It was clear even in the early stages of using lasers for analytical chemistry that improving the selectivity of an analysis was the important frontier of analytical fluorescence spectroscopy. The laser had two properties that were potentially applicable to measurement specificity: the rapid time response of a laser could determine information about relaxation processes characteristic of a sample component (3) while the narrow linewidth of a laser could selectively excite a narrow absorption line of a sample component. This chapter deals with the techniques that take advantage of the laser's narrow linewidth—a set of techniques that has come to be known as site-selective spectroscopy (4).

Site-selective spectroscopy can be used whenever a transition of a spectroscopically active sample component has a homogeneous linewidth that is narrower than its separation from transitions of other sample components. A narrow-line laser can then be tuned to that absorption transition and excite that particular sample component with a higher efficiency than the other components. One can view the laser excitation as labeling the component of interest. The resulting fluorescence spectrum has an enhanced contribution from the labeled component, and the positions and intensities of the fluorescence lines are characteristic of the lower levels of that component. One can also monitor the fluorescence intensity of a characteristic line from the component as a tunable laser is scanned over the region that can excite that fluorescence. The resulting excitation spectrum has an enhanced contribution from the component monitored, and the positions and intensities of the excitation lines are characteristic of the upper levels of that component.

Site-selective spectroscopy can be performed either on samples where there are a finite number of unique components whose individual transitions are resolvable or on samples where there are so many components that it is not possible to resolve individual transitions. The components can be different ions or molecules or the same ion or molecule in different environments or sites. Ions or molecules in amorphous or disordered materials experience a continuous range of environments that together cause an inhomogeneously broadened line. Even in this case, site-selective spectroscopy is possible because a laser tuned to a position within the inhomogeneous line will excite only those components that are resonant with the laser, and the resulting fluorescence can be much narrower. This technique has been called fluorescence line narrowing (FLN).

The success of site-selective spectroscopy depends on the relationships between the energy levels of the labeled component and the other components. In many ways, site-selective spectroscopy is misnamed because it is really an energy-selective method. All components that have an energy level resonant with the laser will be excited. The fluorescence from these sites or components can occur at different frequencies so the nonresonant fluorescence can have contributions from many sites or components (5). Quantitatively, the energy shift of the $i \rightarrow j$ transition from its unperturbed value, ω_{ij}^0, can be described by

$$\omega_{ij} - \omega_{ij}^0 = \alpha \, \Delta\omega_{ij}^\alpha + \beta \, \Delta\omega_{ij}^\beta + \cdots \tag{8.1}$$

where α, β, ... describe the perturbations caused by the environment and $\Delta\omega_{ij}^\alpha$ is the shift of the $i \rightarrow j$ transition caused by the α perturbation. The number of terms required depends on the symmetry of the perturbing environment and the nature of the electronic transition (6). For a finite number of unique components or sites, each would have different values for the α, β, ... parameters. For the sites that form an inhomogeneous distribution, there would be a Gaussian distribution that describes the strains or disorder of the material (6). Considering only α and β as broadening parameters, the overall distribution could be represented as

$$g(\alpha, \beta) = \frac{1}{2\pi\sigma_\alpha\sigma_\beta} e^{-\alpha^2/2\sigma_\alpha^2} e^{-\beta^2/2\sigma_\beta^2} \qquad (8.2)$$

A consequence of Eq. (8.1) is that there is not a unique combination of α, β, ... values for a given transition energy ($\omega_{ij} - \omega_{ij}^0$) so many sites can potentially have the same energy. Thus excitation at any given energy does not determine particular values of α, β, ... This situation is called accidental degeneracy. Since different transitions have different values for $\Delta\omega_{ij}^\alpha$, $\Delta\omega_{ij}^\beta$, ... , the nonresonant fluorescence to other levels can fall at different energies for different sites. Note that the resonant fluorescence must still occur at the same energy because the $\Delta\omega_{ij}^\alpha$, $\Delta\omega_{ij}^\beta$, ... have not changed. However, if there is only one broadening parameter, α, Eq. (8.1) will determine a unique α value for a transition energy ($\omega_{ij} - \omega_{ij}^0$) and an excitation will select a specific site. Similarly, if α, β, ... are highly correlated with each other, the excitation specificity will be high. In the limit of complete correlation, only one parameter is required. The amount of fluorescence line narrowing therefore depends on the number of broadening parameters and the correlation between them. If there is little correlation, the fluorescence resulting from narrow-line excitation within the inhomogeneous line profile can span the entire inhomogeneous width that would be observed with broadband excitation, and fluorescence line narrowing is not observed.

In this chapter, we shall review the applications that site-selective spectroscopy has found in analytical chemistry for systems with finite numbers of components and those with large inhomogeneous broadening. In addition, we shall describe the emerging area of multiresonant, nonlinear, four-wave mixing spectroscopy that promises to extend site-selective techniques to nonfluorescent samples.

2. SITE-SELECTIVE LINEAR SPECTROSCOPY

2.1. Measurement of Fluorescent Inorganic Ions

The first application of site-selective spectroscopy to analytical chemistry was to ultratrace metal analysis (7). Co-precipitation was used to incorporate lanthanide (7, 8) and actinide ions (9) into CaF_2 precipitates. The precipitates are cooled to low temperatures and studied by site-selective spectroscopy. These ions have inherently sharp-line spectra because the transitions occur within inner, unfilled $4f^n$ and $5f^n$ orbitals that are shielded from external broadening effects by outer orbitals. Example fluorescence and excitation spectra are shown in Fig. 8.1a for Er^{3+}, a typical lanthanide ion, and in Fig. 8.1b for U^{6+}, a typical actinide ion. There are broad vibrational lines seen in actinide spectra because of the stronger coupling to the lattice. The vibrational bands appear in both the excitation and fluorescence spectra as is typical for molecular spectra. Lanthanide spectra contain only small contributions (typically 10^3 times lower) from

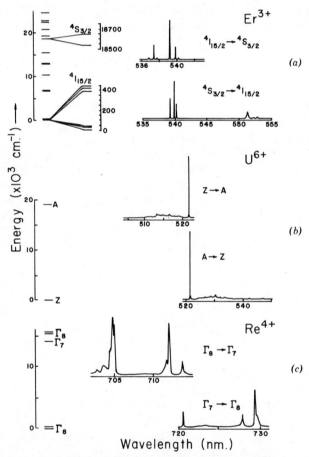

Fig. 8.1. The spectroscopic characteristics of representative lanthanide (Er^{3+}), actinide (U^{6+}), and platinum group transition metals (Re^{4+}) are compared. The electronic energy levels for each ion are given on the left while representative excitation and fluorescence spectra are given on the right.

the vibrational bands and are not visible in the figure. The different transitions correspond to different crystal field states. Recent work in our group has succeeded in extending the co-precipitation techniques to the transition metals as well using MgO (10), K_2SnCl_6, and related materials. The transition metal ions have sharp-line spectra because of transitions within unfilled $3d^n$, $4d^n$, or $5d^n$ orbitals. An example spectrum is shown in Fig. 8.1c for Re^{4+}. The vibrational bands are more important because of the greater interaction with the lattice.

An analysis is performed by dissolving the sample and co-precipitating the analyte ions using the appropriate precipitate. The precipitate is recovered, washed, dried, and cooled to 10 K. The spectra of different lanthanide ions are highly characteristic of the particular ion. The fluorescence intensities can be calibrated against standards prepared by the same precipitation techniques from known concentration solutions. The detection limits for this type of analysis are typically in the low parts per trillion range if a simple N_2 pumped dye laser system is used for excitation. More powerful excitation sources can lower the detection limit by another two orders of magnitude (11). The selectivity of the analysis is very high for two reasons. The lifetimes of the fluorescent levels are long, typically 10^{-4}–10^{-3} s, and pulsed lasers with gated detection can eliminate any fluorescence from organics or other short-lived fluorescent species as well as Raman scatter. Consequently, the background fluorescence is negligible and the detection limits scale inversely with the laser power. The spectral lines of the analyte ions are also very sharp, and there is excellent discrimination between analyte ions in both the excitation step and the fluorescence detection. The selectivity is particularly high for lanthanide ion analysis because the broader vibrational sidebands are so weak.

2.2. Measurement of Nonfluorescent Ions

The procedures can be extended to a wider range of ions by taking advantage of the sensitivity that crystal field splittings of actinide, lanthanide, and transition metals have to the immediate environment. If a probe ion that has sharp-line transitions is introduced to a crystal lattice and another analyte ion foreign to the lattice is associated with the probe ions, the perturbed probe ion site that results can be selectively excited and measured as an indirect method for the analyte ion. The same precipitation methods work well for incorporating both the probe and analyte ions in a suitable matrix. Two physical principles have been used to cause association between the probe and analyte ions—clustering and charge compensation (12–15). In the first, a high concentration of the probe ion is used to force the formation of probe ion dimers. Dimerization typically occurs at concentrations of 0.1 mol % relative to the lattice cation replaced. Analyte ions that are similar to the probe ion can enter the dimers in place of one probe ion and perturb the other probe ions (14). There is a problem, however, in discriminating between the lines from probe ions in dimers containing an analyte and the dimers containing just probe ions. The former are usually present in a small concentration relative to the latter so there is an inherent problem in discriminating between sites over the large dynamic range. This problem was eliminated by choosing the excitation levels and the probe ions judiciously so the fluorescence is quenched by energy transfer between two probe ions in a dimer. Replacement of a probe ion will then remove the energy transfer quenching, and the remaining probe ion in the dimer will fluoresce. These effects are illustrated in Fig. 8.2 for a CaF_2 precipitate with Er^{3+} co-precipitated at a 0.1 mol % level. The energies of two neighboring Er^{3+} ions are shown in the middle of the left-hand side of the figure. If level E is excited, the excitation can relax to the lower level D by the pair of energy transfer processes indicated by the numbered arrows. The excitation first relaxes to a lower level, simultaneously exciting the neighboring Er^{3+} ion to an excited level. The difference

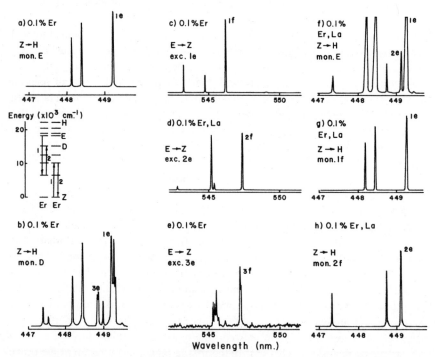

Fig. 8.2. The excitation ($Z \rightarrow H$) and fluorescence ($E \rightarrow Z$) spectra of the different Er^{3+} sites in CaF_2 precipitates are shown. The concentrations of the ions in solutions (either Er^{3+} or La^{3+} relative to Ca^{2+}), the transition scanned, and the transition excited or monitored are indicated in each spectrum. The pair of energy transfer processes that quench the E-level fluorescence are shown on the left for an Er–Er dimer.

in total energy between the initial and final states of the pair is lost to the surroundings as heat. A second energy transfer can occur where the excitation of the second ion is returned to the initial ion, causing excitation of the D level.

This set of energy transfer processes is very efficient because of the close proximity of two ions in a dimer so the E-level fluorescence is quenched. Figure 8.2a shows the excitation spectrum of the H level when E-level fluorescence is monitored. Only the lines of a single Er^{3+} ion with a neighboring oxygen ion charge compensation are observable. If the same scan is performed monitoring fluorescence from level D, new lines appear corresponding to the new dimer site formed. Figure 8.2f shows the analogous scan to Fig. 8.2a for a CaF_2 precipitate containing both La^{3+} and Er^{3+}. Even though the E-level fluorescence is monitored, dimer lines are seen for the Er–La mixed dimers since La^{3+} does not have levels that can quench the E-level fluorescence. If the lines of individual sites are excited, single-site fluorescence spectra can be observed. Setting the dye laser at the lines labeled 1e, 2e, and 3e in Figs. 8.2a, 8.2f, and 8.2b, respectively, produces the fluorescence spectra seen in Figs. 8.2c and 8.2d, and 8.2e, respectively. The fluorescence of the Er–Er dimer in Fig. 8.2e is very weak because of the quenching, but it can be observed with gated detection for a short time after the laser excitation. If a monochromator with 0.05-nm resolution monitors the fluorescence lines labeled 1f and 2f in Figs. 8.2c and 8.2d while the dye laser wavelength is scanned over the same excitation region as before, the single-site excitation spectra shown in Figs. 8.2g and 8.2h are seen. The latter spectrum corresponds to the lines of the Er–La mixed dimer and there is no contribution from other sites.

A variety of different ions can be measured by this technique. Even though the ions may be very similar to each other, the shift of the probe ion's lines is large enough to easily resolve all of the ions. The spectrum at the top of Fig. 8.3 shows the excitation spectrum of the Er^{3+} $^4I_{15/2} \rightarrow {}^4F_{5/2}$ transition in a CaF_2 precipitate containing 0.1 mol % Er^{3+} and 0.1 mol % Sc^{3+}, Zr^{4+}, Ce^{3+}, Th^{4+}, La^{3+}, Gd^{3+}, Lu^{3+}, and Y^{3+}. There are Er^{3+} sites characteristic of the mixed dimer for each different ion. In a similar fashion as in Fig. 8.2, if an individual fluorescence line is monitored and the excitation scan is repeated, only the lines from the site monitored can be seen. The excitation lines below correspond to the decomposition of the overall spectrum at the top into the component sites that make it up. This application illustrates the particularly high selectivity of site-selective spectroscopy.

Charge compensation has also been used to cause association of an analyte and probe ion (12). For example, if a trivalent ion such as Eu^{3+} replaces a divalent cation such as Sr^{2+} in $SrSO_4$, the charge mismatch must be compensated. Typically this compensation would occur by having a Sr vacancy compensate two Eu^{3+} ions. However, if the solution that contained the precipitating $SrSO_4$ also contained PO_4^{3-}, the PO_4^{3-} could compensate the Eu^{3+} directly. The

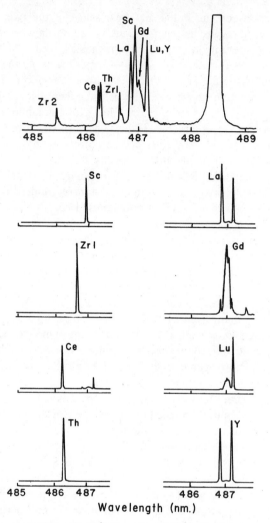

Fig. 8.3. The top spectrum shows the Er^{3+} $Z \rightarrow G$ excitation spectrum monitoring fluorescence from all Er^{3+} sites for an Er^{3+} sites for an Er^{3+}, Zr^{4+}, Ce^{3+}, Th^{4+}, La^{3+}, Sc^{3+}, Gd^{3+}, Lu^{3+}, and Y^{3+} : CaF_2 precipitate. The lower spectra show the single excitation spectra obtained by monitoring a specific fluorescence line from an individual site in the same precipitate.

Coulombic attraction between the net charges of Eu^{3+} and PO_4^{3-} would favor association, and a new site would form characteristic of PO_4^{3-}. These effects have been demonstrated for a number of systems (12, 13, 16). The charge compensation method of achieving association has not been practical for general analysis because the intensity from the charge-compensated sites examined thus far does not vary linearly with concentration but decreases rapidly at lower concentrations. In addition, the kinetics of forming the sites that are examined is not favorable, and there is a basic irreproducibility from sample to sample that has not been solved.

2.3. Measurement of Fluorescent Organic Molecules

Organic molecules can also have sharp-line spectral features if they are frozen in suitable matrices at cryogenic temperatures. This application is discussed in much more detail in Chapter 7. Site-selective spectroscopy was first applied to analytical measurements of such systems by Wehry (17, 18) and Fassel (19). Wehry developed matrix isolation methods in which fluorescent molecules were imbedded in an inert matrix that was formed by freezing gases onto a substrate at cryogenic temperatures. Fassel used the Shpol'skii systems discovered by Shpol'skii (20) in which a polyaromatic hydrocarbon (PAH) molecule is frozen in an n-alkane matrix at cryogenic temperatures. In both cases, the molecular rotational freedom is reduced, and the molecular conformation and environment are sharply restricted. This procedure sharply reduces the inhomogeneous broadening, and the spectra of individual sites and components in a mixture can be resolved by site-selective spectroscopy. A measurement of concentration can be done by comparing the spectral intensities against standards. Figure 8.4 shows typical spectra for a measurement of 6,8 dimethylben[a]anthracene (6,8-DM-B[a]A) and 11-methylbenz[a]anthracene (11-M-B[a]A) mixture in n-octane at 4.2 K by Yang et al. (19). The fluorescence spectrum in Fig. 8.4a was obtained nonselectively so both components contribute. If the fluorescence lines at 385 nm of 6,8 DM-B[a]A are monitored, one can see the single component excitation spectrum in Fig. 8.4d. A similar excitation spectrum can be seen in Fig. 8.4e for 11-M-B[a]A when its fluorescence line is monitored at 384.8 nm. It is clear that the 375.0-nm excitation line of 6,8-DM-B[a]A and the 374.7-nm excitation line of 11-M-B[a]A are resolved. If the dye lasers are tuned to either of these lines, the fluorescence spectra seen in Figs. 8.4b and 8.4c result. Comparing these with Fig. 8.4a shows that there is excellent selectivity in the analysis of these two compounds, both of which are quite similar to each other.

Site-selective methods have also been applied to organic analysis in disordered matrices such as frozen organic glasses (21, 22). There is a great deal of inhomogeneous broadening in amorphous materials because of the wide range of environments for the spectroscopically active molecule. Excitation in the lower energy region of the absorption band where the $0 \rightarrow 0'$ vibrationless transition dominates will selectively excite a specific site, and strong fluorescence line narrowing will be seen, even for the nonresonant transitions. The nonresonant transitions are narrowed because the energy level shifts of molecular levels are highly correlated. The vibrational energies do not change markedly

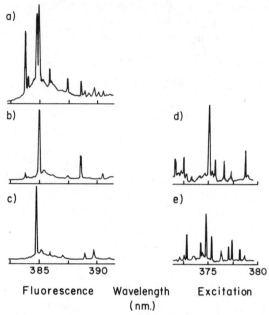

Fig. 8.4. The fluorescence (left hand) and excitation (right hand) spectra of a mixture of 6,8 dimethylbenz[*a*]anthracene and 11-methylbenz[*a*]anthracene in *n*-octane at 4.2 K (19). Single component fluorescence [(b) and (c)] and excitation [(d) and (e)] spectra are obtained by exciting or monitoring transitions from one component. The nonselective fluorescence spectrum is shown in (a).

if the site symmetry changes, so the vibrational levels tend to occur at the same energy above the vibrationless energy state whether it is a ground or excited electronic state. The energies of excited electronic states on the other hand are strongly influenced by the local environment. Selective excitation of the $0 \rightarrow 0'$ transition causes excitation of a subset of sites that have the same $0'$ energy and similar ground-state vibrational levels, v. Thus the $0' \rightarrow v$ fluorescence is sharp. Similarly, monitoring fluorescence at the $0' \rightarrow v$ wavelength with a high-resolution monochromator will fix what $0'$ energy is observed so an excitation scan will locate sharp lines at the $0 \rightarrow v'$ transition. There will still be residual inhomogeneous broadening of these transitions because of the lack of correlation between the shifts of the vibrational and electronic levels. This residual broadening will become very large if one deals with transitions between states that are both strongly affected by environmental effects such as the transition between two excited electronic states. For example, selective excitation of the $S_0 \rightarrow S_2$ transition of azulene does not cause narrowing of the $S_2 \rightarrow S_1$ transition because both S_1 and S_2 are strongly affected by the environment, and the accidental degeneracies are important (5). The spectroscopy of organic systems at cryogenic temperatures is discussed in much more detail in Chapter 7.

2.4. Measurement of Chelates

Site-selective spectroscopy has also been applied to the study of solution chelates. Eu^{3+} is a particularly useful probe ion for solution spectroscopy because its $^7F_0 \rightarrow {}^5D_0$ transition is a singlet-to-singlet transition that can produce only one line for any individual Eu^{3+} environment. Additionally, the $^5D_0$ excited level is well isolated from lower levels and is not as susceptible to quenching by losing energy to the high-energy vibrational modes of water that efficiently quench fluorescence from most other ions. Figure 8.5 shows the relationship between the energy gap below the $^5D_0$ level and the energy of different numbers of H_2O and D_2O vibrational quanta. When Eu^{3+} is chelated with different ligands, site-selective spectroscopy can be used to differentiate between chelates and between different forms of the same chelate. Figure 8.6 shows the excitation spectra of the $^7F_0 \rightarrow {}^5D_0$ transition for a series of β-diketones and a nitrogen

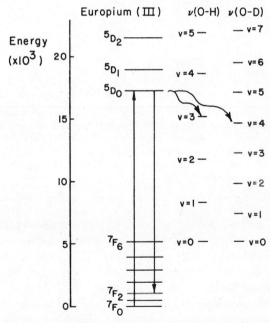

Fig. 8.5. The electronic energy levels of Eu^{3+} are shown along with the energies for different numbers of the O-H or O-D stretching vibrations of the H_2O or D_2O solvent. Since the energy difference between the $^5D_0$ and $^7F_6$ level can be absorbed by 3.6 vibrations of H_2O, some quenching of the $^5D_0$ level occurs. D_2O requires a larger vibrational contribution and is less quenched.

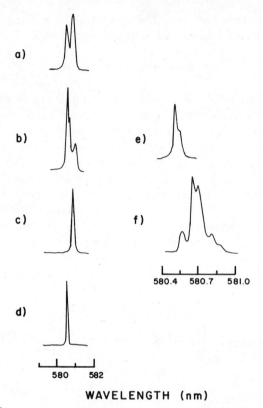

WAVELENGTH (nm)

Fig. 8.6. $^7F_0 \rightarrow {}^5D_0$ excitation spectra of various europium (β-diketone)$_3$(nitrogen donor)$_1$ compounds. $^5D_0 \rightarrow {}^7F_2$ fluorescence is monitored with a low resolution monochromator as the dye laser is scanned. (a) Eu(o-hydroxydibenzoylmethane)$_3$dipyridyl; (b) Eu(di-p-chlorobenzolylmethane)$_3$dipyridyl; (c) Eu(di-p-chlorobenzoylmethane)$_3$1,10-phenanthroline; (d) Eu(1,3-diphenyl-1,3-propanedione)$_3$dipyridyl; (e) Eu(1,3-dimethyl-1,3-propanedione)$_3$1,10-phenanthroline; (f) Eu(1,3-dimethyl-1,3-propanedione)$_3$dipyridyl.

donor ligand of either 2,2′-bipyridyl or 1,10-phenanthroline coordinated with Eu^{3+} in a bidentate fashion. Fluorescence was monitored over a wide bandwidth so all Eu^{3+} environments could be observed. The lines are sharp and well defined. Two of the chelates have unique Eu^{3+} environments because there is only a single line for the Eu^{3+}, but the others have multiple environments. Site-selective spectroscopy works well for resolving individual sites of the chelate. Figure 8.7 illustrates the excellent resolution obtainable from Eu(o-hydroxydibenzoylmethane)$_3$ dipyridyl. Excitation of the two peaks in Fig. 8.6a gives different fluorescence spectra as seen in Figs. 8.7d and 8.7e. If just one of the fluorescence lines is monitored, the resulting excitation scan shown in Figs.

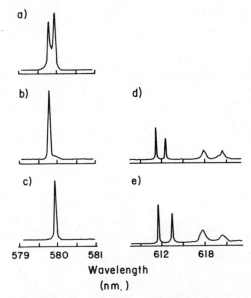

Fig. 8.7. The $^7F_0 \rightarrow {}^5D_0$ excitation spectrum for Eu(o-hydroxydibenzoylmethane)$_3$-dipyridyl is shown monitoring fluorescence from all sites (a) or the fluorescence line at 612.6 nm (b) or 613.5 nm (c). Single-site fluorescence lines result from exciting 579.78 nm (d) or 579.95 nm (e).

8.7b or 8.7c contains only the contribution from the site monitored. The capability for monitoring specific sites of ions in solution has important applications in chemistry because it provides a means for watching the details of the kinetics and equilibria underlying complexation.

2.5. Measurement of Proteins

A particularly attractive application is for monitoring the binding sites in metal binding proteins and enzymes. Proteins that bind Ca^{2+} are particularly amenable to these methods because many times trivalent lanthanides can quantitatively replace the Ca^{2+} but leave the protein or enzyme function intact (23, 24). The lanthanide can provide a direct way to monitor the binding sites as an enzyme participates in biochemical processes.

The first demonstration of site-selective spectroscopy in proteins was made by Valentini and Wright (25) for thermolysin, an enzyme that contains zinc and four calcium binding sites labeled $S(1) \rightarrow S(4)$. Eu^{3+} can be substituted into sites $S(1)$, $S(3)$, and $S(4)$ without destroying the enzymatic activity. Horrocks and co-workers (24) have studied the spectroscopy of the $^7F_0 \rightarrow {}^5D_0$ transition and have identified the binding sites. The line from binding site $S(1)$ is shifted

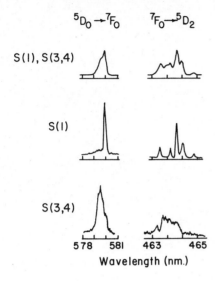

$^5D_0 \rightarrow {}^7F_0$ $^7F_0 \rightarrow {}^5D_2$

S(1), S(3,4)

S(1)

S(3,4)

578   581   463   465

Wavelength (nm.)

Fig. 8.8. The nonselective $^5D_0 \rightarrow {}^7F_0$ fluorescence and the $^7F_0 \rightarrow {}^5D_2$ excitation spectrum are shown at the top for Eu^{3+}-substituted thermolysin. If wavelengths characteristic of site $S(1)$ or sites $S(3, 4)$ are excited or monitored, single-site fluorescence and excitation spectra can be obtained.

and barely resolved from the other sites, but sites $S(3)$ and $S(4)$ are not resolvable. Figure 8.8 shows the results of the site-selective experiments that were performed to resolve the sites. The top left spectrum in Fig. 8.8 shows the fluorescence of the $^5D_0 \rightarrow {}^7F_0$ with non-selective excitation. The line on the right is from $S(1)$ while the shoulder on the left is the unresolved lines of $S(3, 4)$. The $^7F_0 \rightarrow {}^5D_2$ excitation spectrum obtained by monitoring both lines is shown in the upper right of Fig. 8.8. If one monitors fluorescence from the right-hand line or the left-hand shoulder in the $^5D_0 \rightarrow {}^7F_0$ fluorescence, the $^7F_0 \rightarrow {}^5D_2$ excitation spectra shown for $S(1)$ and $S(3, 4)$ are obtained, respectively. Similarly, if the laser is tuned to the $^7F_0 \rightarrow {}^5D_2$ lines at 465 or 463.9 nm, the fluorescence spectra of $S(1)$ and $S(3, 4)$ are obtained, respectively. The transitions from site 1 shown in the middle spectra are completely resolved from the transitions of sites 3, 4 shown on the bottom. It is not possible to resolve the transitions of sites 3 and 4. This failure is probably due to the greater heterogeneity in the local environment for these sites, which causes larger inhomogeneous broadening than site 1.

An interesting possibility of resolving the structure within the inhomogeneous envelope would be to use fluorescence line-narrowing methods on thermolysin. If the lines could be narrowed, one might resolve sites 3 and 4. It was found, however, that narrow-band excitation within the inhomogeneous linewidth of the $^7F_0 \rightarrow {}^5D_0$ transition at 10 K did not appreciably change the width of the nonresonant fluorescence except for the $^5D_0 \rightarrow {}^7F_1$ transition, which changed very slightly (25). The resonance fluorescence transition was strongly narrowed so the failure of nonresonant fluorescence line narrowing is clearly attributable to accidental degeneracy and not to the molecular motion of the protein over the time scale of the experiment. The narrowing of the resonance transition provides information about the homogeneous width, but it does not provide spectral resolution of separate sites because the fluorescence is constrained to occur at the excitation energy.

It is expected that the capabilities of site-selected spectroscopy will have important applications for biological samples, both in fundamental research and assay procedures—particularly fluorescence immunoassay. A typical fluorescence immunoassay results in a sample with several fluorescent components, only one of which is important for the measurement. It is necessary to separate the contributions, either by a physical separation or by spectroscopically distinguishing between them. The use of site-selective methods to provide the discrimination would be an attractive and simple solution to the problem.

3. SITE-SELECTIVE NONLINEAR SPECTROSCOPY

All of the site-selective methods described fail if the sample does not fluoresce. Nonlinear spectroscopy does not require fluorescence, and recent theoretical work suggests that these methods should be capable of being used for site-selective spectroscopy as well (6, 26). Nonlinear spectoscopy is based on the nonlinear relationship between the induced polarization (P) and the driving electric fields (E) of focused lasers when the fields become comparable to the fields within molecules. Since the relationship is unknown, it can be expressed as the power series (neglecting the vectoral and tensorial properties of the variables (1)

$$P = \chi^{(1)}E + \chi^{(2)}E^2 + \chi^{(3)}E^3 + \cdots \qquad (8.3)$$

The second-order susceptibility $(\chi^{(2)})$ describes three-wave mixing where the sum and difference frequencies of two lasers appear in the oscillating polarization. Symmetry arguments show that it vanishes for isotropic materials. The first nonvanishing term comes from the third-order susceptibility $(\chi^{(3)})$ that describes four-wave mixing. In four-wave mixing, the polarization will develop components at all possible combinations of three laser frequencies.

Let us examine the behavior of the particular combination $\omega_1 - \omega_2 + \omega_3 = \omega_S$ where the ω_i represent the frequencies of the three lasers and ω_S represents the signal frequency generated by the polarization. Figure 8.9a sketches this combination for a material that has only virtual levels (dashed lines) at the different combination frequencies. This process is called coherent anti-Stokes Raman spectroscopy or CARS. If a material has real levels at those positions, the efficiency of the nonlinear process is enhanced. The resonance enhancement

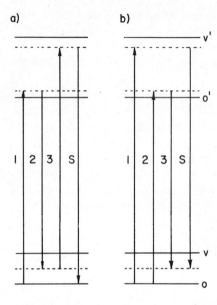

Fig. 8.9. The resonances using three lasers at ω_1, ω_2, and ω_3 to form a signal at $\omega_S = \omega_1 - \omega_2 + \omega_3$. There are virtual levels at the possible resonance positions, but there are real molecular levels at vibrational level v, electronic level $0'$, or the vibronic level v'. (a) The resonances for CARS, a parametric process, (b) the resonances for MENS, a nonparametric process. The arrows indicate the flow of coherence and do not necessarily correspond to absorption or fluorescence.

is typically 10^3 for molecules for each additional resonance. Site-selective spectroscopy can be performed by tuning the lasers to match two resonances on the particular component of interest and scanning the third resonance. When the scan results in a resonance with other components, the efficiency of generating the signal at ω_S will increase by a single resonance enhancement factor, but when a resonance is reached with the labeled component, the total efficiency will reflect three resonance enhancements. The selectivity between components should scale as the difference in the number of enhancements between components.

Scanning over a resonance requires changing two of the devices involved in Fig. 8.9—either the dye lasers at ω_1, ω_2, or ω_3 or the monochromator that monitors the signal at ω_S. The lowest resonance position in Fig. 8.9a can be determined by scanning ω_2 and ω_3 synchronously so ω_S remains constant while ω_1 is left at its resonance position. This scan is called an ω_2/ω_3 scan, and it can provide site-selected information about the ground-state vibrational modes in molecular spectroscopy. Similarly, a ω_1/ω_2 scan will provide information about the next to highest resonance, typically an excited electronic state. The scan needs to be synchronized so the resonances at $(\omega_1 - \omega_2)$ and ω_S are not affected. Finally, an ω_3/ω_S scan can provide information about the highest levels, typically excited-state vibrational modes or vibronics. The resonances at ω_1 and $(\omega_1 - \omega_2)$ are kept constant.

The site-selective capabilities of this type of nonlinear spectroscopy are illustrated in Fig. 8.10 for pentacene doped in a p-terphenyl crystal at 2 K. p-

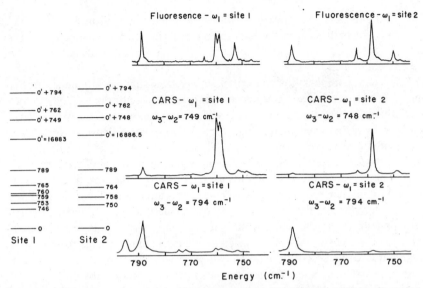

Fig. 8.10. The energy levels of pentacene in sites 1 and 2 of *p*-terphenyl are shown on the left. The excited-state vibronic energies are measured relative to the excited-state energy. The single-site fluorescence spectra exciting site 1 or 2 are shown on the top. Single-site CARS spectra are shown below those.

terphenyl has four crystallographic sites possible for pentacene, two of which have excited electronic states separated by only 3.8 cm^{-1}. This system therefore forms an excellent model for examining whether four-wave mixing can site select the transitions from one of these two closely spaced levels. The pentacene energy levels for the two sites are shown on the left of Fig. 8.10. Conventional site-selective fluorescence spectroscopy can distinguish between the sites. If a laser is tuned to resonance with site 1 at 16,883 cm^{-1}, the fluorescence spectrum at the top left of Fig. 8.10 is obtained, while tuning the laser to resonance with site 2 at 16,886.8 cm^{-1} produces the fluorescence spectrum at the top right of Fig. 8.10. If instead three lasers are used and ω_1 is tuned to the origin of one site, while ω_2 and ω_3 are tuned synchronously so that $\omega_3 - \omega_2$ is always maintained at a constant energy, one can get nonlinear spectra characteristic of the site selected. Spectra for site 1 selected are shown on the left of Fig. 8.10, while those for site 2 are shown on the right. The output signal at $\omega_S = \omega_1 - \omega_2 + \omega_3$ is monitored continuously and since $\omega_3 - \omega_2$ is always a constant, the output signal ω_S will also be a constant. There is additional selection in the nonlinear experiment because of the selection rules on changes in the vibrational quantum numbers. The ground-state vibrational resonance that appears most strongly will be determined by the excited-state vibronic mode that is always resonant with $\omega_1 - \omega_2 + \omega_3$. The vibronic mode selected by $\omega_3 - \omega_2$ is 749

cm^{-1} and 748 cm^{-1} for sites 1 and 2 for the middle spectra in Fig. 8.10 and 794 cm^{-1} for both sites in the bottom spectra. The strongest vibrational lines in the middle spectra correspond to the 759 and 760 cm^{-1} modes of site 1 and the 758 cm^{-1} mode of site 2.

Pentacene:p-terphenyl is an example of a system that has a finite number of unique components. The extension of site-selective nonlinear methods to systems where the line structure is hidden within large inhomogeneously broadened bands is more involved than the analogous fluorescence line-narrowing techniques of site-selective linear methods. The complication arises because nonlinear spectroscopy is coherent and the output signal depends on the square of the susceptibility. Since the susceptibility is determined by the sum of the susceptibilities of each component in a sample, the output signal has cross terms between components that cause constructive or destructive interference. If the resolution of single-site spectral structure within an inhomogeneous band is going to be succesful, the sites selected by the lasers must provide the dominant contribution to the spectrum. However, if there is destructive interference in the cross terms between sites, the large number of sites with nearby energy levels can diminish the enhancement from the selected sites so their contribution is no different than the nearby sites, and line narrowing of the inhomogeneous width is lost. Constructive interference between sites does not cause this effect, and resolution of the spectra for single sites within the inhomogeneous line is possible. The type of interference is determined by the nature of the nonlinear mixing (1, 26). If the coherence that causes the output signal is between two states that were not initially populated, the nonlinear mixing is a nonparametric process. If one of the two states is the initially populated state, the mixing is a parametric process (6). Figure 8.9 shows two schemes for four-wave mixing that are suitable for analytical spectroscopy of organic systems. Figure 8.9a illustrates a parametric process (CARS) while Fig. 8.9b illustrates a nonparametric process that we have called multiple enhanced nonparametric spectroscopy (MENS). Note that both processes can provide outputs that are at higher energy than fluorescence. CARS will always have a higher energy signal while MENS will have a higher energy signal if the separation between state v' and $0'$ is larger than that between v and 0. Parametric and nonparametric processes have different interferences. If the inhomogeneous broadening mechanisms shift the excited levels of a site in the same direction, MENS will have constructive interference between sites and should exhibit narrowing capabilities, while CARS will have destructive interference and should not have narrowing capabilities. If one of the excited levels shifts in the opposite direction from the other two, the type of interference and the narrowing capabilities reverse. An additional complication is caused by the need to consider the excited-state populations that become important because of the resonant lasers. Four-wave mixing can occur from initially populated excited states and cause signals at the same frequency as those from the ground-state population. However, the character of the process can change. For example, an excited-state CARS process is nonparametric and can interfere with the ground-state CARS process. Despite the theoretical work that has been done on this subject (6, 26), the line-narrowing capabilities of nonlinear four-wave mixing spectroscopy have not been demonstrated experimentally, primarily because of the complex nature of the experiments. Nevertheless, the potential for these methods is high, and there is great interest in achieving the ability to perform site-selective spectroscopy in the nonlinear regime because it allows the entire class of nonfluorescent samples to be studied.

ACKNOWLEDGMENTS

The research reported in this chapter was supported by the National Science Foundation under grant number CHE8306084.

REFERENCES

1. J. C. Wright, *Applications of Lasers to Chemical Problems*, T. R. Evans, Ed., Academic Press, New York, 1982.
2. T. G. Matthews and F. E. Lytle, *Anal. Chem.*, **51**, 583 (1979).
3. F. J. Knorr and J. M. Harris, *Anal. Chem.*, **53**, 272 (1981).
4. B. E. Kohler, *Chemical and Biochemical Applications of Lasers*, C. B. Moore, Ed., Academic Press, New York, 1979.
5. H. J. Griesser and U. P. Wild, *J. Chem. Phys.*, **73**, 4715 (1980).
6. J. L. Oudar and Y. R. Shen, *Phys. Rev. A.*, **22**, 1141 (1980).
7. F. J. Gustafson and J. C. Wright, *Anal. Chem.*, **49**, 1680 (1977).
8. F. J. Gustafson and J. C. Wright, *Anal. Chem.*, **51**, 1762 (1979).
9. M. V. Johnston and J. C. Wright, *Anal. Chem.*, **53**, 1050 (1981).
10. M. V. Johnston and J. C. Wright, *Anal. Chem.*, **54**, 2503 (1982).
11. D. L. Perry, S. M. Klainer, H. R. Bowman, F. P. Milanovich, T. Hirschfeld, and S. Miller, *Anal. Chem.*, **53**, 1048 (1981).
12. J. C. Wright, *Anal. Chem.*, **49**, 1690 (1977).
13. J. C. Wright, *Modern Fluorescence Spectroscopy*, Vol. 4, E. L. Wehry, Ed., Plenum Press, New York, 1981.
14. M. V. Johnston and J. C. Wright, *Anal. Chem.*, **51**, 1774 (1979).
15. M. V. Johnston and J. C. Wright, *Anal. Chem.*, **53**, 1054 (1981).
16. L. C. Porter, J. R. Akse, M. V. Johnston, and J. C. Wright, *Appl. Spectr.*, **37**, 360 (1983).
17. R. C. Stroupe, P. Tokousbalides, R. B. Dickinson, E. L. Wehry, and G. Mamantov, *Anal. Chem.*, **49**, 701 (1977).
18. E. L. Wehry and G. Mamantov, *Anal. Chem.*, **51**, 643A (1979).
19. Y. Yang, A. P. D'Silva, and V. A. Fassel, *Anal. Chem.*, **53**, 894 (1981).

20. E. V. Shpol'skii, A. A. Il'ina, and L. A. Klimova, *Dokl. Akad. Nauk SSR*, **87**, 935 (1952).
21. J. C. Brown, M. C. Edelson, and G. J. Small, *Anal. Chem.*, **50**, 1394 (1978).
22. A. P. Marchetti, W. C. McColgin, and J. H. Eberly, *Phys. Rev. Lett.*, **35**, 387 (1975).
23. W. D. Horrocks, *Adv. Inorg. Biochem.*, **4**, 201 (1982).
24. A. P. Snyder, D. R. Sudnick, V. K. Arkle, and W. D. Horrocks, *Biochem.*, **20**, 3334 (1981).
25. M. A. Valentini and J. C. Wright, *Chem. Phys. Lett.*, **100**, 133 (1983).
26. S. A. J. Druet, J. P. E. Taran, and Ch. J. Borde, *J. Phys.*, **41**, 183 (1980).

PART

IV

SELECTED MULTIPHOTON AND MULTIWAVELENGTH METHODS

CHAPTER

9

TWO-PHOTON EXCITED FLUORESCENCE

A. C. KOSKELO and M. J. WIRTH

Department of Chemistry
University of Wisconsin
Madison, Wisconsin

1. INTRODUCTION

The three properties of photons—energy, momentum, and polarization—can be used to advantage in two-photon absorption spectroscopy to provide molecular information unavailable in typical one-photon absorption measurements. In several special cases, two-photon excited fluorescence (TPEF) can be useful for the analytical chemist. For example, the one-photon absorption spectrum of an analyte may be obscured by solvent absorption. Because of selection rules, it is possible that at the frequency of a two-photon transition of the analyte the solvent is transparent (1, 2). In another example, the momenta of two photons can be employed to eliminate the Doppler broadening of gaseous samples in what is known as Doppler-free spectroscopy (3). This technique could be used when two closely related analytes in a gas cannot be resolved spectroscopically because of Doppler broadening, such as complexes of large atoms. The third example, which uses the light polarization to detect small changes in molecular symmetry, is the subject of this chapter. This approach is useful in the analysis of closely related compounds such as the methylnaphthalenes discussed below. The polarization dependence of TPEF also provides information about site symmetry around a probe molecule.

295

Two-photon spectroscopy is sensitive to the symmetry of each of the states involved in the transition. As McClain has shown (4), two parameters can be obtained in a single-laser two-photon experiment that are related to the symmetry of the excited state when the symmetry of the initial state is known. These two parameters, ∂_F and ∂_G, can be determined, from two-photon fluorescence measurements, even for a randomly oriented sample. Thus, measurements that probe excited-state symmetry can be made using room temperature solutions. Previously, such measurements could only be made on single crystals at cryogenic temperatures, conditions which are not common to most chemical systems.

Since two-photon spectroscopy yields information about the symmetries of the states involved in a transition, it also yields information about perturbations on those symmetries. Perturbations on the symmetry of a chromophore can arise from sources external or internal to the molecule. An example of the former is the perturbation caused by a solvent interacting with a solute. An example of the latter is the perturbation caused by substitution of various functional groups on the molecule. In both cases, the manner in which ∂_F and ∂_G are perturbed yields information about the symmetry of the perturbation. An explanation of the basis for TPEF symmetry measurements, the experimental requirements, and several applications of the method as a probe of symmetry perturbations follows.

2. MOLECULAR SYMMETRY AND TRANSITIONS

The selection rules for two-photon transitions are different from those of one-photon transitions. The classic case is for molecules with an inversion center. Whereas the one-photon selection rule is $g \rightarrow u$, the two-photon selection rule is $g \rightarrow g$. Other selection rules occur and are easily obtained from the character table for the point group of the molecule. In one-photon spectroscopy, only transitions to states that transform as x, y, or z or some linear combination of these are allowed. Alternately, in two-photon spectroscopy the excited state must transform as a quadratic, such as x^2. The reason for this is, quite simply, the process involves two photons instead of one. Qualitatively, two-photon absorption may be thought of as two sequential one-photon absorptions. The first photon induces a transition to an intermediate state, either virtual or real. The symmetry of this intermediate state is limited by the one-photon selection rules. The second photon induces a transition from the intermediate state to the final state and is also subject to the one-photon selection rules. Clearly, the two-photon process requires a transformation involving two coordinates; thus, the final state must transform as a quadratic. For example, for a D_{2h} molecule, transitions to states of A_g, B_{1g}, B_{2g}, or B_{3g} symmetry are two-photon allowed, and states that are of B_{1u}, B_{2u}, or B_{3u} symmetry are one-photon allowed. States of A_u symmetry can only be excited by a three-photon process (5).

The utility of TPEF lies in its ability to provide information about the symmetry of the states involved in the transition for randomly oriented samples. In a real sample, there is an ensemble of molecules, all undergoing transitions. The net signal is dependent on the average of the orientations of each of these absorbers. The signal for a single molecule is proportional to $|E_1 \cdot \partial \cdot E_2|^2$ where ∂ is the intrinsic two-photon absorption tensor and E_1 and E_2 are the electric fields of the two photons. As shown by Monson and McClain (6), the net experimental two-photon absorption coefficient, when averaged over a random ensemble, becomes

$$\partial = F\partial_F + G\partial_G + H\partial_H \qquad (9.1)$$

where ∂_F, ∂_G, and ∂_H are independent of orientation and contain only molecular quantities along with the energy of the photons. They are defined as

$$\partial_F = \left| \sum_a S_{aa} \right|^2 \qquad (9.2)$$

$$\partial_G = \sum_a \sum_b |S_{ab}|^2 \qquad (9.3)$$

$$\partial_H = \sum_a \sum_b S_{ab} S_{ba}^* \qquad (9.4)$$

where S is the two-photon transition tensor and a and b run over the three symmetry coordinates of the molecule. For molecules of a rectangular point group, such as D_{2h}, these are Cartesian coordinates. F, G, and H depend only on the polarization of the light and are defined as

$$F = 4|e_1 \cdot e_2|^2 - 1 - |e_1 \cdot e_2^*|^2 \qquad (9.5)$$

$$G = -|e_1 \cdot e_2|^2 + 4 - |e_1 \cdot e_2^*|^2 \qquad (9.6)$$

$$H = -|e_1 \cdot e_2|^2 - 1 + 4|e_1 \cdot e_2^*|^2 \qquad (9.7)$$

where e_1 and e_2 are the polarizations of photon 1 and photon 2, respectively. By varying the polarization of each photon, F, G, and H can be changed. For example, when both photons are polarized linearly, with parallel polarizations the values of F, G, and H are all 2; and with perpendicular polarizations $F = H = -1$ and $G = 4$. When both photons are circularly polarized in the same sense, $F = -2$ and $G = H = 3$ (Table II, ref. (6) gives other examples). Then by measuring ∂ for each of three sets of F, G, and H, three independent equations of the form of Eq. (9.1) are obtained from which the values of ∂_F, ∂_G, and ∂_H can be determined. In two-photon spectroscopy, one can therefore determine three constants characteristic of the transition, whereas in one-photon spectroscopy only one constant is obtained. These three constants aid in determining the symmetry of the excited state of the molecule.

Of importance is that ∂_F involves only diagonal elements of the two-photon tensor—only transitions to states that transform as a^2 will have a nonzero ∂_F. For all point groups these states must belong to the totally symmetric representation; that is, the states are symmetric with respect to all symmetry elements of the point group. Thus a nonzero ∂_F indicates, in principle, that the transition is to a totally symmetric excited state.

For transitions to a non-totally symmetric excited state, either the transition tensor is zero for all components or it has a zero trace. The first case occurs in transitions to states that are symmetry forbidden or to states that do not overlap with the ground state, that is, where Franck–Condon overlap is forbidden. The second case is two-photon allowed but has ∂_F equal to zero. Thus a measurement of ∂_F, in principle, distinguishes totally symmetric excited states from nontotally symmetric states.

McClain (4) also showed how many of the various types of excited states in a given point group can be distinguished using ∂_F, $\partial_G - \partial_H$, and $\partial_G + \partial_H$. For molecules of D_{2h} or C_{2v} symmetry, however, three different B_g representations all show the same behavior with respect to the symmetry indicators. Thus, a two-photon experiment can only distinguish between a totally symmetric excited state and a nontotally symmetric one. A measurement of ∂_F and ∂_G is sufficient to establish this difference; ∂_H adds no information. This is important since in a single-laser two-photon experiment, it can be shown that ∂_H equals ∂_G necessarily; that is, only ∂_F and ∂_G can be measured. As discussed below, the two-laser experiment is more difficult and more sensitive to equipment noise than the one-laser experiment. Therefore, for molecules of the D_{2h} or C_{2v} point groups, the maximum amount of information about a transition can be obtained in the experimental configuration that has the optimum experimental requirements.

In practice, ∂_F is a difficult quantity to determine. Source power must be free of noise and accurately known. Since two-photon absorption is sensitive to the photon statistics of the source, the latter must be carefully controlled. The interaction length and cross section of the beam over this length must also be known. In a fluorescence experiment, the collection efficiency must be constant and the quantum yield known. One method of eliminating these problems is to use an internal standard; ∂_G works excellently. Its sensitivity to all of the above parameters is the same as ∂_F and it arises from the same molecule. Thus the ratio ∂_F/∂_G eliminates the need for controlling all of the above parameters.

Table 9.1. Summary of the Effect of Different Perturbations on ∂_F/∂_G

Transition Symmetry[a]	Perturbation Symmetry	Change in ∂_F/∂_G[b]
TS	TS	+ or −
TS	NTS	+ or −[c]
NTS	TS	+
NTS	NTS	Remains 0

[a]TS = totally symmetric; NTS = nontotally symmetric. By totally symmetric transition is meant a two-photon transition between states of the same symmetry; by a nontotally symmetric transition is meant a two-photon transition where the initial and final states are of different symmetries.
[b]+ = ∂_F/∂_G increases as a result of the perturbation; − = ∂_F/∂_G decreases as a result of the perturbation.
[c]For D_{2h} and C_{2v} molecules, ∂_F/∂_G must decrease.

All the information contained in ∂_F about the symmetry of the excited state is in ∂_F/∂_G.

Since two-photon spectroscopy yields information about the symmetry of the states involved in a transition for a randomly oriented ensemble, the forms of perturbations on these symmetries are determinable. Just like the transition tensor of a molecule, a perturbation can be decomposed into parts that transform as the irreducible representations of the molecular point group. A summary of the effects of a perturbation on a transition is given in Table 9.1. The perturbations on ∂_F/∂_G of a chromophore occur both within a molecule and external to it. Examples of each will be given below.

3. EXPERIMENTAL CONSIDERATIONS

From Eq. (9.1) and (9.5)–(9.7) it can be shown that, for a single-laser experiment,

$$\frac{\partial_F}{\partial_G} = \frac{3(\partial_L/\partial_C) - 2}{(\partial_L/\partial_C) + 1} \tag{9.8}$$

where ∂_L is the molecular cross section for linearly polarized light and ∂_C is that for perfectly circularly polarized light.

In order to obtain the ratio ∂_L/∂_C, numbers proportional to ∂_L and ∂_C are each obtained from a measurement of the two-photon fluorescence signal, I, resulting from excitation by a beam of the corresponding polarized light. For both cases, I can be expressed as:

$$I = \frac{KQlcP^2 \partial}{A} \qquad (9.9)$$

Here l is the path length of excitation, c is the sample concentration, A is the cross-sectional area of the beam, ∂ is the molecular two-photon absorption coefficient defined in Eq. (9.1), and P is the power. K allows for the collection efficiency and Q the quantum yield in a fluorescence experiment. ∂ is small, being on the order of $10^{-50} \cdot cm^4 \cdot photon^{-1} \cdot molecule^{-1}$. Thus, in order to get a measurable signal, K, l, and P must be as large as possible and A must be small. To achieve a signal of reasonable size, the power should be on the order of 1 kW. Generating power of that magnitude in a single wavelength requires a pulsed laser.

The experimental configuration used in the TPEF experiments in our lab is illustrated in Fig. 9.1. The laser system consists of a cavity-dumped dye laser synchronously pumped by a mode-locked argon ion laser (7). The light generated is passed through a Glan–Thompson prism. Although the dye laser polarization is fairly clean, the polarizing prism is used to guarantee linearly polarized light to 1 ppm. The beam is then passed through a mica quarter-wave plate that allows the polarization of the light to be varied from linear to circular. The mica quarter-wave plate is chromatic, and the effect of this on the light polarization must be accounted for when calculating ∂_F/∂_G. Neglect of nonideal polarization can lead to large errors in calculating ∂_F/∂_G as illustrated in Fig. 9.2.

A deviation in phase angle drives ∂_F/∂_G toward 0.5. For ∂_F/∂_G equal to 0.5, the value observed is independent of polarization error. The farther the true value of ∂_F/∂_G is from 0.5, the greater the error in calculating ∂_F/∂_G when

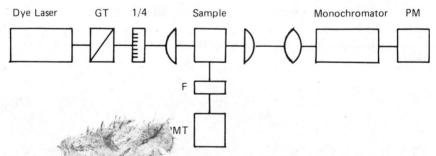

Fig. 9.1. Diagram of the TPEF experiment. GT is a Glan–Thompson prism with an air gap. 1/4 is a mica quarter-wave plate retarder. The next element is a plano-convex lens with a 4-in. focal length; a collimating lens follows the sample chamber. A lens is used to focus the beam into a monochromator and PM is a power meter used to detect the point of maximum transmission by the monochromator. F is a filter box; filters are used to isolate the fluorescence from the scatter. PMT is a photomultiplier tube.

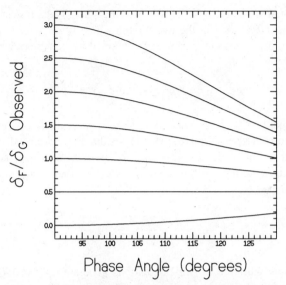

Fig. 9.2. Error in ∂_F/∂_G versus the true light polarization. The true value of ∂_F/∂_G is found at $\epsilon = 90°$. The values calculated when the light is not truly circularly polarized are plotted.

assuming that the light is perfectly polarized. This fact, taken with the chromaticity of the quarter wave plate, must be considered in determining ∂_F/∂_G.

For some experiments it is useful to measure ∂_F and ∂_G separately. It can be shown that experiments requiring the measurement of ∂_F must have better control of the light polarization than those using only ∂_G (8). This suggests the use of ∂_G as the more quantitative indicator of the presence of an absorption band rather than ∂_F.

4. COMPARISON OF SOURCE NOISE BETWEEN ONE- AND TWO-LASER EXPERIMENTS

Experiments that require only ∂_F and ∂_G and not ∂_H can be done with either one or two lasers. Since one-laser experiments are cheaper than two-laser experiments, it is desirable to compare the differences in the effect of source noise in the two experiments. Consider the effect of power fluctuations of a single dye laser on the observed two-photon signal. It is readily shown from Eq. (9.1) that the relative noise is given by

$$\frac{d\partial}{\partial} = 2\frac{dP}{P} \tag{9.10}$$

where $d\partial$ is the absolute error in the two-photon signal. Thus, power fluctuations of 1% generate a 2% error in the two-photon signal.

The error in a two-laser experiment is obtained by considering the instantaneous two-photon signal,

$$\partial(t) = M|E_1(t) + E_2(t)|^4 \tag{9.11}$$

where M is equal to KQ/clA in Eq. (9.9) and $E_n(t)$ is the instantaneous amplitude of the field from laser n. Equation (9.11) results since the total electric field amplitude is the sum of the amplitudes from the individual lasers. Evaluating Eq. (9.11) and assuming that the relative power fluctuation of a single laser is independent of power:

$$\frac{d\partial}{\partial} = \frac{\sqrt{10}}{2} \frac{dE}{E} \cong 0.79 \frac{dP}{P} \tag{9.12}$$

Thus, two lasers with 1% power fluctuations generate only a 0.79% error in the two-photon absorption signal. Compare this with the single-laser experiment of the same total power. The two-laser experiment therefore contains a signal-to-noise ratio advantage over the one-laser experiment.

The maximum advantage is obtained when the power fluctuations between the two dye lasers are uncorrelated; when the power of one laser drifts up, the power of the second laser may drift either up or down thereby augmenting or balancing the effect of the first laser's drift. In a one-laser experiment no cancellation of power fluctuations can take place. For the same reason, if the power fluctuations in a two-laser experiment are correlated, the canceling of one laser's drift by the other is not as effective as when the drifts are uncorrelated. Depending on whether the correlation is 0 or 100%, the two-photon signal exhibits a fluctuation between 0.395 and 1 times that in a single-laser experiment. Correlation of the noise between two dye lasers can occur if both are driven by the same pump laser and the noise is determined by pump laser fluctuations. For jet stream dye lasers, noise is thought to be primarily determined by jet stream fluctuations; thus, the two dye lasers should be uncorrelated in their noise properties. The relative error due to power fluctuations in a two-laser experiment is up to 2.5 times better than in a one-laser experiment.

Although a two-laser experiment is less sensitive to power fluctuations, it has a source of noise not present in the one-laser experiment, namely, noise due to changes in the temporal overlap of the pulses from the two lasers. The pulse output from a single synchronously pumped dye laser will jitter in time as a result of small gain and loss fluctuations in the laser cavity. Two dye lasers having different cavities will jitter in time independently of one another; thus, the temporal overlap between the pulses from the two lasers will vary with time.

For pulses of equal amplitude but originating in different lasers, the ratio of the two-photon signal when the pulses overlap completely to the signal when they do not overlap at all is $3:1$ as in an autocorrelation. Thus, the two-photon signal can vary by as much as $\pm 50\%$.

The effect of temporal jitter on the observed two-photon signal depends on the repetition rate of the laser. For low repetition rate lasers, such as N_2-pumped dye lasers, the jitter will result in large changes in the two-photon signal from shot to shot. For a synchronously pumped system, where the repetition rate is on the order of megahertz, the pulse-to-pulse jitter is effectively signal averaged. The temporal jitter in this system therfore appears only as a reduced magnitude of signal. Typically, the two-photon signal will be reduced by $1.5/1.16 = 1.29$ times from the case where there is no jitter. The no-jitter case is the same as a one-laser experiment of the same total power. Thus, the two-laser experiment produces a slightly lower signal than does the one-laser experiment. The net increase in signal-to-noise ratio of the two-laser experiment over the one-laser experiment is then $2.5/1.29 = 2$ times. In order to gain the advantage of using two lasers, the wavelengths of the two must be matched to within a few angstroms. As shown below, two-photon spectra are sensitive to wavelength to this degree. Because of the difficulty of matching the wavelengths of two dye lasers to within this range, along with the rather meager increase in signal-to-noise ratio in the two-laser experiment, a one-laser experiment is preferable in practice when either method results in the same amount of spectroscopic information.

TPEF can yield quantitative as well as qualitative information about a sample if absolute measurements of ∂_F and ∂_G are made. Since ∂_F and ∂_G are essentially absorption coefficients, the normal means of obtaining quantitative concentration information by the use of calibration curves or standard additions can be employed. A measurement of ∂_F/∂_G, which can be used in qualitative analysis and does not require absolute intensity measurements, is not easily interpretable in terms of the separate concentrations of the solutes. This is a result of the following equation for ∂_F/∂_G for a sample containing two components, 1 and 2:

$$\partial_F = c_1\partial_{F1} + c_2\partial_{F2} \tag{9.13}$$

$$\partial_G = c_1\partial_{G1} + c_2\partial_{G2} \tag{9.14}$$

$$\frac{\partial_F}{\partial_G} = \frac{c_1\partial_{F1} + c_2\partial_{F2}}{c_1\partial_{G1} + c_2\partial_{G2}} \tag{9.15}$$

A simple measurement of ∂_F/∂_G does not allow c_1 and c_2 to be determined. However, if ∂_F and ∂_G are measured separately, Eqs. (9.13) and (9.14) show that c_1 and c_2 can be obtained if ∂_{F1}, ∂_{F2}, ∂_{G1}, and ∂_{G2} are known, which can be obtained by measurements of standards. Therefore, in an analysis of a mixture using TPEF, ∂_F/∂_G identifies the components; a measurement of ∂_F and ∂_G can then be used to obtain the concentrations. Thus, TPEF permits both a qualitative and quantitative analysis simultaneously without the need for physical separation.

5. IDENTIFICATION OF METHYL DERIVATIVES OF NAPHTHALENE USING TWO-PHOTON EXCITED FLUORESCENCE

The sensitivity of two-photon spectroscopy to symmetry perturbations suggests its ability to discern the position of substitution on a chromophore. The perturbation due to the substituent breaks the symmetry of the molecule. The effectiveness of this perturbation as well as the manner in which the symmetry of the molecule is reduced depends on where the alkyl group is located. For ex-

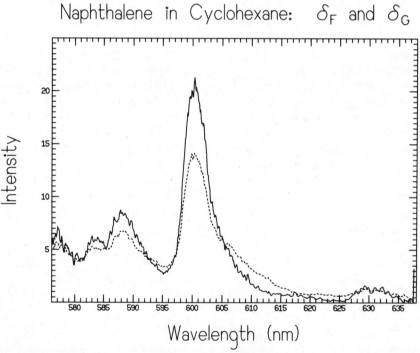

Fig. 9.3. Naphthalene in cyclohexane. Figures 9.3 through 9.9 show the plot of ∂_F and ∂_G vs. wavelength; the ∂_F spectrum is the solid line and the ∂_G spectrum is the dashed line. Each spectrum has been corrected for background, which is due to solvent scatter and photomultiplier dark current. In addition, correction has been made for the laser gain profile using bis-orthomethylstyrylbenzene as a reference. The wavelength axis is accurate to 0.5 nm.

ample, methyl substitution at both the 2 and 6 positions of naphthalene maintains the inversion center of the naphthalene chromophore, whereas substitution at both the 1 and 4 positions destroys the center of symmetry. Thus, 2,6-dimethylnapthalene and 1,4-dimethylnaphthalene are expected to exhibit different values for the two-photon symmetry indicator ∂_F/∂_G. Substitution at other positions also produces changes in ∂_F/∂_G. As shown in this section these expectations are fulfilled (8). The results indicate that TPEF can provide analytical information about a complex mixture of methyl-substituted naphthalenes. Clearly, the method can be extended to other mixtures of closely related molecules.

Throughout this section the methylnaphthalenes will be referred to by their respective positions of substitution. For example, 1-methylnaphthalene will be referred to as 1-.

The TPEF spectra of the methylnaphthalenes are all similarly structured as shown in Figs. 9.3–9.9. This verifies the integrity of the naphthalene chromophore and that methyl substitution provides a weak perturbation (9). The large peak between 600 and 610 nm is the $A_g \rightarrow A_g$ vibronically coupled peak

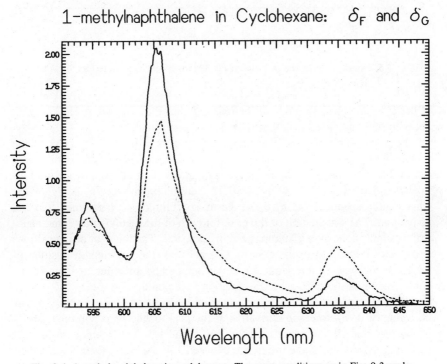

Fig. 9.4. 1-methylnaphthalene in cyclohexane. The same conditions as in Fig. 9.3 apply.

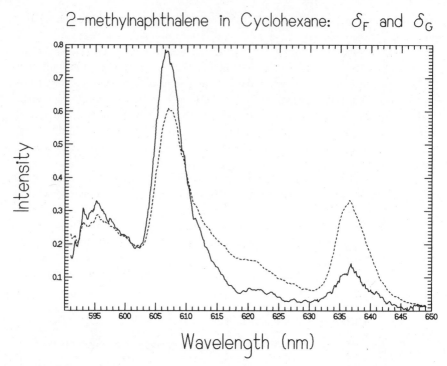

Fig. 9.5. 2-methylnaphthalene in cyclohexane. The same conditions as in Fig. 9.3 apply.

of the $S_0 \rightarrow S_1$ transition and will be referred to as the main vibronic. The peak between 630 and 640 nm is the $S_0 \rightarrow S_1$ 0–0 band.

The main vibronic of all the compounds occurs at or near the peak of the rhodamine 6G dye laser gain profile, and thus the greatest signal-to-noise ratio achievable with this common laser dye is obtained. This allowed concentrations in the region of 10^{-3} M to be used. A careful point-by-point study of the wavelength dependence of ∂_F/∂_G of the main vibronic was performed for each compound. The wavelength of the peak for each of the methyl-substituted naph-thalenes in cyclohexane is summarized in Table 9.2. The range of peak positions is not very large, and in the case of 1,3- and 1,4- the wavelength position of the main vibronic is the same for both compounds. In order to identify the compounds in a mixture, more information is required.

TPEF provides additional information in the form of ∂_F/∂_G. Table 9.3 lists the value of ∂_F/∂_G for each derivative at the peak of the main vibronic, and it is observed that all compounds except 2,6-,2- and 1,3- have a unique ∂_F/∂_G. Of particular importance is the difference between 1,4- and 1,3-. Even though their spectra are very similar in this region, their ∂_F/∂_G values are discernibly different;

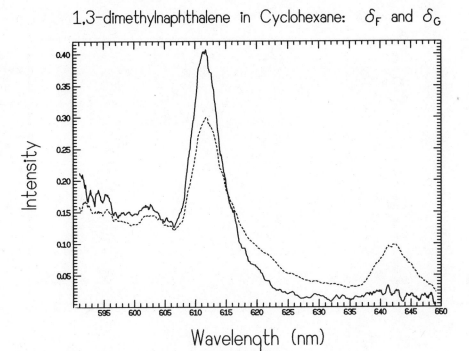

Fig. 9.6. 1,3-dimethylnaphthalene in cyclohexane. The same conditions as in Fig. 9.3 apply.

thus, 1,4- and 1,3- are uniquely identified by their values of ∂_F/∂_G. The derivatives 2,6-, 2- and 1,3- have the same value of ∂_F/∂_G, but the peak of the main vibronic occurs at different wavelengths for each. Thus, a measurement of the peak position and the value of ∂_F/∂_G for the main vibronic uniquely identifies a specific methyl-substituted naphthalene and can be used to obtain qualitative information about a mixture of these compounds. This is important since their physical separation is difficult (10).

With some loss in sensitivity, the 0–0 bands of each of the compounds can be used in identification. For the molecules with a center of symmetry, 2,6- and naphthalene, the 0–0 band is formally not allowed because a two-photon transition from a gerade ground state to an ungerade final state is symmetry forbidden. For these two molecules the 0–0 band is weak with respect to the main vibronic. The other derivatives perturb the naphthalene chromophore, reducing the symmetry, thereby making the 0–0 band more allowed, as can be seen by examining Table 9.3. Thus, the 0–0 band is strong enough in most cases to be of use in an analytical measurement. Unlike the main vibronic, the 0–0 band has a very different value of ∂_F/∂_G for every methyl derivative, as shown in

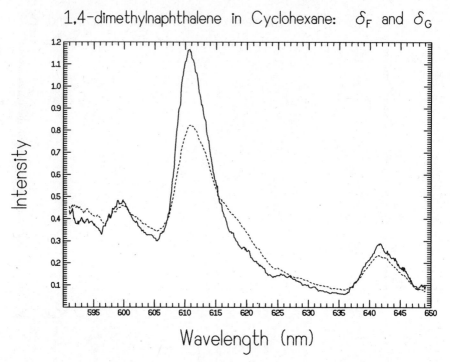

Fig. 9.7. 1,4-dimethylnaphthalene in cyclohexane. The same conditions as in Fig. 9.3 apply.

Table 9.3. The only compounds with the same ∂_F/∂_G are 1,4- and naphthalene; however, these two compounds can be distinguished because the 0–0 bands are at two measurably different wavelengths.

In summary, qualitative analysis of methylnaphthalenes is possible using TPEF of the main vibronic or the 0–0 band. The main vibronic allows greater sensitivity than the 0–0 band, but a measurement of ∂_F/∂_G is not as discriminating as for the 0–0 band. Therefore, the choice of which band to use depends on the particular requirements of sensitivity versus selectivity.

6. SOLVENT STRUCTURE AND TWO-PHOTON EXCITED FLUORESCENCE

The interaction between two molecules perturbs the energy of each. For a solute molecule immersed in a collection of solvent molecules, all of the molecules interact. To fully describe the system, the time-dependent Schrödinger equation must be solved using a Hamiltonian that includes all particle interactions. This

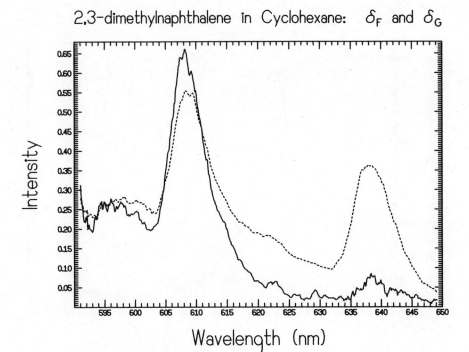

Fig. 9.8. 2,3-dimethylnaphthalene in cyclohexane. The same conditions as in Fig. 9.3 apply.

is a very difficult problem and considerable theoretical effort is underway (11–21). Many chemical processes are influenced by solvent–solute interactions; the many forms of chromatography, solvent effects on reactions, and solubility are examples. Understanding how these processes work essentially involves understanding the shape of the potential energy surface of the solute–solvent system.

In the frame of the solute molecule, the potential energy surface is determined by the local field generated by the surrounding solvent molecules. The contours of this surface are determined by the structure of the solvent cage and the shape and properties of the individual solvent molecules. The use of the term *solvation sphere* as occurs commonly in the literature is purposely avoided here. To call the solvent structure around the solute a solvation sphere requires that the solvent cage be spherically symmetric. For dense systems where the molecules are in close proximity, it is expected that the potential energy across a given molecule will be sharply modulated by the presence of a neighboring nonspherical solvent molecule. The potential energy modulations contributed by a number of solvent molecules around a solute may partially cancel. To a first approximation the solvent cage may be treated as a sphere; however, the degree to which the

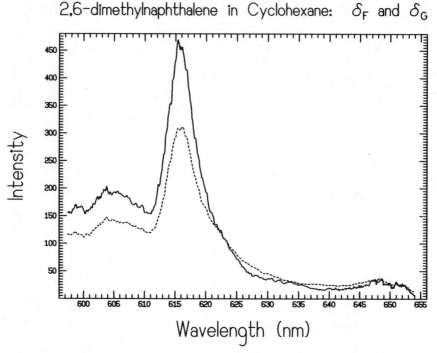

Fig. 9.9. 2,6-dimethylnaphthalene in cyclohexane. The same conditions as in Fig. 9.3 apply.

individual solvent perturbations collectively cancel depends on the local order in the region of the solute. In the reaction field theory (22), the solvent cage is taken to be an average over all possible orientations of the solvent with respect to the solute. Since this is a variational calculation, the radius of the sphere is adjusted until experimental data are matched (23, 24). The result may predict solvent shifts reasonably accurately, but it does so through the use of a non-physical mechanism. No correlation of molecular positions or orientations is assumed. In order to fully understand solvation, the effects of correlation must be considered. To provide an experimental understanding of solvent structure on a molecular scale, methods sensitive to this structure, which also reliably provide more than radial information, must be found.

TPEF can lead to insight into the shape of the solvent cage. With knowledge of the symmetry of a spectroscopic transition of a solute, the effect of a pertur-bation on that transition by the solvent allows determination of the form of the potential energy surface in terms of the point group of the solute molecule. A solvent cage that has all the symmetry elements of the solute, and which is centered about the solute will have a totally symmetric interaction with the

Table 9.2 Observed Parameters in TPEF of the Main Vibronic of the $S_0 \rightarrow S_1$ Transition in Methyl-Substituted Naphthalenes

Compound	Peak Position[a]	∂_F/∂_G[b]	∂_G[c]
Naphthalene	600.3	1.50	49
1-methylnaphthalene	605.8	1.47	180
2-methylnaphthalene	607.1	1.33	505
2,3-dimethylnaphthalene	608.1	1.24	74
1,4-dimethylnaphthalene	610.9	1.37	185
1,3-dimethylnaphthalene	611.9	1.33	220
2,6-dimethylnaphthalene	616.0	1.34	255

[a] Peak position is given in nanometers for a single-laser experiment. The error in wavelength is ±0.5 nm.
[b] The 95% confidence interval of ∂_F/∂_G is ±0.01.
[c] ∂_G is in arbitrary units. The value of ∂_G for all compounds was referenced to the same bis-MSB solution.

solute. In this case the solvent cage must perturb all symmetrically equivalent sites on a solute molecule in the same manner. A site is an atom, a bond, a region of charge, or any other part of the solute molecule one chooses to consider. A totally symmetric solvent cage would most likely occur when the solute determines the shape of the solvent cage. In this limit the solvent–solvent molecular interactions are much weaker than the solvent–solute interactions, and the solute creates a cavity for itself in the solvent. However, if the solvent–solvent interaction is comparable to or greater than that between the solute and solvent, then the solvent cage will not necessarily be totally symmetric. The

Table 9.3 Observed Parameters in TPEF of the 0–0 band of the $S_0 \rightarrow S_1$ Transition in Methyl-Substituted Naphthalene

Compound	Peak Position[a]	∂_F/∂_G[b]	∂_G[c]
Naphthalene	630.3	1.4	4
1,4-dimethylnaphthalene	641.7	1.4	22
2,6-dimethylnaphthalene	648.2	0.95	58
1-methylnaphthalene	634.8	0.4	150
2-methylnaphthalene	636.6	0.3	51
2,3-dimethylnaphthalene	637.9	0.2	80
1,3-dimethylnaphthalene	642.1	0.15	57

[a] Peak position is given in nanometers for a single-laser experiment. The error in wavelength is ±0.5 nm.
[b] The error in ∂_F/∂_G is estimated to be ±0.5.
[c] ∂_G is in arbitrary units. The same reference used for the data of Table 9.1 was used here.

solvent–solvent interaction may be steric in that the solvent molecules may not fit around the solute in a symmetric manner. The interaction may also be attractive if the solvent tends to form long-range structures as in hydrogen-bonding solvents. For both cases, asymmetry may result in the solute–solvent interaction potential. Restricted according to Table 9.1, a measurement of ∂_F/∂_G indicates the presence of an antisymmetric component of the solvent cage with respect to some reference. Ideally, this reference would be the value of ∂_F/∂_G in the gas phase after allowing for band overlap in the liquid.

It has been established that TPEF is sensitive to the symmetry of the local environment of a solute (25, 26). TPEF can therefore aid in interpreting how a solvent induces transition strength in a symmetry forbidden band (8). Using one-photon absorption spectroscopy, Mukhopadhyay and Georghiou observed an intensity enhancement of the 0–0 band of naphthalene in dimethylformamide, which they attributed to complex formation between solute and solvent (27). The complexation reduces the symmetry of naphthalene thereby making the $S_0 \rightarrow S_1$ 0–0 transition allowed. TPEF show that DMF does perturb the symmetry of naphthalene. Measurements of ∂_F/∂_G for the 0–0 band in DMF and in cyclohexane yield values of 0.7 and 1.0, respectively, which indicates that DMF induces more B_{1g} transition strength in the 0–0 band than does cyclohexane. Taking the long axis of naphthalene to be the B_{3u} axis and the short in plane axis to be B_{2u}, the DMF solvation cage is more effectively asymmetric with respect to the short axis of naphthalene than the cyclohexane solvation cage. That the solvent has a greater effect on the short axis of naphthalene is also shown by the increase in absorption coefficient of the short axis polarized $S_0 \rightarrow S_2$ one-photon transition.

Another example of the use of TPEF in probing solvation structure can be found in ref. 25. TPEF indicates that methanol has structure around fluorene, phenanthrene, naphthalene, and chrysene (25).

7. CONCLUSION

TPEF can aid the analytical chemist in both qualitative and quantitative analyses as illustrated by the example of methylnaphthalenes. The use of TPEF in identifying and determining local site symmetry around a probe molecule has greater potential for broad applicability. The examples cited here have been for the probing of solvent structure; concrete identification of structures await gas-phase measurements and theoretical calculation to fit the observed data. While local solvent structure was discussed here, TPEF could also be used as an aid in establishing the structures of complexes and in establishing the presence of symmetries in biological molecules.

REFERENCES

1. M. J. Wirth and F. E. Lytle, *Two Photon Excited Molecular Fluorescence*, New Applications of Lasers to Chemistry, Gary Hieftje, Ed., American Chemical Society Washington, D.C., 1978.
2. M. J. Wirth and F. E. Lytle, *Anal. Chem.*, **49,** 2054 (1977).
3. E. Riedle, R. Moder, and H. J. Neusser, *Opt. Comm.*, **43,** 388 (1982).
4. W. M. McClain, *J. Chem. Phys.*, **55,** 2789 (1971).
5. D. M. Friedrich, *J. Chem. Phys.*, **75,** 3258 (1981).
6. P. R. Monson and W. M. McClain, *J. Chem. Phys.*, **53,** 29 (1970).
7. M. J. Wirth, M. J. Sanders, and A. C. Koskelo, *Appl. Phys. Lett.*, **38,** 295 (1981).
8. A. C. Koskelo, Ph.D. thesis, University of Wisconsin-Madison, 1983.
9. M. J. Wirth, A. C. Koskelo, C. E. Mohler, and B. L. Lentz, *Analyt. Chem.*, **53,** 2045 (1981).
10. M. J. Wirth, D. A. Hahn, and R. A. Holland, *Analyt. Chem.*, **55,** 787 (1983).
11. E. W. Knapp and S. F. Fischer, *J. Chem. Phys.*, **74,** 89 (1981).
12. C. N. R. Rao, S. Singh, and V. P. Senthilnathan, *Chem. Soc. Rev.*, **5,** 297 (1976).
13. M. F. Nicol, *Appl. Spec. Rev.*, **8,** 183 (1974).
14. A. T. Amos and B. L. Burrows, in *Advances in Quantum Chemistry*, Vol. 7, Academic Press, New York, 1973.
15. A. Warchel, *J. Phys. Chem.*, **83,** 1640 (1979).
16. S. Miertus, E. Scrocco, and J. Tomasi, *Chem. Phys.*, **55,** 117 (1981).
17. O. Tapia, *Theoret. Chim. Acta (Berl.)*, **47,** 157 (1978).
18. O. Tapia and O. Goscinski, *Mol. Phys.*, **29,** 1653 (1975).
19. M. F. Herman and B. J. Berne, *J. Chem. Phys.*, **78,** 4103 (1983).
20. K. S. Schweitzer and D. Chandler, *J. Chem. Phys.*, **78,** 4118 (1983).
21. F. Kohler, *Pure Appl. Chem.*, **51,** 1637 (1979).
22. L. Onsager, *J. Am. Chem. Soc.*, **58,** 1486 (1936).
23. I. A. Arev, *Opt. Spectrosc.*, **40,** 10 (1976).
24. I. A. Arev, *Russ. J. Phys. Chem.*, **53,** 335 (1979).
25. M. J. Wirth, A. C. Koskelo, and C. E. Mohler, *J. Phys. Chem.*, **87,** 4395 (1983).
26. D. M. Friedrich, J. Van Alsten, and M. A. Walters, *Chem. Phys. Lett.*, **76,** 504 (1980).
27. A. K. Mukhopadhyay and S. Georghiou, *Photochem. Photobio.*, **31,** 407 (1980).

CHAPTER

10

RAMAN AND RELATED METHODS IN CHEMICAL ANALYSIS

EDWARD S. YEUNG

Department of Chemistry and Ames Laboratory
Iowa State University
Ames, Iowa

1. INTRODUCTION

After its prediction 60 years ago (1) and its demonstration shortly thereafter (2, 3) Raman spectroscopy has established itself as an important spectroscopic technique. In particular, the complementary nature of Raman and infrared techniques makes them valuable tools for structural determination. For chemical analysis, this suggests a functional-group-specific method. Many developments have occurred to boost the potential of Raman spectroscopy in chemical analysis. Of these, the development of the laser must be considered the most significant. In this chapter, we shall examine the various ways through which lasers have contributed to the success of Raman spectroscopy. We shall first outline the principles and instrumentation in the various variations of Raman spectroscopy

315

and then show how the individual features can be used to guide our choice of a technique for a particular application.

2. VARIOUS RAMAN TECHNIQUES

2.1. Conventional Raman Scattering (RS)

Even though normal RS is not flashy compared to the newer types of laser Raman spectroscopy, there is much to be gained from recent developments in laser technology. It is fair to say that lasers have brought on the renaissance of Raman spectroscopy (4), which would otherwise have stayed under the shadow of infrared methods for studying molecular structures. A few examples are given below to show that certain properties of the laser beam are particularly advantageous for RS, providing information that cannot be obtained using conventional light sources.

Lasers can provide spectral output much narrower than conventional light sources. For gas-phase RS, this is important even if only Doppler-limited linewidths are desired. One notes that in the forward scattering geometry, Doppler widths in RS are comparable to those in infrared spectroscopy, but are much narrower than excitation widths for conventional light sources. Using a single-mode argon ion laser, the Q-branch structure of O_2 has been studied to a resolution of 0.05 cm^{-1} (5), using photographic detection. Similar results have been obtained using photoelectric detection (6). Spectral calibration is conveniently provided by hollow cathode lamps to obtain relative line positions to 0.002 cm^{-1} (7) and absolute line positions to 0.02 cm^{-1} (8).

The high-power, collimated beam leads to efficient collection of the signal and efficient excitation. The larger signal level is the key to the success of multichannel image intensifiers coupled to television cameras for detection (9). The pure rotational spectrum of O_2 at 1 bar has been studied with good signal-to-noise ratios using either 50 mW of radiation at 442 nm for 3 s or 2 mJ pulses of 532-nm radiation at 1 Hz for 30 min with gating. The collimated beam can be focused to a small diameter. This is ideal for studying small samples (10). A commercial version of this Raman microprobe can be used to study local regions of samples to a few micrometers in diameter, to scan the entire sample area while monitoring one line, or to study temporal changes in composition. The small beam size also makes laser Raman suitable for detection in liquid chromatography (11). A cell volume of 5 μL is used for detecting the order of 10^{-3} M of mesogenic azoxybenzenes injected, with a time constant of 5 s. The resonance-enhanced RS signal is strong enough for multichannel detection at an excitation power of 700 mW.

The extremely high instantaneous photon flux available in pulsed lasers can provide discrimination from background radiation. An intracavity arrangement (12) based on a free-running ruby laser at 694 nm is operated at 20 J per pulse with a duration of 270 μs for the pulse train, and is shown in Fig. 10.1. A H_2–O_2 flame with added N_2 is placed in the focal region, where power levels of 10^8 W/cm^2 are expected. The Stokes emission from the N_2 fundamental Raman transition at 828 nm is monitored by L1, F1, and P1. The anti-Stokes emission from the same transition at 598 nm is monitored by L2, F2, and P2. This way, the number density as well as the vibrational temperature can be measured in a single pulse. The high powers also vaporize and eject any particles that may be in the optical path and reduce the background signals substantially.

The characteristics of lasers can lead to various methods for rejecting background due to Rayleigh scattering or luminescence from the sample. Techniques for conveniently modulating the polarization of the laser are known and can be used to distinguish strongly polarized Raman lines from the isotropic background luminescence (13). It is also relatively easy to modulate the frequency of the excitation laser over a region on the order of Raman linewidths (14). Since fluorescence excitation generally has a much weaker dependence on frequency, it can be effectively suppressed. Modulation of the laser frequency is superior to the analogous technique of modulating the detection wavelength (15) when the linewidths are narrow. The reason is that the former allows the use of much higher frequencies, thus reducing $1/f$ noise. The short durations of laser pulses can be used in RS to discriminate against the longer-lived fluorescence emission from the samples (16). The extent of discrimination is proportional to the ratio of the fluorescence lifetime to the gate width. At present, temporal rejection of the fluorescence background in RS is limited by the response time of the photodetector (about 150 ps) and not the duration of the laser pulse.

Other promising developments in RS include studies of polymer films, short-lived radicals, and hyper-Raman spectroscopy. To study submicron films, one has to increase the interaction length to compensate for the low number density.

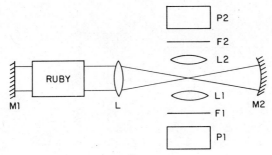

Fig. 10.1. Intractivity arrangement for Raman scattering. M1, M2, cavity mirrors; L, L1, L2, lenses; F1, F2, filters; and P1, P2, phototubes.

One approach is to form an optical waveguide (17) from the film of polymers and a high refractive-index glass layer, 1 μ thick, which is in turn deposited on Pyrex glass. This is possible because of the highly collimated laser light. A monolayer of squarylium dye is sufficient to record a Raman spectrum. The time resolution of the Raman probe can be on the order of the laser pulse width. This has been applied to the study of the lifetime of the p-terphenyl anion radical after its generation by a field emission accelerator (18). The hyper-Raman effect is based on simultaneous excitation by two laser photons of frequency ω and the emission of a single photon at $2\omega \pm \omega_R$, where ω_R is the Raman frequency. This process is shown in Fig. 10.2. The cross section is much lower than normal RS, but the background is almost negligible since the emitted light is at about twice the laser frequency. The information is distinct from RS, but the sensitivities (19) preclude any serious analytical applications.

2.2. Raman Remote Sensing

Lidar is a general term for the optical frequency (light) equivalent of radar, which is concerned with radio frequencies. (See also Chapter 11.) In the typical case, light from a laser beam is sent toward the atmospheric region of interest, and backscattered radiation is collected and detected at the same location as the source. The collimated laser beam provides spatial selectivity along the line of sight, and the temporal information of the scattered photons is used to determine the range of the region probed. Direct backscattering at a single wavelength can be used to measure concentrations and sizes of aerosols. If the signal is observed as a function of laser wavelength, absorption along the optical path can be determined. If a wavelength different from the laser wavelength is monitored, Raman scattering can be measured. The main consideration is that the Raman signal is generally four orders of magnitude lower in strength than the above-mentioned signals. This puts extreme demands on the instrumentation.

The early Raman lidar used a ruby laser and a monochromator to observe

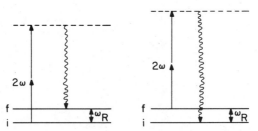

Fig. 10.2. Transition scheme for hyper-Raman spectroscopy. ω, input photons; ω_R Raman frequency; i, initial state; and f, final state.

N_2 to a distance of 3 km (20). To monitor trace gases, 160 mJ of frequency-doubled ruby radiation at 2 Hz is used (21) so that the Raman cross section can be increased 16-fold. The optical arrangement is shown in Fig. 10.3. When the 90 cm Cassegrain telescope is used at the 200–400-m range, concentrations as low as 1 ppm can be detected. To further improve detectability, resonance Raman scattering can be used. However, beam attenuation will eventually limit the utility of resonance Raman.

The advantage of Raman lidar is that all species can be monitored simultaneously by, for example, a multichannel detector (22). By appropriate gating, it is also possible to have spatial resolution on a single pulse. Another important application is the determination of temperatures from the rotational Raman band profiles, which can be measured remotely. Accuracy can be as high as $\pm 1°$ K within a distance of a few kilometers (23), something few other techniques are capable of achieving.

2.3. Spectroscopy by Inverse Raman Scattering (SIRS)

Although SIRS was predicted much earlier, the first observation was made only after lasers had been invented (24). The transition scheme for SIRS is no different from normal Raman scattering (RS) or stimulated Raman scattering (SRS), as shown in Fig. 10.4. Only the mode of observation is different. In SRS the gain in stimulated ω_S photons is monitored, while in SIRS the loss in ω_L excitation photons is monitored. If the radiation field is quantized, one can consider the scattering process as one that changes the occupation numbers of the two kinds of radiation modes, at ω_L and ω_S, such that the first mode decreases by one and the second mode increases by one. Using conventional raising and lowering operators for the harmonic oscillators, one can see that the transition probability for the Raman process is proportional to the quantity $N_L(N_S + 1)$,

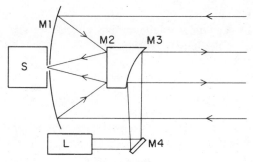

Fig. 10.3. Raman remote sensing apparatus. M1, M2, mirrors of Cassegrain telescope; M3, M4, collimation mirror for laser; S, spectrometer; and L, laser.

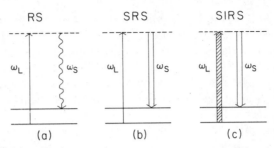

Fig. 10.4. Transition schemes in normal Raman scattering (RS), stimulated Raman scattering (SRS), and inverse Raman scattering (SIRS): straight arrow, excitation; wavy arrow, spontaneous scattering; double arrow, stimulated scattering; hatched double arrow indicates beam absorption is monitored.

where N_L and N_S are the occupation numbers for photons at ω_L and ω_S, respectively. More explicitly, the transition probability W is proportional to two terms,

$$W \propto \rho(\omega_L) \, [\rho(\omega_S) \, + \, 8\pi h \omega_S^3 n_S^3] \qquad (10.1)$$

where ω_L and ω_S are the frequencies of the photons shown in Fig. 10.4, ρ is the photon energy per unit volume per unit frequency, and n_S is the index of refraction at the Stokes frequency. The two terms in parentheses represent stimulated scattering and spontaneous scattering, respectively. In normal Raman scattering, $\rho(\omega_S)$ is essentially zero, so that the intensity of the scattering is proportional to the intensity of the incident photons. When $\rho(\omega_S)$ becomes large enough, either due to the large flux of exciting photons in SRS or due to the presence of a second laser beam at frequency ω_S in SIRS, the transition probability becomes proportional to the product of the two photon densities. It is now clear why there is a threshold in SRS. The photon density at ω_S must build up and become large compared to the second term, and losses in the medium at ω_S must be overcome. In SIRS, the presence of the second laser ensures that these conditions apply.

To determine the magnitudes of the various effects, we start out with the conversion efficiency of the Raman process,

$$\eta = \frac{\text{number of Stokes photons}}{\text{number of exciting photons}} = \frac{d\sigma}{d\Omega} \, 4\pi N l \qquad (10.2)$$

where $d\sigma/d\Omega$ is the absolute Raman scattering cross section per polarization in units of $cm^2 \, sr^{-1}$, N is the number density of the molecules or atoms, and l is the total interaction length of the excitation source with the sample. The typical value for η is 10^{-8} for the most favorable cases. In SRS, the conversion efficiency can be as high as 75% (25). In SIRS, the conversion efficiency can be likewise in the tens of percent. However, since one does not measure the increase in intensity of the laser beam at ω_S, but rather the decrease in the intensity of the laser beam at ω_L, it is more appropriate to think of absorbances of the order of unity in SIRS.

We can define a quantity γ as the relative enhancement of the Raman conversion efficiency in the presence of the intense radiation at ω_S. This is simply the ratio of the two terms in parenthesis in Eq. (10.1). In units commonly used for laser experiments,

$$\gamma = \frac{10^7 \, P_S}{8\pi hc^2 n_S^2 \, \Delta\omega \, \omega_S^3} \tag{10.3}$$

where P_S is the laser power density (W/cm^2) at ω_S (cm^{-1}), $\Delta\omega$ is the frequency spread of the Raman transition, and all the physical constants are in cgs units. The product $\eta\gamma/l$ thus represents the conversion efficiency of ω_L photons to ω_S photons per unit length. Since in ordinary Raman, the observable is $\Delta P_S = h\omega_S W$ and in SIRS the observable is $\Delta P_L = h\omega_L W$, one needs to further introduce a factor ω_L/ω_S to compare the two. Combining all these, one can derive the inverse Raman absorption coefficient g:

$$P_L(l) = P_L(0) \, e^{-gl} \tag{10.4}$$

$$g = \frac{10^7 \, P_S \omega_L N(d\sigma/d\Omega)}{2hc^2 n_S^2 \, \Delta\omega \, \omega_S^4} \tag{10.5}$$

We note that the stimulated Raman gain coefficient G is related to g such that

$$P_S(l) = P_S(0) \, e^{Gl} \tag{10.6}$$

$$G = \frac{\omega_S}{\omega_L} g \tag{10.7}$$

Equation (10.7) implies that if in Fig. 10.4c the increase in intensity in ω_S is measured rather than the decrease in intensity in ω_L, similar information is obtained. This technique is called stimulated Raman–gain spectroscopy (SRGS) (26).

We can estimate the degree of enhancement of the Raman effect for SRS and SIRS over normal Raman scattering. A typical giant pulse laser used for these experiments has P_S on the order of 100 MW/cm^2. For a strong Raman scatterer, such as liquid CCl$_4$ (459 cm^{-1} line), with a linewidth of 1 cm^{-1}, Eq. (10.3) gives a value of 10^6 for γ. This is a very significant enhancement. For any noticable enhancement, that is, for $\gamma > 1$, P_S needs to be 94 W/cm^2, a power density certainly readily achievable with lasers.

Equations (10.2), (10.5), and (10.7) show that the magnitudes of the three scattering processes are proportional to the same quantity. This implies that there should be a one-to-one correspondence in spectra obtained using any one of the three techniques. In actual experiments, normal Raman spectra do show this correspondence with inverse Raman spectra. However, the presence of a threshold in SRS preferentially selects the stronger Raman modes to go into oscillation. The relative intensities in SRS spectra thus show up quite differently than in the other two processes. This affects also the corresponding linewidths of the transitions, which tend to be more narrow in SRS spectra. Higher-order mixing can also occur in SRS, so that the anti-Stokes scattering can achieve comparable intensities as the Stokes scattering.

There is no theoretical threshold in SIRS because of the presence of the radiation field at ω_S. There is, however, a minimum power requirement on P_S that is related to the minimum observable absorbance in the particular experiment. To measure an absorption of 0.1%, for a 10-cm interaction length and pure liquid CCl$_4$, one needs a P_S of 15 kW/cm^2 to observe SIRS.

Because relative intensities are measured, Eq. (10.4) does not show any limitation on the power P_L. However, as has been pointed out (27), when P_L becomes very small, the normal anti-Stokes Raman scattering reaching the detector will become a source of interference. One can quickly see that at this limit

$$P_L = \frac{P_S(d\sigma/d\Omega)\ NB\Phi}{g} \tag{10.8}$$

where Φ is the beam divergence at ω_L and B the Boltzmann factor governing the intensity of the anti-Stokes scattering. For room temperature experiments involving Raman shifts of 1000 cm^{-1}, this limit is 0.6 mW/cm^2. For typical laser sources, this limit is well exceeded.

Again, there is no theoretical limit as to the maximum values of P_L and P_S. In practice, one needs to avoid breakdown of the medium at high powers and any heating effects because of absorption of either of the photons. One must also avoid the generation of SRS from either laser beam in the sample. Not only will the values of P_L and P_S have complicated dependences on the experimental parameters, but the other properties of the two beams (e.g., polarization) will also be affected by higher-order interactions.

Many different experimental arrangements have been tried for SIRS. The first demonstration uses liquid toluene to generate the anti-Stokes emission line at 1003 cm^{-1} as a source for ω_L and the residual ruby laser light as ω_S (24). Since temporal coincidence is guaranteed, the broad emission can be used to probe

SIRS absorption in nitromethane by photographic detection. Another arrangement uses fluorescence from dyes pumped by the frequency-doubled ruby laser as the probe and the ruby fundamental as the pump (28). The broad emisison allows a substantial portion of the Raman spectrum to be recorded. To obtain a larger range of the Raman spectrum, one can use a broader source like self-phase modulated emission (29). The above systems suffer from one common characteristic: the emission is too weak to allow observation of the Raman spectrum on a single laser shot. Photographic integration over several shots must be used, which makes quantitation very difficult.

Alternately, a broadband dye laser can be used (30). An optical arrangement based on a laser-pumped dye laser is shown in Fig. 10.5. A giant-pulse ruby laser is frequency-doubled in a KDP crystal. The second harmonic is separated from the fundamental ruby radiation by M1 and pumps a dye laser after passing through M2 and M3. The dye laser cavity is made up of a partially transmitting mirror M3 and a totally reflecting mirror M6. The dye laser output is directed to the sample cell by M2. The ruby fundamental output is combined with the dye laser by M4 and M5 so that spatial overlap can be achieved. Since the dye laser pulse is only 7 ns in duration, compared to the ruby laser pulse of 25 ns, temporal overlap of the two pulses is assured. The dye laser typically gives output that is $1-200 \text{ cm}^{-1}$ wide. By photographing this in a spectrograph, various portions of the Raman spectrum can be covered, depending on the dye and its concentration.

To cover the Raman region below 800 cm^{-1}, a different arrangement must be used. This is because lasers in the 670–700-nm range pumped by doubled ruby are not reliable and because one would have difficulties with the optical coating for M5. The alternative is shown in Fig. 10.6. In this case, ω_S is chosen

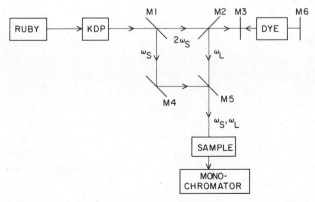

Fig. 10.5. Experimental scheme for SIRS using two pulsed lasers for moderate Raman shifts. KDP, frequency doubler; M1 through M6, dichroic mirrors; and ω_L and ω_S are defined in Fig. 10.4.

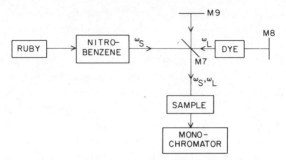

Fig. 10.6. Experimental scheme for SIRS using two pulsed lasers for small Raman shifts. M7, M8, dichroic mirrors; and ω_L and ω_S are defined in Fig. 10.4.

to be the Stokes SRS in nitrobenzene liquid generated from the ruby laser. An infrared dye laser pumped by the ruby fundamental is used as a source for ω_L. The dye laser cavity is formed by total reflectors M8 and M9. The pump laser enters the cavity through M7 and the dye laser output also exits via M7.

A fairly serious problem with laser-pumped lasers is one of mode competition in the lasing process. Any two of the optical surfaces in the laser cavity, including the cell windows, the mirrors, and even the multiple-layer dielectric coatings on the mirrors, can maintain standing waves for particular wavelengths. The broad spectral output of the dye laser is therefore not truly continuous, but has unpredictable maxima and minima in intensity. The inverse Raman spectrum can thus be obscured by the intensity variations. To avoid this problem the dye solvent can be index matched to the windows, and off-angles can be used for all except the cavity mirrors.

Since it is not necessary to have high intensities at ω_L, one can use a continuous-wave (cw) dye laser in conjunction with a giant-pulse laser to study SIRS. The first such quasi-cw or ac-coupled inverse Raman spectrometer is shown in Fig. 10.7 (31). The Q-switched ruby laser is directed into the sample by M2. To monitor the energy of each laser pulse, the exiting beam is directed by a second M2 into a ballistic thermopile. The cw dye laser is expanded 10 times to sample a major portion of the ruby laser. The cw dye laser passes through an optical delay line and then goes into the sample cell via M1. After the sample cell, the dye laser intensity is monitored using a phototube through a monochromator to reject scattered ruby light. The two lasers travel in opposite directions to provide the best discrimination against scattered light. The dye laser is controlled by a Pockels cell so that light only reaches the phototube slightly before and slightly after the ruby pulse. This is to avoid saturation of the phototube. The dye laser is affected by the ruby radiation reaching it; hence there is an optical delay line to isolate the two lasers. The output of the phototube is displayed on an oscilloscope and photographed.

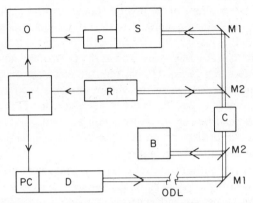

Fig. 10.7. Experimental scheme for SIRS using one pulsed and one cw laser. O, oscilloscope; T, trigger generator; PC, Pockels cell; R, ruby laser; D, dye laser; ODL, optical delay line; M1, M2, dichroic mirrors; B, ballistic thermopile; C, sample cell; S, spectrometer; and P, phototube.

A typical inverse Raman event is shown in Fig. 10.8. Before the dye laser switches on, no light reaches the phototube. The dot at the upper left-hand corner of the oscilloscope trace represents this zero-intensity level. When the dye laser switches on just before the arrival of the ruby pulse, the signal corresponds to the very early part of the trace. The oscilloscope is not triggered until later, so that the dye intensity has time to stabilize. The interaction of the ruby laser and the dye laser produces a drop in the dye intensity, and this is the inverse Raman absorption. The second, much larger drop in the dye intensity is due to the interference of the ruby laser inside the dye laser cavity, and has no significance in this experiment.

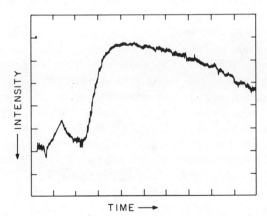

Fig. 10.8. SIRS signal as recorded on an oscilloscope. Horizontal scale, 50 ns/div.

To determine the magnitude of the inverse Raman absorption, one can simply measure off the corresponding quantities on the oscilloscope trace. The quantity $P_L(l)/P_L(0)$ in Eq. (10.4) is simply the ratio of the heights (dye laser intensities) immediately before the ruby laser pulse and at the peak of the absorption. Since l can be reliably measured in Eq. (10.4), one can calculate the quantity g. To have a meaningful interpretation of the absorption, one needs to normalize g with respect to the laser power P_S. This is necessary for each event because parameters such as peak power and pulse width are typically not reproducible for giant-pulse lasers. Fortunately, all the necessary information is contained in Fig. 10.8 so that this can be done properly.

The total energy per pulse of the ruby laser is conveniently obtained by the ballistic thermopile in Fig. 10.7. The losses due to M2 and the exit window of the sample cell can be determined experimentally. Since the power levels are low enough to avoid SRS and other higher-order processes in the sample, further losses can be neglected. To obtain P_S as a function of time, one then needs to know the pulse width and pulse shape for each event. We note that if g in Eq. (10.3) is small,

$$\frac{P_L(0) - P_L(l)}{P_L(0)} = gl \qquad (10.9)$$

Since the right-hand side of Eq. (10.9) is proportional to P_S, according to Eq. (10.5), the time dependence of P_S can be determined unambiguously.

Instead of tuning the ω_L beam outlined in Fig. 10.7, one can tune the ω_S beam (32, 33) by combining a pulsed dye laser with a cw Ar ion laser (ω_L). If the dye laser has a high repetition rate, one can scan through the Raman region from 450 cm^{-1} to 2000 cm^{-1} at a rate of 366 cm^{-1}/min (34). Alternately, two dye lasers, one cw and one pulsed, can be used (35). In this arrangement, a single-mode cw ring dye laser is used at ω_L, chopped mechanically to produce 100-μs pulses at 10 Hz (to reduce the average light level on the detector). A second cw single-mode dye laser is amplified in three stages in dye cells pumped by a frequency-doubled Nd:YAG laser (36). The result is a Fourier transform limited bandwidth in the ω_S beam at 10 Hz and over 1 MW in power for 12-ns pulses. High spectral resolution can thus be obtained in SIRS with good sensitivity. For SRGS, one can use a frequency-doubled Nd:YAG laser for ω_L and a cw dye laser for ω_S (37). By combining the two beams in a Jamin interferometer, the SRGS signal can be used to study the real (nonlinear dispersion) or the imaginary (Raman gain) parts of the interaction simply by altering the relative retardations in the two arms of the interferometer.

When a cw pump beam is used, the lower powers imply that the SIRS or SRGS signal is decreased. But, if one can improve the detectability in absorption, cw SRGS (two cw lasers) is feasible (38). A sophisticated version of this is based on two single-mode cw lasers at 1 W and 10 mW, respectively (39). The success of such a system is due to an increased duty cycle to partially offset the lower power levels, the low noise levels of the probe laser at the modulation frequency, by a noise reduction system, a reference channel to the lock-in amplifier to provide compensation, and a multipass gas cell that allows for a gain of 50 for 97 passes. By monitoring the absorption in the argon ion laser rather than the gain in the dye laser, SIRS can be studied in the same apparatus. However, the power level in the dye laser has to be increased substantially to provide similar signal levels.

An important feature of SIRS is its quantitative reliability. Since the two laser beams accurately define the interaction region, the uncertainties in collection efficiency in normal Raman spectroscopy are avoided. The measurement is also relative, that is, the ratio $P_L(0)/P_L(l)$, so that calibration of the detector response is not necessary. There is also an averaging effect over the cross section of the beam if the mode structures of the beams are poor (30). The experimental line shapes are identical to those in normal Raman spectra, and no nonlinear background effects are present. That SIRS and normal Raman parameters are equivalent is shown in Table 10.1. In fact, Raman cross sections determined by SIRS are expected to be more reliable than other methods. For example, the nitrobenzene line at 1345 cm^{-1} has been determined to be $(1.38 \pm 0.27) \times 10^{-29}$ cm^2 sr^{-1} (31). Similarly, SRGS has been used to compare the 992-cm^{-1} Raman line of benzene to the normal Kerr effect in CS_2, to obtain both the peak cross section and the nonlinear dispersion (37). Other quantitative applications include the determination of the Raman cross section of the Q branch of the fundamental vibration of N_2 (40).

The high spectroscopic resolution that can be achieved with SIRS and SRGS is an attractive feature. In the normal 90° geometry of Raman scattering, the full Doppler width is observed for the emitted photon, ω_S. In the collinear geometry for SIRS, if the beams are co-propagating, the Doppler width is reduced to the order of $\omega_L - \omega_S$. For pure rotational spectra, this reduction is substantial. It is even possible to obtain sub-Doppler spectra with SIRS (41).

Table 10.1. CW Raman Experiments

Type	P_1 (mW)	P_2 (mW)	Gas	Pressure (torr)	Reference
RS	150	—	O_2	100	9
SRGS	1,000	10	CH_4	35[a]	39
RIKES	450	15	H_2	3	52
PARS	10,400	1,000	CH_4	800	56
CARS	640	10	CH_4	46	81

[a] Multiple pass.

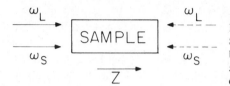

Fig. 10.9. Saturated inverse Raman absorption. Solid arrows, high-intensity saturating lasers; broken arrows, lower-intensity probe lasers; and z, molecular velocity vector relevant to Doppler effect.

This can be seen in the optical arrangement in Fig. 10.9. Under the influence of strong pump and probe beams in SIRS (solid arrows), the molecules in the sample with a certain velocity distribution in the z direction will take part in the SIRS process. Now, the counterpropagating pump and probe beams (broken arrows) will be able to experience this saturation effect when their frequency difference allows them to interact with the same set of molecules. This is identical to normal saturation spectroscopy except that the excitation mechanism is based on the Raman effect. The widths of such saturation spectra are determined by the molecular homogeneous linewidth, or the width of the laser, whichever is broader. Using saturating intensities of 2 MW in ω_S and 100 kW in ω_L, and a cw probe laser for monitoring, the effect was demonstrated in D_2 gas at a pressure as low as 2 torr (41). The spectroscopic utility of SIRS and SGRS has also been demonstrated in 1.3 torr of $^{13}CH_4$ (35) at a resolution of 100 MHz, in 3.8 torr of SF_6 with good resolution of the very complex vibrational-rotational structure (42), and in CH_4 in a molecular jet to give rotationally cooled spectra (43).

The discrimination of SIRS against fluorescence can be clearly seen in the spectrum obtained in a 10^{-2} M solution of rhodamine 6G in ethanol (44). The detectability of SIRS depends on the ability to measure small absorptions, and can be quite impressive with respect to the actual number density of the analyte present, as evidenced by the above-mentioned gas-phase studies. SIRS does not have a nonlinear background term as in CARS and does not have Kerr effect terms like RIKES. So, the fundamental limit in detectability for trace analysis is the Raman scattering of the major component. Such background scattering is present even far away from the Raman-active fundamentals of typical molecules. By scanning across the Raman line, better discrimination against Raman scattering of the major component is possible, but the latter still contributes to detectability since it magnifies the amplitude noise in the pump beam. At present, the limit is in the 0.1% range for the minor component (32).

Finally, there are two interesting applications that are based on these techniques. By monitoring SRGS of N_2 at the exit of a subsonic nozzle at high resolution, the flow velocity can be determined from the spectral shifts (45). This then has the potential for diagnostics in analytical flames and plasmas, as well as in combustion chambers. A transient stimulated Raman signal is produced when a longer, third laser pulse is used to interrogate the molecules produced by two shorter pulsed lasers in the normal SIRS configuration after some time delay (46). Since collisions change the phase of the coherent excitation, this third pulse only probes the molecules that did not suffer any collisions. If the pulse length of the third laser is longer than 1.4 times the dephasing time, a higher spectroscopic resolution will be achieved. This way, weak Raman transitions normally hidden at the wings of stronger Raman lines can be observed (46).

2.4. Raman-Induced Kerr Effect Spectroscopy (RIKES)

The optical Kerr effect is an intensity-dependent birefringence caused by a plane-polarized pump beam at one frequency, resulting in a rotation of the plane of polarization of a probe beam at a different frequency, as long as the two planes of polarization are not orthogonal to each other (47). When the frequencies of the two beams correspond to a Raman resonance, the effect is greatly enhanced. RIKES is analogous to inverse Raman spectroscopy (Figure 10.4) and requires two lasers to interact with the sample. When a plane-polarized probe beam is used as ω_L in Figure 10.4, one can decompose it into two electric fields with vectors, one in the plane of the linearly polarized pump beam, ω_S, and one perpendicular to this plane. Since Raman transition probabilities in general are different for parallel and perpendicular photons, inverse Raman absorption will be different in magnitude for each of these electric fields. After the sample, the resultant electric field will therefore be rotated in its polarization direction, and can be conveniently detected as a transmitted intensity through crossed polarizers. An equation can be obtained for the intensity of the field passing through the second polarizer. Figure 10.10 shows the relationship between the two sets of coordinate axes used. The unit vectors $\hat{p}_1$ and $\hat{p}_2$ are parallel to the polarization directions of the fields passed by the first and second polarizers, respectively. These polarizers are crossed, that is:

$$\hat{p}_1 \cdot \hat{p}_2 = 0 \tag{10.10}$$

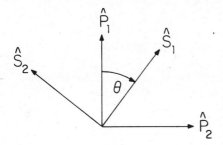

Fig. 10.10. Coordinate system for polarization rotation in SIRS. P_1, P_2, axes for polarization direction of probe beam; and S_1, S_2, axes for polarization direction of pump beam.

The electric vector of the probe field in the sample $\mathbf{E}_L$ (having passed through the first polarizer), is

$$\mathbf{E}_L = E_L \hat{p}_1 \tag{10.11}$$

The electric vector of the pump field in the sample, $\mathbf{E}_S$, is

$$\mathbf{E}_S = E_S \hat{s}_1 \tag{10.12}$$

where the frequency functionality of the fields has been suppressed. The intensity of a field will be given by

$$P = \mathbf{E} \cdot \mathbf{E} \tag{10.13}$$

To treat the rotation, we reexpress $\mathbf{E}_L$ in the coordinate system of the pump field, that is

$$\begin{aligned} \mathbf{E}_L &= E_L \{ (\hat{p}_1 \cdot \hat{s}_1) \, \hat{s}_1 + (\hat{p}_1 \cdot \hat{s}_2) \, \hat{s}_2 \} \\ &= E_L(1) \, \hat{s}_1 + E_L(2) \, \hat{s}_2 \end{aligned} \tag{10.14}$$

We thus consider the excitation field to be composed of two fields of identical frequency, propagation direction and phase, polarized orthogonally with intensities $\{E_L(1)\}^2$ and $\{E_L(2)\}^2$.

After passing through the sample, each component field will be attenuated by the absorption induced by the pump field. Thus,

$$\begin{aligned} P_{L1}(l) &= P_{L1}(0) \exp \{ -g_1 l \} \\ P_{L2}(l) &= P_{L2}(0) \exp \{ -g_2 l \} \end{aligned} \tag{10.15}$$

where

$$g_1 = \frac{g}{(1 + \rho)} \qquad g_2 = \frac{g\rho}{(1 + \rho)} \tag{10.16}$$

and g is given by Eq. (10.5) and ρ is the depolarization ratio for linearly polarized light of the Raman transition.

Using Eqs. (10.13)–(10.15), one deduces that the electric vector of the excitation field after passing through the sample, $\mathbf{E}'_L$, can be written

$$\mathbf{E}'_L = E_L \{ e^{-\alpha} (\hat{p}_1 \cdot \hat{s}_1) \, \hat{s}_1 + e^{-\beta} (\hat{p}_1 \cdot \hat{s}_2) \, \hat{s}_2 \} \tag{10.17}$$

where

$$\alpha = \tfrac{1}{2} g_1 l \qquad \beta = \tfrac{1}{2} g_2 l \tag{10.18}$$

The intensity of the excitation field leaving the sample cell, $P_L(l)$, can be obtained from Eq. (10.17).

$$P_L(l) = \mathbf{E}_L' \cdot \mathbf{E}_L' = P_L(0) \{ e^{-2\alpha} \cos^2 \Theta + e^{-2\beta} \sin^2 \Theta \} \tag{10.19}$$

The intensity of the field passed by the second polarizer, P_L'', is readily given by

$$P_L'' = (\mathbf{E}_L' \cdot \hat{p}_2)^2 = P_L(0)\, (e^{-\alpha} - e^{-\beta})^2 \sin^2 \Theta \cos^2 \Theta \tag{10.20}$$

which for low absorption becomes

$$P_L'' \simeq P_L(0) \left(\frac{gl}{4} \right)^2 \left(\frac{1 - \rho}{1 + \rho} \right)^2 \sin^2 2\Theta \tag{10.21}$$

Application of simple differential calculus to Eq. (10.20) shows that maximum sensitivity occurs when $\Theta = 45°$. In actual measurements, however, there is a nonresonant contribution from the normal optical Kerr effect. This produces an interference that shows up as a dispersion-like line shape, very similar to an analogous interference in CARS.

One can also use a circularly polarized pump beam, ω_S, for RIKES. There, one can consider the plane-polarized probe beam, ω_L, as being composed of equal amounts of left- and right-circularly polarized light propagating in phase. Since Raman transitions in general show different probabilities for same-sense versus opposite-sense circularly polarized photons, the intensities of the two components of ω_L will no longer be identical after the sample. This is detected again as a transmitted intensity through crossed polarizers. If we define x as the transmission axis of the first polarizer and y as the transmission axis (perpendicular to x) of the second polarizer, the probe can be decomposed into two components:

$$(\mathbf{E}_y)_1 = \mathbf{E}_x e^{-i\pi/2}$$

$$(\mathbf{E}_y)_2 = \mathbf{E}_x e^{i\pi/2} \tag{10.22}$$

where 1 and 2 refer to the sense of circular polarization. Since the amplitudes are identical, the net intensity in the y direction, $P_y(0) = |(\mathbf{E}_y)_1 + (\mathbf{E}_y)_2|^2 = 0$.

In the presence of inverse Raman absorption, the two intensities decrease by different amounts similar to Eq. (10.15) and (10.16), where ρ now refers to the depolarization ratio for circularly polarized light of the Raman transition. So, after the sample, we have

$$(\mathbf{E}_y')_1 = \mathbf{E}_x e^{-i\pi/2} e^{-\alpha}$$

$$(\mathbf{E}_y')_2 = \mathbf{E}_x e^{i\pi/2} e^{-\beta} \tag{10.23}$$

with α and β having analogous meanings as in Eq. (10.18). The square of the vector sum of the two components is then the transmitted intensity

$$P_L(l) = P_L(0) \, (e^{-\alpha} - e^{-\beta})^2 \tag{10.24}$$

for low absorptions, this becomes

$$P_L(l) = P_L(0) \, \frac{gl^2}{4} \left\{ \frac{1 - \rho}{1 + \rho} \right\}^2 \tag{10.25}$$

Because the pump radiation is circularly polarized, there is no optical Kerr effect. The line shape is then predictable from that in normal Raman scattering, that is, from the wavelength dependence of g. For the general case of a plane-polarized probe beam and an elliptically polarized pump, the RIKES signal shows a more complicated, but predictable, dependance (48).

The first demonstration of RIKES (49) relies on a broadband dye laser that is directed through two crossed polarizers, as shown in Fig. 10.11. Excitation is by a narrow-band dye laser, synchronized in time since the two lasers are pumped by the same nitrogen laser, circularly polarized, and intersecting the probe beam at a shallow angle in the sample cell. Using neat liquids and about 30 kW of power in each of the 8-ns long beams, about 4 shots are needed to record a spectrum on ASA-400 film. The advantage of recording a substantial portion of the Raman spectrum at the time resolution of the laser is offset by the poor mode structure of the dye laser and the problems of calibrating the photographic emulsion for quantitation.

Since pulse-to-pulse intensity variation is the main contributor of noise in RIKES using pulsed lasers, one can replace one or both of the lasers by cw radiation. When the probe beam is a cw source, it is possible to introduce heterodyne detection (50). If after the second polarizer, there exists a small amount of probe radiation polarized in the y plane, with amplitude $(\mathbf{E}_y)_0$, that is derived from the slight ellipticity in the field described by Eq. (10.22), this can interfere with the RIKES signal. The detector will then respond to an intensity

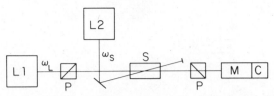

Fig. 10.11. Experimental scheme for RIKES. L1, linearly polarized probe beam; L2, circularly polarized pump beam; P, polarizers; S, sample; M, monochromator; and C, camera.

$$P_L(l) = |(\mathbf{E}_y)_0 + (\mathbf{E}_y')_1 + (\mathbf{E}_y')_2|^2 \qquad (10.26)$$

In the case where $(\mathbf{E}_y)_0$ is much larger than the other fields, the dominant terms in Eq. (10.26) give rise to

$$P_L(l) = I_y + \sqrt{I_y P_L(0)}\,(e^{-\alpha} - e^{-\beta}) \qquad (10.27)$$

where $I_y = |(\mathbf{E}_y)_0|^2$ is a dc signal. If ac detection is used, then

$$P_L(l) = \sqrt{I_y P_L(0)}\,\frac{gl}{4}\frac{1-\rho}{1+\rho} \qquad (10.28)$$

The linear dependence here implies that the signal will be larger than that given in Eq. (10.25), and that the line shape will be identical to that in the normal Raman spectrum. Naturally, there is an optimum value for I_y, since the noise in I_y also increases with I_y, and will eventually become large compared to the RIKES signal.

The optical arrangement for heterodyne-RIKES is similar to Fig. 10.11 (50). C and M are replaced by a photomultiplier tube, L1 is replaced by a 0.5-W single-mode cw argon ion laser, with about 0.3 mW in the heterodyne beam, I_y, and L2 is a 50-kW nitrogen-pumped dye laser. So, a fair amount of light is incident on the detector all the time. An ac amplifier extracts the RIKES signal, which is then gated with a boxcar integrator. The sodium benzoate peak (water solution) at 1005 cm^{-1} was recorded with a signal-to-noise ratio (S/N) of 5 when the concentration was 0.2%. Even better S/N is obtained using two cw lasers with proper design (51). The apparatus is shown in Fig. 10.12. The good detectability is a result of the high modulation frequency and the stable probe beam. Using about 100 mW in the pump and 10 mW in the probe laser, a detection limit of 0.2% (S/N = 3) is found for C_6H_6 in CCl_4. The limit is from residual Raman scattering from the major component, even though resonances are quite far away. For a single component, for example, gas-phase H_2, the background is absent, and a detection limit of 3 torr of H_2 has been demonstrated (52).

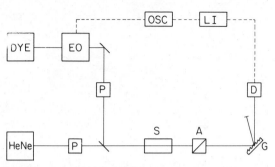

Fig. 10.12. Experimental scheme for cw RIKES. EO, electro-optical modulator at 25 kHz; P, polarizers; A, analyzer; S, sample; G, grating; D, photodiode; OSC, frequency generator; and LI, lock-in amplifier.

The dependence of the S/N ratio and the optimum heterodyne intensity on the various instrumental characteristics have been explained (53). It is clear that the lower the background radiation due to birefringence in the system, the lower the heterodyne power that is needed, provided that the system is not shot-noise or electronic-noise limited. And, the lower these two light intensities become, the better the S/N ratio. In our laboratory, we have developed the techniques needed to reduce background birefringence substantially, and have applied this to measure small optical rotations in liquids (54). By incorporating a pump laser, we can then study RIKES with little modification in the system, as shown in Fig. 10.13. Using about 150 mW in the probe beam at 488 nm and about 11-mJ pulses (7 ns wide) of pump beam at 571 nm, we have determined the detection limit of CH_3CN (2965-cm^{-1} line) in H_2O to be 0.5% by volume (S/N = 3). On switching to D_2O as the solvent, the detection limit was 0.1% The limit is due to the background Raman scattering of the solvent, and increasing the pump power did not help. It is clear that to detect trace components in a mixture, a different approach should be used. A possibility is to introduce a reference cell in series in the optical path to allow compensation of the signal from the major component—the solvent. If now two pump beams (derived from the same laser) that are linearly polarized enter the reference cell and the sample cell with polarization vectors perpendicular to each other but 45° relative to the probe beam, the normal optical Kerr effect and any background scattering from the solvent will cause rotations in opposite directions for the probe beam. The result is mutual cancellation if the reference cell is properly matched with the sample cell. This concept was tested on benzyl alcohol in the two cells by modifying Fig. 10.13. The RIKES signal observed at around 2965 cm^{-1} was 51 units and 52 units, respectively, when the pump laser was introduced into one or the other cell. When the pump laser was introduced into both cells at perpendicular polarizations at identical power levels, the signal became 11 units.

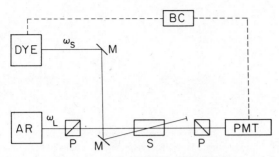

Fig. 10.13. Experimental scheme for quasi-cw RIKES. DYE, YAG-pumped dye laser; AR, argon ion laser; M, mirrors; P, polarizers; S, sample; PMT, phototube; and BC, boxcar integrator.

The cancellation is then 80% of the background in the sample cell. The inhomogeneous nature of the pump beam and imperfect beam overlap in the two cells are the main reasons for incomplete cancellation. The utility of this optical-null technique thus depends on having available better optics.

2.5. Photoacoustic Raman Spectroscopy (PARS)

Simultaneous with the Raman process (Fig. 10.4a) is the excitation of the molecule to an excited rovibronic state. Relaxation of this excited state results in translational energy deposited in this molecule and in the surrounding atoms or molecules. This is then the Raman analog of optoacoustic spectroscopy (Chapter 5). The only difference is that excitation is not through a simple absorption process. The microphone still functions to detect the excited molecules. In normal Raman spectroscopy, the number of Raman events is so small that the photoacoustic signal is negligible compared to acoustic noise in the room or from cell window absorption. In the case of resonance Raman, the signal is enhanced, but the probability for normal absorption to an electronic state is also increased, thus masking the acoustic signal from the Raman event. So, realistically, only when nonlinear Raman methods are used to drive the Raman process can one observe PARS. In stimulated Raman events (Fig. 10.4b), an acoustic signal is generated, but the utility of such a signal is poor because little spectroscopic information is present. The excitation beam can be tuned over a large range of frequencies without altering the signal substantially. To introduce spectroscopic information, one must use inverse Raman excitation (Figure 10.4c). Excitation then depends on the presence of both beams, the frequency difference of which provides the Raman spectrum. It should be noted that the CARS process (Figure 10.14) takes a molecule from the ground state to the ground state, and no vibrational excitation results. PARS is then unrelated to CARS. It turns out that in CARS experiments, the mere presence of the two laser beams

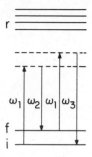

Fig. 10.14. Transition scheme for CARS. i, initial state; f, Raman level; r, coupling intermediate states; ω_1, ω_2, pump lasers; and ω_3, CARS output.

will produce a PARS signal via inverse Raman excitation, but the CARS emission actually competes with PARS for excited molecules.

A brief derivation of the equation for the pressure change in a gas resulting from inverse Raman absorption will be presented here. In the regime where the p-V-T behavior of a gas can be satisfactorily represented by the ideal-gas relationship, we find that

$$\Delta p = \frac{(\gamma - 1)U}{V} \tag{10.29}$$

where γ is the heat capacity ratio (C_p/C_v) of the gas, V is the cell volume, and U is the thermal energy deposited in the gas (adiabatic conditions are assumed). If more than one gas is present, γ can be calculated using a partial pressure weighted average C_p/C_v.

For example, if we consider an ideal monatomic gas where $\gamma = \frac{5}{3}$ and $U = \Delta E_T$, that is, there are no internal degrees of freedom and all of the thermal energy is channeled into translational energy E_T; we find using Eq. (10.29):

$$\Delta p = \frac{2\Delta E_T}{3V} \tag{10.30}$$

a familiar result for ideal monatomic gases. For real molecular gases the ($\gamma - 1$) term will correct for the loss of energy to internal modes where it will not contribute to a pressure change.

To use Eq. (10.29) it is necessary to determine U for inverse Raman absorption. The total energy, U_L (in joules), removed from the excitation beam as a result of inverse Raman absorption is

$$U_L = A\tau P_L(0)(1 - e^{-gl}) \tag{10.31}$$

where A (cm^2) is the cross-sectional area of the interaction region, l (cm) is the length of the interaction region, τ (s) is the interaction time, $P_L(0)$ is expressed in watts/cm^2, and g (cm^{-1}) is given by Eq. (10.5). In the regime of low absorption Eq. (10.31) can be approximated by

$$U_L = A\tau P_L(0)gl \qquad (10.32)$$

The energy that appears as thermal energy in the gas is not U_L, however, but rather the energy that results from the relaxation of the vibrational level populated by the inverse Raman process. Assuming all of the energy in the Raman-excited vibrational level degrades to thermal energy, we find

$$U = U_L \Delta\sigma_R/\sigma_L \qquad (10.33)$$

where $\Delta\sigma_R$ (cm^{-1}) is the Raman shift of the vibrational level and σ_L (cm^{-1}) is the energy of an excitation photon. Thus,

$$\Delta p = \frac{(\gamma - 1)\, A\tau P_L(0)\, gl\, \Delta\sigma_R}{V\sigma_L}$$

$$= \frac{K(\gamma - 1)\, A\tau l P_L(0)\, \Delta\sigma_R N(d\sigma/d\Omega)}{V n_s^2 \sigma_s^4 \Delta\sigma} \qquad (\text{J/cm}^3) \qquad (10.34)$$

where $K = 5.6 \times 10^{11}$. When a modulated cw source is used, τ is the reciprocal of the modulation frequency.

Rigorously, Eq. (10.34) is valid only for collimated beams of uniform irradiance in spatial coincidence at all points in the interaction region. The use of focused Gaussian beams with different waist diameters will lead to a mismatch in beam overlap that can be corrected for by performing radial and axial integration over the two beams in the interaction region. For two Gaussian beams, it was found theoretically that the maximum excitation is achieved when the two beam waists are identical in diameter (55). Unless the mismatch is severe, Eq. (10.34) can be used to provide a good indication of the magnitude of the pressure change that can be expected. Of interest is that unlike CARS, no phase matching is required for PARS, and beam overlap can be readily maximized.

The first demonstration of PARS is based on two cw lasers. By proper rearrangement of a cw dye laser cavity, as shown in Fig. 10.15, one can obtain a spatial region where the dye laser beam and an external Ar ion laser beam overlap with closely matching beam waists (56). This allows 10.4 W of 514.5-nm radiation (double-pass) to interact with 1 W of 605.4-nm radiation. When 800 torr of CH$_4$ gas is put into the acoustic gas cell at the common beam waist, the 2916.7-cm^{-1} Raman line is excited to produce a PARS signal. The estimated pressure change of 1.3×10^{-3} μbar during one chopper cycle at 573 Hz is in good agreement with the measured value. There is an acoustic background due to the cell windows from the modulated Ar ion laser, which is about 10 times

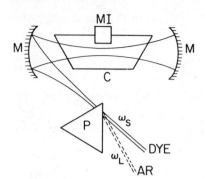

Fig. 10.15. Experimental scheme for PARS using cw lasers. M, cavity mirrors of dye laser; AR, argon laser; P, prism; C, gas cell; and MI, microphone.

that of the actual PARS signal. However, by scanning the dye laser through the Raman resonance, the Raman-induced effect is clearly demonstrated.

The incentive for using PARS is one of sensitivity. In contrast to absorption measurements, one is looking at the increase of response on top of an ideally low background when no excitation occurs, as opposed to the detection of a small change on top of a large intensity. The advantages are common to all indirect absorption measurements based on the optoacoustic effect (57). In practice, environmental noise, window absorption, and laser flicker noise at acoustic frequencies seriously degrade the detectability. It may be possible to use resonant acoustic cells (58) to overcome some of these problems in the gas phase. In liquids, solvent absorption and inefficient coupling of the acoustical wave to the transducer are the main experimental limitations. So, the more promising analytical applications are all in the gas phase.

To increase the detectability, one can use high-peak-power pulsed lasers (59). A schematic diagram of a system based on the Nd:YAG laser is shown in Fig. 10.16. The frequency-doubled output is used for PARS excitation as well as to pump a dye laser, which provides the other excitation beam. Typical power levels are 1.5 MW (15 mJ) and 1.0 MW (10 mJ) at 532 and 574 nm, respectively, for exciting the CO_2 line at 1388 cm^{-1} (59). The amount of energy deposited in the gas per pulse is about 10^6 more than the cw experiments. Even accounting for the lower duty cycle, the gain is still substantial. Moreover, the short pulses allow one to use a gated detector, such as a boxcar integrator, to monitor the initial pressure wave from the interaction region. This then discriminates against the acoustic wave generated at the cell windows, environmental noise, and the weaker pressure waves derived from reflections in the cell.

The response in PARS for a minor component in gas mixtures has been investigated (60). In general, when the minor component is present at levels less than 1%, a linear PARS response is obtained with respect to the concentration. This is in contrast to the PARS response for a single gaseous component at different pressures (61). The latter shows a nonlinear response with respect

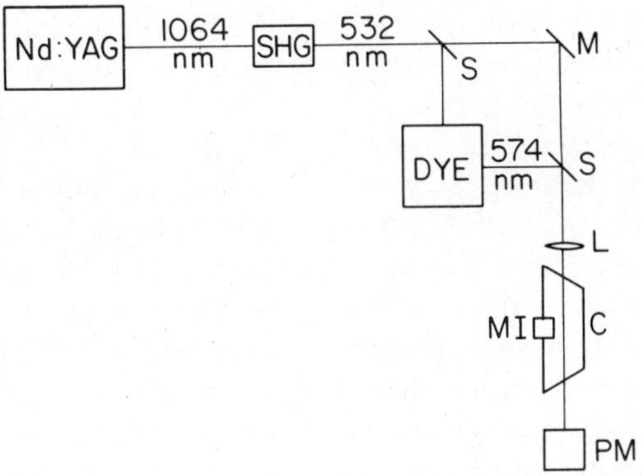

Fig. 10.16. Experimental scheme for PARS using pulsed lasers. SHG, second-harmonic generator; S, beam splitters; M, mirror; L, lens; C, gas cell; PM, power meter; and MI, microphone.

to number density due to a combination of varying linewidth, degree of saturation, efficiency of vibrational relaxation, and coupling efficiency with the transducer. Since the PARS excitation is exactly analogous to normal Raman transitions, there are no contributions from the nonresonant terms that are inherent to, for example, CARS. So, the major component in the gas mixture has less influence on the PARS detectability. The only exception is if the major component shows absorption at one of the laser wavelengths, such as the case of CH_4 (56). There, the green laser must be the one that is modulated to avoid the weak absorption at the red laser wavelength. Naturally, such discrimination is not possible using two pulsed lasers, and one must then rely on a more judicious choice of the laser wavelengths. PARS is inherently more sensitive than inverse Raman spectroscopy, for exactly the same reasons that optoacoustic spectroscopy is more sensitive than conventional absorption measurements. Using a high-power pulsed laser system similar to that in Fig. 10.16, a detection limit of 3 ppm (S/N = 3) of CH_4 in N_2 was found. Comparable detection limits are expected for other gases for the same power levels. These are better than those obtained for CARS (62), where the nonlinear terms from the background determine the fundamental limit. An attempt to improve the detectability of PARS using a multiple-pass cell has been reported (61), but the larger cell required results in a lower acoustic detection sensitivity, and negates any gain from multiple passing.

As a Raman spectroscopic tool, PARS has several important advantages. Since two lasers are required for excitation, the inherent spectral resolution is determined by the laser linewidths. This is particularly useful for studying gas-phase vibrational-rotational transitions. Another unique feature is the absence of a Rayleigh line in PARS, since Rayleigh scattering returns the molecules to the ground state with no energy exchange. So, transitions at very low frequencies can be easily studied. There is also a curious situation with the anti-Stokes transition. Since molecules are taken from the excited state to the ground state, a *decrease* in effective temperature occurs. This is then a negative pressure wave, and the microphone response (using a lock-in amplifier or a boxcar integrator) should be negative. In the actual experiment, however, one cannot dictate which of the two laser beams is ω_L and which is ω_S in Fig. 10.4b, and both Stokes and anti-Stokes transitions will occur. This predicts that there is some cancellation of the PARS signals at very low frequencies, when the excited level is almost equal in population to the ground state (after accounting for degeneracies). A final feature is that the power levels used in PARS are low compared to stimulated Raman studies. This allows weak spectral features to be recorded and not be suppressed by the laser action on the stronger Raman lines (63).

2.6. Coherent Anti-Stokes Raman Spectroscopy (CARS)

Although CARS is not the first of the various nonlinear Raman techniques to be discovered, it is the most popular technique at present. It certainly has an attractive acronym, but its popularity may be due to the generation of a third color of light that is easily detected visually, at least in typical neat liquids used for demonstration. The energy scheme for CARS is shown in Fig. 10.14. When two photons of frequencies ω_1 and ω_2 interact with a molecule, the Raman resonance is driven coherently. This coherence can be probed by allowing another photon of ω_1 to interact with the system, so that a photon at ω_3 is generated. By energy conservation, it is easy to see that $\omega_3 = 2\omega_1 - \omega_2$, and the emission is anti-Stokes by convention. Naturally, the coherence can also interact with ω_2 once more, but the resulting emission is at frequency ω_1 and cannot be conveniently monitored. On the other hand, other frequencies can be used to probe the coherence, if desired. The important point is that the four-photon event is simultaneous and is different from a stepwise SRS excitation and the anti-Stokes generation.

Neglecting any one or two photon absorption processes, the probability for CARS is related to a susceptibility for the transition from state i to state f back to state i (64).

$$\chi_{\alpha\beta\gamma\delta}^{CARS} = \chi_{NR} + (N_i - N_f)\, L[24\hbar(\omega_{if} - \omega_1 + \omega_2 - i\Gamma_{if}]^{-1}$$
$$\times\; [\langle i|\alpha_{\alpha\beta}^A|f\rangle\langle f|\alpha_{\gamma\delta}^S|i\rangle + \langle i|\alpha_{\alpha\gamma}^A|f\rangle\langle f|\alpha_{\beta\delta}^S|i\rangle] \qquad (10.35)$$

where α, β, γ, δ are Cartesian coordinates, L is a local field correction factor, χ_{NR} is a nonresonant contribution, N_i is the number density of the state i, Γ_{if} is the half width at half height for the spontaneous Raman line, ω_{if} is the frequency of the Raman line, and the Stokes (S) and anti-Stokes (A) transition elements are:

$$\langle i|\alpha_{\alpha\beta}^{A}|f\rangle = \frac{1}{\hbar}\sum_{r}\left[\frac{(M_{\alpha})_{ir}\,(M_{\beta})_{rf}}{\omega_{r}-\omega_{3}+i\Gamma_{r}} + \frac{(M_{\beta})_{ir}\,(M_{\alpha})_{rf}}{\omega_{r}+\omega_{1}+i\Gamma_{r}}\right]$$

$$\langle f|\alpha_{\gamma\delta}^{S}|i\rangle = \frac{1}{\hbar}\sum_{r}\left[\frac{(M_{\gamma})_{fr}\,(M_{\delta})_{ri}}{\omega_{r}+\omega_{2}+i\Gamma_{r}} + \frac{(M_{\delta})_{fr}\,(M_{\gamma})_{ri}}{\omega_{r}-\omega_{1}+i\Gamma_{r}}\right] \quad (10.36)$$

where Γ is a damping constant, $(M_{\alpha})_{ir} = \langle i|M_{\alpha}|r\rangle$ is a dipole transition moment, and the summation is over all real states r in the molecule. For plane waves, the intensity (power/area) of the emission, time averaged, after the two initial laser beams have traversed a distance l is

$$I_{3} = \frac{256\pi^{4}}{n_{1}^{2}n_{2}n_{3}c^{4}}\,\omega_{3}^{2}\,|\chi^{CARS}|^{2}\,I_{1}^{2}I_{2}l^{2}\left[\frac{\sin\,(\Delta kl/2)}{\Delta kl/2}\right]^{2} \quad (10.37)$$

where χ^{CARS} is a bulk susceptibility and Δk is a measure of the mismatch of the wave vectors:

$$\Delta k = 2k_{1} - k_{2} - k_{3} \quad \text{and} \quad k_{i} = \frac{n_{i}\omega_{i}}{c} \quad (10.38)$$

For exact phase matching, $\Delta k = 0$ and the last term in Eq. (10.37) becomes unity. The CARS signal thus increases as l^{2}. For mismatched beams, the maxima occurs at $l_{c} = \pi/\Delta k$, with a reduction in intensity of a factor of $(2/\pi)^{2}$. For gases, l_{c} is about 100 cm at 1 atm pressure and is inversely related to pressure. For condensed phases, l_{c} is only a few millimeters. It is possible to increase the signal by reducing the phase mismatch by crossing the laser beams at an angle Θ given in Fig. 10.17. The angle is typically 1–3° and limits the CARS signal because the beams can only overlap in a short distance.

In practice, one has focused Gaussian beams rather than plane waves. The effective interaction region is then a cylinder of plane waves with a diameter of $w_{0} = 4\lambda f/\pi w$ and a length of $l = \pi w_{0}/2\lambda$, where λ is the wavelength of the laser and f is the focal length of the lens used to image the beam of diameter w ($1/e^{2}$ intensity points). The result is a CARS signal power of

$$P_{3} = \left(\frac{2}{\lambda}\right)^{2}\frac{256\pi^{4}\omega_{3}^{2}}{n_{1}^{2}n_{2}n_{3}c^{4}}\,|\chi^{CARS}|^{2}P_{1}\,^{2}P_{2} \quad (10.39)$$

At this level of approximation, there is no dependence on focal length of the

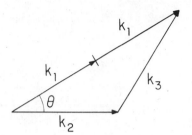

Fig. 10.17. Phase-matching diagram for CARS. k_1, k_2, and k_3, momentum vectors for photons at ω_1, ω_2, and ω_3.

lens or the initial diameter of the beams. More sophisticated treatment (65) shows that most of the signal is generated over a distance of $6l$ and that there is a dependence on the focal length.

To relate these microscopic susceptibilities to the normal Raman scattering cross sections, an orientation average must be performed. It can be shown that

$$\chi^{CARS} = 3\chi^{CARS}_{\alpha\beta\gamma\delta} = \frac{(N_i - N_f)\, n_1 c^4}{2hn_2\omega_2^4} \left(\frac{d\sigma}{d\Omega}\right) \frac{1}{\omega_{if} - \omega_1 + \omega_2 - i\Gamma_{if}} \quad (10.40)$$

For a single Raman resonance, this can be written as

$$\chi^{CARS} = \frac{A}{\delta\omega^2 + \Gamma^2} (\delta\omega + i\Gamma) \quad (10.41)$$

with $\delta\omega = \omega_{if} - \omega_1 + \omega_2$. This shows that CARS contains an imaginary part that is identical to the normal Raman spectrum, but also contains a real part that is the result of interference between χ_{NR} and the Raman cross section. The latter gives rise to a dispersion line shape that makes CARS spectra more difficult to interpret. Additional interferences occur when several Raman lines are present in the same spectral region. Of particular importance to trace analysis is that the solvent (major component) also contributes to χ_{NR}. At high concentrations, the Raman resonance of the analyte dominates so that the signal, according to Eq. (10.39), depends on the square of the number density. At low concentrations, the background χ_{NR} dominates, and the signal depends linearly on the number density. Eventually, fluctuations on the high nonresonant background (e.g., flicker noise in the lasers) limit detectability.

One way to suppress the nonresonant background in CARS is to use ellipsometry (66). The optical arrangement is exactly the same as CARS, except that the polarizations of the laser beams are carefully chosen, as shown in Fig. 10.18. If the photons ω_1 and ω_2 are plane polarized in the directions shown in Fig. 10.18, then the induced emission at ω_3 for an arbitrary polarization direction will be composed of a resonant and a nonresonant term as before. However,

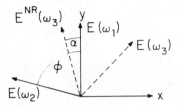

Fig. 10.18. Polarization discrimination of CARS nonresonant background. $E(\omega_i)$, polarization direction for photon at ω_i.

this can be decomposed into a nonresonant component E^{NR} and its orthogonal partner, respectively. So, if the emission is analyzed by a linear polarizer at the direction $E(\omega_3)$ in Fig. 10.18, the background is suppressed. The exact angle is given by maximizing the quantity $\sin \alpha \cos \phi$ and is different for different systems. Even though the signal is also decreased in this geometry, the S/N ratio is improved substantially (67).

Another method for suppressing the nonresonant CARS background is through temporal discrimination (68). The dephasing times for the nonresonant background is usually much faster than the corresponding vibrational dephasing times. So, one can use picosecond laser pulses ω_1 and ω_2 to coherently excite the Raman level, and delay the ω_1 pulse, which probes the coherence by generating ω_3, until the resonant process dominates. Applications of this technique depends on having very short laser pulses and long vibrational dephasing times so the signal level can be preserved.

It is also possible to cancel out the nonresonant contributions by using a third laser frequency to probe the coherent Raman excitation in a technique called ASTERISK (69). The relative polarizations of the three input laser photons are shown in Fig. 10.19, and the transition scheme for the molecule is shown in Fig. 10.20. Qualitatively, the nonresonant term generated by different combinations of the three input lasers in the polarization direction of ω_4 can be made to cancel each other, through the proper choice of Θ and ϕ. The phase-matching condition must still be satisfied, as seen in Fig. 10.21, because that separately determines the interaction length. This cancellation has been demonstrated in a

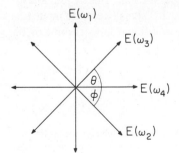

Fig. 10.19. Polarization directions for ASTERISK. $E(\omega_i)$, polarization direction for photon at ω_i.

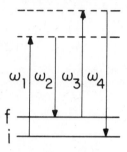

Fig. 10.20. Transition scheme for ASTERISK. i, initial state; and f, Raman level.

$0.3M$ solution of benzene (992-cm^{-1} line) in CS_2 (69), providing a factor of 1000 decrease in the nonresonant background and a band shape that is Lorentzian rather than dispersion-like.

Because the CARS signal, according to Eqs. (10.37) and (10.39), depends on the product of the power squared in one beam and the power in the other beam, the simplest arrangement is to use two pulsed lasers synchronized in time. Since this four-wave mixing process was first observed in stimulated Raman experiments (70), it is not surprising that the early CARS apparatus (71) is based on a ruby laser (ω_1) and its SRS in various liquids (ω_2). By using the ruby-laser-pumped infrared dye laser (ω_2), continuous CARS spectra can be obtained (72). Since proper phase matching is required for CARS in condensed phases, the experimental arrangement is slightly different from those for SIRS. A straightforward method for crossing two laser beams at a predefined angle Θ is to first produce collimated, collinear laser beams separated by a distance d, as shown in Fig. 10.22. Then, if a lens of focal length f is used to combine the two beams at a common point, and if the two beams are symmetrically placed relative to the axis of the lens, then a crossing angle of $\Theta = \tan^{-1} (d/2f)$ is obtained. It should be noted that on entering the condensed phase, refraction occurs and the crossing angle changes. However, this can be accounted for quite readily. Also, the crossing angle changes as ω_2 is scanned, and a correction must be made during the scan. Another design feature is that ω_3 exits in a well-defined angle dictated by the phase matching in Eq. (10.38). Apertures can thus be judiciously placed to spatially discriminate the collimated CARS signal from the input lasers. Normally, the signal level requires that an additional monochromator be

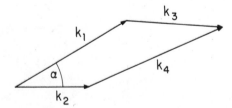

Fig. 10.21. Phase-matching diagram for ASTERISK. k_i, momentum vector for photon at ω_i.

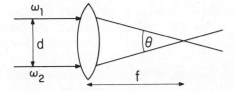

Fig. 10.22. Focusing of the pump beams in CARS separated by a distance d using a lens with focal length f so that the crossing angle θ equals the phase-matching angle.

used to further reject the pump beams. This places a burden on the design to be able to vary the observation direction as one scans over the CARS spectrum.

To provide more flexibility in laser frequencies, two dye lasers pumped by the same nitrogen laser can be used (73). The lower power levels available in laser-pumped dye lasers (10 kW typical) is still adequate for condensed-phase systems. Very narrow linewidths can be obtained (0.01 cm^{-1}), and the laser frequencies can be chosen to take advantage of resonance enhancement. A sophisticated version of this instrument is shown in Fig. 10.23 (74). To control the crossing angle, it is necessary to translate the mirror directing the ω_2 beam relative to the ω_1 beam. The crossing angle can be calculated knowing the refractive indices of the solvent at various wavelengths. However, to make sure that they still cross at the same spatial location, the mirror must also be adjusted slightly by rotating in the xy plane. The second adjustment is made in a servo mode to maximize the CARS signal. To scan the CARS spectrum, one needs to control the two dye lasers independently. When UV wavelengths are used, frequency-doubling crystals must be angle tuned to provide maximum power in each beam. Finally, the angular position and the wavelength of the monochromator must be scanned synchronously. So, a total of nine computer-controlled stepping motors are needed, plus various computer algorithms for predicting the

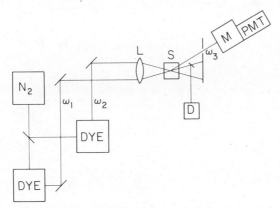

Fig. 10.23. Experimental scheme for CARS using two pulsed dye lasers. L, lens; S, sample; D, photodiode; M, monochromator; and PMT, phototube.

phase-matching angle. This points to the complexity of the CARS apparatus compared to other nonlinear Raman techniques. Recent advances in excimer lasers should allow even higher peak powers in the pump lasers, without sacrificing monochromaticity.

Another convenient high-power laser source is the 532-nm frequency-doubled output of a Nd:YAG laser. When this is used as ω_1, a laser-pumped dye laser derived from the same system can be used as ω_2. The resolution is essentially limited by the 0.1 cm^{-1} width of the ω_1 beam. Typical powers (75) are 10 MW for ω_1 (7-ns pulses at 15 pps) and up to 500 kW in ω_2 (oscillator-amplifier) in an arrangement shown in Fig. 10.24. Since the sample is in the gas phase, a collinear geometry is used. The presence of fluorescence is unlikely, and dichroic mirrors and filters are sufficient to isolate the CARS signal at ω_3. A similar apparatus can be used based on an optical parametric oscillator using either the 1064-nm Nd:YAG fundamental or the 532-nm second harmonic (76). The difficulty there is the inability of the parametric oscillator to generate very narrow spectral outputs.

If ω_2 is a broadband source rather than monochromatic, then a whole series of ω_3 output can be derived from the sample, corresponding to each Raman resonance. With an optical multichannel analyzer at the exit of a monochromator, the entire spectrum can be recorded on a single shot (77). One difficulty is that for normal broadband lasers, the intensity varies substantially over the gain profile. So, a reference intensity profile must be obtained for ω_2, for example, by imaging on a different portion of the same detector the nonresonant CARS signal from a gas such as argon. The sensitivity of the detector is only slightly poorer than a photomultiplier tube. For example, the gas-phase spectrum of CH_4 has been reported at a pressure of 600 μ with a good S/N ratio for a 200-kW ω_1 pulse and a 2-kW ω_2 pulse (78).

If a lower signal can be tolerated, cw lasers can be used (79). The advantages

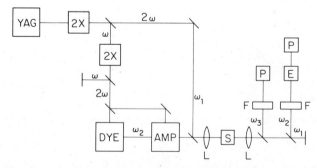

Fig. 10.24. Experimental scheme for high-resolution CARS using pulsed lasers. 2X, frequency doubler; AMP, dye amplifier; L, lens; S, sample cell; F, filters; P, phototube; and E, etalon.

are the higher spectral resolution possible and the improved amplitude stability of the beams. A single-mode argon ion laser at 640 mW and a single-mode cw dye laser at 10 mW have been combined to study at high resolution the Q branch of CH_4 (80). Because of the reduced signal, extreme care must be taken to minimize stray light, for example, argon laser plasma lines and scattering at ω_1 and ω_2. An improved version uses prisms to filter the light (81), which allows the CARS spectrum to be recorded at a gas pressure of 0.06 atm. The signal can also be increased by producing a common focal point inside both the Ar and dye laser cavities to achieve powers of 0.5 kW (82), and the order of 10^7 photons per second can be obtained from N_2 gas.

CARS is an attractive alternative to normal Raman scattering because of the enhanced signal level. A consideration of Eq. (10.39) and (10.40) shows that the relative power of the scattered beam in CARS exceeds that of normal Raman by a factor of $10^{-5} P_1 P_2 p$ for a gas at a pressure p (64). With pulsed lasers, $P_1 P_2 = 10^{10}$ W^2, and the enhancement is 10^5 for gases at 1 torr. Moreover, the CARS power is all in a narrow beam compared to the 4π steradians of normal scattering. In fact, for pure gases, concentrations as low as 10^{-10} atm should be detectable (71). In analytical applications, however, the major component will contribute toward the nonresonant susceptibility. Laser flicker noise will introduce fluctuations in this background signal to the order of $P_1^2 P_2$. The signal from the analyte will then be buried in this noisy background. In the most favorable case of H_2 in N_2, a detection limit of 10 ppm is reported (83). In solutions, the number densities are higher so that the signals are also higher. Still, the same background interference exists. For example, at below 1% concentration, benzene dissolved in toluene cannot be detected (84). Similarly, sodium benzoate in water cannot be detected below the $0.3M$ concentration range (85). It is apparent that some of the background suppression techniques discussed earlier must be used to improve detectability. The higher sensitivity of CARS implies that fewer total laser photons can be used to obtain Raman spectra at the same signal-to-noise level, even though the actual powers during the pulse are higher. For absorbing samples, this results in less heating, and sample spinning techniques may not be necessary. The only precaution is that the laser beams must not be depleted substantially or else the signal will degrade rapidly.

The collimated CARS signal makes it suitable to discriminate against fluorescence, which is the major problem in normal Raman scattering. Impressive spectra were obtained in laser dyes such as Rhodamine 6G and Rhodamine B (86). Further discrimination results from observing emission on the anti-Stokes side of the input laser beams. It is also possible to discriminate against Rayleigh scattering to produce Raman spectra at low frequencies in a variation of AS-TERISK without the polarization features (87). This permits large angular separations between the laser beams while still preserving phase matching at low Raman frequencies. Even at zero frequency, stray light is reduced to only 2%

of the total signal (86). Such discrimination against scattering and fluorescence is critical to applications of resonance-enhanced CARS to improve detectability.

The high sensitivity of CARS allows it to be used to study the small amounts of molecules in the multilayers on a metal surface (88). The optical arrangement is shown in Fig. 10.25. This is essentially bulk CARS with surface boundary conditions, so that four-wave mixing is possible via the plasmon waves at the silver surface. The actual CARS interaction length is estimated to be 10 μm and about 100 layers of benzene on the surface participate in the process. The small interaction length rules out any enhancement mechanism due to plasmon excitation, which is one of the theories proposed to explain SERS (see Section 2.8). In fact, the Raman frequencies of such a CARS process of pyridine on a silver surface were found to be nearly identical to those in the bulk, in contrast to SERS frequencies (89).

An interesting application of CARS to solutions is its combination with liquid chromatography. The advantage is a functional-group-specific detector that can provide structural information as the solutes elute from the column. The small crossing region in liquids is directly compatible with the small-volume flow cells required for liquid chromatography. For typical concentrations of materials eluting, however, the normal CARS signal is too weak to be useful and resonance enhancement is necessary. The eluting materials are first passed through an absorption detector to determine its absorption spectrum. The CARS system can then be set to take advantage of the resonance enhancement under computer control. In fact, a complete system can provide during the same chromatographic run detection based on absorption, fluorescence, fluorescence excitation, and CARS spectra (90).

CARS can provide spatially resolved concentration profiles in the sample. In gases, the collinear geometry is the simplest. Since most of the CARS signal is generated within 1–20 mm of the beam waist, depth profiling is possible. For still higher spatial resolution, one can use a crossed-beam geometry like ASTERISK to define the interaction region to 1–2 mm, with some loss of signal. A mapping of the local concentrations of H_2 in an air-natural gas flame has been

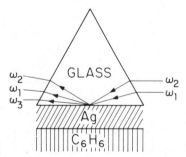

Fig. 10.25. CARS for studying the liquid-solid interface.

obtained by scanning the lasers radially as well as vertically, and H_2 in the range of 10–2000 ppm is found (83). By comparing the intensities of the $Q(0)$ through $Q(4)$ lines in the fundamental band of H_2, the rotational temperature can be determined in a H_2 discharge (91). Also, by comparing the $Q(1)$ lines of the fundamental and two overtone vibrations, the vibrational temperature can be established (91). These measurements are not readily performed using other techniques. For more complex vibrational-rotational structures, such as in N_2, a spectral fitting procedure must be employed to account for the odd line shapes inherent to CARS, but concentrations and temperatures can also be obtained (92). CARS is thus a valuable tool for characterizing the major components in analytical flames and plasmas, but probably not suitable for the trace components.

CARS has been suggested as a method for obtaining Doppler-free spectra by saturation (93). By counter-propagating pairs of ω_1 and ω_2 beams, the CARS output can become saturated. This saturation, however, is not due to the CARS process, since in Fig. 10.14 we can see that the CARS process returns the molecules to the same initial state. Rather, it is due to a SRS process similar to SIRS or SRGS, which actually depletes the ground-state population of a particular velocity component during Raman excitation. Fewer such ground-state species are then available to the counter-propagated pair of pump beams, thus creating saturation. It is interesting to note that if instead an excited Raman transition is probed by CARS (94), say from ω_1 and another frequency-matched laser in the opposite direction, the same saturation effect will be observed. However, the latter arrangement provides an increasing signal on top of a small background of CSRS (coherent Stokes Raman scattering), and is more desirable. So, saturated CARS is really a misnomer, and should be called SRS–CARS double-resonance spectroscopy. Another possibility for Doppler-free CARS is obtained by analyzing the linewidth of the coherent emission (95). In Fig. 10.14, one can think of the second ω_1 photon as a probe of the coherently generated Raman resonance in the molecule. This coherent excitation has a finite lifetime, and thus a finite spectral width, which should be reflected in ω_3 for completely monochromatic pump beams. This is equivalent to the picosecond CARS experiment (96), where the information is the Fourier transform of the linewidth.

2.7. Resonance Raman (RR)

The real effect of any of the above nonlinear Raman methods is to increase the transition probability of the Raman line for a given molecular system with a fixed $(d\sigma/d\Omega)$. The alternative is to increase the transition cross section $(d\sigma/d\Omega)$ characteristic of the Raman line. From quantum theory, we can derive an explicit relationship between the Raman cross section and electronic transition dipoles connecting states shown in Fig. 10.26 (97). If μ_{ir} is the transition moment

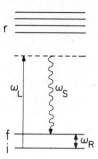

Fig. 10.26. States involved in RR. i, initial level; f, final level; r, coupling intermediate states; and ω_R, Raman frequency.

connecting the initial state i and the intermediate (real) state r at an energy E_r, then

$$\left(\frac{d\sigma}{d\Omega}\right)_{if} = \frac{2^5\pi^4}{3^2 C^4}\,\omega_s^4\,\sum|\alpha|^2 \tag{10.42}$$

$$\alpha = \sum_r\left(\frac{\mu_{rf}\mu_{ir}}{E_r - E_i - \hbar\omega_L + i\Gamma_r} + \frac{\mu_{ir}\mu_{rf}}{E_r - E_f + \hbar\omega_L + i\Gamma_r}\right) \tag{10.43}$$

where f refers to the final state and Γ is the damping constant. Qualitatively, the Raman transition "borrows" intensity from the allowed electronic transitions to state r. Eq. (10.43) shows that when the excitation photon at ω_L matches the energy difference for the real electronic levels, resonance occurs and the Raman cross section can be greatly increased. In fact, both adiabatic Condon type and vibronically induced transitions to r can be enhanced (98), as well as nonadiabatic vibronic interactions (99). Actually, at true resonance, it becomes difficult to distinguish between RR and resonance fluorescence (100), which points to the extremely large enhancements that are possible.

In normal RS in liquids, the incorporation of RR is to enhance the signal from the minor component relative to the solvent, which is the ultimate limit of detectability. This is clearly demonstrated in a methanol solution of pseudo-isocyanine, where even parts-per-million levels can be observed (101). The intensity of the band at 1380 cm^{-1} excited by a Hg arc is about 5×10^5 that of the normal RS of CCl$_4$ at 313 cm^{-1}. The diphenyldodecahexene bands at 1142 cm^{-1} and 1552 cm^{-1} excited by Hg arc show even larger enhancements of 3.5×10^6 and 5.3×10^6, respectively, in acetone solution (102). So, the increased utility of resonance enhancement is not restricted to laser excitation. The incorporation of RR simplifies the design problems in laser RS instrumentation, by providing a larger inherent signal. On the other hand, to take advantage of RR without relying on accidental coincidence in ω_L, lasers will have to be tunable and will have to work in the UV region, both of which require sophisticated technology.

The resonance enhancement for nonlinear Raman methods follows a more complicated dependence than Eq. (10.42), since ω_S is also from an input beam. For SIRS, there exists another frequency factor (103) of the form $(\omega_R - \omega_L + \omega_S + i\Gamma_R)^{-1}$, where R refers to the Raman level. The imaginary part of such a modification of Eq. (10.42) is proportional to the Raman intensity. So (104),

$$I_R \propto \left\{ \frac{\Gamma_R(\omega_R - \omega_L)^2 - \Gamma_R\Gamma_r^2 + 2\Gamma_r(\omega_r - \omega_L)(\omega_R - \omega_L + \omega_S)}{[(\omega_R - \omega_L)^2 + \Gamma_r^2]^2 \, [(\omega_R - \omega_L + \omega_S)^2 + \Gamma_R^2]} \right\} \quad (10.44)$$

Far from electronic resonance, the second term in the denominator predicts a normal Lorentzian line shape identical to RS, while the first term in the denominator leads to a simple increase in intensity. At exact electronic resonance, the second term in the numerator predicts a negative-going Lorentzian line shape. In the intermediate region where $\omega_L = \omega_r \pm \Gamma_r$, a pure dispersion-type line shape is present. Since electronic resonance is relevant to ω_L only, if the Raman spectrum is obtained by scanning ω_S for a fixed ω_L, then all lines should show identical line shapes. On the other hand, if ω_L is scanned while ω_S is fixed, different Raman lines may show very different line shapes. This is clearly demonstrated in the SIRS of acridine orange (104).

The main concern in resonance-enhanced SIRS is that the sample must be relatively transparent at both ω_L and ω_S. If the linear absorption coefficient of the species (or other species in the sample) at ω_S is ϵ_S, then one must keep $N\epsilon_S l$ smaller than unity to avoid depletion of the pump beam significantly. This implies that as one increases the concentration (or interaction length), there will be a point where any further increase in concentration (length) will not lead to additional inverse Raman absorption. We can call this point N_{max}. The lowest detectable concentration N_{min} is inversely proportional to g, which is now much larger because of the resonance enhancement. These two determine the useful range for SIRS. Linear absorption at ω_L puts a further restriction on the inverse Raman process. One can in principle measure the absorption with and without ω_S to correct for this interference. Another complication from absorption is the appearance of thermal lensing, as described in Chapter 13. If there are no limiting apertures in the optical system, including the finite size of the photodetector, thermal lensing will not contribute to the signal. Also, one can measure SIRS with very short laser pulses to avoid the lens that is formed in the 10–50 ns time scale. The possibility of efficient excitation also implies that ground-state depletion can occur, so that the laser powers used must be sufficiently low for resonance SIRS. Finally, one must also consider the possiblity of excited-state vibrations rather than ground-state vibrations dictating the SIRS transition and complicating the spectrum.

In normal resonance Raman studies, any fluorescence from the enhancing electronic state will confuse the results, for example, in the case of gaseous molecular iodine (105). The added feature of fluorescence rejection in SIRS should make it a better tool in some cases.

In RIKES, the resonance case is even more interesting. RIKES essentially measures the difference in absorption of the same- versus opposite-sense circularly polarized light in a probe beam. It can be seen (106) that the former is proportional to $3G^2/2$ and the latter is proportional to $G^0/2 + G^2/4$, where G^0 and G^2 are the trace and symmetric scattering tensors (squared), respectively. Depending on the symmetry types of the Raman transition and the electronic state in resonance, one of the tensors will gain in magnitude. This implies that resonance enhancement can either increase or decrease the degree of rotation of the probe beam, and thus a corresponding increase or decrease in the observed RIKES signal. There is also an interference effect if a light wave $(E_y)_0$ is used to enhance the RIKES signal according to Eqs. (10.26)–(10.28), and absorption occurs at this wavelength. The optical arrangement in RIKES is very similar to those of polarization spectroscopy (107). At electronic resonance, it is difficult to separate out these individual contributions to the signal.

In PARS, resonance enhancement does not provide any advantage. Since any absorption is likely to generate an acoustic signal, one loses the advantage by increasing the signal and the background at the same time. This is in contrast with SIRS and RIKES, where the background absorption can be accounted for by fixing ω_L and scanning ω_S.

In CARS, Eq. (10.43) shows that resonance enhancement occurs if ω_1 or ω_3 matches the energy of an electronic state. Furthermore, two different electronic states can affect the two photons. The line shapes are thus complicated, and interference between the two can occur. For a single electronic state r, Eq. (10.43) reduces to give (108):

$$\chi^{\text{CARS}} = \frac{1}{(U^2 + V^2)} (V + iU)$$

$$U = \Gamma_r(\omega_1 - \omega_2)$$

$$V = (\omega_r - \omega_1)(\omega_r - 2\omega_1 + \omega_2) + \Gamma_r^2$$

(10.45)

For ordinary CARS and preresonance CARS, the real part of Eq. (10.45) dominates, as expected. At one of the two electronic resonances, the ratio of the imaginary part to the real part is $(\omega_1 - \omega_2)/\Gamma_r$, and can be very large. In such cases, a Lorentzian line shape similar to RS is present except that the peak is negative going (109). In between the two resonances, the dispersion-shaped CARS signal changes signs. For minor components in a mixture, additional terms involving χ_{NR} for the major component(s) must be included, and complex band shapes will result. The importance of resonance CARS is not so much to increase the signal, which is typically plentiful, but to increase the fractional contribution of the minor component to the overall CARS process, so that the detectability can be extended. For example, a detection limit of $5 \times 10^{-7}M$ of β-carotene has been reported in benzene solution (110). Compared to conventional RR, CARS has the advantage of negligible interference from fluorescence, which can be substantial due to the real electronic transition. On the other hand, depletion of the ground state due to actual absorption or due to other nonlinear Raman processes present can limit the usefulness of resonance enhancement. And, unlike other Raman events, observation in CARS is at a frequency higher than those of the exciting beams, increasing the probability that the signal will be decreased due to absorption.

2.8. Raman Spectroelectrochemistry

One of the many links between spectroscopy and electrochemistry is the use of RS for characterization of species on or adjacent to the electrode surface. The structural information in aqueous solutions is more readily accessible by RS than by infrared. Depending on the experimental geometry, one can study the species in the bulk solution, the diffusion layer, or on the electrode surface.

An experimental setup for studying the bulk solution is shown in Fig. 10.27 (111). The species being studied is typically generated by controlled potential coulometry. As long as the species is stable in the time scale of the Raman measurement, it can be studied. Because the sensitivity of RS is generally low, RR is almost always used to provide a sufficient signal level. An application is the potentiometric titration of cytochrome c using RR to monitor the oxidized and the reduced forms, so that the formal reduction potential can be determined (112).

The RS of species in the diffusion layer can be obtained in a geometry shown in Fig. 10.28 (111). A square-wave potential is applied to the working electrode to create a steady-state concentration in the diffusion layer so that the reaction product is formed at a diffusion-controlled rate and depleted at the same rate on

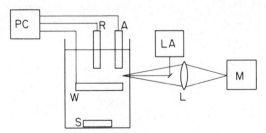

Fig. 10.27. Experimental scheme for studying bulk electrolyte using RS. PC, potentiostat and coulometer; R, reference electrode; A, auxiliary electrode; W, working electrode; S, stirrer; LA, laser; L, lens; and M, monochromator.

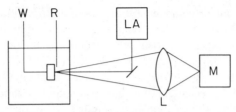

Fig. 10.28. Experimental scheme to study the diffusion layer using RS. W, working electrode; R, reference electrode; LA, laser; L, lens; and M, monochromator.

the reverse step. It is also possible to orient the laser tangent to the electrode surface to determine the concentration profiles as a function of distance from the electrode (113), but such experiments have not been too successful.

To record the Raman spectra of species adsorbed on the electrode surface, a geometry such as Fig. 10.29 can be used (114). The working electrode is placed close to the cell wall to avoid contributions from the species in the bulk solution. It is in the attempt to observe RR from the surface layer of electrodes that the surface enhanced phenomenon (SERS) was recognized (115). On depositing a monolayer of pyridine onto a clean silver electrode by a single oxidation–reduction cycle, a large Raman signal is obtained. This signal level cannot be explained simply by considering the RR effect. It also cannot be explained as the result of increased surface area on the electrode due to roughening. In any case, the enhancement factor is on the order of $10^3–10^6$ (115). In fact, if not for this enhancement process, the study of surface species with RS is very difficult because of the low signal levels.

Many theories have been suggested to explain this enhancement effect, but the debate still continues since no one theory seems to be able to explain all of the experimental observations. There is evidence that more than one monolayer is present on the surface (116). The presence of surface carbon may also be

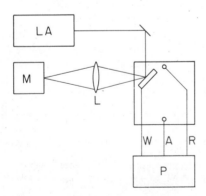

Fig. 10.29. Experimental scheme to study surface species at the electrode using RS. LA, laser; L, lens; M, monochromator; W, working electrode; A, auxiliary electrode; R, reference electrode; and P, potentiostat.

related to the SERS signal (117). It has been suggested that molecular electronic levels actually couple with surface plasmons to create new intermediate states that increase the RR effect (118). The enhancement can also be caused by the image dipole in the metal when the radiation induces an oscillating dipole moment in the absorbing species on the surface (119). Resonant excitation of the conduction electrons in the roughened part of the metal surface that is covered by the molecular species generates a large local field, which in turn enhances the scattering process (120). Another mechanism is the recombination of the electron-hole pairs formed after resonant excitation of the molecules coupled to the metal (121). At present, it seems that more than one mechanism may be responsible for SERS.

Theory suggests that SERS does not occur on arbitrary electrode surfaces and does not occur with arbitrary surface species. The types of species likely to exhibit SERS are oxygenated species, heterocycle bases, pseudohalides, and reduced organic compounds. The types of electrode materials suitable for SERS include silver, gold, copper, tunnel junctions, and binary metal/silver. Without surface enhancement, the normal RR signal is too weak to be broadly useful for studying surface species. For example, for the platinum electrode, which is the best understood in electrochemistry, SERS is not observed. With normal RR, species such as I_3^- (122), p-nitrosodimethylaniline (123), phenylhydrazine, and phenol (124) have been reported on the platinum electrode. On the other hand, the SERS process is due to some special interaction with the surface that may make the environment atypical for the electroactive species. The most interesting applications of SERS so far are the Raman spectroscopic studies of biological molecules such as cytochrome c (125), and the study of time-dependent processes on surfaces (126).

3. COMPARISON OF FEATURES

In addition to the differences in resonance spectra, there are many features inherent to each Raman method that must be considered when designing an experiment.

3.1. Discrimination Against Fluorescence

One of the most obvious advantages in nonlinear Raman effects is the possibility of discriminating against molecular fluorescence in Raman spectroscopy. This applies equally to fluorescence (or phosphorescence) of the molecule under investigation, or any other species that happens to be in the sample. There are three mechanisms for discrimination in SIRS. First, the probe beam is on the anti-Stokes side of the high-power laser. Fluorescence at these wavelengths is due to hot-band absorption and is considerably less intense because of the Boltzmann factor for the initial state populations. If the excited state is structurally not too different from the ground state in the fluorescing species, that is, there is no substantial geometry change or frequency change in the vibrational modes, Franck–Condon factors will become even less favorable for hot-band absorption. In other cases, the Boltzmann factor will be the determining one. Second, fluorescence is generally distributed isotropically over 4π sr while the probe beam in SIRS has a smaller angular spread. It can be easily shown that if the half-angular spread of the probe beam is b rad, the extent of discrimination is then $b^2/4$. For a laser collimated to the order of 1 mrad, the rejection for fluorescence can be better than 1 ppm. The fluorescence intensity naturally depends on the intensity P_S. What the above shows is that the probe beam needs only be somewhat larger than $10^{-6}P_S$ in intensity to reject fluorescence completely even for the case where the fluorescence quantum yield is unity, and the exciting beam is completely absorbed. This, of course, is only a limiting case because one cannot perform SIRS studies when P_S decreases so drastically inside the sample anyway. Third, since a constant fraction of ω_L is absorbed for a given P_S, discrimination depends on P_L for $P_L \ll P_S$. We can conveniently calculate the probe power P_L that is needed to give a signal-to-noise ratio of unity in such an experiment, that is, when the change in P_L due to inverse Raman absorption is equal to the fluorescence flux at the detector. Considering all the possible parameters, one finds that for small linear absorption at ω_S and small inverse Raman absorption at ω_L

$$P_L = P_S \frac{\alpha}{g} \frac{l'}{l''} \frac{c'}{c''} \frac{b^2}{4} \Phi_f K \qquad (10.46)$$

where α is the linear absorption coefficient at ω_S, g is as defined in Eq. (10.5), l' is the effective interaction length of the fluorescence, l'' is the interaction length of the inverse Raman process, c' and c'' are the concentrations of the fluorescing species and the Raman active species, respectively, Φ_f is the fluorescence quantum yield of the species, and K is the fraction of fluorescence passed because of the limited spectral bandpass of the spectrometer. In SRGS, the situation is worse. Observation is on the Stokes side of the exciting laser. There, fluorescence is the most intense because one does not require hot-band absorption. In principle, at least, one can use very high probe laser powers to enhance the discrimination level. Further, there is an unfavorable Boltzmann factor—ground-state absorption–fluorescence versus excited-state Raman process.

In PARS, discrimination against fluorescence is absolute since no observation of light is involved. However, since absorption is the precursor to fluorescence, it can be assumed that in fluorescing samples the PARS background is large.

Also, there can be secondary effects such as the absorption of the fluorescence by other molecules or parts of the cell to create additional acoustic signal.

For RIKES, the sample is placed between crossed polarizers. Since fluorescence is generally randomly polarized, half of it will pass through the second polarizer and be recorded. Rejection comes in mainly because of the narrow divergence of the probe laser. Again, an equation of the form of Eq. (10.46) will be found, but P_L is replaced with the power of the rotated probe beam, which is the part that passes through the second polarizer. In general, the rotated beam is much lower in power than the original probe beam. This is why RIKES will not have as high a discrimination level as SIRS against fluorescence.

For CARS, two photons are supplied to the system and the output to the anti-Stokes side of both of these is monitored. There is therefore a minimum power for the coherent anti-Stokes output with the form of Eq. (10.46). The right-hand side must be modified to include contributions from both photons supplied to the system, that is, two separate P_S and two separate α. To achieve this power level in the CARS output, the input lasers must be intense enough. Since the fluorescence intensity increases linearly with input power, and the CARS output is proportional to the square of the laser power in one of the beams, discrimination is possible. This is in contrast to SIRS, where P_L is a probe laser that can be arbitrarily intense (to the limit of generating other nonlinear processes). For typical CARS experiments, the output is much lower in power than the typical probe beam powers in SIRS. This means that generally SIRS has a higher discrimination level than CARS against fluorescence.

In addition to these specific features of the various Raman techniques, one can simultaneously introduce temporal discrimination (16). So, the optical arrangements for nonlinear Raman spectroscopy based on pulsed lasers are the most favorable for rejecting fluorescence.

3.2. Applicability to Small Volumes

In many analytical measurements, for example in liquid chromatography (127), a small sample volume is desirable. The ability of Gaussian laser beams to be focused down to the diffraction limit is therefore an advantage. For Raman methods based on a single laser, the desirable sample volume can be calculated following the procedure in Chapter 17. The result is that a volume of $16z^2\lambda$ is required for a laser beam of wavelength λ interacting with the sample over a distance of $2z$. In principle, one can make the sample volume as small as desired by proper focusing. However, as the volume is made smaller, the useful interaction length also decreases, leading to a real loss in signal in RS.

For SIRS, the signal depends on pathlength according to Eq. (10.4.) The signal also depends on the power density P_S, so P_S should be maximized. Since P_S is inversely proportional to the size of the beam waist, ω_0^2, and the useful length

z is proportional to ω_0^2, no change in SIRS signal should occur on focusing if collinear beams are used with matched beam waists. So, for the same concentration, the volume can be reduced without loss of sensitivity, that is, until dielectric breakdown or other nonlinear processes become important. In RIKES arrangements where the beams are almost collinear, the signal is a function of the product gl, as shown in Eqs. (10.21), (10.25), and (10.28). Again, g increases with ω_0^2 and $l = z$ decreases with ω_0^2, and focusing does not affect the signal level. PARS excitation follows the same dependence. However, PARS involves the conversion of the Raman excitation to heat and thus a pressure wave; the smaller the volume the heating effect is confined to, the larger the temperature increase. So, it is in fact desirable to use as small a volume as possible, provided that vibrational to translational relaxation is complete in that region and the microphone is small enough to benefit from this increased pressure level.

The case of CARS is more complicated. χ depends linearly on pathlength and on ω_0^2 to show a net ω_0^4 dependence. The three intensity terms in Eq. (10.37) each depends inversely on ω_0^2. The first approximation is that there is no dependence on focusing, as shown in Eq. (10.39). However, more exact treatments (65, 128) show a slight increase of CARS signal with tighter focusing. In practice, thermal lensing prevents very tight focusing. In liquids, the phase-matching condition requires the beams be crossed, so that the actual sample volume must be larger than the interaction volume to provide the proper geometry. On the other hand, a crossed-beam geometry for any nonlinear Raman effect allows the selection of a localized interaction region, with some loss of sensitivity because one cannot optimize the volume relative to both beam waists simultaneously.

A further consideration is related to the powers necessary for signal generation, as discussed in the next section. This is because, in general, low-power, cw lasers have good Gaussian beam profiles that minimize the volume. If pulsed lasers are used as one or both of the beams, the interaction volume will be larger than those mentioned above.

3.3. Other Considerations

The need for high peak powers for efficient interaction between two laser beams is easily understood. But, depending on the mode of detection, sometimes cw lasers can be used. The important point is that high peak powers can lead to undesirable effects in the molecules, such as saturation and the ac Stark effect (129), or in the environment, such as dielectric breakdown (130). One way to compare the power requirements is to consider the performance of cw versions of the various Raman techniques in gas samples. This is shown in Table 10.1. One can see that PARS requires the most power. Conventional RS and the other nonlinear techniques are somewhat comparable. Table 10.1 is only a rough picture since signal-to-noise ratios have not been included. Also, the situation will be different depending on the presence of interfering materials, the desired spectral resolution, and fluorescence. The distinction between RS and the others is that the former cannot be enhanced by using higher peak powers for a given average power.

The spectral shape of the Raman lines influences interpretation and the degree of interference from other components. In this respect, SIRS–SRGS, PARS, and certain versions of RIKES are better because they resemble RS in spectral information. Quantitative accuracy is thus also better for the above techniques. The effect on detectability is more complex. On the one hand, the nonresonant background in CARS and some versions of RIKES can be the limiting factor for trace analysis. On the other hand, the background produces a signal that is linear rather than quadratic with respect to concentration, and may provide better signal levels for the same number density.

Finally, the complexity of the instrumentation must be considered. The phase-matching condition in solution CARS makes it the most difficult to align and to keep aligned while scanning. The reduction of background birefringence in RIKES is also a difficult problem, but is probably easier to solve (54). PARS and SIRS are thus the simpler experimental arrangements involving two beams. In these situations, the detection and modulation electronics are more complex. The simplest experiment is still the normal RS, that is, if one can sacrifice fluorescence discrimination, temporal resolution, spectral resolution, and spatial resolution.

Obviously, no one Raman technique will solve all analytical problems. Despite the complex instrumentation involved, many new laser Raman techniques are available for specific applications where normal RS cannot be used. The popularity of these techniques depends very much on laser technology, but the prospect of broad applications is extremely bright.

ACKNOWLEDGMENTS

The author thanks the many co-workers in his laboratory who have contributed to many parts of this work, particularly L. J. Hughes and D. R. Bobbitt. Funding from the Research Corporation and the U.S. Department of Energy, Office of Basic Energy Sciences, Division of Chemical Sciences is gratefully acknowledged. The Ames Laboratory is operated by Iowa State University for the U.S. Department of Energy under contract No. W-7405-eng-82.

REFERENCES

1. A. Smekal, *Naturwiss.*, **11**, 873 (1923).
2. C. V. Raman and K. S. Krishnan, *Nature*, **121**, 501 (1928).

3. C. V. Raman, *Nature*, **121**, 619 (1928).
4. S. P. S. Porto and D. L. Wood, *J. Opt. Soc. Am.*, **52**, 251 (1962).
5. M. Loëte and H. Bergen, *J. Mol. Spectrosc.*, **68**, 317 (1977).
6. W. H. Fletcher, J. S. Rayside, and W. B. McLendon, *J. Raman Spectrosc.*, **7**, 205 (1978).
7. S. Brodersen and J. Bentsen, *J. Raman Spectrosc.*, **3**, 207 (1975).
8. S. Brodersen, *Topics in Current Physics*, **11**, A. Weber, Ed., Springer, Berlin, 1979.
9. W. Klockner, W. J. Schmid, and H. W. Schrotter, *Proceedings V Int. Conference on Raman Spectrosc.*, E. D. Schmid, J. Brandmüller, W. Kiefer, B. Schrader, and H. W. Schrotter, Eds., Schulz, Freiburg, 1976.
10. M. Delhaye and P. Dhamelincourt, *J. Raman Spectrosc.*, **3**, 33 (1975).
11. M. D'Orazio and U. Schimpf, *Anal. Chem.*, **53**, 809 (1981).
12. M. Pealat, R. Bailly, and J. P. E. Taran, *Optics Comm.*, **22**, 91 (1977).
13. C. A. Arguello, G. F. Mendes, and R. C. C. Leite, *Appl. Optics*, **13**, 1731 (1974).
14. F. L. Galeener, *Chem. Phys. Lett.*, **48**, 7 (1977).
15. Y. Yacoby, I. Wagner, J. Bodenheimer, and W. Low, *Phys. Rev. Lett.*, **27**, 248 (1971).
16. R. P. Van Duyne, D. L. Jeanmaire, and D. F. Shriver, *Anal. Chem.*, **46**, 213 (1974).
17. J. F. Rabolt, R. Santo, and J. D. Swalen, *Appl. Spectrosc.*, **34**, 517 (1980).
18. P. Pagsberg, R. Wilbrandt, K. B. Hansen, and K. V. Weisberg, *Chem. Phys. Lett.*, **39**, 538 (1976).
19. M. J. French and D. A. Long, *J. Raman Spectrosc.*, **3**, 391 (1975).
20. J. A. Cooney, *Appl. Phys. Lett.*, **12**, 40 (1968).
21. T. Hirschfeld, E. R. Schildkraut, H. Tannenbaum, and D. Tanenbaum, *Appl. Phys. Lett.*, **22**, 38 (1973).
22. Y. G. Vainer, M. Y. Kuzin, L. P. Malyavkin, E. G. Sil'kis, K. V. Tanana, and U. D. Titov, *Sov. J. Quant. Electron.*, **9**, 296 (1979).
23. R. Gill, K. Geller, J. Farina, J. A. Cooney, and A. Cohen, *J. Appl. Meteoro.*, **18**, 225 (1979).
24. W. J. Jones and B. P. Stoicheff, *Phys. Rev. Lett.*, **13**, 657 (1964).
25. J. B. Grun, A. K. McQuillan, and B. P. Stoicheff, *Phys. Rev.*, **180**, 61 (1969).
26. I. Reinhold and M. Maier, *Opt. Commun.*, **5**, 31 (1972).
27. V. L. Strizhevskii and E. I. Kondilenko, *Opt. Spectrosx. (USSR)*, **30**, 127 (1970).
28. R. A. McLaren and B. P. Stoicheff, *Appl. Phys. Lett.*, **16**, 140 (1970).
29. R. R. Alfano and S. L. Shapiro, *Chem. Phys. Lett.*, **8**, 631 (1971).
30. E. S. Yeung, *J. Mol. Spectrosc.*, **53**, 379 (1974).
31. L. J. Hughes, L. E. Steenhoek, and E. S. Yeung, *Chem. Phys. Lett.*, **58**, 413 (1978).
32. J. P. Haushalter, G. P. Ritz, D. J. Wallan, K. Dien, and M. D. Morris, *Appl. Spectrosc.*, **34**, 144 (1980).
33. Y. Taira, K. Ide, and H. Takuma, *Chem. Phys. Lett.*, **91**, 299 (1982).
34. C. Buffett and M. D. Morris, *Appl. Spectrosc.*, **35**, 203 (1981).
35. R. S. McDowell, C. W. Patterson, and A. Owyoung, *J. Chem. Phys.*, **72**, 1071 (1980).

36. P. Drell and S. Chu, *Opt. Commun.*, **28**, 343 (1979).
37. A. Owyoung and P. S. Peercy, *J. Appl. Phys.*, **48**, 674 (1977).
38. A. Owyoung and E. D. Jones, *Opt. Lett.*, **1**, 152 (1977).
39. A. Owyoung, C. W. Patterson, and R. S. McDowell, *Chem. Phys. Lett.*, **59**, 156 (1978).
40. B. E. Kincaid and J. R. Fontana, *Appl. Phys. Lett.*, **28**, 12 (1976).
41. A. Owyoung and P. Escherick, *Opt. Lett.*, **5**, 421 (1980).
42. A. Owyoung and P. Escherick, *Proceedings of the VII International Conference on Raman Spectroscopy*, W. F. Murphy, Ed., North-Holland, New York, 1980, p. 656.
43. J. J. Valentini, P. Escherick, and A. Owyoung, *Chem. Phys. Lett.*, **75**, 590 (1980).
44. A. Lau, W. Werncke, M. Pfeiffer, K. Lenz, and H. J. Weigmann, *Sov. J. Quant. Electron.*, **6**, 402 (1976).
45. G. C. Herring, W. M. Fairbank, Jr., and C. Y. She, *IEEE J. Quant. Electron.*, **QE17**, 1975 (1981).
46. W. Zinth, M. C. Nuss, and W. Kaiser, *Chem. Phys. Lett.*, **88**, 257 (1982).
47. G. Mayer and F. Gires, *Comp. Rend.*, **25**, 2039 (1964).
48. M. D. Levenson and J. J. Song, *J. Opt. Soc. Am.*, **66**, 641 (1976).
49. D. Heiman, R. W. Hellwarth, M. D. Levenson, and G. Martin, *Phys. Rev. Lett.*, **36**, 189 (1976).
50. G. L. Eesley, M. D. Levenson, and W. M. Tolles, *IEEE J. Quant. Electron.*, **14**, 45 (1978).
51. A. Owyoung, *IEEE J. Quant. Electron.*, **14**, 192 (1978).
52. A. Owyoung, *Opt. Lett.*, **2**, 91 (1978).
53. M. D. Levenson and G. L. Eesley, *Appl. Phys.*, **19**, 1 (1979).
54. E. S. Yeung, L. E. Steenhoek, S. D. Woodruff, and J. C. Kuo, *Anal. Chem.*, **52**, 1399 (1980).
55. J. J. Barrett and D. F. Heller, *J. Opt. Soc. Am.*, **71**, 1299 (1981).
56. J. J. Barrett and M. J. Berry, *Appl. Phys. Lett.*, **34**, 144 (1979).
57. E. L. Kerr and J. G. Atwood, *Appl. Opt.*, **7**, 915 (1968).
58. T. Kritchman, S. Shtrikman, and M. Slatkine, *J. Opt. Soc. Am.*, **68**, 1257 (1978).
59. G. A. West, D. R. Siebert, and J. J. Barrett, *J. Appl. Phys.*, **51**, 2823 (1980).
60. D. R. Siebert, G. A. West, and J. J. Barrett, *Appl. Opt.*, **19**, 53 (1980).
61. J. J. Barrett, in *Chemical Applications of Nonlinear Raman Spectroscopy*, A. B. Harvey, Ed., Academic Press, New York, 1981.
62. P. R. Regnier and J. P. E. Taran, *Laser Raman Gas Diagnostics*, M. Lapp and C. M. Penney, Eds., Plenum, New York, 1974.
63. C. K. N. Patel and A. C. Tam, *Appl. Phys. Lett.*, **34**, 760 (1979).
64. J. W. Nibler and G. V. Knighten, *Topics in Current Physics*, **11**, A. Weber, Ed., Springer-Verlag, New York, 1979.
65. G. C. Bjorklund, *IEEE J. Quant. Electron.*, **QE11**, 287 (1975).
66. S. A. Akhmanov and N. I. Koroteev, *Sov. Phys. Usp.*, **20**, 899 (1977).
67. J. L. Oudar, R. W. Smith, and Y. R. Shen, *Appl. Phys. Lett.*, **34**, 758 (1979).
68. F. M. Kamga and M. G. Skeats, *Opt. Lett.*, **5**, 126 (1980).
69. J. J. Song, G. L. Eesley, and M. D. Levenson, *Appl. Phys. Lett.*, **29**, 567 (1976).
70. P. D. Maker and R. W. Terhune, *Phys. Rev.*, **137**, A801 (1965).

71. P. Regnier and J. P. E. Taran, *Appl. Phys. Lett.*, **23**, 240 (1973).
72. F. Moya, S. A. J. Druet, and J. P. E. Taran, *Opt. Comm.*, **13**, 169 (1975).
73. I. Chabay, G. K. Klauminzer, and B. S. Hudson, *Appl. Phys. Lett.*, **28**, 27 (1976).
74. L. A. Carreira, L. B. Rogers, L. P. Goss, G. W. Martin, R. M. Irwin, R. Von Wandruska, and D. A. Berkowitz, *Chem. Biomed. Environ. Instrumentation*, **10**, 249 (1980).
75. J. W. Nibler, J. R. McDonald, and A. B. Harvey, *Opt. Comm.*, **18**, 134 (1976).
76. S. A. Akhmanov, N. I. Koroteev, and A. I. Kholodnykh, *J. Raman Spectrosc.*, **2**, 239 (1974).
77. W. E. Roh, P. Schreiber, and J. P. E. Taran, *Appl. Phys. Lett.*, **29**, 174 (1976).
78. J. W. Nibler, W. M. Shaub, J. R. McDonald, and A. B. Harvey, *Vibrational Spectra and Structure*, J. R. Diring, Ed., **6**, Elsevier, New York, Chapter 3.
79. J. J. Barrett and R. F. Begley, *Appl. Phys. Lett.*, **27**, 129 (1975).
80. M. A. Henesian, L. Kulevskii, and R. L. Byer, *J. Chem. Phys.*, **65**, 5530 (1976).
81. V. I. Fabelinsky, B. B. Krynetsky, L. A. Kulevsky, V. A. Mishim, A. M. Prokhorov, A. D. Savel'ev, and V. V. Smirnov, *Opt. Comm.*, **20**, 389 (1977).
82. A. Hirth and K. Volrath, *Opt. Comm.*, **18**, 213 (1976).
83. P. Regnier, F. Moya, and J. P. E. Taran, *AIAA J.*, **12**, 826 (1974).
84. R. F. Begley, A. B. Harvey, and R. L. Byer, *Appl. Phys. Lett.*, **25**, 387 (1974).
88. B. S. Hudson, W. Hetherington, S. Cramer, I. Chabay, and G. Klauminzer, *Proc. Natl. Acad. Sci.*, **73**, 3798 (1976).
86. L. A. Carreira and M. L. Horovitz, *Nonlinear Raman Spectroscopy and Its Chemical Applications*, W. Kiefer and D. A. Long, Eds. Reidel, Boston, 1982, p. 429.
87. J. A. Shirley, R. J. Hall, and A. C. Eckbreth, *Opt. Lett.*, **5**, 380 (1980).
88. C. K. Chen, A. R. B. deCastro, Y. R. Shen, and F. DeMartini, *Phys. Rev. Lett.*, **43**, 946 (1979).
89. F. W. Schneider, *Nonlinear Raman Spectroscopy and its Chemical Applications*, W. Kiefer and D. A. Long, Eds., Reidel, Boston, 1982.
90. G. D. Boutilier, R. M. Irwin, R. R. Antcliff, L. B. Rogers, L. A. Carreira, and L. Azarraga, *Appl. Spectrosc.*, **35**, 576 (1981).
91. M. Pealat, J. P. E. Taran, J. Taillet, M. Bacal, and A. M. Bruneteau, *J. Appl. Phys.*, **52**, 2687 (1981).
92. A. C. Eckbreth and R. J. Hall, *Combust. Flame*, **36**, 87 (1979).
93. J. Moret-Bailly, *Nonlinear Raman Spectroscopy and its Chemical Applications*, W. Kiefer and D. A. Long, Eds., Reidel, Boston, 1982.
94. S. M. Gladkov, M. G. Karimov, and N. I. Koroteev, *Opt. Lett.*, **8**, 298 (1983).
95. I. C. Khoo and E. S. Yeung, *Opt. Comm.*, **22**, 83 (1977).
96. G. Marowsky, A. Anliker, and Q. Munir, *Opt. Comm.*, **45**, 183 (1983).
97. J. Behringer, *Z. Elektrochem.*, **62**, 906 (1958).
98. E. S. Yeung, M. Heiling, and G. J. Small, *Spectrochim. Acta*, **31A**, 1921 (1975).
99. G. J. Small and E. S. Yeung, *Chem. Phys.*, **9**, 379 (1975).
100. J. M. Friedman and R. M. Hochstrasser, *Chem. Phys.*, **6**, 155 (1974).
101. W. Maier and F. Dorr, *Appl. Spectrosc.*, **14**, 1 (1960).
102. P. P. Shorygin and T. M. Ivanova, *Optic Spektrosk.*, **15**, 176 (1963).
103. H. Lotem, R. T. Lynch, Jr., and N. Bloembergen, *Phys. Rev.*, **14A**, 1748 (1976).
104. J. P. Haushalter and M. D. Morris, *Anal. Chem.*, **53**, 21 (1981).

105. P. F. Williams, D. L. Rousseau, and S. H. Dworetsky, *Phys. Rev. Lett.*, **32**, 196 (1974).

106. G. Placzek, *Handb. Radiol.*, **6**(2) (1934).

107. C. Wieman and T. W. Hansch, *Phys. Rev. Lett.*, **34**, 1120 (1976).

108. B. Hudson, W. Hetherington III, S. Cramer, I. Chaby, and G. K. Klauminzer, *Proc. Natl. Acad. Sci. (USA)*, **73**, 3798 (1976).

109. L. A. Carriera, L. P. Goss, and T. B. Malloy, Jr., *J. Chem. Phys.*, **66**, 2762 (1977).

110. L. B. Rogers, J. D. Stuart, L. P. Goss, T. B. Malloy, Jr., and L. A. Carriera, *Anal. Chem.*, **49**, 949 (1977).

111. D. L. Jeanmaire, M. R. Suchanski, and R. P. Van Duyne, *J. Am. Chem. Soc.*, **97**, 1699 (1975).

112. J. L. Anderson and J. R. Kincaid, *Appl. Spectrosc.*, **32**, 356 (1978).

113. M. R. Mahony, M. W. Howard, and R. P. Cocney, *Chem. Phys. Lett.*, **71**, 59 (1980).

114. A. J. McQuillan, P. J. Hendra, and M. Fleischmann, *J. Electroanal. Chem.*, **65**, 933 (1975).

115. D. L. Jeanmaire and R. P. Van Duyne, *J. Electroanal. Chem.*, **84**, 1 (1977).

116. G. Blondeau, J. Zerbino, and N. Jaffrezic-Renault, *J. Electroanal. Chem.*, **112**, 127 (1980).

117. M. W. Howard, R. P. Cooney, and A. J. McQuillan, *J. Raman Spectrosc.*, **9**, 273 (1980).

118. R. M. Hexter and M. G. Albrecht, *Spectrochim Acta*, **A35**, 233 (1979).

119. F. W. King, R. P. Van Duyne, and G. C. Schatz, *J. Chem. Phys.*, **69**, 4472 (1978).

120. M. Moskovits, *J. Chem. Phys.*, **69**, 4159 (1978).

121. E. Burstein, Y. J. Chen, C. Y. Chen, S. Lundquist, and E. Tosatti, *Solid State Comm.*, **29**, 567 (1979).

122. R. P. Cooney, E. S. Reid, P. J. Hendra, and M. Fleischmann, *J. Am. Chem. Soc.*, **99**, 2002 (1977).

123. G. Hagen, B. Simic Galavaski, and E. Yeager, *J. Electroanal. Chem.*, **88**, 269 (1978).

124. J. Heitbaum, *Z. Phys. Chem. Wiesbaden*, **105**, 307 (1977).

125. T. M. Cotton, S. G. Schultz, and R. P. Van Duyne, *J. Am. Chem. Soc.*, **102**, 7960 (1980).

126. J. E. Pemberton and R. P. Buck, *J. Electroanal Chem.*, **136**, 201 (1982).

127. E. S. Yeung, *Adv. in Chromatogr.*, **23**, 1 (1984).

128. W. M. Shaub, A. B. Harvey, and G. C. Bjorklund, *J. Chem. Phys.*, **67**, 2547 (1977).

129. P. Esherick and A. Owyoung, *Nonlinear Raman Spectrosc. and Its Chemical Applications*, W. Kiefer and D. A. Long, Eds., Reidel, Boston, 1982.

130. M. W. Dowley, K. B. Eisenthal, and W. L. Peticolas, *Phys. Rev. Lett.*, **18**, 531 (1967).

METHODS BASED ON SPECIAL CHARACTERISTICS OF LASERS

CHAPTER

11

REMOTE SENSING WITH LASERS

R. M. MEASURES

University of Toronto
Institute for Aerospace Studies
Toronto, Ontario, Canada

1. INTRODUCTION

Environmental issues such as the influence of fluorocarbons and nitric oxide on the earth's protective shield of ozone, the effect of carbon dioxide and volcanic dust on the climate, the formation of photochemical smog, oil pollution, and acid rain have drawn to our attention the fragile nature of the biosphere. Lasers have provided us with important new methods of strengthening our understanding of the delicate ecosystems involved and determining the extent to which we are perturbing them.

The thermal balance between the incident solar radiant flux and the emission from the earth leads to the thermal structure of the atmosphere. The lowest, and therefore most dense, part of the atmosphere is termed the *troposphere*. Its boundary is defined to coincide with the lowest temperature minimum and ranges from about 10 km at the poles to about 15 km in the tropics. Above the troposphere lies the *stratosphere*—a region owing its characteristics to its ozone content.

The density profiles for most of the important species within the atmosphere are described in the U.S. Standard Atmosphere (1). The density in general falls off exponentially with height. However, the scale height changes at around 100 km so that the decrease is less severe above 100 km. The density of some constituents (primarily those that are photoactive like O and O_3) cannot be described in this simple manner. Since the molecular weights of N_2 and O_2 are 28 and 32 kg $kmol^{-1}$, respectively, the mean molecular weight of air is 28.96 kg $kmol^{-1}$. The volume of an ideal gas at "standard temperature and pressure" (STP) is 22.4 m^3 $kmol^{-1}$. In this condition the average mass density of air is 1.29 kg m^{-3}.

Many of these trace constituents can be classified as "pollutants"; that is to say, they can be regarded as harmful to mankind. Some affect man directly—carbon monoxide for example—while the influence of others is much more subtle. Chlorofluorocarbons (CFC, also known by their trade name Freon) were once thought to be completely innoxious due to their chemical inertness; however, the stability of these molecules permits them to be carried into the stratosphere, where intense solar UV can break them up and release chlorine atoms, which are capable of participating in the catalytic annihilation of ozone molecules.

Another long-term problem facing mankind is the so-called greenhouse effect, whereby solar radiation passes relatively unimpeded through the atmosphere while the thermal radiation from the earth's surface is trapped by the ever increasing amount of carbon dioxide (CO_2) being produced as a result of burning fossil fuels. This can be understood in terms of the optical properties of CO_2 for it is transparent to visible radiation but relatively opaque to thermal infrared radiation. Other molecules can also contribute to this phenomenon—for example, Varanasi and Ko (2) suggested that the chlorofluorocarbons (CFC-11 and CFC-12) could raise the global surface temperature by as much as 0.9 K when present in a concentration as low as a few parts per 10^9 (ppb) due to their strong absorption bands in the 8–12-μm interval, which includes the region of strongest emission from the earth.

Both nitric oxide and sulfur dioxide have been implicated in the growing environmental problem of acid rain, for it is suspected that NO is converted to nitric acid (HNO_3) and SO_2 to sulfuric acid (H_2SO_4) through some chain of chemical reactions within the atmosphere (3). SO_2 has, of course been a pollutant over the centuries, back to the start of the industrial revolution in Europe. Most European countries today have opted for an SO_2 standard[†] of 100 μg m^{-3} annual level (4). Around the end of the forties, the effect of a new kind of air pollution,

---

[†]The molecular weight of SO_2 is 64 kg $kmol^{-1}$. Consequently, its density is 2.86 kg m^{-3}, as 1 kmol occupies 22.4 m^3. Thus, the volume taken up by 100 μg (or 10^{-7} kg) of SO_2 is 3.5 $\times$ 10^{-8} m^3, and the concentration corresponding to 100 μg m^{-3} is then 35 ppb.

causing eye irritation, plant damage, and visibility degradation, became evident in Los Angeles. Subsequently, this form of oxidizing pollution has been experienced in many large cities throughout the world; it is attributed to the action of solar ultraviolet radiation upon mixtures of hydrocarbons (HC) and oxides of nitrogen (NO_x) (5).

The principal constituents of the troposphere (apart from N_2, O_2, H_2O, and the inert gases) are listed in Table 11.1 together with some idea of their sources and an estimate of their emission rates, background concentrations, and characteristics residence times (6). Besides the atmospheric constituents listed in Table 11.1 there are many others that are present in very small quantities. Some of these have an importance that is way out of line with their concentration. The hydroxyl free radical OH is an excellent example for it has been implicated in the global conversion of CO to CO_2 and as a key intermediary in the photochemical formation of smog, yet its concentration has been measured to be less than a few parts per 10^{12} (ppt) (7).

The scope of lasers in environmental remote sensing will be shown to be extensive as they can be used to undertake:

1. Concentration measurements including both major and minor constituents, and are therefore well suited to pollution surveillance and monitoring.
2. Evaluation of thermal, structural, and dynamic properties.
3. Threshold detection of specific constituents, and are therefore well suited for alarm purposes.
4. Mapping of effluent plume dispersal as a function of time.
5. Spectral fingerprinting of a specific target, such as an oil slick.

Furthermore, these observations can be made remotely from the ground or from ships, helicopters, aircraft, or satellites with both spatial and temporal resolution in most instances. A broad based monograph, *Laser Remote Sensing*, has recently been published by the author (8).

2. REMOTE-SENSING TECHNIQUES USING LASERS

The range of processes amenable to laser remote sensing includes Rayleigh scattering, Mie scattering, Raman scattering, resonance scattering, fluorescence, absorption, and differential absorption and scattering (DAS). The range of cross sections observed for each process is schematically presented in Fig. 11.1 from which it is evident that the cross section for Mie scattering can be so large that just a few appropriate-size scatters could give rise to a scattered signal that

Table 11.1. Characteristics of Atmospheric Trace Gases (6)

Pollutant	Major Sources		Estimated Emission Rates		Polluted-Atmosphere Concentrations	Atmospheric Background Concentrations	Calculated Atmospheric Residence Time
	Anthropogenic	Natural	Anthropogenic (10^9 kg yr^{-1})	Natural (10^9 kg yr^{-1})			
CO_2	Combustion	Biological decay; release from oceans	13,000	10^6	350 ppm	320 ppm	2–4 yr
CO	Auto exhaust; combustion	CH_4 oxidation	250	3000	5 ppm	0.1 ppm	0.1 yr
SO_2	Combustion of coal and oil	Volcanoes	133	Small	1 ppm	0.2–2 ppb	4 days
H_2S	Chemical processes; sewage treatment	Volcanoes, biological action in swamps	2.7	90	4 ppb	0.2 ppb	2 days
O_3	Photochemical smog	Photolysis of O_2 (25–50 km)	Uncertain	Uncertain	0.3 ppm	0.01 ppm	1 day
NO	Combustion	Bacterial action in soil	48	460	0.2 ppm	0.2–2 ppb	1 day
NO_2	None	Conversion of NO	Negligible	Negligible	0.1 ppm	0.5–4 ppb	5 days
N_2O	None	Biological action in soil	Small	540	0.25 ppm	0.25 ppm	4 yr
NH_3	Waste treatment	Biological decay	3.6	1000	0.01 ppm	6–20 ppb	7 days
Hydrocarbons	Combustion exhaust; chemical processes	Biological processes	CH_4:80 Others: uncertain	CH_4:2000 Others: uncertain	CH_4: 3 ppm Others: 2 ppm	CH_4: 1.5 ppm Others: < 1 ppb	CH_4: 1 yr Others: unknown
HCHO	Combustion exhaust; atmospheric reactions	Biological processes	100	1000	0.05 ppm	1 ppb	1–5 days
HCl	Chemical processes; rocket engine exhaust	Unknown	Uncertain	Unknown	1–5 ppm	Unknown	Unknown

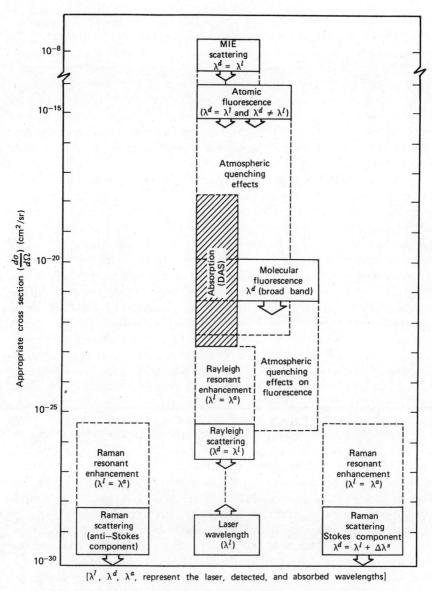

Fig. 11.1. Optical interactions of relevance to laser environmental sensing.

$[\lambda^l, \lambda^d, \lambda^a,$ represent the laser, detected, and absorbed wavelengths]

would completely swamp any other signal. This implies that quite low concentrations (or changes in concentration) of dust or aerosols can be detected.

Although resonance scattering, sometimes referred to as atomic (or resonance) fluorescence, also has an inherently large cross section, collision quenching with the more abundant atmospheric species generally ensures that the detected signal is small; consequently, this technique is used to best effect in studies of the trace constituents in the upper atmosphere. The influence of collision quenching on molecular fluorescence can be equally detrimental and the broadband nature of molecular fluorescence invariably leads to an even smaller signal.

In the case of hydrographic work, laser-induced fluorescence does play an important role because many contaminants of natural bodies of water fluoresce when excited by short wavelength (<400 nm) lasers. This realization led to the development of a new form of remote sensor termed a *laser fluorosensor* (9–13). Detailed spectroscopic studies (11, 12, 14) of both crude oils and petroleum products indicated that an airborne laser fluorosensor with high spectral resolution may indeed be capable of classifying an oil slick and measuring its thickness (15). The fluorescence of chlorophyll has long been known, and the possibility of employing a laser fluorosensor to remotely map the chlorophyll concentration of natural bodies of water has also been studied (16–21).

Raman scattering is an inelastic scattering process wherein the laser radiation may be thought of as raising the molecule to a virtual level from which it immediately decays (in $<10^{-14}$s), with the subsequent emission of radiation having a different wavelength. The difference in energy between the incident and emitted photons is a characteristic of the irradiated molecule and usually corresponds to a change of one vibrational quantum.

The frequency shifts of the Q-branch of vibrational-rotational Raman spectra have been summarized for a large group of molecules by Inaba and Kobayasi (22). A particularly attractive feature of Raman scattering relates to the ease with which the concentration of any species relative to some reference species, such as nitrogen, can be evaluated from the ratio of the respective Raman signals, provided the relevant cross section ratio is known (23–26).

The cross section for absorption of radiation is in general much greater than either the effective (quenched) fluorescence cross section or the cross section for Raman scattering. Consequently, the attenuation of a beam of suitably tuned laser radiation is a sensitive method of evaluating the mean density of a given constituent. In order to separate absorption by the molecule of interest from other causes of attenuation, a differential approach is usually adopted. In this instance two frequencies are employed, one centered on a line within the absorption band of interest, the other detuned into the wing of the line. With a few notable exceptions, most of the absorption bands of interest lie in the infrared and correspond to vibrational-rotational transitions (27–36).

High sensitivity with good spatial resolution can be achieved by the combination of differential absorption and scattering (DAS). This technique was first suggested by Schotland (37) for the purpose of remotely evaluating the water vapor content of the atmosphere. In this approach a comparison is made between the atmospheric backscattered laser radiation monitored when the frequency of the laser is tuned to closely match that of an absorption line (within the molecule of interest) and when it is detuned to lie in the wing of the line. In this way the large Mie scattering cross section is employed to provide spatial resolution and to ensure a strong return signal at both frequencies, while the ratio of the signal yields the required degree of specificity due to differential absorption. These advantages appear to bestow upon the DAS technique the greatest sensitivity for long-range monitoring of specific molecular constituents (38–48).

Although detector sensitivity makes both fluorescence and the DAS techniques more amenable to those molecules that possess an absorption band in the visible or near-ultraviolet part of the spectrum, recent improvements in infrared detector sensitivity has given the DAS approach more universal appeal (49–51). The acronym DIAL, standing for differential-absorption lidar, has gained considerable popularity recently for all of the laser remote-sensing techniques that rely on differential absorption.

The functional elements and manner of operation of most remote laser environmental sensors are schematically illustrated in Fig. 11.2. (The acronym

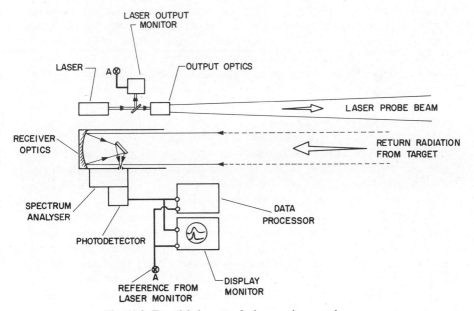

Fig. 11.2. Essential elements of a laser environmental sensor.

lidar, which can be thought to stand for laser identification, detection and ranging, will be used to encompass all forms of laser remote sensor.) An intense pulse of optical energy emitted by a laser is directed through some appropriate output optics toward the target of interest. The function of the output optics can be threefold: to improve the beam collimation, provide spatial filtering, and block the transmission of any unwanted broadband radiation, including emission that arises from some lasers. Often a small fraction of this pulse is sampled to provide a zero-time maker (a reference signal with which the return signal can be normalized in the event the laser's output reproducibility is inadequate) and a check on the laser wavelength where this is important.

The radiation gathered by the receiver optics (usually a Newtonian or Cassegrainian telescope) is passed through some form of spectrum analyzer on its way to the photodetection system. The spectrum analyzer serves to select the observation wavelength interval and thereby discriminate against background radiation at other wavelengths. It can take the form of a monochromator, a polychromator, or a set of narrow-band spectral filters together with a laser wavelength-blocking filter (unless elastically scattered light is of interest). The choice of photodetector is often dictated by the spectral region of interest, which in turn is determined by the kind of application and the type of laser employed.

The signal from the photodetector may be processed via analog or digital techniques. The development of very fast dual-waveform digitizers, such as the Biomation 4500 and Tektronix 7612D, make real-time data processing possible. Uthe and Allen (52) provided a brief review of the data-handling techniques employed in atmospheric probing, while Fredriksson et al. (48) and Hoge and Swift (53) have detailed representative lidar real-time data-recording systems.

The basic characteristics instrumental in determining the choice of a photodetector include the spectral response, quantum efficiency, frequency response, current gain, and dark current. Sometimes, other considerations such as physical size, ruggedness, and cost may also be important. In most instances the wavelength of the signal to be detected constitutes the primary factor in selecting the class of photodetector to be employed in any application. For wavelengths that lie between 200 nm and 1 μm (ultraviolet to near infrared), photomultipliers are generally preferred because of their high gain and low noise. Indeed, the single-photon detection capability of these devices has led to low-light-level detection schemes based on counting of individual photons (54).

In general, the performance of a photomultiplier is determined by (1) the spectral response of its photocathode, (2) the dark-current characteristics of its photocathode, (3) the gain of the dynode chain, (4) time dispersal effects of the electrons moving through the dynode chain, and (5) the transit time of the electrons between the last dynode and the anode. Photomultipliers with high gain attain almost ideal quantum-noise-limited sensitivity for the detection of weak light signals. The large range of photocathode materials currently available offer a wide choice of spectral response characteristics.

Infrared detectors can be divided, broadly speaking, into two classes—photodetectors and thermal detectors. The most sensitive infrared detectors are semiconductors in which the incident radiation creates charge carriers via a quantum interaction. These photodetectors may be further divided into photovoltaic and photoconductive devices. Of these, photovoltaic devices (photodiodes) are the more popular for environmental sensing. Although some photodiodes can be used in the visible, they come into their own at longer wavelengths, where their high quantum efficiency (30–80%) becomes important. Unfortunatley, the output must be externally amplified, so that its sensitivity is often limited by thermal noise. Altmann et al. (55) developed a "fast current amplifier" that enables a photovoltaic InSb detector, operated at zero bias voltage (for optimum detectivity), to achieve background-limited performance.

Some photodiodes, when operated at a high reverse bias, develop internal gain through a process of carrier multiplication. These avalanche photodiodes are similar to photomultipliers in the sense that their sensitivity is no longer determined by thermal noise of the detector and output circuit (56). One of the most widely used and most sensitive infrared detectors for the 1–5.5-μm range appears to be the liquid-nitrogen-cooled InSb photodiode (34, 57). A useful overview of infrared detectors has been prepared by Emmons et al. (58), and Lussier (59) has compiled several tables that summarize various infrared detector characteristics.

The spectrum analyzer (absorption filters, interferometric elements or dispersive systems) is used to select the wavelength interval of observation and to provide adequate rejection of all off-frequency radiation, whether this be laser-scattered radiation, solar background radiation, or any other form of radiation having a wavelength different from that of the signal. In general, this is accomplished with the aid of one or more spectral components.

The best, but most expensive, narrow-band filters are made of birefringent materials and are called Lyot filters (60). The wide field of view of this kind of narrow-band filter is particularly useful in lidar work. This filters can also be tuned. Reviews of this capability of birefringent filters have been prepared by Title and Rosenberg (61) and Gunning (62).

For those applications that require high spectral resolution, the choice often lies between a Fabry–Perot interferometer and a grating monochromator (63). Of these, the Fabry–Perot etalon is usually the cheaper, can provide the higher resolving power, and has the greater light throughput. Indeed, it would be ideal for many applications if it were not for its major drawback—many overlapping orders. This difficulty can be overcome by prefiltering with an interference filter, with a second, wider-passband Fabry–Perot interferometer, or with a dispersive element. Each approach has its own limitations. A detailed comparison of prism, grating, and Fabry–Perot etalon spectrometers has been given by Jacquinot (64). The passband of a Fabry–Perot etalon can be scanned by varying the pressure of the gas between the interferometer plates (65) or by displacing one plate relative to the other (66).

In those situations where measurement of a spectral profile is important or where many wavelengths are of interest, a grating monochromator offers some advantages. Of the wide array of monochromator systems available, the Czerny–Turner arrangement appears to be one of the most popular, being typically capable of providing a stray-light rejection ratio of 10^{-6}. For applications where this feature is of critical importance (such as those involving Raman backscattered signals), double monochromators are often employed. Stray-light rejection ratios of 10^{-6} (single) and 10^{-12} (double) can be achieved for displacements of about 60 cm^{-1} for the exciting line.

The range of lasers available is extensive, but only those capable of emitting pulses of very high peak power, narrow bandwidth, short duration, and that propagate with a low degree of divergence are ideal for probing the environment. In addition they should also be capable of operating at a high repetition rate for most airborne missions and for those atmospheric applications in which the return signal is very weak. An overview of the types of lasers appropriate for remote sensing is provided in Table 11.2. One of the most popular lasers used in lidar systems is the neodymium–yttrium aluminum garnet (or Nd:YAG). This laser not only meets the criteria suggested above but its second or third harmonic can be used to pump one or more dye lasers and thereby provide a tunable output (67).

3. LIDAR EQUATION AND ATMOSPHERIC ATTENUATION

The form of equation to be used in any given situation depends on the kind of interaction and the nature of the measurement to be undertaken. For those applications in which backscattering (elastic or inelastic) of the laser beam is utilized, the form of the lidar equation is fairly simple. Most atmospheric probing, including those instances where differential absorption is employed, is covered by this equation.

For those situations involving laser-induced fluorescence, a more complex form of the lidar equation that depends on the optical depth of the target media, detector integration period, and laser pulse duration is somewhat appropriate (68). In the limit of large optical depth this form of the lidar equation becomes identical to the "laser fluorosensor" equation, which was specifically developed to cover airborne lidars that probe natural bodies of water.

Table 11.2. Types of Lasers Relevant to Remote Sensing (8)

	Solid State	Gas	Liquid	Semiconductor
Representative examples	Ruby Neodymium (YAG) Alexandrite	XeCl (rare-gas halide) N_2 (transient) $HgBr_2$/HgBr (dissociation) CO_2 (molecular)	Organic dyes such as: Rhodamine 6G Coumarin Cresyl violet	GaAs GaAsP InAs $Pb_{1-x}Sn_xSe$
Primary pumping technique	Flashlamp	Intense electrical discharge in gas	Flashlamp or laser	High current injection leading to n, p radiative annihilation at an np junction
Range of wavelengths and tuning	Ruby (694.3 nm)—thermal tuning ±0.4 nm. Nd: YAG (1.06 μm) Alexandrite—tunable (701-818 nm) Second or third harmonic generation possible with all three kinds	H_2 (116, 160 nm) Xe_2 (170 nm) KrF (249 nm) XeCl (308 nm) N_2 (337 nm) $HgBr_2$/HgBr (502-504 nm) DF or HF (2.7-4.0 μm) CO (5.0-5.7 μm) CO_2 (9.0-11 μm) HCN (337 μm)	Large range of dyes provide wavelengths from 340 nm to 1.1 μm Typical tuning range per dye $\approx$ 40 nm with widths of 0.1-0.01 nm possible with grating or prism (+ etalon) arrangement	$GaAsP\text{-}Pb_{1-x}Sn_xSe$ (550 nm to 32 μm) Tuning possible by changing current, applying pressure or magnetic field
Modes of operation and pulse duration	Q-switching leads to 10-100-ns pulses Mode locking can yield 10-ps pulses	Fast discharges lead to pulses that typically range from 1 ns to 1 μs Q-switching possible with certain molecular gas lasers, cavity dumping with others	When N_2 laser-pumped pulses are ~ 5-10 ns When flashlamp pumped 0.3-1 μs pulses Cavity dumping of latter can yield 30-ns pulses	Current pulsed but requires cooling and efficient heat sink 10 ns to 1 μs possible
Peak power and energy/pulse attainable	For ruby and Nd: YAG 10^6-10^8 W and 1-10 J when Q switched; for Alexandrite lasers 10^7 W and 500 mJ	10^4-10^7 W and 1 mJ to 1 J	10^4-10^6 W in narrow, tunable bandwidth; 0.1-3 J	100 W possible from laser diode arrays

377

In general, interpretation of the lidar signal is further complicated by geometrical considerations that include the degree of overlap between the laser beam and the field of view of the receiver optics as well as the details of the telescope. A detailed treatment of the lidar equation is provided by Measures (8). Fortunately, it is possible to use one simplified form of the lidar equation under a fairly wide range of conditions.

3.1. Scattering Form of the Lidar Equation

The total power received at the wavelength, λ, by the lidar photodetector at the instant $t(=2R/c)$ corresponding to the time taken for the leading edge of the laser pulse to propagate (at the velocity of light, c) to range R and the returned radiation to reach the lidar, can often be expressed by an equation of the form

$$P(\lambda, t) = P_L \xi(\lambda) \ \xi(R) \ \frac{A_0}{R^2} \ \beta(\lambda_L, \lambda, R) \ T(\lambda_L, R) \ T(\lambda, R) \ \frac{c\tau_L}{2} \quad (11.1)$$

where P_L represents the transmitted laser power, $\xi(\lambda)$ the receiver's spectral transmission factor at λ, $\xi(R)$ the geometrical overlap factor (between the laser beam and the field of view of the receiver optics), A_0/R^2 the acceptance solid angle of the receiver optics, $\beta(\lambda_L, \lambda, R)$ the unit-volume backscattering coefficient at wavelength λ (for laser excitation at wavelength λ_L) and range R. The form of this equation implicitly assumes that the spatial extent of the laser pulse $c\tau_L$ is small compared to the range of interest. Lastly, $T(\lambda_L, R)$ and $T(\lambda, R)$ represent the single-pass atmospheric transmittances at λ_L and λ, respectively. If the laser radiation is weak enough to avoid perturbing the intervening atmosphere, then Beer–Lambert's law applies and we can write

$$T(\lambda_L, R) = \exp\left\{ -\int_0^R \kappa(\lambda_L, R) \ dR \right\} \quad (11.2)$$

and

$$T(\lambda, R) = \exp\left\{ -\int_0^R \kappa(\lambda, R) \ dR \right\} \quad (11.3)$$

where $\kappa(\lambda_L, R)$ and $\kappa(\lambda, R)$ represent the atmospheric attenuation coefficients at the laser and detected wavelengths, respectively. Combining these leads to the total (two-way) atmospheric transmittance (sometimes called the transmission factor)

$$T(R) = T(\lambda_L, R) \ T(\lambda, R) \quad (11.4)$$

Although the instantaneous power falling upon the photodetector is a useful quantity to evaluate, more often than not the increment of radiative energy at wavelength λ received during the interval (t, τ_d)

$$E(\lambda, R) = \int_{2R/c}^{2R/c + \tau_d} P(\lambda, t)\, dt \qquad (11.5)$$

is the more pertinent entity, where τ_d is the photodetector's response time. If E_L is the transmitted energy of the laser, Eq. (11.5) yields the basic lidar equation,

$$E(\lambda, R) = E_L \xi(\lambda)\, T(R)\, \xi(R)\, \frac{A_0}{R^2}\, \beta(\lambda_L, \lambda, R)\, \frac{c\tau_d}{2} \qquad (11.6)$$

Implicit in this derivation are the assumptions that the laser pulse is approximately rectangular and that $\tau_d \ll 2R/c$. The effective range resolution of such a system is limited to $c(\tau_d + \tau_L)/2$ as is seen by reference to the range–time diagram presented as Fig. 3.

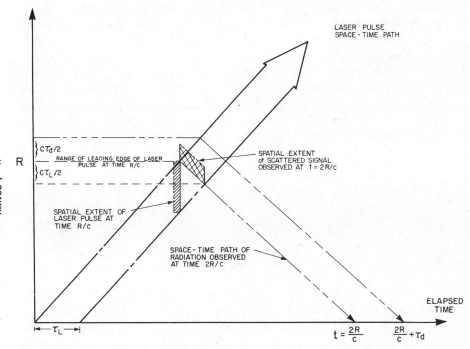

Fig. 11.3. Spatial resolution for scattering phenomena as seen from space–time diagram of propagating rectangular-shaped laser pulse (68).

In the case of inelastic scattering by one dominant species having an isotropic scattering cross section $\sigma_S(\lambda_L, \lambda)$, we can write

$$E(\lambda, R) = E_L \xi(\lambda)\, T(R)\, \xi(R)\, \frac{A_0}{R^2}\, \frac{N(R)\,\sigma_S(\lambda_L, \lambda)}{4\pi}\, \frac{c\tau_d}{2} \qquad (11.7)$$

where $N(R)$ represents the number density of this species at range R.

3.2. Differential Absorption Lidar (DIAL) Equation

Differential absorption of laser radiation by a particular molecular species represents both a selective and a sensitive method of measuring specific atmospheric constituents. There are two ways in which such measurements can be undertaken. Both involve using two laser pulses of slightly different wavelength (one chosen to coincide with a strong absorption feature of the specific constituent of interest, the other detuned into the wing of this feature) and comparing the attenuation of the two pulses. The difference in the techniques stems from the mechanism chosen to return the laser radiation to the lidar receiver system. In one case elastic scattering from atmospheric aerosols and particulates is employed, while the other approach relies on scattering of the laser radiation from some conveniently located topographical target. An extreme example of this uses a strategically positioned retroreflector.

In the first approach, two laser wavelengths, λ_0 and $\lambda_0 + \delta\lambda$, are selected such that λ_0 corresponds to the center wavelength of some prominent absorption line of the molecule of interest, while $\lambda_0 + \delta\lambda$ lies in the wing of this line. If we write λ_W for $\lambda_0 + \delta\lambda$, and use the elastic-scattering form of lidar Eq. (11.1) above, then we can write

$$\int_0^R N(R)\,\sigma_A(\lambda_0:\lambda_W)\,dR = \frac{1}{2}\ln \frac{P(\lambda_W, R)\,\xi(\lambda_0)\,\beta(\lambda_0, R)}{P(\lambda_0 R)\,\xi(\lambda_W)\,\beta(\lambda_W, R)}$$

$$- \int_0^R \{\bar\kappa(\lambda_0, R) - \bar\kappa(\lambda_W, R)\}\,dR \qquad (11.8)$$

where we have introduced the "differential absorption cross section,"

$$\sigma_A(\lambda_0:\lambda_W) \equiv \sigma^A(\lambda_0) - \sigma^A(\lambda_W) \qquad (11.9)$$

and have assumed that in general the total attenuation (or extinction) coefficient (m^{-1}) is given by

$$\kappa(\lambda, R) = \bar\kappa(\lambda, R) + N(R)\,\sigma^A(\lambda) \qquad (11.10)$$

where $\bar{\kappa}(\lambda, R)$ is the attenuation coefficient (m^{-1}) exclusive of the absorption contribution from the molecular species of interest, $N(R)$ represents the number density of these molecules (m^{-3}) at range R, and $\sigma^A(\lambda)$ their absorption cross section (m^2) at wavelength λ.

In differential form Eq. (11.8) becomes

$$N(R) = \frac{1}{2\sigma_A(\lambda_0 : \lambda_W)} \left| \frac{d}{dR} \left\{ \ln \left[\frac{P(\lambda_W, R)}{P(\lambda_0, R)} \right] - \ln \left[\frac{\beta(\lambda_W, R)}{\beta(\lambda_0, R)} \right] \right\} \right.$$

$$\left. + \bar{\kappa}(\lambda_W, R) - \bar{\kappa}(\lambda_0, R) \right| \qquad (11.11)$$

where we have assumed that the receiver's spectral transmittance is effectively independent of wavelength over the small interval $\delta\lambda$:

$$\xi(\lambda_0) \approx \xi(\lambda_W)$$

Additional simplification can be attained if we also assume that the volume backscattering coefficient β and the residual attenuation coefficient $\bar{\kappa}$ are independent of wavelength over this small interval $\delta\lambda$.

A considerable improvement in sensitivity can be achieved if this differential absorption technique is used in conjunction with a "topographical" scatterer. However, this gain in sensitivity is achieved at the expense of range resolution, so that this technique is only applicable in situations where the integrated concentration of the trace constituent along the path of the laser beam is worth evaluating.

In the case of a topographical scatterer the increment of radiative energy received within the detector's integration period τ_d at wavelength λ_0 is

$$E(\lambda_0, R_T) = E_L \frac{A_0}{R_T^2} \xi(\lambda_0) \xi(R_T) \frac{\rho^S \tau_d}{\pi \tau_L} \exp \left| -2 \int_0^{R_T} \kappa(\lambda_0, R) \, dR \right| \qquad (11.12)$$

where ρ^S represents the scattering efficiency of the target and R_T its range, provided $\tau_d \le \tau_L$. In the event that $\tau_d > \tau_L$, the factor τ_d/τ_L is replaced with unity. Values of ρ^S can range from 0.1 in the visible to 1 in the infrared (69–71). In order to have optimum temporal discrimination against any solar background illumination of the target, τ_d should be chosen to be as close to τ_L as possible. At locations of known pollution emission, a retroflector might be positioned so as to maximize the system sensitivity. Under these conditions the factor $\rho^S A_0/\pi R_T^2$ in Eq. (11.12) is replaced with ξ_0, the receiver collection efficiency. This can amount to an improvement of several orders of magnitude, depending primarily upon the range.

3.3. Lidar Equation in the Case of a Fluorescent Target

If the laser is capable of exciting fluorescence within the target, the finite relaxation effects have to be taken into consideration. Kildal and Byer (72) were the first to recognize the significance of this effect and to illustrate its influence on the return signal. A more detailed analysis of this problem was undertaken by Measures (8, 68) and reveals that the appropriate equation can be written in a form that is very similar to Eq. (11.7) for scattering.

For a fluorescent target the increment of radiative energy in the spectral interval $(\lambda, \Delta\lambda_0)$ received by the photodetector in the time interval (t, τ_d) is

$$E(\lambda, R) = E_L K_0(\lambda)\, T(R)\, \xi(R)\, \frac{A_0}{R^2}\, \frac{N_F(R)\, \sigma^F(\lambda_L, \lambda)}{4\pi}\, \frac{c\tau_d}{2}\, \gamma(R) \quad (11.13)$$

where $\Delta\lambda_0$ is the total bandpass and $K_0(\lambda)$ the filter function of the receiver optics and photodetector. $N_F(R)$ is the number density and $\sigma^F(\lambda_L, \lambda)$ the isotropic fluorescence cross section for the fluorescing species. Introduction of the filter function,

$$K_0(\lambda) \equiv \int_{\Delta\lambda_0} \xi(\lambda)\, d\lambda \quad (11.14)$$

implies that the spectral sampling window of the receiver system is narrow compared to the fluorescence linewidth.

The factor $\gamma(R)$ appearing in Eq. (11.13) takes account of many of the complexities associated with the fluorescence case. Fortunately, this factor tends to unity for target penetration greater than a few laser pulse lengths, provided the medium is optically thin and the laser pulse duration is greater than the lifetime of the laser-excited molecule, (8, 68). In the case of an optically thick fluorescent target Eq. (11.13) becomes

$$E(\lambda, R) = E_L K_0(\lambda)\, T(R)\, \xi(R)\, \frac{A_0}{R^2}\, \frac{N_F(R)\, \sigma^F(\lambda_L, \lambda)}{4\pi\{\kappa(\lambda_L) + \kappa(\lambda)\}} \quad (11.15)$$

which turns out to be quite appropriate for some hydrographic applications and is sometimes termed the laser fluorosensor equation.

As we have seen, the appropriate lidar equation provides a means of relating the radiation returned from a probing laser beam to the relevant optical properties and through them to the related physical properties (such as the density of some specific constituent) of the target medium. In order to evaluate these optical properties from the return signal, the appropriate lidar equation has to be solved (73, 74).

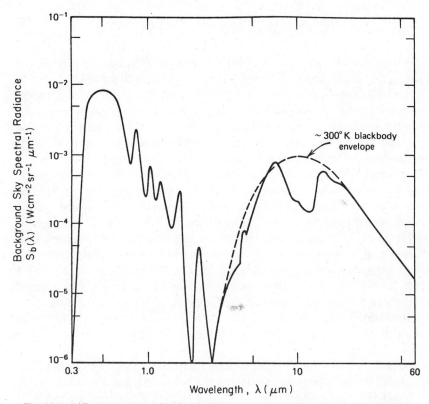

Fig. 11.4. Diffuse component of typical background spectral radiance from sea level (75).

The correct interpretation of these lidar return signals requires careful analysis and adequate information. It is also necessary that background sources of noise be correctly appreciated in order that they can either be avoided or taken into consideration. One of the most important sources of noise in the field of laser remote sensing is solar background radiation. The spectral radiance of a clear sky is presented as Fig. 11.4 and can be seen to peak in the visible with a sharp cutoff at about 300 nm, due to the ozone shield in the earth's atmosphere. The slower decline toward the infrared is punctuated by many absorption bands (75). The second hump, centered at about 10 μm, represents the thermal background radiation. In the case of downward-pointing airborne lidars this background is augmented by reflected and scattered solar radiation (76).

3.4. Atmospheric Attenuation

In the case of radiation propagating through the atmosphere the attenuation coefficient (m^{-1})

$$\kappa(\lambda) = \sum_i \{\kappa_E^i(\lambda) + \kappa_R^i(\lambda) + \kappa_A^i(\lambda)\} + \kappa_M(\lambda) \qquad (11.16)$$

where the sum extends over each of the atmospheric molecular constituents and $\kappa_E(\lambda)$, $\kappa_R(\lambda)$, $\kappa_A(\lambda)$, and $\kappa_M(\lambda)$ represent the elastic (Rayleigh), Raman absorption, and Mie volume attenuation coefficients, respectively. In regard to attenuation, the contribution arising from inelastic (Raman) scattering is negligible. When the wavelength of the radiation coincides with that of a relatively strong absorption line or band of even a minor constituent of the atmosphere, $\kappa_A(\nu)$ can dominate and appreciable attenuation result.

For wavelengths less than about 200 nm, the atmosphere is totally opaque as a result of the Schumann–Runge bands of molecular oxygen (O_2). The absorption due to O_2 decreases with increasing wavelength so that beyond 250 nm it is unimportant and likely to be exceeded by the effect of small quantities of ozone (O_3) (77). In the infrared part of the spectrum many atmospheric constituents contribute to the absorption, leaving only a few spectral windows through which optical probing is possible (27, 78). Reference to Fig. 11.5 reveals that water vapor (H_2O) and carbon dioxide (CO_2) are the principal absorbers in the unpolluted atmosphere. Between 300 nm and 1 μm there are few absorption bands, and under clear sky conditions it is Rayleigh–Mie scattering that determines the attenuation characteristics of the atmosphere in this portion of the spectrum.

The electronic absorption bands of most molecules, other than O_3, SO_2, and NO_2, lie in the far ultraviolet at wavelengths below 185 nm. The most intense and broad vibration-rotation band of "water vapor" is centered at about 6.27 μm and completely absorbs electromagnetic radiation in an interval that extends from 5.5 μm to around 7.5 μm. Other vibration-rotation band centers are at 2.73, 2.66, 1.87, 1.38, 1.10, 0.94, 0.81, and 0.71 μm. The large dipole moments of the water molecule and its isotopes give rise to an intense rotational spectrum that runs from about 8 μm through to the far infrared.

The "carbon dioxide" molecule has two main infrared absorption bands centered at about 4.3 and 15 μm (the third is optically inactive). In addition to the main bands, CO_2 has overtone combination bands and hot bands with centers near 10.4, 9.4, 5.2, 4.3, 2.7, 2.0, 1.6, and 1.4 μm. The strong absorption observed beyond 14 μm is also primarily due to CO_2.

Electronic transitions in the "ozone" molecule produce Hartley and Huggins bands located in the ultraviolet at wavelengths shorter than 340 nm. Weaker Chappius bands are found between 450 and 740 nm. The three main vibration-rotation bands of O_3 have centers at 9.0, 14.1, and 9.6 μm, with additional weaker bands at 5.75, 4.75, 3.59, 3.27, and 2.7 μm.

Additional information on the infrared absorption characteristics of these and other molecules is provided by Zuev (80) and by Pressley's *Handbook of Lasers*

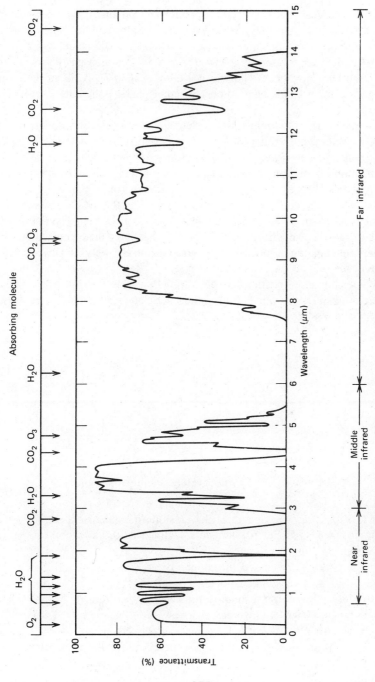

Fig. 11.5. Transmittance through the earth's atmosphere (horizontal path at sea level, length 1828 m) (79).

385

(81), and detailed calculations of atmospheric transmittance have been undertaken by McClatchey et al. (82, 83), LaRocca (84), Selby and McClatchey (85), and Roberts et al. (86). Specific atmospheric transmittance calculations for a number of laser lines emitted by a few of the more common gas lasers (CO_2, CO, HF, and DF) have been reported by Kelley et al. (87). A list of the more important absorption bands of a number of atmospheric pollutants is provided in Table 11.3.

As indicated earlier, the transmission characteristics of the atmosphere in the visible portion of the spectrum (400–700 nm approximately) is determined under clear-sky conditions by Rayleigh–Mie scattering of particulates and aerosols. The attenuation of electromagnetic radiation over a wide range of wavelengths becomes progressively more severe with the formation of haze.

Unfortunately, the complexity of this scattering and the local fluctuations in the density and shape of the scatterers makes $\kappa_M(\lambda)$ both highly variable and difficult to evaluate. Nevertheless, Nilsson (89) has attempted to calculate the atmospheric attenuation associated with aerosols for a variety of weather conditions over the 0.2–40-μm wavelength range.

A considerable simplification has been found possible (90) in certain situations by relating the mean value of this atmospheric attenuation coefficient $\kappa_M(\lambda)$ to visibility through the empirical relation

$$\kappa_M(\lambda) = \frac{3.91}{R_0} \left\{ \frac{550}{\lambda} \right\}^q \ \mathrm{km}^{-1} \tag{11.17}$$

where

$$q = 0.585 R_0^{1/3} \quad \text{for} \quad R_0 \leq 6 \ \mathrm{km}$$

$$\approx 1.3 \quad \text{for average seeing conditions}$$

In Eq. (11.17) λ is the wavelength of the radiation in nanometers and R_0 is the meteorological visual range in kilometers (defined as the horizontal range at which the transmission at 550 nm is 2%).

Woodman (91) has discussed the reliability of Eq. (11.17) and concluded that although its accuracy is questionable in the infrared ($\lambda > 2$ μm), it can probably be used to provide a rough estimate of $\kappa_M(\lambda)$ from R_0 in the visible wavelength range. However, Twomey and Howell (92) and Fenn (93) have also questioned the reliability of Eq. (11.17) in the case of laser radiation. Clay and Lenham (94) have measured the attenuation coefficient $\kappa_M(\lambda)$ for fogs at several wavelengths.

Table 11.3. Measured Absorption Cross Sections for Gaseous Species[a]

Molecule	(cm^{-1})	λ	$\sigma^A (\lambda_1)$ $(10^{-18}\ cm^2)$	$\kappa_A (\lambda)$ at STP $[(ppm\ cm)^{-1}]$
Acetylene, C_2H_2	719.9	13.89 μm	9.2	2.48 × 10^{-4}
Ammonia, NH_3	1,084.6	9.220 μm	3.6	9.68 × 10^{-5}
Benzene, C_6H_6	1,037.5	9.639 μm	0.09	2.42 × 10^{-6}
1,3-Butadiene, C_4H_6	1,609.0	6.215 μm	0.27	7.26 × 10^{-6}
1-Butene, C_4H_8	927.0	10.787 μm	0.13	3.50 × 10^{-6}
Carbon monoxide, CO	2,123.7	4.709 μm	2.8	7.53 × 10^{-5}
Carbon tetrachloride, CCl_4	793.0	12.610 μm	4.8	1.29 × 10^{-4}
Ethylene, C_4H_4	949.5	10.531 μm	1.34	3.60 × 10^{-5}
	950	10.526 μm	1.70	4.57 × 10^{-5}
Fluorocarbon-11, CCl_3F (Freon-11)	847	11.806 μm	4.4	1.18 × 10^{-4}
	1,084.6	9.220 μm	1.24	3.34 × 10^{-5}
Fluorocarbon-12, CCl_2F_2 (Freon-12)	920.8	10.860 μm	11.0	2.96 × 10^{-4}
	923.0	10.834 μm	3.68	9.90 × 10^{-5}
Fluorocarbon-113, $C_2Cl_3F_3$	1,041.2	9.604 μm	0.77	2.07 × 10^{-5}
Methane, CH_4	2,948.7	3.391 μm	0.6	1.61 × 10^{-5}
	3,057.7	3.270 μm	2.0	5.38 × 10^{-5}
Nitric oxide, NO	1,900.1	5.265 μm	0.6	1.61 × 10^{-5}
	1,917.5	5.215 μm	0.67	1.80 × 10^{-5}
Nitrogen dioxide, NO_2	1,605.4	6.229 μm	2.68	7.21 × 10^{-5}
	22,311.0	448.2 nm	0.2	5.38 × 10^{-6}
Ozone, O_3	1,051.8	9.508 μm	0.9	2.42 × 10^{-5}
	1,052.2	9.504 μm	0.56	1.51 × 10^{-5}
	39,425.0	253.6 mm	12.0	3.23 × 10^{-4}
Perchloroethylene, C_2Cl_4	923.0	10.834 μm	1.14	3.07 × 10^{-5}
Propane, C_3H_8	2,948.7	3.391 μm	0.8	2.15 × 10^{-5}
Propylene, C_3H_6	1,647.7	6.069 μm	0.09	2.42 × 10^{-6}
Sulfur dioxide, SO_2	1,108.2	9.024 μm	0.25	6.73 × 10^{-6}
	1,126.0	8.880 μm	0.2	5.38 × 10^{-6}
	2,499.1	4.001 μm	0.02^b	5.38 × 10^{-7}
	33,330.0	300.1 nm	1.0	2.69 × 10^{-5}
Trichloroethylene, C_2HCl_3	944.2	10.591 μm	0.56	1.51 × 10^{-5}
Vinyl chloride, C_2H_3Cl	940.0	10.638 μm	0.4	1.08 × 10^{-5}

[a] Based on the data presented by Hinkley et al. (88).
[b] Recent measurements (36) indicate an absorption cross section of 0.416 × 10^{-18} cm^2 for SO$_2$ at 3.9843 μm, corresponding to the $P_4(6)$ line of a DF laser.

387

4. ATMOSPHERIC LIDAR APPLICATIONS

The development of laser remote sensing holds the promise of substantially improving our understanding of the environment in which we are immersed by providing new methods of measuring some of the most important parameters and constituents. In many instances lidar techniques enable us to uniquely assess the extent to which anthropogenic activity may be perturbing our atmosphere.

4.1. Atmospheric Studies

The oxygen–nitrogen balance of the atmosphere represents an important measurement in view of the potential reduction in the sources of free oxygen (namely, decline in marine life and vegetation) and the increase in the rate of oxygen consumption (population growth and corresponding increased fossil fuel combustion). Schwiesow and Derr (95) have indicated that a precise measurement of the O_2–N_2 balance in the atmosphere should be possible using laser Raman scattering. They point out that with this technique a two-order-of-magnitude improvement in precision over other techniques should be possible (leading to an accuracy of 0.3 ppm for the O_2-N_2 ratio and 0.006 ppm for the CO_2-N_2 ratio). This would be adequate to determine the magnitude of any long-term drifts in the oxygen–carbon dioxide–nitrogen balance of the atmosphere.

Leonard (96) was the first to use the nitrogen laser (operating at 337 nm) to observe Raman backscattering from nitrogen (at 365.9 nm) and oxygen (at 355.7 nm) at a range of around 1 km. The first Raman measurement of the gas density profile was undertaken by Cooney (97) using a 25-MW, Q-switched ruby laser. The Raman vibrational-rotational return from nitrogen was observed at night, up to an altitude of 3 km. Cooney used a combination of a 694.3-nm blocking filter and a 15-nm passband interference filter to provide a net spectral rejection of 10^7. This was more than adequate to overcome the intensity factor of 500 between the elastically backscattered return at 694.3 nm and the nitrogen Raman return at 828.5 nm. Garvey and Kent (98) extended the range of these Raman investigations of atmospheric nitrogen well into the stratosphere (to a height of at least 40 km) and obtained good agreement between their observations, balloon-mounted radiosonde measurements, and the U.S. Standard Atmosphere (99).

The first remote laser measurement of the vertical water vapor profiles in the atmosphere was based on differential absorption and scattering and was made by Schotland (37) using a thermally tuned ruby laser. The earliest laser Raman measurements to yield spatial distribution of water vapor in the atmosphere were performed by Melfi et al. (100) and Cooney (101). Each used a frequency-doubled, Q-switched ruby laser and normalized their water vapor return with the nitrogen vibrational Raman return. Melfi et al. (100) and Cooney (102) were also able to demonstrate good agreement between their lidar-evaluated profiles and measurements undertaken by radiosondes. Similar results have also been obtained by Pourny et al. (103).

Although the general agreement obtained in the above work was excellent, the comparisons were made rather far apart in space and time. Strauch et al. (104) avoided this problem by making a direct comparison of the Raman lidar measurements with that of a standard humidity meter mounted on a tower some 30 m above the ground. Their results show an excellent correlation between the two measurements and indicate that their system should be capable of determining the water vapor profile to a range of about 4 km.

Most Raman-based lidar measurements are restricted to nighttime operation because of the strong daytime sky background radiance. One way of avoiding this form of noise is to operate between 230 and 300 nm. Stratospheric ozone absorbs the incoming solar radiation within this spectral interval, and consequently this is termed the *solar blind* region of the spectrum. Unfortunately, operation at these wavelengths is something of a double-edged sword—for the absorption by ozone that is responsible for the solar-blind region also causes attenuation of both the outgoing laser pulse and the Raman-backscattered return. This problem is further aggravated by the strong wavelength dependence (Hartley bands) of this ozone absorption. Renaut et al. (105) and Petri et al. (106) have both attempted to use multiwavelength, solar-blind Raman lidars to remotely measure atmospheric water vapor content and temperature.

To date all measurements of the atmospheric water vapor content based on Raman scattering have been limited to an operational range of less than 3 km. Although this could be extended with higher-power lasers and more sophisticated detection schemes (particularly better spectral rejection of background radiation), eye-safety considerations (especially in the solar-blind region) will probably limit the improvement attained in practice. On the other hand, elastic (Rayleigh and Mie) backscattering is much more intense (typically 10^5 to 10^7 times greater) than the H_2O Raman backscattering and lends itself to a much better method of long-range concentration mapping through differential absorption and scattering (37). A useful comparison of these techniques has been made by Cooney (107).

Murray et al. (49) used a 1-J/pulse CO_2 TEA laser to measure the water vapor content of the atmosphere over a horizontal range of about 1 km using the differential absorption and scattering DAS technique. The differential absorption between the $R(20)$, $R(12)$, and $R(18)$ rotational-vibrational transitions of the CO_2 laser is clearly seen in Fig. 11.6.

A differential absorption lidar (DIAL) system based on a continuously tunable Nd:YAG-pumped optical parametric oscillator (lithium niobate) could greatly extend the potential range of atmospheric constituents amenable to range-resolved measurement. Brassington (51) has demonstrated that, although such a

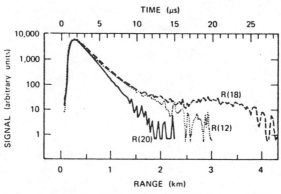

Fig. 11.6. Lidar backscatter signal for $R(12)$, $R(18)$, and $R(20)$ lines on the 10-μm band of a CO_2 laser. The $R(20)$ is more strongly absorbed by ambient H_2O vapor than is $R(12)$ or $R(18)$ (49).

system is possible, large errors can arise if tight control of the excitation wavelength is not maintained.

A more sensitive differential absorption lidar operating in the near infrared—on the 724.37-nm H_2O absorption line—was developed by Browell et al. (108). Their system employed a 1.5-J Q-switched (Holobeam) ruby laser to provide simultaneously both the off-line laser pulse (at 694.3 nm) and the on-line laser pulse (at 724.37 nm). This was accomplished through the use of a beam splitter that allowed 0.125 J to be transmitted and the remaining 1.25 J to pump a dye oscillator-amplifier arrangement. An important feature of this DIAL system was the incorporation of a multipass H_2O absorption cell for exact calibration of the dye laser's spectral output on *each shot*. This determination of the H_2O absorption for each dye laser pulse eliminates a major source of uncertainty in such DIAL measurements. A representative vertical atmospheric H_2O profile (with comparison radiosonde data) is presented as Fig. 11.7. This successful demonstration of a ground-based system represented the initial step in the development of an airborne and ultimately Shuttle-borne H_2O DIAL. Although these results are impressive, Zuev et al. (109) have been able to extend such water vapor measurements to an altitude of 17 km using a temperature-tuned ruby laser to probe the 694.38-nm H_2O absorption line and a spectrophone cell for fine tuning the laser.

An aircraft-mounted H_2O DIAL would allow important studies of macro- and micrometeorology, air-mass modification over bodies of water, aerosol growth, and tropospheric-stratospheric exchange mechanisms; Shuttle-derived H_2O profiles would provide inputs for meteorological forecasting models and improve our understanding of atmospheric radiative processes. Browell (110) has reported on the successful preliminary trials of an airborne DIAL system

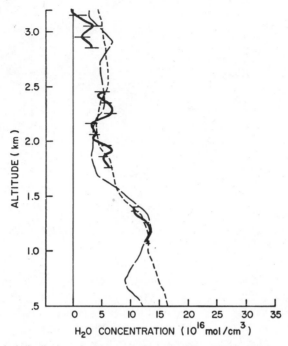

Fig. 11.7. Vertical distribution of water vapor determined by a DIAL system (—), involving 100 laser firings, a 100-m range cell, and a H_2O absorption cross section of 5.2×10^{-23} m^2. Data from radiosondes launched before (---) and after (— — —) DIAL measurements are also displayed (108).

that has measured vertical water vapor profiles with a flight range resolution of about 4 km.

Temperature profiles are obviously important in both climate modeling and weather forecasting. For example, the severity of storms is often related to the temperature lapse rate, and the cloud ceiling and visibility are influenced by a combination of humidity and temperature profiles. It has also been shown that the formation of a temperature inversion over an urban area is often responsible for the most severe pollution episodes. The idea of using a light beam to determine the atmospheric molecular-density profile and thereby derive the temperature profile was first proposed by Elterman (111, 112). With a searchlight he attempted to evaluate the temperature between 10 and 67 km by assuming that beyond 10-km altitude the return was determined principally by Rayleigh scattering and that the ideal-gas law could be combined with a hydrostatic relation to give the temperature in terms of the density change. Sandford (113) also used elastic scattering, but from a powerful laser, to evaluate the density change and thereby the temperature. Figure 11.8 presents the upper-atmospheric

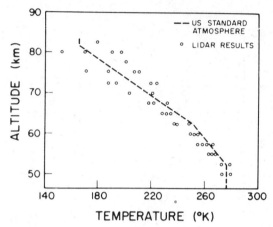

Fig. 11.8. Upper atmospheric temperature profile deduced from lidar elastic backscattering measurements at Wakefield, England (113).

temperature profile deduced by Sandford (113) and found to be in reasonable agreement with the U.S. Standard Atmosphere.

Cooney (114) proposed that rotational Raman scattering from N_2 might serve as a more convenient vehicle for ascertaining the temperature profile of the atmosphere. He based his argument on the stronger scattering intensity and the temperature sensitivity of the rotational Raman spectrum. He also suggested that a differential technique might be the best way of ensuring adequate temperature sensitivity and also canceling out most of the extraneous factors, such as atmospheric transmission and the spectral response of the photodetector. Salzman (115) demonstrated the feasibility of this approach by making temperature measurements from −20 to 30°C at a range of 100 m (indoors) with a resolution of 5 m and an accuracy of ±3°C. He used two interference filters with circular apertures. These sampled the rotational anti-Stokes spectrum out in the wing and close to the exciting line. The ratio of the resulting intensities is, in principle, soley dependent on the gas temperature.

The temperature of the atmosphere can also be ascertained from the molecular-level populations by means of differential absorption measurements. Mason (116) first proposed that the DAS technique could be applied to such temperature evaluations, and Schwemmer and Wilkerson (117) undertook an initial error analysis.

Endermann and Byer (118, 119) were able to show (through simulation) that the temperatures should be proportional to the logarithm of the ratio of the absorbances over the range of −10 to 30°C. This would be adequate for work in the lower troposphere. Their simulation involved the wavelengths $\lambda_1 = 1.7695$ μm, $\lambda_2 = 1.7698$ μm, and $\lambda_3 = 1.7696$ μm, and was based on absorption within the wing of the 1.9-μm band of water vapor. These wavelengths were selected after careful examination of the Air Force Cambridge Research Laboratory tapes of McClatchey et al. (83). In a preliminary set of experiments, Endermann and Byer (119) used a Nd:YAG laser to pump an optical parametric oscillator. This provided a tuning range from 1.4 to 4.0 μm and an output energy of 5 mJ in a 10-ns pulse. Their measurements relied on scattering of these laser pulses from a building located 775 m from the telescope receiver and therefore only provided the average values of the temperature and humidity over that range. Endemann and Byer (119) suggest that an appreciable improvement could be expected if all three laser pulses were fired within the atmosphere correlation time of a few milliseconds (120).

Kalshoven et al. (121) have considered the case of an atmospheric molecular constituent for which the density can be assumed to be independent of location. Under these circumstances the temperature can be evaluated from an absorption measurement on a single temperature-sensitive line—this is, taken to mean the lower level of the transition is elevated above the ground level by about the mean thermal energy of the atmospheric constituents. Oxygen was chosen by Kalshoven et al. (121) because of its virtually uniform mixing ratio throughout the atmosphere and the availability of convenient high-resolution high-energy lasers that are capable of being tuned to the temperature-sensitive oxygen A-band at 770 nm. The DIAL system used by Kalshoven et al. (121) comprised two dye lasers that were pumped by a continuous-wave (cw) krypton laser. One dye laser was tuned to the 768.38-nm line of O_2 and had a linewidth of better than 5.9×10^{-5} nm; the other dye laser provided the reference laser radiation and was displaced by about 5.9×10^{-2} nm from the absorption line center. Its linewidth was about 5.9×10^{-3} nm. Their results indicate that the average temperature of a 1-km path can be determined to better than 1.0°C with a noise level of 0.3°C.

A temperature measurement technique that is particularly applicable to the stratosphere has been proposed by McGee and McIlrath (122). Their approach involves comparing the ratio of laser-induced fluorescence on two lines of the OH radical and is based on the fact that the chemical lifetime of this trace constituent of the stratosphere is long enough for its ground vibronic-rotational state populations to be in thermodynamic equilibrium. The OH radical was chosen because it is readily excited by current high-power tunable lasers and its emission is strong and well defined.

The atmosphere contains a number of naturally occurring minor (trace) constituents that play an important role in the scheme of things. Some of the best-known examples are CO_2, O_3, and OH. The first has a strong influence on the thermal balance of the atmosphere, the second forms an extremely effective shield against the life-damaging short-wavelength ($\lambda < 300$ nm) radiation from the sun, and the last appears to play a crucial role in many of the atmosphere's photochemical reactions. Since man's activities could disturb the natural balance of these important constituents, it is evident that we must have adequate means of monitoring their concentrations in order to detect such perturbations before they can become significant. The review of Schofield (123) provides an extensive evaluation of the potential of atomic and molecular fluorescence for in situ monitoring of the concentration of a large number of minor species within the stratosphere.

Although laser-induced fluorescence can, in principle, be used for the troposphere, the large quenching rates associated with the high density in that region make it rather impractical for remote-sensing applications. In the stratosphere the severity of the quenching rates is greatly reduced and fluorescence lidar looks attractive for the detection of a number of species. McIlrath (124) has provided a list of potential species, their respective wavelengths for excitation and detection, and their appropriate cross sections.

The hydroxyl free radical OH is considered to be one of the most chemically active trace constituents of the atmosphere and is believed to control the worldwide conversion of CO to CO_2 as well as being an important intermediary in

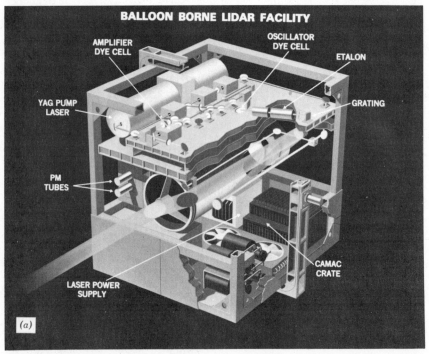

Fig. 11.9. Balloon-borne lidar facility: (a) artist's representation of the layout and (b) schematic diagram of transmitter system (all wavelengths are in microns) (127).

the photochemical formation of smog. In the stratosphere OH takes part in catalytic ozone destruction processes that regulate the concentration of O_3 in both the lower mesosphere and the upper stratosphere. Laser-induced fluorescence has been shown to be capable of detecting OH at the low concentrations of importance ($\approx 10^6$ cm^{-3}) (7, 125). Although the latter work involved an aircraft-mounted system, both groups were involved with in situ techniques. McGee and McIlrath (122) and McIlrath (124) have suggested that laser-induced fluorescence could be used to remotely monitor the stratospheric concentration of OH from a balloon. Furthermore, the developments in areas of narrow-band tunable excimer lasers, high-resolution, high-rejection optical filters, and wavelength measurement devices could make such measurements even more sensitive (126).

Recently, Heaps et al. (127) have successfully undertaken such a mission, and their results reveal a temporal variation of the hydroxyl-radical concentration over the 34–37 km altitude range, from 40 ppt shortly after noon to around 5 ppt 2 hr after sunset. The balloon-borne lidar facility of Heaps et al. (127) is illustrated in Fig. 11.9a, while Fig. 11.9b reveals the layout of the laser system

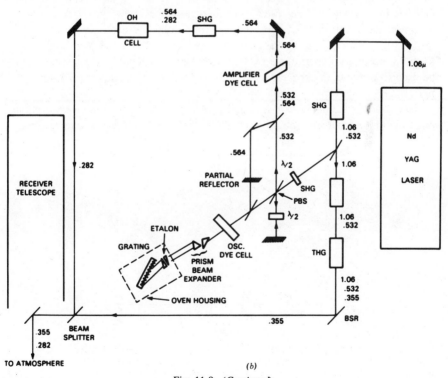

(b)

Fig. 11.9. (Continued)

and transmission optics employed. A 400-mJ Nd:YAG laser constitutes the primary source, and a 30-cm-diameter Cassegrain telescope serves as the collector for the receiver subsystem.

In addition to the OH measurements undertaken with this facility, Heaps et al. (127) also used it to evaluate the distribution of ozone in the stratosphere. For this work the system was operated in a differential absorption mode. However, the uniform density of backscatterers in the stratosphere enabled observations to be made using only one emission wavelength (i.e., 282 nm). Figure 11.10 shows the ozone profile determined during the ascent of the balloon. The lidar results are plotted as crosses with the horizontal crossbar extending twice the standard deviation on either side of the measured value. The averaging period for each measurement was 6–7 min, during which the balloon ascended between 500 and 700 m. The dashed line is a plot of the value obtained by a Dasibi in situ ozone analyzer located aboard the gondola. Range-resolved ozone profiles taken on the ascent at an altitude of about 23 km revealed an unexpected horizontal structure. This structure had more or less disappeared at an altitude of 37 km.

Another balloon-borne approach at measuring stratospheric trace gas constituents is currently being developed at the Jet Propulsion Laboratory. In this instance absorption of laser radiation propagating between the gondola and a retroreflector suspended 0.5 km below the gondola is used as an indicator of the in situ density of various trace constituents such as NO, and NO_2, and H_2O. In the preliminary experiments two infrared lead salt tunable diode lasers will be used to provide an expected minimum detectable concentration of about 0.1 ppbv (128).

The coincidence of the 308-nm emission of a XeCl laser with an absorption band within O_3 enabled Uchino et al. (129) to undertake the first experimental

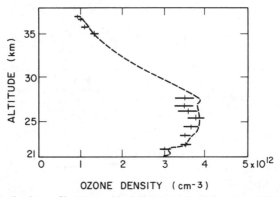

Fig. 11.10. Ozone density profile measured by balloon-mounted lidar (+) and Dasibi in situ ozone analyzer (---) during the ascent (error bars represent two standard deviations (127).

measurements of the ozone distribution in the 15–30-km section of the strato-sphere with a ground-based lidar system.

The XeCl laser used by Uchino et al. (130) transmitted an energy pulse of about 50 mJ, spread over the three vibrational bands at 307.6, 307.9, and 308.2 nm, with a duration of about 16 ns and a beam divergence of 1 mrad. Back-scattered laser radiation was collected by a 50-cm-diameter Newtonian tele-scope. A comparison between the lidar observations taken a Fukuoka and ozonesondes launched the same day from adjacent locations of Tateno and Kagoshina showed good agreement over the 15–25-km range of altitudes. Also a high correlation was obtained between the mean ozone density at 17.25 km and the total vertical column density of atmospheric ozone observed by a Dobson spectrophotometer at Tateno.

Recently, Uchino et al. (131) have extended their measurements down into the troposphere using a two-wavelength lidar system that employed both a XeCl laser at 308 nm and a KrF–Raman hybrid laser (at 209.4 nm). Their measure-ments indicated that the density of O_3 centered around 10^{12} cm^{-3} in the 4–12-km altitude range.

An airborne DIAL system developed at the NASA Langley Research Center has the capability to investigate the spatial distribution of many tropospheric gases and aeroticulates (132). This system has the flexibility to operate in the UV for temperature and pressure measurements of O_3 or SO_2, in the visible for NO_2, and in the near IR for H_2O. Also, aeroticulate backscattering investigations in the visible and the near IR can be conducted simultaneously with the DIAL measurements. This lidar system employs two frequency-doubled Nd:YAG la-sers to pump two dye lasers. The "on" and "off" (or wing) wavelength laser pulses at 286 and 300 nm, respectively, are sequentially produced and separated by less than 100 μs. Dielectric-coated steering optics are used to direct the dye laser outputs through a 40-cm-diameter quartz window (used for high UV trans-mittance) in the bottom or top of the aircraft. The receiver system is composed of a 36-cm-diameter Cassegrain telescope and gatable photomultiplier tubes. A photograph and schematic view of this airborne DIAL system are presented in Fig. 11.11.

The first remote measurement of tropospheric ozone profiles from an aircraft was obtained with this system in May, 1980, and the results compared with in situ measurements obtained from an instrumented Cessna 402 aircraft. A major field experiment with the U.S. Environmental Protection Agency was directed at studying the large-scale pollution events in the northeast sector of the United States. The objectives of this program included the characterization of persistent elevated pollution episodes (PEPE) and the evaluation of a four-layer regional oxidant model. A comparison of O_3 measurements made with the airborne DIAL system at an altitude of 3200 m and the in situ instruments on the Cessna is presented as Fig. 11.12. The lidar-observed variation in O_3 concentration, from

AIRBORNE DIAL SYSTEM SCHEMATIC

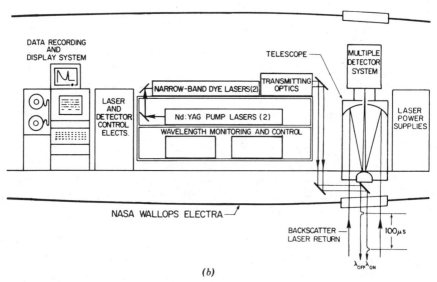

(b)

Fig. 11.11. (a) The NASA–DIAL system aboard the Wallops Flight Centre *Electra* aircraft and (b) schematic of the system (110).

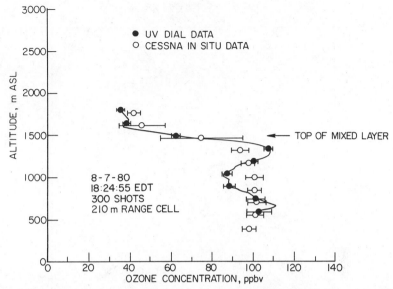

Fig. 11.12. Comparison of DIAL and in situ O_3 data obtained in the vicinity of Salisbury, Maryland, during EPA field experiments (132).

42 ppb above the mixed layer to 100 ppb within the mixed layer, is in fairly good agreement with the in situ Cessna measurements.

An even more interesting result is displayed in Fig. 11.13, where the NASA UV airborne DIAL system reproduces fairly faithfully the narrow double layer of enhanced ozone concentration detected with an ozonesonde launched from Caribou at 2209 EDT (110). The altitude for these layers is within 150 m, which is very good, considering that the DIAL data represent the average O_3 profile obtained from 300 lidar measurements along a 6-km flight path. A more detailed description of this potentially multipurpose airborne DIAL system, with additional O_3 measurements, have been provided recently by Browell et al. (133).

The development of high-energy, flashlamp-pumped, tunable dye lasers made it possible to map certain trace constituents in the upper atmosphere through the process of laser-induced (resonance) fluorescence. Bowman et al. (134) undertook the first ground-based measurement of the sodium atom concentration in the tenuous outer regions of the atmosphere. More detailed measurements, including the seasonal variations of this sodium concentration, were obtained by Gibson and Sandford (135). Hake et al. (136) observed a fourfold increase in the sodium layer content during the maximum of the Geminids meteor shower. This observation, presented as Fig. 11.14, lends support to the idea that meteor ablation represents an important source of this material. The laser used by Hake et al. is representative of those used in these kinds of experiments and comprised

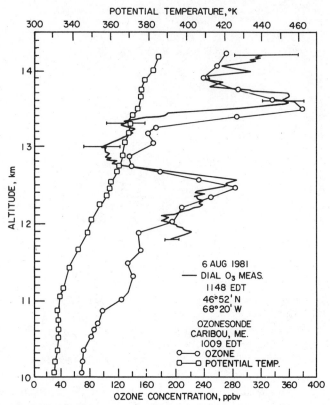

Fig. 11.13. Comparison of DIAL and ozonesonde measurements of ozone layers in the vicinity of the tropopause (110).

a two-stage (oscillator-amplifier) Rhodamine 6G dye laser with an output of 0.5 J in 300 ns at 589.0 nm. The spectral linewidth was less than 0.005 nm, and the final beam was collimated to better than 0.5 mrad. Additional confirmation of the meteor production theory was obtained by Aruga et al. (137). Although most observations of this sodium layer were made at night, Gibson and Sandford (138) were able to modify their system sufficiently to map the spatial distribution of the sodium during the day and thereby dispel the notion of daytime enhancement.

Nighttime studies by Blamont et al. (139) revealed a stratification of the sodium layer. They also employed an (oscillator-amplifier) flashlamp-pumped dye laser, the output of which was about 0.3 J in a 0.013-nm linewidth. The pulse duration was 1 μs, and beam was collimated by an afocal system to 0.3 mrad. The receiver system comprised an 80-cm-diameter telescope, an afocal

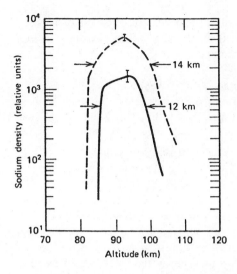

Fig. 11.14. Altitude profiles of free atomic sodium density obtained at 2315-2400 PST (solid line) and 0255-0310 PST (dashed line), before and slightly after transit of the radiant of the Geminids meteor shower on December 11-14, 1971. Reprinted from ref. (136) with permission.

optical arrangement for reducing the beam diameter to that of a 2-nm linewidth interference filter, and a photomultiplier. A sodium vapor scattering cell was used to check the tuning of the laser and provide a measurement of its energy. More recently Megie and Blamont (140) have reported on an extensive program of nighttime atmospheric sodium measurements. They have developed a dynamic photochemical model of the upper atmosphere and used it to simulate the behavior of the atmospheric sodium. The experimental results were then compared with this simulation and seem to confirm that photochemical equilibrium between

$$Na + O_3 = NaO + O_2$$

and

$$NaO + O = Na + O_2$$

exists during the night. The behavior of this sodium layer is then surmised to be determined by dynamical processes related essentially to eddy diffusion mixing and to the strength of a permanent source of sodium. The stratifications in the density profiles first observed by Blamont et al. (139) were also confirmed and correlated with the propagation of gravity waves at mesospheric heights. Although sporadic enhancement of the sodium content of the atmosphere seems to have been related to meteoritic showers, Megie and Blamont (140) stated that the influence of strong eddy mixing makes it impossible to draw a conclusion concerning the meteoric origin of *all* the atmospheric sodium. They contended that a simultaneous measurement of the sodium and potassium concentrations in the upper atmosphere would permit the influence of vertical transport to be separated from that of meteoric deposition, and later reported on such observations (141). These observations led them to suggest that there are, in fact, two sources for the alkali content of the upper atmosphere:

1. A meteoritic source that is constant over the year and responsible for the low value of the abundance ratio in the summer. The correlation between sporadic meteoritic showers and increases in the alkali content bears witness to this source.

2. A terrestrial source due to the vertical transport of salt particles at high latitudes, which works only in winter, when the circulation pattern of the polar stratosphere breaks down.

More recently, Granier and Megie (142) reported on daytime measurements of the mesospheric sodium layer with an improved lidar system and conclude that the dynamical effects such as horizontal transport and organized vertical motions dominate the regular variations induced by the solar diurnal cycle as predicted by the photochemical models of the sodium layer (140).

4.2. Atmospheric Pollution Surveillance

The adverse effect of air pollution on human health is well established (143, 144), and many government agencies have established maximum levels of exposure for a wide range of atmospheric contaminants. In the following sections we shall look at laser techniques that are capable of providing information on the general level of pollution and those better suited for effluent source monitoring. Grant and Menzies (145) have recently provided a useful survey of laser and selected optical systems for remote measurement of pollutant gas concentrations.

The possibility of using multiple-line gas lasers to detect pollution over an extended path was first considered toward the end of the sixties by Hanst and Morreal (146). They showed that it was possible to detect gaseous pollutants, such as CO, NO, SO_2, and O_3, down to concentrations of a few parts per million over a 1-km path using either a CO_2 or and I_2 laser and a retroreflector. In order to avoid an absolute calibration of the system, a differential absorption approach was adopted.

Tunable infrared lasers clearly increase the scope of pollution detection since they provide a wider choice of wavelengths. If this tuning can be accomplished rapidly, additional opportunities present themselves. Ku et al. (147) demonstrated that fast modulation of the laser frequency about an absorption line can lead to the elimination of signal noise arising from atmospheric turbulence. This "fast-derivative spectroscopy" involved the use of a cryogenically cooled $PbS_{0.82}Se_{0.18}$ semiconductor diode laser operating around the CO 4.7-μm absorption band with an output power of less than 1 mW. The laser frequency was modulated over a band of about 0.5 cm^{-1} at 10 kHz by a 10% oscillation in the injection current. Detection was accomplished with a liquid-nitrogen-cooled InSb infrared detector.

Hinkley (148) reported on a U.S. Environmental Protection Agency's Regional Air Pollution Study that involved the use of this technique for monitoring the CO concentration over a 1-km path in downtown St. Louis. In addition he presented the results of monitoring the nitric oxide concentration at a busy traffic rotary in Cambridge, Massachusetts, using resonance absorption and second-derivative spectroscopy. Figure 11.15 illustrates this work and shows both calibrations and sudden increases in NO concentration due to individual vehicles. In principle, much greater sensitivity can be obtained by either a multipath technique or positioning the retroreflector much further away. Indeed, the limit of sensitivity could easily be extended by a factor of 50, which would mean in some instances a sensitivity of better than a few parts per billion (149). Additional improvement can be achieved if heterodyne detection is employed (150, 151). A further point worth making stems from the very modest laser energy required for long-path absorption. This allows the use of a broad spectrum of lasers, including tunable infrared laser diodes and Raman spin-flip lasers (152, 153). Unfortunately, the range of situations amenable to installation of a retroreflector will invariably be restrictive. Topographical targets may be used to relax this constraint but at a cost of increases in laser energy due to the return signal's range dependence.

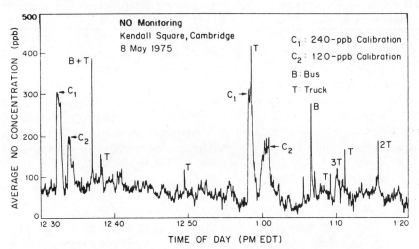

Fig. 11.15. Nitric oxide monitoring by resonance absorption (second derivative) at busy traffic rotary in Cambridge, Massachusetts. Calibrations are shown, and large increases due to diesel-fueled buses (B) and trucks (T) are evident. Time constant was 1 s. Reprinted from ref. (148) with permission.

Menzies and Shumate (154) measured the average concentrations of nitric oxide, ozone, and ethylene over a 0.8- and a 3.75-km path using differential absorption with either a retroreflector or a solid scattering surface to provide the return signals. CO and CO_2 lasers supplied the respective 5.2-, 9.5- and 10.5-μm wavelength radiation. Direct detection of the laser radiation was employed with cube-corner retroflectors, while heterodyne was employed when the return signal was laser radiation scattered from the rough surface. In principle, heterodyne detection of laser radiation can be several orders of magnitude more sensitive than direct photodetection in the 5–12-μm infrared wavelength region. This holds out the prospect of detection from topographical surfaces at a range of several kilometers, even when the emitted laser power is less than 1 W. Recently, Menzies and Shumate (155) measured tropospheric ambient ozone concentrations using an airborne differential absorption lidar system.

Henningsen et al. (30) and Guagliardo and Bundy (156, 157) were among the first to demonstrate that single-ended resonance absorption using topographical backscattering was practical. Henningsen et al. (30) were able to detect CO over a range of 107 m using backscattered radiation at 2.3 μm from foliage; Guagliardo and Bundy (156, 157) developed an airborne system to measure the column concentration of O_3 using the earth as a topographical target. They employed two grating-tuned TEA CO_2 lasers that were fired within a 20-μs interval in order to both minimize atmospheric scintillation and ensure that both beams struck, as far as possible, the same spot on the ground.

Ethylene is an important urban air pollutant in that it is directly emitted in the exhaust of motor vehicles, yet many plants exhibit symptoms of ethylene toxicity at concentrations as low as 10 ppb. Laser backscattering from foliage on foothills located at a range of around 5 km has been used to monitor the ambient concentrations of ethylene (35). These experiments were conducted with 1-J (100-ns) pulses on the $P(14)$ and $P(16)$ lines of a CO_2 TEA laser with a beam divergence of around 1.8 mrad. The receiver system comprised a 31.75-cm-diameter telescope having a 3.3-mrad field of view, and a HgCdTe detector with a detectivity D^* of 1.1×10^{10} cm $Hz^{1/2}$ W^{-1}. Two extreme examples of ethylene measurements undertaken by Murray and van der Laan (35) are presented as Fig. 11.16. The spectral interference from water vapor was estimated to be equivalent to about 7.6 ppb of ethylene. The results shown in Fig. 11.16 were corrected for this interference.

Killinger et al. (158) have also used a line-tunable CO_2 TEA laser to monitor the concentration of carbon monoxide near a major traffic roadway. They frequency doubled the CO_2 output in a crystal of $CdGeAs_2$ to provide pulses of about 1 mJ at around 4.65 μm. The return radiation was collected by a 30-cm-diameter Cassegrain telescope and directed onto a InSb pyroelectric detector. The frequency-doubled $R(18)$ line is fairly strongly absorbed by CO and is relatively free from spectral interference of other atmospheric constituents. The diurnal variation of the average atmospheric CO concentration over a 13-hr

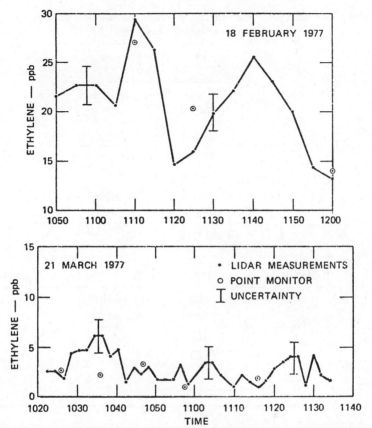

Fig. 11.16. Lidar-measured ethylene concentrations corrected for water vapor interference and compared with a point monitor. Reprinted from ref. (35) with permission.

period for a 500-m path was observed to range between 0.35 and 1.0 ppm. The peaks tended to correlate with periods of heavy traffic on a nearby major road-way.

This same system was also used to monitor the concentration of nitric oxide over a well-traveled road (159). These measurements involved the spectral co-incidence between the frequency-doubled CO_2 laser radiation and the NO ab-sorption lines near 5.3 μm. Return signals were obtained from a topographical target at a range of 1.4 km, and significant NO concentrations (of about 250 ppb) above ambient were recorded even under high-humidity conditions. Re-cently, Menyuk et al. (160) have reported that this system has also been used to detect about 100 ppb of highly toxic rocket fuels over ranges between 0.5 and 5 km using a topographical backscattering target (see Table 11.4).

The first remote measurement of the ambient background concentration of

Table 11.4. Absorption Parameters for Laser Remote Sensing of Several Hydrazine Rocket Fuels (160)

CO$_2$ Laser Transition	Wavelength λ (μm)	Absorption Coefficients κ_A (cm^{-1} atm^{-1})			Atmospheric Attenuation κ_ϵ (km^{-1})
		Hydrazine	Interfering Species		
		N$_2$H$_4$	NH$_3$	C$_2$H$_4$	
P(22)	10.611	4.77	0.045	1.09	0.1142
P(28)	10.675	2.06	0.36	1.30	0.0976
		Unsymmetrical Dimethylhydrazine	Interfering Species		
		(CH$_3$)$_2$N$_2$H$_4$	NH$_3$	C$_2$H$_4$	
P(30)	10.696	2.22	0.86	1.63	0.0907
R(10)	10.318	0.18	0.78	1.51	0.1142
		Monomethyl-hydrazine	Interfering Species		
		CH$_3$N$_2$H$_4$	NH$_3$	C$_2$H$_4$	
R(30)	10.182	1.69	0.029	0.56	0.1137
R(18)	9.282	0.31	0.13	0.61	0.1418

N$_2$O using a DF laser-based lidar was reported by Altmann et al. (36). They used the coincidence of the DF P_3(7) laser line, at around 3.8903 μm, with a strong absorption (1.33 $\times$ 10^{-6} cm^{-1} ppm^{-1}) in N$_2$O. The adjacent P_3(6) laser line at 3.855 μm provided the reference signal, having a much weaker absorption (8.1 $\times$ 10^{-8} cm^{-1} ppm^{-1}). Return signals from topographical targets at distances up to 8 km were detected, and their N$_2$O concentration measurement of about 290 ppb was found to be in reasonable agreement with in situ measurements. Altmann and Pokrowsky (36) measured the absorption coefficient of SO$_2$ for twenty of the DF laser lines. Only the P_4(6) line at 3.9843 μm was found to have a strong absorption coefficient (4.4 $\times$ 10^{-7} cm^{-1} ppm^{-1}).

A Nd:YAG-pumped LiNbO$_3$ optical parametric oscillator (OPO) represents an important alternative to line-tunable gas lasers as a source of intense tunable infrared radiation (161, 162). Baumgartner and Byer (163, 164) have developed a differential absorption lidar based on such a transmitter. Their LiNbO$_3$ parametric oscillator provided an output of about 20 mJ over the range 1.4–4.2 μm with a 1.0-cm^{-1} linewidth. A 41-cm-diameter telescope collected the return signal and focused it onto a 1-mm^2 InSb photovoltaic detector (cooled to 77 K). The output of the detector was amplified through a computer-controlled gain-switchable amplifier having a 1-MHz bandwidth. Preliminary observations were undertaken with this system using a topographic reflector for SO_2 at 4.0 μm, CH_4 at 3.3 μm (and 1.66 μm), and H_2O at 1.7 μm. During the CH_4 studies at 3.3 μm, absorption scans of the atmosphere were made to determine the possible interference effects of water vapor (165).

Recently, Aldén et al. (166) have reported the first lidar measurement of atomic mercury in the atmosphere. Differential absorption and topographical scattering were employed. An anti-Stokes-shifted, frequency-doubled Nd:YAG-pumped dye laser with an output of about 0.7 mJ at the on and off wavelengths (253.65 and 253.68 nm, respectively) of the resonance transition of mercury was used in conjuction with a 25-cm, $f/4$ Newtonian telescope. A detection limit of about 8 μg m^{-3}m was achieved with a laser linewidth of around 0.15 cm^{-1}. This corresponds to an average concentration of 4 ng m^{-3} over a pathlength of 2 km, which is representative of background concentrations and typically one-tenth of that expected near chloroalkali plants.

Occasionally, the ambient concentration of the constituent of interest can be large enough to limit the range of observations based on resonance absorption through excessive attenuation of the laser radiation. This was first pointed out by Measures and Pilon (38). Detuning of the on-wavelength laser can relax this constraint.

Byer and Garbuny (32) have shown by similar arguments that where an excess density of some pollutant, ΔN_i, is to be detected against a background density N_i of the same constituent, then the cross section appropriate to a minimum in the laser energy is given by

$$\sigma_i^A(\lambda) = \frac{1}{2L\Delta N_i}\ln\left\{1 + \frac{L\Delta N_i}{RN_i}\right\} \qquad (11.18)$$

where L is the extent of the pollutant plume and R its range.

As we have seen from Table 11.3, most pollutants have absorption bands in the infrared. However, a few—notably O_3, NO_2, SO_2, and a number of metals—have healthy absorption features that lie in the visible or near-ultraviolet part of the spectrum. Hoell et al. (45) and Thompson et al. (167) attained a measurement sensitivity of 10 ppb at a range of 0.8 km for SO_2 in field experiments. A frequency-doubled, flashlamp-pumped tunable dye laser having an output of 100 μJ, in a spectral bandwidth of less than 0.03 nm, and a duration of 1.3 μs was used for these observations. More recently, Baumgartner et al. (168) have also used a flashlamp-pumped dye laser to measure the ambient NO_2 concentration over Redwood City, California. Their laser emits a 10-mJ, 700-ns pulse that alternates in wavelength between 448.1 and 446.5 nm (to correspond to the on and off absorption bands of NO_2) with a divergence of about 1.3 mrad. The laser repetition rate was 5 pps, and the output linewidth was 0.2 nm. Similar measurements for SO_2 over the city of Göteborg (Sweden) have been undertaken by Fredriksson et al. (46). Their system employed a frequency-doubled flash-lamp-pumped dye laser with an output of 0.4 mJ in a 1-μs pulse around 300 nm. The laser repetition rate was 25 pps, and a 25-cm-diameter $f/4$ Newtonian telescope served as the optical receiver. An example of their results is displayed in Fig. 11.17. A comparison with the day average SO_2 concentration obtained by the local Public Health Board, using conventional chemical methods, is also provided.

4.3. Pollution Source Monitoring

As we have seen, the differential absorption approach is unlikely to be rivaled for range and sensitivity. Nevertheless, the sophistication of these DIAL systems and the complexity of their signal interpretation provides considerable incentive to develop alternative approaches where the bounds on range and concentration are conducive. Such a situation is likely to be encountered when the spatial and temporal distributions of specific gaseous contaminants are to be monitored as they emerge from a pollution source.

An approach based upon laser-induced fluorescence may appear to be attractive for a number of constituents such as SO_2, NO_2, I_2, O_3, various hydrocarbon

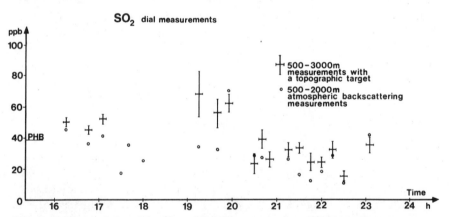

Fig. 11.17. Time variation of the SO_2 concentration over the city of Göteborg. The average concentration over a distance of 3 km, as determined in DIAL measurements against a topographic target, is shown with indications of the time periods of measurements and the estimated maximal error. In addition, mean values over a distance of 2 km are given by (o), as obtained in DIAL measurements using atmospheric backscattering. The day average value as measured by the local Public Health Board (PHB) with conventional chemical methods is also given. This value was obtained at a point along the measurement path. Reprinted from ref. (46) with permission.

vapors, and certain kinds of aerosol pollutants. An early analysis of the fluorescence return signal expected from a localized source of pollution was undertaken by Measures and Pilon (38). The results of this study revealed that above a certain peak concentration a distortion of the returned signal could lead to a misinterpretation of both the range and concentration of the source. Unfortunately, in the case of atmospheric work, collision quenching and the broadband nature of the emission, combined with the concomitant high aeroticulate background associated with such sources of pollution, tend to restrict the remote-sensing potential of this approach.

Although the extraordinary small cross section associated with Raman scattering represents a considerable impediment to its use in remote sensing, it possesses several desirable characteristics that make it very attractive for pollution source monitoring:

1. The spectral shift of the Raman-backscattered raditation is specific to each molecule.

2. The intensity of a given Raman signal is directly proportional to the density of the appropriate scattering molecule and independent of the others. Consequently, a direct measure of the concentration of a pollutant relative to nitrogen can be obtained without calibration problems.

3. The narrow spectral width and shift of the Raman return are conducive to spectral discrimination against both solar background radiation and elastically scattered laser radiation.

4. The inherent short duration of the Raman process can also be used to discriminate against solar background radiation when a small range interval is of interest.

5. Only a single, fixed-frequency laser is required to produce the simultaneous Raman spectra of all the pollutants within the region being probed. The advantage of this for multiplexing is obvious.

6. Good spatial and temporal resolution is possible since a backscattering process is involved.

There is, however, a serious problem that can arise in pollution monitoring. The Raman signal from a trace constituent of a plume could be masked by the O- or S-branch Raman signal from a major component. This can be appreciated by reference to Fig. 11.18. Inaba and Kobayasi (22, 169), who were the first to undertake a comprehensive study of Raman scattering for pollution monitoring, addressed this problem and found that in many instances spectral interference can be avoided by the use of narrow spectral filters.

An early foretaste of what can be achieved was provided by Hirschfeld et al. (170), who designed and built one of the most powerful lidars to date. This

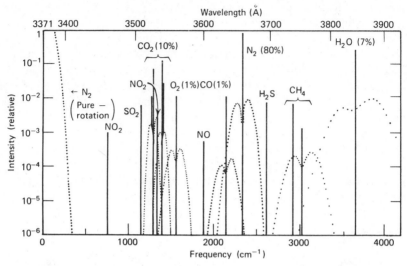

Fig. 11.18. Theoretical distribution of Raman volume backscattering coefficient due to a molecular mixture contained in a typical oil smoke as a function of Raman-shifted frequency (22).

system incorporated a frequency-doubled, Q-switched, 2-J (2-pps) ruby laser, a 91.4-cm-diameter $f/6.8$ Dall–Kirkham Cassegrain telescope, a polychromator, and an array of photomultipliers to provide multiplex detection. The high sensitivity achieved with this lidar is attested by the daylight Raman spectra obtained from a controlled plume of SO_2 and kerosene at a range of 200 m, with a 10-m range resolution. DeLong (171) extended these measurements and demonstrated that sensitivities on the order of 100 ppm for a wide variety of constituents should be achieved, even under unfavorable daytime weather conditions for a range of several hundred meters. Melfi et al. (172) also reported the detection of Raman scattering by SO_2 in the plume of a 200-MW coal-burning electrical generating plant at a slant range of 210 m. Poultney et al. (173) demonstrated that an improved version of this system was capable of measuring SO_2 at a concentration of about 10^3 ppm with 12% accuracy at a range of 300 m in about 15 min with a range resolution of 6 m.

In spite of the complexity of both equipment and data processing required for the differential absorption and scattering technique, its sensitivity enables it to do more than just monitor the pollutant concentration within the plume at the source of emission. In particular, a DIAL system can be used to map the dispersion of these pollutants, day or night. This, combined with the growing concern regarding the international problem of acid rain, has prompted the development of many operational DIAL systems throughout the industrialized world. We shall examine several representative systems.

One of the first such operational DIAL systems was developed by Rothe et al. (43, 44) in Germany. They used a 1-mJ, 300-ns, flashlamp-pumped, tunable dye laser operating between 455 and 470 nm with a 1-pps repetition rate to demonstrate that NO_2 concentrations down to 200 ppb could be detected with spatial distribution over a chemical factory obtained with the lidar situated some 750 m from the chimney stack. It is reproduced here as Fig. 11.19 and represents one of the most striking examples of the potential of DAS. Similar concentration maps of water vapor in the vicinity of a cooling tower and the ethylene distribution over a refinery have also been obtained by Rothe (174).

In the United States the Electric Power Research Institute (EPRI) has pro-

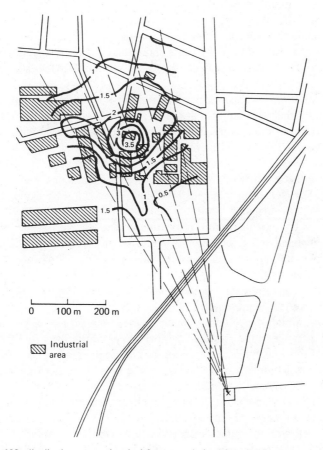

Fig. 11.19. NO_2 distribution over a chemical factory as derived from DAS measurements performed in the indicated directions at an altitude of about 45 m. The concentrations are given in ppm, parts per million (44).

moted the development of a mobile DIAL system for monitoring the emission and dispersion of SO_2 and NO_2 released from the tall stacks commonly used by utilities burning fossil fuels. At the heart of this system, which was designed and built at SRI International, California, lie two independently controlled Nd:YAG lasers (67). Their outputs are frequency doubled and used to pump two dye lasers, which are in turn frequency doubled to provide outputs at 300.0 and 299.5 nm for SO_2. The 448.1- and 446.5-nm wavelengths for NO_2 detection can be obtained by pumping a pair of coumarine dye lasers with the third-harmonic outputs of the Nd:YAG lasers. The sensitivity of this DIAL system for SO_2 was about 2 parts per million meter (ppm m) when the output energy was around 10 mJ and a 51-cm-diameter telescope was employed. The integration period was around 2 min. This means that a concentration of only 220 ppb of SO_2 could be detected for a smokestack plume of 10-m diameter. Moreover, such measurements could be undertaken at a range of 3 km. A photograph of the van, laser, and receiver telescope is shown in Fig. 11.20a. The twin lasers and other electro-optical equipment that constitute this DIAL system are dramatically illustrated in Fig. 11.20b.

In early 1979 the Swedish Space Corporation funded the development of a powerful, fully mobile lidar system that would be used for both research and routine atmospheric monitoring. The lidar was built around a 250-mJ, 7-ns, 10-pps Nd:YAG laser that could be frequency doubled or tripled to yield 100 and 50 mJ, respectively. These shorter-wavelength outputs are used to pump a tunable dye laser that consists of a grating-tuned oscillator and one or two amplifier stages (48). For NO_2 measurements 448.1 and 446.5 nm were again chosen as the on and off wavelengths, while for SO_2 the dye laser is frequency doubled to provide 300.05 and 299.30 nm, respectively. The receiver system used a 30-cm $f/3.3$ Newtonian telescope to focus the backscattered laser radiation onto an EMI 9817 photomultiplier after transmission through a set of narrow-band optical filters. The first 20 μs of signal (corresponding to backscattered radiation out to 3 km) is captured with a Biomation 8100 transient digitizer. A schematic of this DIAL system is presented as Fig. 11.21.

It is worth noting that Fredriksson et al. (48) avoided having to match the large dynamic range of the return signal (arising primarily from its $1/R^2$ falloff) to the limited dynamic range of the fast-transient digitizer through a combination of applying a synchronized ramp voltage to the dynode chain of the photomultiplier and using geometrical compression in the optical design (175). Their results indicated that the NO concentration emanating from the chimney of the saltpeter plant at a range of 1350 m was about 122 mg m^{-3} or roughly 100 ppm and agreed well with the value determined by conventional means.

Chlorinated hydrocarbons are today extensively used in a variety of industrial

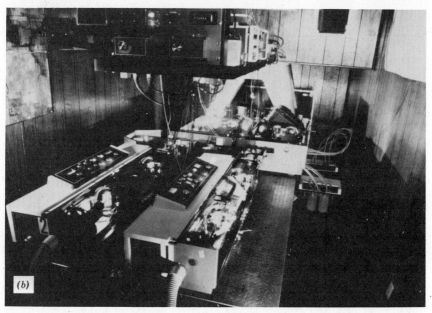

Fig. 11.20. SRI International mobile lidar (a) shows the van with receiver telescope, (b) reveals dual laser system and receiver electronics. Reprinted from ref. (67) with permission.

413

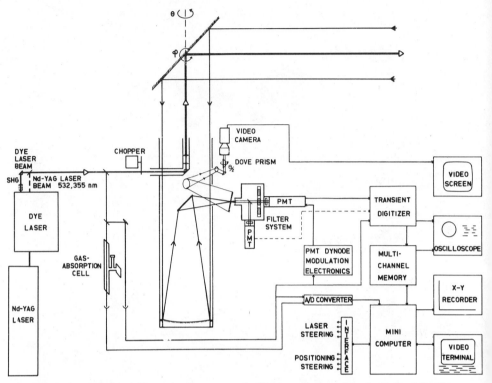

Fig. 11.21. Optical and electronic arrangement of a lidar system operated by the National Swedish Environmental Protection Board (48).

processes. Indeed, organochloride wastes in excess of 100,000 tons are produced annually by the industrialized nations. Because of the established carcinogenic, mutagenic, teratogenic, and acute toxic effects of certain components of this waste, its disposal presents considerable problems. Destruction by incineration is regarded as the best means of disposing of these substances with minimal impact on the biosphere. If this is done at sea, expensive scrubbers (for the removal of the hydrogen chloride produced by combustion) are not required, provided the incineration is conducted far enough from populated areas. However, the further out to sea, the greater the cost, so the question arises—how far and in what direction? What is needed to answer this question is an investigation of the plume concentration of HCl under different weather conditions and a mathematical model for describing its disappearance. Weitkamp et al. (176) and Weitkamp (50) have developed a ship-mounted DIAL system based on a pulsed TEA deuterium fluoride laser for this purpose.

5. HYDROGRAPHIC LIDAR APPLICATIONS

Remote sensing of the oceans, lakes, and rivers of our planet are possible with visible, infrared, and microwave radiation. Indeed, an enormous wealth of data has been collected from ships, aircraft, and satellites. However, the introduction of the laser into hydrographic work not only extended and complemented these measurements, it can truly be said to have added another dimension for it permits a degree of depth resolution and subsurface interrogation that was unattainable through other remote-sensing techniques. The primary reasons for this are that infrared and microwave radiation have negligible penetration in water, and visible observations were essentially passive in nature prior to the introduction of the hydrographic lidar systems.

Toward the end of the sixties concern over the environmental impact of oil spills on the oceans, lakes, and rivers, combined with inadequate means of detecting intentional spills at night, provided the incentive for the development of a new class of remote airborne sensor termed the *laser fluorosensor* (9, 10). Fluorescence had been used in many disciplines as a powerful analytical tool which enabled laboratory detection of substances in concentrations as low as a few parts per million. The invention of the high-power laser operating in the near-ultraviolet made it possible to extend this capability to the field of remote sensing. Indeed, we saw earlier that laser-induced fluorescence had an important role in studies of the upper atmosphere.

Although the laser fluorosensor was originally conceived for airborne oil-spill detection, Measures et al. (177) indicated that from the onset this new form of active remote sensor was conceived to be capable of undertaking a broad class of missions, some of which had never previously been considered to be within the realm of airborne surveillance. The prototype laser fluorosensor developed by Measures et al. (14) involved a 100-kW, 10-ns, 337-nm nitrogen laser. Its principle of operation is illustrated in Fig. 11.22. A more sophisticated version was developed by Fantasia and Ingrao (11) for the U.S. Coast Guard.

The Canada Centre for Remote Sensing Mk III laser fluorosensor is a representative state-of-the-art airborne hydrographic lidar (178). A rugged and reliable N_2 laser, operating at 337 nm irradiates the target of interest while the return fluorescence in the 380–700-nm range is observed by a 16-channel photodetection system. A schematic of this system is presented as Fig. 11.23. The first channel is centered at the OH-stretch water Raman line at 381 nm, while the next 14 channels have center wavelengths between 400 and 660 nm and are each 20 nm wide. The last channel extends from 670 to 720 nm and has been chosen to respond to the chlorophyll-*a* fluorescence band at 680 nm.

In preliminary flights the observed suppression of the water Raman return in the 380-nm channel at the location of an oil spill led to the development of a

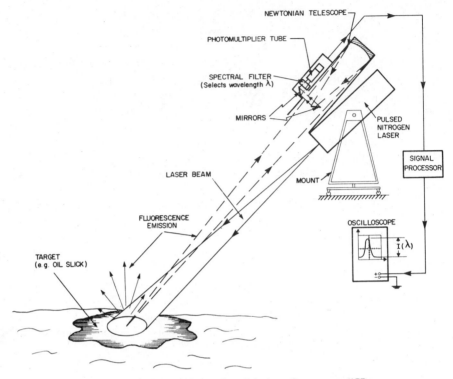

Fig. 11.22. Principle of operation of the laser fluorosensor (177).

method of determining the thickness of the oil slick at its boundaries. O'Neil et al. (178) showed that if the total attenuation coefficient for the oil is known, the thickness of the layer can be evaluated from the ratio of the laser fluorosensor signal obtained while over the oil to that obtained from an adjacent water mass where no oil slick is present (178). In essence it is the *suppression* of the return at the water Raman wavelength resulting from attenuation within the oil film that is used to determine the thickness of the oil. If the oil fluorescence at the water Raman wavelength is not negligible, then its contribution can be taken into account by using the laser-fluorosensor signal at another wavelength (where the water Raman contribution is negligible) (179). Recently, Burlamacchi, et al. (180) have shown that the 308-nm wavelength of a XeCl laser is optimal for those measurements.

Photosynthesis is the ultimate source of all food and the oxygen we breathe. It is the process by which all plants, including the aquatic varieties, convert solar energy into chemical energy. Chlorophyll-*a* plays a leading role in this

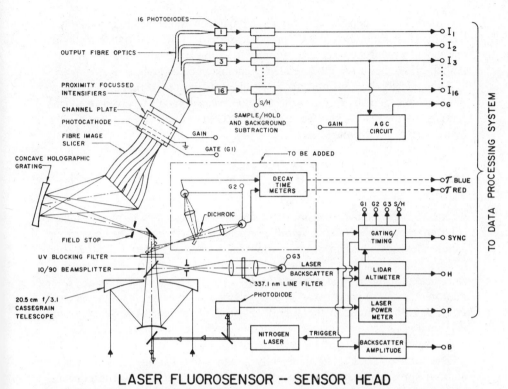

LASER FLUOROSENSOR -- SENSOR HEAD

Fig. 11.23. Block diagram of the Canada Centre for Remote Sensing airborne laser fluorosensor showing the major electrooptical components. In addition, the entire fluorosensor system requires a laser support pallet (consisting of a power supply, a nitrogen supply, and a vacuum pump) and a microprocessor-based data acquisition system that includes a real-time display and a computer-compatible tape transport. Reprinted from ref. (178) with permission.

process, and so it should come as no surprise to find that Sorenzen (181) and El-Sayed (182) discovered a highly significant correlation between chlorophyll-a concentration and the primary productivity in marine surface waters.

Hickman and Moore (16) first suggested, on the basis of laboratory studies, that chlorophyll measurements with a sensitivity of better than 10 mg m^{-3} might be undertaken using an airborne pulsed neon laser (output of about 20 kW) from an altitude of 100 m. An improved system, incorporating a 250-mJ, 300-ns dye laser, was shown by Kim (18) to be capable of detecting chlorophyll-a concentrations down to a fraction of a mg m^{-3} at an altitude of 30 m. Mumola et al. (17) showed that accurate measurements of the chlorophyll-a concentration can only be undertaken if allowance is made for the different color groups that may be present in any natural mixture of algae. Furthermore, the excitation spectrum was found to vary considerably from one phytoplankton color group to another. Consequently, Mumola et al. (17) proposed a four-wavelength excitation scheme that would exploit these features in order to evaluate the concentration of chlorophyll-*a* in an arbitrary mixture of phytoplankton color groups.

The optically thick form of the lidar equation [Eq. (11.15)] is appropriate to this situation, and so we can write the fluorescence return received in a band centered on the chlorophyll wavelength λ_f (685 nm in all cases) when excited by a laser of wavelength λ_L^i as

$$E_S(\lambda_L^i) = \frac{E_L K_0(\lambda_f) T(R_A) A_0 \phi_w \xi(R_A)}{4\pi R_A^2 [\kappa(\lambda_L^i) + \kappa(\lambda_f)] n^2} \sum_j N_j \sigma_j^F(\lambda_L^i) \qquad (11.19)$$

where N_j is the number density of chlorophyll-*a* molecules in phytoplankton color group j; $\sigma_j^F(\lambda_L^i)$ is the in vivo fluorescence cross section of the color group j molecules, for emission per unit wavelength interval centered about λ_f (the 685-nm chlorophyll-*a* band) with excitation of λ_L^i; ϕ_w is the two-way transmission factor for the air–water interface; $K_0(\lambda_f)$ is the filter function of the laser fluorosensor; $T(R_A)$ is the two-way atmospheric transmittance; A_0 is the receiver area; R_A is the altitude of the aircraft; E_L is the transmitted laser energy; and $\kappa(\lambda_L^i)$ and $\kappa(\lambda_f)$ are the effective attenuation coefficients at the laser and fluorescence wavelengths, λ_L^i and λ_f, respectively.

The form of Eq. (11.19) inherently assumes that the altitude of the laser fluorosensor is far greater than the effective penetration depth of the laser beam. This would be reasonable in any operational airborne mission. We have also assumed that τ_d (the detector integration time) $\gtrsim \tau_L$ (the laser pulse duration) in order to capitalize on the return signal (at the expense of spatial resolution), and that the fluorescence lifetime is much less than the laser pulse duration. This is a reasonable assumption since the lifetime of the excited chlorophyll-*a* molecule is known to be very short [close to 1 ns (12)].

In principle, the densities N_j can be evaluated by matrix inversion, given all of the other factors. A prototype four-wavelength (454.4, 539.0, 598.7, and 617.8 nm) laser fluorosensor was developed by Mumola et al. (17). This system operated with a minimum energy of 0.6 mJ per pulse and was able to detect < 1 mg m^{-3} of chlorophyll-*a* from an altitude of 100 m during daytime operating conditions.

Browell (19) undertook an analysis of the errors likely to be incurred in such measurements. His results indicate that remote quantification of chlorophyll-*a* requires optimum excitation wavelengths and careful measurement of the marine attenuation coefficients. Difficulties in determining these attenuation coefficients constituted a major limitation on the implementation of this form of remote sensing.

Fortunately, in most situations of interest this does not occur, and the solution to the problem of the unknown attenuation coefficients lies in the recognition that the laser-excited water Raman signal could serve as an indicator of changes in optical attenuation. This Raman-scattered signal is a property of the water alone and consequently its magnitude will be primarily determined by the attenuation coefficients at the laser and Raman wavelengths, $\kappa(\lambda_L)$ and $\kappa(\lambda_R)$, respectively. For a given laser wavelength λ_L, the fluorescence signal observed in the 685-nm chlorophyll band can be expressed in the form

$$P_F = \frac{E_A^* K_0(\lambda_f) N_c \sigma_C^F(\lambda_L)}{K_0(\lambda_f)[\kappa_L + \kappa_f]} \qquad (11.20)$$

where E_A^* contains most of the system parameters, see Eq. (11.19), we have also introduced the total chlorophyll-a molecule density N_c and the appropriate mean fluorescence cross section $\sigma_c^F(\lambda_L)$. In a similar manner the laser fluorosensor signal observed in the Raman band at wavelength λ_R (displaced 3418 cm^{-1} to the longer-wavelength side of λ_L) can be written

$$P_R = \frac{E_A^* K_0(\lambda_R) N_W \sigma_W^R(\lambda_L)}{K_0(\lambda_f)[\kappa_L + \kappa_R]} \qquad (11.21)$$

where N_W represents the density of water molecules and σ_W^R is the Raman-scattering cross section for the OH stretching vibrational mode of liquid water. Division of the fluorescence signal by the Raman signal eliminates uncertainties and fluctuations associated with E_L, R_A, $\xi(R_A)$, ϕ_W, n, and $T(R_A)$ and yields

$$\frac{P_F}{P_R} = \frac{K_0(\lambda_f)}{K_0(\lambda_R)} \left(\frac{\kappa_L + \kappa_R}{\kappa_L + \kappa_f}\right) \frac{N_c \sigma_c^F(\lambda_L)}{N_W \sigma_W^R(\lambda_L)} \qquad (11.22)$$

It should be noted that Exton et al. (183) have used a similar equation for Mie scattering and have thereby suggested that this ratio could be used to indicate the total suspended solids (TSS) burden. They also propose that the dissolved organic material (DOM) content could be evaluated in estuaries by this means.

Clearly, for Eq. (11.22) to be useful for evaluating the chlorophyll-a concentration, the attenuation ratio $(\kappa_L + \kappa_R)/(\kappa_L + \kappa_f)$ must remain constant, or at most change slowly for significant changes in κ_L, κ_R, and κ_f. This appears to be true for a water body in which dissolved and particulate materials change only in concentration but not in character, according to the observations of Bristow et al. (20). Their relatively simple laser fluorosensor employed a 200-kW, 250-

ns (FWHM) flashlamp-pumped dye laser, a 30-cm-diameter $f/1.8$ refracting telescope (using an acrylic Fresnel collector lens), and the two-channel photo-detection system. The repetition rate for the laser was 1 Hz, and its wavelength was chosen to be 470 nm. This wavelength was selected on the basis of a compromise. In order to minimize the background water fluorescence, ensure that the attenuation ratio was reasonably constant, and maximize both the penetration depth of the laser pulse and the output energy available, a long wavelength of excitation was desired, while to maximize both the Raman and fluorescence signals, a short wavelength was desired. Results of an airborne flight test conducted over the Las Vegas Bay are shown in Fig. 11.24 in the form of smoothed profiles of the chlorophyll-a fluorescence pulse energy P_F, the water Raman pulse energy P_R, and P_F/P_R as functions of flight time. In addition, the location of the sampling sites and an approximate horizontal distance scale are indicated. Also shown in Fig. 11.24 are the ground truth measurements of the

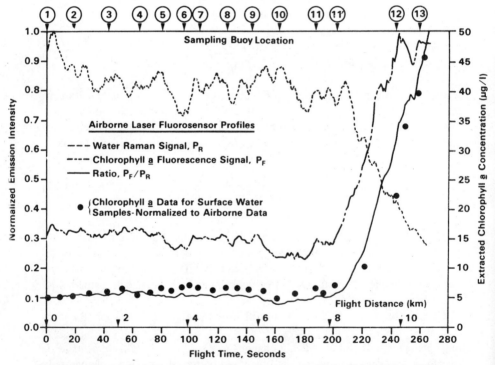

Fig. 11.24. Airborne laser fluorosensor profiles of P_F, P_R, and P_F/P_R for flight of 7 June 1979 over Las Vegas Bay. Profiles smoothed using a moving five-point average scheme. Also shown are the chlorophyll-a values for the 28 ground truth samples, which have been normalized to the laser fluorosensor profile for P_F/P_R (20).

extracted chlorophyll-*a* concentration for grab samples obtained at the 28 sampling sites at the time of the laser fluorosensor overflights. From this figure it is immediately apparent that P_F/P_R and the chlorophyll-*a* data are highly correlated over the extent of the flight path.

A NASA airborne oceanographic lidar has also been used for the remote sensing of chlorophyll-*a* in the oceans (21). Unfortunately, the low spectral resolution of their 40-channel receiver, combined with a laser wavelength of 532 nm, led to problems separating the chlorophyll fluorescence profile at 685 nm and the water Raman band, which in this instance occurred at 645 nm. Another problem caused by the poor choice of laser wavelength was competitive absorption and fluorescence by organic material (often referred to as *Gelbstoff*) carried to the sea by rivers (184). In spite of these difficulties, Hoge and Swift (21) were able to map the relative concentration of chlorophyll-*a* over an extensive area of the North Sea and at the intersection of Chesapeake Bay with the Potomac River in the United States. This work has recently been complemented by airborne laser fluorosensor mapping of phytoplankton photopigments in a Gulf Stream Warm Core Ring where sharp boundaries in the chlorophyll fluorescence have been observed corresponding to the thermal boundaries of the Warm Core Ring (53). They also compared excitation at 427 nm (from a XeCl laser-pumped dye laser) which directly stimulates chlorophyll-*a* fluorescence with that at 532 nm (from the second harmonic of a Nd:YAG laser) which involves fluorescence via indirect transfer from accessory pigments.

It is possible that the ultimate limit on the precision of chlorophyll-*a* measurements undertaken from an airborne laser fluorosensor may well be determined by the fact that the quantum efficiency of phytoplankton fluorescence depends on the ambient light level (photoinhibition). This could mean that accurate measurements can only be made at night and with very short pulse lasers.

The oceans of our planet are recognized as one of our greatest resources, not only for the food they yield, the oxygen they provide, and their possible future use as a source of energy but also for the influence they have on our climate. Basic among the ocean's properties, certainly from these points of view, is its thermal structure down to about 100 m. Currently, there is no operational means of remotely monitoring the subsurface temperature profile. At present satellites can map the surface temperature globally. A remote subsurface temperature measurement capability would greatly benefit weather forecasting and aid in our understanding of the climate and the air–sea interface. It would also facilitate the development of more reliable models for thermal waste-disposal studies and for the design of ocean thermal-energy conversion (OTEC) power plants.

Chang and Young (185) were among the first to consider using Raman backscattering for ocean temperature measurements, and the experimental work of Slusher and Derr (186) led to a Raman differential scattering cross section of $4.5 \ 10^{-29}$ cm^2 sr^{-1} for the O–H stretching band of liquid water. Their work also indicated that ice temperatures might be determined by measuring the ratio of Stokes to anti-Stokes Raman lines.

More recently, Leonard et al. (187) have undertaken the first remote ocean temperature measurements based on Raman backscattering from a ship-mounted lidar system. The physical basis of the Raman temperature measurement stems from the coexistence of both monomer and polymer forms of liquid water. The relative concentration of these two species depends primarily on the temperature, and since the O–H Raman stretching frequency is different for each, it is possible to infer the water temperature from the resulting Raman backscattering spectra (188). This model is shown in a heuristic manner by means of the Raman spectrum displayed in Fig. 11.25 (187). As is seen, each species has its individual Raman spectrum, proportional to its concentration, so that a simultaneous measurement of the Raman spectra at the two wavelengths λ_1 and λ_2 provides a measure of the concentration ratio and, through an equilibrium constant, the temperature.

The preliminary ocean tests of this technique were undertaken by Leonard et al. (187) using the Raman lidar system originally designed and used by Leonard and Caputo (189) for airport transmissometer work. A 100-kW, 10-ns, 337-nm pulsed nitrogen laser with a repetition rate of 500 Hz and a 2-mrad divergence was used as the transmitter. The output of the laser is passed through an interference filter, which passes the 337.1-nm laser line but blocks associated broadband spontaneous emission. A water solution of 2,7-dimethyl-3,6-diazacyclohepta-1,6-dieneperchlorate in a quartz cell acts as a very effective filter for blocking backscattered UV laser radiation, while permitting the Raman signal at about 375 nm to be transmitted quite efficiently into a double 0.25-m-focal-

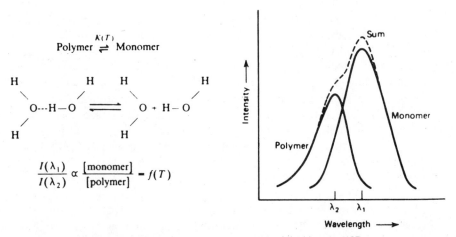

Fig. 11.25. Raman spectrum and structure of liquid water (187).

length scanning spectrometer having a 0.15-nm spectral resolution. This system was mounted aboard the afterdeck of the unique twin-hulled research vessel *Hayes*.

The laser Raman spectra of the ocean waters clearly confirmed the temperature-dependent effects (187). The ground truth measurements were 4.7 and 15.7°C, respectively. Both spectra were obtained for water in the first 10 m of depth. Analysis of the two band shapes with the standard two-color technique reveals that the polymer component decreased by 13% between the Labrador Current and the Gulf Stream. This suggests a temperature difference of 13°C, compared to the measured difference of 11°C. In attempting to undertake depth-resolved Raman temperature measurements, differential attenuation was observed to cause an apparent increase in the spectral temperature as the depth increased.

Recently, Leonard and Caputo (190) have suggested that this difficulty can be circumvented by using the ratio of water Raman-scattered signal-polarized parallel and perpendicular to the laser pulse.

REFERENCES

1. *U.S. Standard Atmosphere*, U.S. Government Printing Office, Washington, D.C. (1962).
2. P. Varanasi and F-K. Ko, *J. Quant. Spectosc. Radiat. Transfer*, **17,** 385–388 (1977).
3. G. E. Likens, *Conference on Emerging Environmental Problems, Rensselaerville, New York*, EPA Report 902/9-75-001 (1975).
4. A. Middleton, *New Scientist*, **4,** Nov., 279 (1976).
5. J. N. Pitts, Jr., *Env. Sci. Tech*, **11,** 456–461 (1977).
6. M. L. Wright, E. K. Proctor, L. S. Gasiorek, and E. M. Liston, *A Preliminary Study of Air-Pollution Measurement by Active Remote Sensing Techniques*, NASA CR-132724, 1975.
7. D. D. Davis, W. S. Heaps, D. Philen, M. Rodgers, T. McGee, A. Nelson, and A. J. Moriarty, *Rev. Sci. Instrum.*, **50,** 1505–1516 (1975).
8. R. M. Measures, *Laser Remote Sensing*, Wiley-Interscience, New York, 1984.
9. R. M. Measures and M. Bristow, *Can. Aeron. Space J.*, **17,** 421–422 (1971).
10. J. F. Fantasia, T. M. Hard, and H. C. Ingrao, *An Investigation of Oil Fluorescence as a Technique for Remote Sensing of Oil Spills*, Report No. DOT-TSC-USCG-71-7, Transportation Systems Center, Dept. of Transportation, Cambridge, MA (1971).
11. J. F. Fantasia, and H. C. Ingrao, *Proc. of the 9th Intern. Symp. on Remote Sensing of the Environment, 15–19 April 1974*, Paper 10700-1-X, 1711–1745 (1974).
12. W. R. Houston, D. G. Stephenson, and R. M. Measures, "LIFES: Laser Induced Fluorescence and Environmental Sensing," *NASA Conference on the Use of Lasers for Hydrographic Studies*, NASA SP-375, 153–169, 1973.

13. R. M. Measures, H. R. Houston, and D. G. Stephenson, *Optical Eng.*, **13**, 494–450 (1974).

14. R. M. Measures, W. R. Houston, and M. Bristow, *Can. Aeron. Space J.*, **19**, 501–506 (1973).

15. F. E. Hoge, R. N. Swift, and E. B. Frederick, *Appl. Optics*, **19**, 871–883 (1980).

16. G. D. Hickman and R. B. Moore, *Proc. 13th Conf. Great Lakes Res.*, Int. Assoc. Great Lakes Res., 1–4 (1970).

17. P. B. Mumola, O. Jarrett, Jr., and C. A. Brown, Jr., *NASA Conference on the Use of Lasers for Hydrographic Studies*, NASA SP-375, 137–145 (1973).

18. H. H. Kim, *Appl. Optics*, **12**, 1454–1459 (1973).

19. E. V. Browell, *Analysis of Laser Fluorosensor Systems for Remote Algae Detection and Quantification*, NASA TN D-8447 (1977).

20. M. Bristow, D. Nielsen, D. Bundy, and R. Furtek, *Appl. Optics*, **20**, 2889–2906 (1981).

21. F. E. Hoge and R. N. Swift, *Appl. Optics*, **20**, 1191–1202 (1981).

22. H. Inaba and T. Kobayasi, *Opto-electronics*, **4**, 101–123 (1972).

23. W. R. Fenner, H. A. Hyatt, J. M. Kellan, and S. P. S. Porto, *J. Opt. Soc. Am.*, **63**, 73–77 (1973).

24. D. A. Stephenson, *J. Quant. Spectrosc. Radiat. Transfer*, **14**, 1291–1301 (1974).

25. D. G. Fouche and R. K. Chang, *Appl. Phys. Lett.*, **18**, 579–580 (1971).

26. D. G. Fouche, A. Herzenberg, and R. K. Chang, *J. Appl. Phys.*, **43**, 3846–3851 (1972).

27. R. A. Smith, F. E. Jones, and R. P. Chasmar, *The Detection and Measurements of Infrared Radiation*, Oxford London. Univ. Press, 1968.

28. J. A. Hodgeson, W. A. McClenny, and P. L. Hanst, *Science*, **182**, 248–258 (1973).

29. K. Asai and T. Igarashi, *Opt. Quantum Elect.*, **7**, 211–214 (1975).

30. T. Henningsen, M. Garbuny, and R. L. Byer, *Appl. Phys. Lett.*, **24**, 242–244 (1974).

31. E. D. Hinkley and P. L. Kelly, *Science*, **17**, 635–639 (1971).

32. R. L. Byer and M. Garbuny, *Appl. Optics*, **12**, 1496–1505 (1973).

33. W. Schnell and G. Fischer, *Appl. Optics*, **14**, 2058–2059 (1975).

34. J. Shewchun, B. K. Garside, E. A. Ballik, C. C. Y. Kwan, M. M. Elsherbiny, G. Hogenkamp, and A. Kazandjian, *Appl. Optics*, **15**, 340–346 (1976).

35. E. R. Murray and J. E. van der Laan, *Appl. Optics*, **17**, 814–817 (1978).

36. J. Altmann and P. Pokrowsky, *Appl. Optics*, **19**, 3449–3452 (1980).

37. R. M. Schotland, *Proc. 4th Symposium on Remote Sensing of the Environment 12–14 April 1966*, Univ. of Michigan, Ann Arbor, 273–283, 1966.

38. R. M. Measures and G. Pilon, *Opto-electronics*, **4**, 141–153 (1972).

39. R. L. Byer, *Opt. Quantum Elect.*, **7**, 147–177 (1975).

40. W. B. Grant, R. D. Hake, Jr., E. M. Liston, R. C. Robbins, and E. K. Proctor, Jr., *Appl. Phys. Lett.*, **24**, 550–552 (1974).

41. C. O'Shea and L. G. Dodge, *Appl. Optics*, **13**, 1481–1486 (1974).

42. S. A. Ahmed, *Appl. Optics*, **12**, 901–903 (1973).

43. K. W. Rothe, U. Brinkman, and H. Walter, *Appl. Phys.*, **3**, 115–119 (1974).

44. K. W. Rothe, U. Brinkman, and H. Walter, *Appl. Phys.*, **4**, 181–182 (1974).

45. J. M. Hoell, Jr., W. R. Wade, and R. T. Thompson, Jr., "Remote Sensing of Atmospheric SO_2 Using the Differential Absorption Lidar Technique," Int. Conf. on Environ. Sens. & Assessment, Las Vegas, 14 Sept., 1975.

46. K. Fredriksson, B. Galle, K. Nystrom, and S. Svanberg, *Appl. Optics*, **18**, 2998–3003 (1979).

47. R. A. Baumgartner, J. G. Depp, W. E. Evans, W. B. Grant, J. G. Hawley, R. G. March, E. R. Murray, and E. K. Proctor, *Characterization of the EPRI Differential Absorption Lidar* (DIAL) *System*, EA-1267, Project Report 862-14, 1979.

48. K. Fredriksson, B. Galle, K. Nystrom, and S. Svanberg, *Appl. Optics*, **20**, 4181–4189 (1981).

49. E. R. Murray, R. D. Hake, Jr., J. E. van der Laan, and J. G. Hawley, *Appl. Phys. Lett.*, **28**, 542–543 (1976).

50. C. Weitkamp, "The Distribution of Hydrogen Chloride in the Plume of Incineration Ships: Development of New Measurements Systems," in *Wastes in the Ocean*, Vol. 3, Wiley, New York, 1981.

51. D. J. Brassington, *Appl. Optics*, **21**, 4411–4416 (1982).

52. E. E. Uthe and R. J. Allen, *Opti. Quantum Elect.*, **7**, 121–129, (1975).

53. F. E. Hoge and R. N. Swift, *Appl. Opt.*, **22**, 2272–2281 (1983).

54. S. K. Poultney, *Adv. Elect. Electron Phys.*, **31**, 39–117 (1972).

55. J. Altmann, S. Kohler, and W. Lahmann, *J. Physics. 6: Sci. Instrum.*, **13**, 1275–1279 (1980).

56. R. J. Keyes and R. H. Kingston, *Phys. Today*, Mar., 48–54 (1972).

57. C. P. Wang, *Acta Astronaut.*, **1**, 105–123 (1974).

58. R. B. Emmons, S. R. Hawkins, and C. F. Cuff, *Opt. Eng.*, **14**, 21–30 (1975).

59. F. M. Lussier, *Laser Focus*, Oct., 66–71 (1976).

60. H. Walther and J. L. Hall, *Appl. Phys. Lett.*, **17**, 239 (1970).

61. A. M. Title and W. J. Rosenberg, *Optical Eng.*, **20**, 815–823 (1981).

62. W. J. Gunning, *Opt. Eng.*, **20**, 837–845 (1981).

63. M. Born and E. Wolf, *Principles of Optics*, 2nd ed., Pergamon Press, New York, 1964.

64. P. J. Jacquinot, *J. Opt. Soc. Am.*, **44**, 761 (1954).

65. A. Girard, and O. Jacquinot, Chapter 3 in *Advanced Optical Techniques* (A. van Heel, Ed.), North-Holland, Amsterdam (1967).

66. J. V. Ramsay, *Appl. Optics*, **1**, 411–413 (1962).

67. J. G. Hawley, *Laser Focus*, Mar., 60–62 (1981).

68. R. M. Measures *Appl. Optics*, **16**, 1092–1103 (1977).

69. W. L. Wolfe, Ed., *Handbook of Military Infrared Technology*, ONR Cat. No. 65-62266, U.S. Government Printing Office, Washington, D.C., 1966.

70. M. S. Shumate, S. Lundqvist, V. Persson, and S. T. Eng, *Appl. Optics*, **21**, 2386–2389 (1982).

71. W. B. Grant, *Appl. Optics*, **21**, 2390–2394 (1982).

72. H. Kildal and R. L. Byer *Proc. IEEE*, **59**, 1644–1663 (1971).

73. J. D. Klett, *Appl. Optics*, **20**, 211–220 (1981).

74. V. E. Zuev and I. E. Naats, *Inverse Problems of Lidar Sensing of the Atmosphere*, Springer-Verlag, Berlin, 1983.

75. W. K. Pratt, *Laser Communication Systems*, Wiley, New York, 1969.

76. G. N. Plass, G. W. Kattawar, and J. A. Guinn, Jr., *Appl. Optics,* **15,** 3161–3165 (1976).
77. A. E. S. Green, *The Middle Ultraviolet,* Wiley, New York, 1966.
78. H. S. Gebbie, W. R. Harding, C. Hilsum, A. W. Pryce, and V. Roberts, *Proc. R. Soc. A,* **206,** 87–107 (1951).
79. R. D. Hudson, Jr. and J. W. Hudson, *Proc. IEEE.,* **63,** 104 (1975).
80. V. E. Zuev, "Laser-Light Transmission through the Atmosphere," in *Laser Monitoring of the Atmosphere,* E. D. Hinckley, Ed., Springer-Verlag, 1976.
81. R. J. Pressley, *Handbook of Lasers,* Chemical Rubber Co., Cleveland, 1971.
82. R. W. McClatchey, R. W. Fenn, J. E. A. Selby, F. E. Volz, and J. S. Garing, *Optical Properties of the Atmosphere,* Air Force Cambridge Research Laboratories 71-0279, 1971; also 72-0497, 1972.
83. R. A. McClatchey, W. S. Benedict, S. A. Clough, D. E. Burch, R. F. Calfee, K. Fox, L. S. Rothman, and J. S. Garing, *AFCRL Atmospheric Absorption Line Parameters Compilation,* AFCRL-TR-73-0096, 1973.
84. A. J. LaRocca, *Proc. IEEE,* **63,** 75–94 (1975).
85. J. E. A. Selby and R. A. McClatchey, *Atmospheric Transmittance from 0.25 Microns to 28.5 Microns: Computer Code III,* AFCRL-TR-0255, 1975.
86. R. E. Roberts, J. E. A. Selby, and L. M. Biberman, *Appl. Optics,* **15,** 2085–2090 (1976).
87 P. L. Kelly, R. A. McClatchey, R. K. Long, and A. Snelson, *Opt. Quantum Elect.,* **8,** 117–144 (1976).
88. E. D. Hinkley, R. T. Ku, and P. L. Kelly, "Techniques for Detection of Molecular Pollutants by Absorption of Laser Radiation," in *Laser Monitoring of the Atmosphere,* Springer-Verlag, New York, 1976.
89. B. Nilsson, *Appl. Optics,* **18,** 3457–3473 (1979).
90. P. W. Kruse, L. D. McGlauchlin, and R. B. McQuistan, *Elements of Infrared Technology,* Wiley, New York, 1963.
91. D. P. Woodman, *Appl. Optics,* **13,** 2193–2195 (1974).
92. S. Twomey and H. B. Howell, *Appl. Optics,* **4,** 501–505 (1965).
93. R. W. Fenn, *Appl. Optics,* **5,** 293–295 (1966).
94. M. R. Clay and A. P. Lenham, *Appl. Optics,* **20,** 3831–3832 (1981).
95. R. L. Schwiesow and V. E. Derr, *J. Geophys. Res.,* **75,** 1629–1632 (1970).
96. D. A. Leonard, *Nature,* **216,** 142–143 (1967).
97. J. A. Cooney, *Appl. Phys. Lett.,* **12,** 40–42 (1968).
98. M. J. Garvey and G. S. Kent, *Nature,* **248,** 124–125 (1974).
99. *U.S. Standard Atmosphere,* U.S. Government Printing Office, Washington, D.C. (1962).
100. S. H. Melfi, J. D. Lawrence, Jr., and M. P. McCormick, *Appl. Phys. Lett.,* **15,** 295–297 (1969).
101. J. A. Cooney, *J. Appl. Meteor.,* **9,** 182–184 (1970).
102. J. A. Cooney, *J. Appl. Meteor.,* **10,** 301–308 (1971).
103. J. C. Pourny, D. Renaut, and A. Orszag, *Appl. Optics,* **18,** 1141–1148 (1979).
104. R. G. Strauch, V. E. Derr, and R. E. Cupp, *Remote Sensing of Envir.,* **2,** 101–108 (1972).
105. D. Renaut, C. Pourney, and R. Capitini, *Optics Letters,* **5,** 233–235 (1980).

106. K. Petri, A. Salik, and J. Cooney, *Appl. Optics*, **21**, 1212–1218 (1982).

107. J. A. Cooney, *Opt. Eng.*, **22**, 292–301 (1983).

108. E. V. Browell, T. D. Wilkerson, and T. J. McIlrath, *Appl. Optics*, **18**, 3474–3483 (1979).

109. V. V. Zeuv, V. E. Zuev, Yu. S. Makushkin, V. N. Marichev, and A. A. Mitsel, *Appl. Optics*, **22**, 3742–3746 (1983).

110. E. V. Browell, "Remote Sensing of Tropospheric Gases and Aerosols with an Airborne Dial System," *Proc. of Workshop on Optical and Laser Remote Sensing, Monterey, Feb. 9–11, 1982* (1982).

111. G. Elterman, *J. Geophys.*, **58**, 519 (1953).

112. G. Elterman, *J. Geophys.*, **59**, 351 (1954).

113. M. C. W. Sandford, *J. Atm. Terres. Phys.*, **29**, 1657–1662 (1967).

114. J. A. Cooney, *J. Appl. Meteor.*, **11**, 108–112 (1972).

115. J. A. Salzman, "Low Temperature Measurements by Rotational Raman Scattering," in *Laser Raman Gas Diagnostics*, M. Lapp and C. M. Penney, Eds. Plenum Press, New York, pp. 179–188, 1974.

116. J. B. Mason, *Appl. Optics*, **14**, 76–78 (1975).

117. B. K. Schwemmer, B. K. and T. O. Wilkerson, *Appl. Optics*, **18**, 3539–3541 (1979).

118. M. Endermann and R. L. Byer, *Optics Letters*, **5**, 452–454 (1980).

119. M. Endermann and R. L. Byer, *Appl. Optics*, **20**, 3211–3217 (1981).

120. N. Menyuk and D. K. Killinger, *Optics Letters*, **6**, 301–303 (1981).

121. J. E. Kalshoven, Jr., C. L. Korb, G. K. Schwemmer, and M. Dombrowski, *Appl. Optics*, **20**, 1967–1971 (1981).

122. T. J. McGee and T. J. McIlrath, *Appl. Optics*, **18**, 1710–1714 (1979).

123. K. Schofield, *J. Quant. Spectosc. Radiat. Transfer*, **17**, 13–51 (1977).

124. T. J. McIlrath, *Opt. Eng.*, **19**, 494–502 (1980).

125. E. L. Baardsen, and R. W. Terhune, *Appl. Phys. Lett.*, **21**, 209–211 (1972).

126. S. McDermid, J. B. Laudenslager, and T. T. Pacala, *Appl. Optics*, **22**, 2586–2591 (1983).

127. W. S. Heaps, T. J. McGee, R. D. Hudson, and L. O. Caudill, *Appl. Optics*, **21**, 2265–2274 (1982).

128. R. T. Menzies, C. R. Webster, and E. D. Hinkley, (1983), *Appl. Opt.*, **22**, 2655–2664 (1983).

129. O. Uchino, M. Maeda, J. Kohno, T. Shibata, C. Nagasawa, and M. Hirono, *Appl. Phys. Lett.*, **33**, 807–809 (1978).

130. O. Uchino, M. Maeda, T. Shibata, M. Hirono, and M. Fujimara, *Appl. Optics*, **19**, 4175–4181 (1980).

131. O. Uchino M. Tokunaga, M. Maeda, and Y. Miyazoe, *Optics Letters*, **8**, 347–349 (1983).

132. E. V. Browell, *Optical Eng.*, **21**, 128–132 (1982).

133. E. V. Browell, A. F. Carter, S. T. Shipley, R. J. Allen, C. F. Butler, M. N. Mayo, J. H. Siviter, Jr., and W. M. Hall, *Appl. Optics*, **22**, 522–534 (1983).

134. M. R. Bowman, A. J. Gibson, and M. C. W. Sandford, *Nature*, **221**, 456 (1969).

135. A. J. Gibson and M. C. W. Sandford *J. Atmos. Terr. Phys.*, **33**, 1675–1684 (1971).

136. R. D. Hake, Jr., D. E. Arnold, D. W. Jackson, W. E. Evans, B. P. Ficklin, and R. A. Long, *Geophys. Res.*, **77**, 6389–6848 (1972).

137. T. Aruga, H. Kamiyama, M. Jyumonji, T. Kobayashi, and H. Inaba, *Report Ionosphere Space Research Japan*, **28**, 65–68 (1974).

138. A. J. Gibson and M. C. W. Sandford, *Nature*, **239**, 509–511 (1972).

139. J. E. Blamont, M. L. Chanin, and G. Megie, *Appl. Optics Ann. Geophys.*, **28**, 833–838 (1972).

140. G. Megie and J. E. Blamont, *Planet. Space Sci.*, **25**, 1093–1109 (1977).

141. G. Megie, F. Bos, J. E. Blamont, and M. L. Chanin, *Planet. Space Sci.*, **26**, 27–35 (1978).

142. C. Granier and G. Megie, *Planet. Space Sci.*, **30**, 169–177 (1982).

143. A. C. Stern, *Air Pollution*, Academic Press, New York, 1968.

144. S. J. Williamson, *Fundamentals of Air Pollution*, Addison-Wesley, Toronto, 1973.

145. W. B. Grant and R. J. Menzies, *J. Air Pollution Contr. Ass.*, **33**, 187–194 (1983).

146. P. L. Hanst and J. A. Morreal, *J. Air Poll. Control Assoc.*, **18**, 754–759 (1968).

147. R. T. Ku, E. D. Hinkley, and J. D. Sample, *Appl. Optics*, **14**, 854–861 (1975).

148. E. D. Hinkley, *Optical Quantum Electronics*, **8**, 155–167 (1976).

149. J. Reid, J. Schewchun, B. K. Garside, and E. A. Ballik, *Appl. Optics*, **17**, 300–307 (1978).

150. R. T. Menzies, *Opto-electronics*, **4**, 178–186 (1972).

151. R. T. Menzies, "Laser Heterodyne Detection Techniques," in *Laser Monitoring of the Atmosphere*, E. O. Hinkley, Ed., Springer-Verlag, Berlin, 1976.

152. E. D. Hinkley, *Opto-electronics*, **4**, 69–86 (1972).

153. K. W. Nill, *Opt. Eng.*, **13**, 516–522 (1974).

154. R. T. Menzies and M. S. Shumate, *Appl. Optics*, **15**, 2080–2084 (1976).

155. R. T. Menzies and M. S. Shumate, *J. Geophy. Res.*, **83**, 4039 (1978).

156. J. L. Guagliardo and D. H. Bundy, "Differential Monitoring of Ozone in the Troposphere Using Earth Reflected Differential Absorption," International Conference (Oct.) 1974.

157. J. L. Guagliardo and D. H. Bundy, "Earth Reflected Differential Absorption Using TEA Lasers: A Remote Sensing Method for Ozone," 7th International Laser Radar Conference, Palo Alto, Calif., 1975.

158. D. K. Dillinger, N. Menyuk, and W. E. DeFeo, *Appl. Phys. Lett.*, **36**, 402–405 (1980).

159. N. Menyuk, D. K. Killinger, and W. E. DeFeo, *Appl. Optics*, **19**, 3282–3286 (1980).

160. N. Menyuk, D. K. Killinger, and W. E. DeFeo, *Appl. Optics*, **21**, 2275–2286 (1982).

161. A. Yariv, *Introduction to Optical Electronics*, 2nd ed., Holt, Rinehart and Winston, New York, 1976.

162. R. A. Baumgartner and R. L. Byer, *IEEE J. Quant. Electr.*, **QE-15**, 432 (1975).

163. R. A. Baumgartner and R. L. Byer, *Optics Letters*, **2**, 163 (1978).

164. R. A. Baumgartner and R. L. Byer *Appl. Optics*, **17**, 3555 (1978).

165. E. R. Murray and R. L. Byer, *Remote Measurements of Air Pollutants*, SRI International report. Jan. 1980.

166. M. Aldén, H. Edner, and S. Svanberg, *Optics Letters*, **7**, 221–223 (1982).

167. R. T. Thompson, Jr., J. M. Hoell, Jr., and W. R. Wade, *J. Appl. Phys.*, **46**, 3040–3043 (1975).

168. R. A. Baumgartner, L. D. Fletcher, and J. G. Hawley, *ARCO J.*, **29**, 1162–1165 (1979).

169. H. Inaba and T. Kobayasi, *Nature*, **224**, 170–172 (1969).

170. T. Hirschfeld, E. R. Schildkraut, H. Tannenbaum, and D. Tannenbaum, *Appl. Phys. Lett.*, **22**, 38–40 (1973).

171. H. P. DeLong, *Optical Engineering*, **13**, 5–9 (1974).

172. S. H. Melfi, M. L. Brumfield, and R. M. Storey, Jr., *Appl. Phys. Lett.*, **22**, 402–403 (1975).

173. S. K. Poultney, M. L. Brumfield, and J. H. Siviter, Jr. *Appl. Optics*, **16**, 3180–3182 (1977).

174. K. W. Rothe, *Radio Electr. Eng.*, **50**, 567–572 (1980).

175. J. Harms, W. Lahmann, and C. Weirkamp, *Appl. Optics*, **17**, 1131–1135 (1978).

176. C. Weitkamp, H. J. Heinrich, W. Herrmann, W. Michaelis, V. Lenhard, and R. N. Schindler, "Measurement of Hydrogen Chloride in the Plume of Incineration Ships," 5th International Clean Air Congress, 20–26 Oct., Buenos Aires, Argentina; also GKSS 80/E/55, 1980.

177. R. M. Measures, J. Garlick, W. R. Houston, and D. G. Stephenson, *Can. J. Remote Sensing*, **1**, 95–102 (1975).

178. R. A. O'Neil, L. Buje-Bijunas, and D. M. Rayner, *Appl. Optics*, **19**, 863–870 (1980).

179. F. E. Hoge and R. N. Swift, *Appl. Optics*, **12**, 3269–3281 (1980).

180. P. Burlamacchi, G. Cecchi, P. Mazzinghi, and L. Pantani, *Appl. Opt.*, **22**, 48–53 (1983).

181. C. J. Sorenzen, *Limnology Oceanogra.*, **15**, 479–480 (1970).

182. S. Z. El-Sayed, "Phytoplankton Production of the South Pacific and the Pacific Sector of the Antarctic," in *Scientific Exploration of the South Pacific*, National Academy of Sciences, Washington, D.C., 1970.

183. R. J. Exton, V. M. Houghton, W. Esaias, R. C. Harris, F. H. Farmer, and H. H. White, *Appl. Optics*, **22**, 54–64 (1983).

184. C. S. Yentsch, "The Fluorescence of Chlorophyll and Yellow Substance in Natural Waters: A Note on the Problems on Measurements and Their Importance to Remote Sensing," in *NASA Conference on the Use of Lasers for Hydrographic Studies, Wallops Island*, NASA SP-375, 147–151, 1973.

185. C. H. Chang, and L. A. Young, "Remote Measurement of Ocean Temperature from Depolarization in Raman Scattering," in *NASA Conference on the Use of Lasers for Hydrographic Studies*, NASA SP-375, 105–112, 1973.

186. R. B. Slusher and V. E. Derr, *Appl. Optics*, **14**, 2116–2120 (1975).

187. D. A. Leonard, B. Caputo, and F. E. Hoge, *Appl. Optics*, **18**, 1732–1745 (1979).

188. G. E. Walrafen, *J. Chem. Phys.*, **47**, 114–126 (1967).

189. D. A. Leonard and B. Caputo, *Opt. Eng.*, **13**, 10–14 (1974).

190. D. A. Leonard and B. Caputo, *Opt. Eng.*, **22**, 288–291 (1983).

CHAPTER

12

INTRACAVITY-ENHANCED SPECTROSCOPY

EDWARD H. PIEPMEIER

Department of Chemistry
Oregon State University
Corvallis, Oregon

1. INTRODUCTION

A weak absorber placed inside a laser cavity (Fig. 12.1) can produce large changes in the laser output. Analytical intracavity-enhanced spectroscopy takes advantage of this phenomenon to determine minute amounts of absorbing species and to observe weakly absorbing species, often in the presence of intense background emission. Absorption coefficients of 10^{-8}–10^{-9} cm^{-1} have been detected in some circumstances (1–5). As pointed out in a comparison (6), this makes intracavity spectroscopy one of the most, if not the most, sensitive technique for the detection of optical absorption.

The ability to detect small absorbances occurs in part because of an enhance-

431

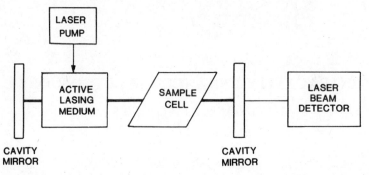

Fig. 12.1. Basic components of an intracavity absorption spectrophotometer showing the sample cell inside the cavity of the laser. A partially transmitting mirror allows a small fraction of the beam to reach the detector.

ment in the absorbance *signal* that occurs. The enhancement factor is the ratio of the relative change in laser intensity *at the absorption wavelength* produced by the absorber in the laser cavity to the relative change in the intensity of a light beam outside of the cavity caused by the absorber (i.e., in a conventional absorption determination). Enhancement factors range from 10 to 10^6, depending on the type and quality of the instrument system and whether the laser bandwidth is narrow or broad relative to the absorption peak of the analyte.

The enhancement is in part caused by the multiple-pass character of the absorption cell. Each time the laser beam passes through the cell, an additional loss occurs. Another effect is caused by the fact that a laser exhibits a threshold below which lasing ceases. If a laser is operated near threshold, small losses cause large changes in laser intensity. For example, even a small loss caused by absorption could cause the laser intensity to go from a high value to zero. Enhancements caused by these two effects total only a few hundred or less (7) for a laser that has a spectral bandwidth that is narrow relative to the absorption peak. A third effect takes place when only one mode (one wavelength) of a multiple-mode laser is affected in these ways by the absorber. As we will see in the next section, the dynamic interplay of the absorbed mode with the other modes causes an additional enhancement of several decades of magnitude. This case requires the spectral bandwidth of the absorber to be narrow compared to the spectral bandwidth of the laser. It is in this situation that the greatest enhancement factors are obtained.

The next section considers the theory of the intracavity enhancement and provides the background for the discussion of instrument systems that follows. Analytical applications are then considered and some hopes for the future presented.

2. THEORETICAL CONSIDERATIONS

Intracavity spectroscopy can be subdivided into cases where the spectral bandwidth of the laser is (a) narrow and (b) broad relative to the absorption spectrum of the analyte in the laser cavity. We shall prepare for a discussion of the second case by considering the simpler situation where the laser bandwidth is narrow relative to the absorption band.

2.1. Narrow-band Laser

For the case where the laser bandwidth is narrow relative to the absorption band, the effect of absorption loss on the output power can be determined by considering the case of a single-mode laser. For a single-mode laser, the output power P is proportional to the ratio of the cavity (unsaturated) gain G (photons gained per single pass of the beam through the cavity) to the intracavity losses L (photons lost per single pass) (8),

$$P = K\left(\frac{G}{L} - 1\right) \tag{12.1}$$

where K is a proportionality factor that depends on the laser. Following the lead of Harris (7), we take the derivative of P with respect to L and divide the corresponding sides of the resulting equation by Eq. (12.1) to obtain

$$\frac{dP}{P} = \left[\frac{G/L}{G - L}\right] dL \tag{12.2}$$

This shows that the relative loss in laser power is proportional to the incremental loss dL, at least to a first approximation where G, L, and their difference are assumed constant. Equation (12.2) has an analogy in the derivation of Beer's law (9) in which case the term in brackets is 1, and dL is the product of absorptivity, concentration, and incremental pathlength, dx. The term in brackets is thus an enhancement factor by which the conventional absorbance of the absorber in the intracavity cell appears to be increased when the *intracavity absorbance* is calculated as $\ln(P_0/P)$, the logarithm of the ratio of the laser powers with the blank P_0 and with the analyte P. This indicates that intracavity absorbance is proportional to concentration to a first approximation where the term in brackets (the enhancement factor) is assumed to be constant.

The significance of the enhancement factor $(G/L)/(G - L)$ may be clarified by considering two limiting cases. For high pumping power, $G \gg L$, and the enhancement factor approaches $1/L$, and $dP/P = dL/L$; that is, the relative change in laser power is only equal to the relative change in photons lost from the cavity. At very low power the gain and loss are close to each other, $G - L$ becomes very small, and the enhancement factor becomes very large. In this case the enhancement is a result of the laser operating close to threshold. A small additional intracavity loss could cause lasing to cease altogether, which is a very large effect. The ability to take full advantage of this effect is limited by the difficulty of obtaining reproducible operation of a laser very near threshold where even very minor cavity fluctuations (often caused by imperfections) produce large changes in laser power. Consequently, enhancements of only a few hundred have been obtained using this method, where all modes or wavelengths of the laser are subjected to absorption by the absorber.

2.2. Broad-band Multimode Laser

The highest enhancements are obtained when the absorption peak is narrow compared to the laser spectral profile of a multimode laser, that is, only the photons in one or a relatively few modes of a multimode laser are absorbed by the absorber. Although multimode operation of lasers is well known, general considerations (8) predict that a laser should not be able to operate in a *steady-state* condition at more than one wavelength (mode) if the active gain medium is homogeneously broadened (i.e., so that *any* photon can stimulate emission from *any* gain site). In such a laser all modes compete with each other for the limited number of excited states in the gain medium. Upon each pass through the cavity, the mode with the highest irradiance (because it also has the highest gain or lowest losses) continues to take photons from a larger and larger fraction of excited states, while the weaker modes then find fewer and fewer excited states available. The weaker modes eventually die out, and at steady state only the mode with the highest gain lases. Why, then, does a continuous-wave (cw) dye laser lase in hundreds or thousands of modes instead of just one?

There have been two theoretical responses to the question. One approach, taken by Hansch et al. (10), assumes that cw dye lasers operate at steady state, but that they are not homogeneously broadened. They adopt an inhomogeneity model (11) whereby dye molecules located at the nodes of the standing-wave pattern of a particular mode are not stimulated to emit into that mode. These molecules are available to act as the gain medium for other modes. Using standard laser rate equations and assuming steady state, they found approximate analytical solutions to the equations. In the limit of a large number of modes all having the same gain and loss, except for one mode subject to absorption by the analyte, they find the enhancement factor to be

$$\text{Enhancement factor} = \frac{2M(G/L)}{G - L} \tag{12.3}$$

where M is the number of modes. This is the same enhancement factor as in Eq. (12.2) except for the factor $2M$. An intuitive reason for the factor $2M$ is that with more modes there is greater competition for the excited dye molecules. A small loss (caused by the absorber) in one mode will make more molecules available for gain in the other modes, which will in turn increase their irradiances and make fewer excited molecules available for the weakened mode, making it even weaker, and thereby increasing the enhancement factor.

Figure 12.2 (7) compares experimentally determined enhancement factors for different pump powers with a line having the shape of Eq. (12.3). Although the shapes agree, it was necessary to substantially scale the magnitude of the theoretical predictions.

To determine whether the approximations of Hansch et al. (10) were valid, Harris (12) solved the rate equations numerically and obtained time-dependent solutions. The inclusion of spatial inhomogeneity allows less than 10 modes to

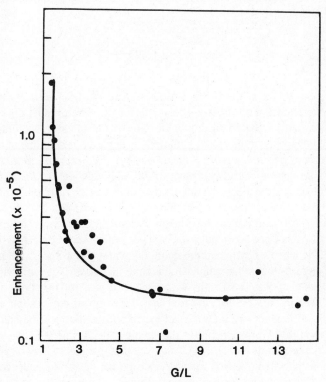

Fig. 12.2. Enhancement factor versus gain-to-loss ratio G/L for water vapor overtone spectra. The laser threshold occurs when $G/L = 1$. The dots show experimental results, while the solid line gives the prediction of the theory of Hansch et al. (10); reproduced from ref. (7) with permission.

survive by the time steady state is reached. This is because such a small number of modes is all that is necessary to produce enough overlap of the standing-wave patterns to uniformly saturate the entire region of the active medium and prevent other modes from lasing.

Not only do these results show that these equations cannot explain broad-band steady-state operation for more than 10 modes but, when they are solved with an intracavity absorber for one mode, that mode dies out in several hundred microseconds (12). Therefore, it seems that something other than a steady-state condition must exist in a real cw dye laser.

The theory of Belikova et al. (13, 14) explains broad-band laser operation by assuming that steady state is never reached. Instead, individual modes are continually and randomly quenched and regenerated. Regeneration is initiated by spontaneous emission. Approximate analytical solutions to the multimode rate equations that explicitly omit spatial inhomogeneity predict an enhancement factor equal to the number of times that the laser beam passes through the absorption cell during the lifetime of the laser mode being absorbed by the analyte (13):

$$\text{Enhancement factor} = \frac{ct}{X} \qquad (12.4)$$

where c is the speed of light, t is the duration of the laser mode (which they call the generation time), and X is the cavity length. The product ct is the total distance traveled by the beam during the life of the mode. According to Eq. (12.4), the enhancement increases with time throughout the lifetime of the mode being absorbed, *and time-resolved measurements are necessary to take full advantage of the greater enhancement available near the end of the pulse.* This temporal behavior has been quantitatively confirmed by a number of experimental studies using pulsed and chopped laser beams (13–15). Figure 12.3 shows four spectra of water vapor taken at four different times following the initiation of a chopped laser pulse (15). The vertical axis represents the laser intensity. The narrow downward peaks are the water vapor lines, and they grow larger and approach zero laser intensity as time increases. The narrowing of the spectral profile of the laser beam with time is typical of this type of pulsed operation. Theory and observation find that the width of the spectral profile of the laser beam is inversely proportional to the square root of time and that the height of the peak is proportional to the square root of time (15).

Experimental results by Harris (12) suggest that this theory may be applicable to cw dye lasers when a way is found to incorporate power dependence into the theory. The Fourier transform of a single mode in a broad-band cw dye laser operating at pumping rates three to four times threshold shows that the mode turns on and off with frequencies as high as 200 kHz. At low pumping rates the

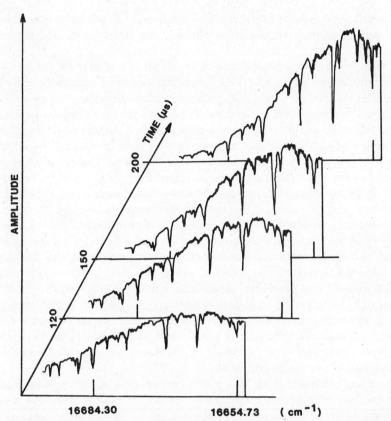

Fig. 12.3. Four time-resolved intracavity-enhanced absorption spectra of H_2O vapor lines. The vertical axis represents the laser beam intensity. The overall spectral peak of the laser beam (only part of which is shown) becomes narrower and increases in height with time. The narrow downward absorption peaks become stronger as the enhancement factor increases with time; reproduced from ref. (15) with permission.

number of modes is considerably reduced, and the remaining modes have much longer lifetimes. Equation (12.4) predicts that there should be greater enhancement at these longer lifetimes that accompany the lower pump powers. This could help explain, in addition to the threshold effect, the observation that greater enhancement occurs at lower pumping rates for cw dye lasers (7, 16).

The transient nature of each mode in a multiple-mode laser is taken one step further by a rate equation model by Brink (17). For each mode Brink adds a quenching cycle during which the mode n grows from a background intensity level B_n until a time τ, after which the mode is exponentially quenched (toward zero intensity) with a characteristic time T, until it reaches the background level again. The mode growth time τ and the decay characteristic time T depend upon the operating parameters of the laser. The loss in gain of each mode is assumed to be directly proportional to the intensities of the other modes. The proportionality constant for each mode is called the mode coupling and saturation parameter and is assumed, for example, to fall off as one over the distance of the mth mode from the zeroth mode (where the absorption peak is located). It is assumed that the main influence of the intracavity absorber is to perturb the background radiation level of those modes that overlap the absorption peak.

The equations are integrated over a typical quench cycle, and the signal is calculated as the difference ΔP in the time-integrated intensity of the zeroth mode with and without the absorber present in the cavity. Numerical solutions to the equations show (for some assumed parameter values) that ΔP is linearly related to absorber concentrations over several orders of magnitude, after which ΔP tends to saturate (because the mode totally disappears when the absorption is high enough). At first this may appear to be distinctly different from the other theories just discussed, which predict (at least over a limited range of absorbances) that the logarithm of the ratio of the laser mode intensities, and not their difference, is proportional to concentration. However, this new relationship is valid for only low single-pass absorbances where the magnitude of the difference (the absorption) in single-pass transmittances is (to a first approximation) proportional to absorbance and, therefore, also to concentration. When the single-pass transmittance difference is proportional to absorbance, the same enhancement factor applies to the enhanced absorbance and the enhanced transmittance because of the proportionality.

Numerical predictions were also made of the spectral line shape of the laser when an absorber is present. Results are shown in Fig. 12.4. The absorption minimum occurs at the center of the absorption profile, and the intensities of the nearby modes are enhanced. These enhanced wings have been experimentally observed in many studies and show that the intracavity spectral profile does not have the same shape as the conventional absorption profile. Similarly, reversed wings in the laser spectrum are predicted by Harris (12), who also considers mode coupling.

Although the small signal gain and the laser rise time τ strongly influence the signal size and enhancement factor (enhancement factors on the order of 10^6 were calculated), and the mode coupling parameters have not yet been characterized experimentally, this study found that the magnitudes of these parameters have little effect on the *shape* of the line.

In summary, some theories predict that the intracavity absorbance is proportional to the conventional single-pass absorbance, and another theory predicts for small single-pass absorbances that the mode intensity difference is proportional to the single-pass absorbance. In any case the proportionality factor (the intracavity enhancement factor) is predicted to be highest at low pump powers. For pulsed operation the intracavity absorbance is directly proportional to time.

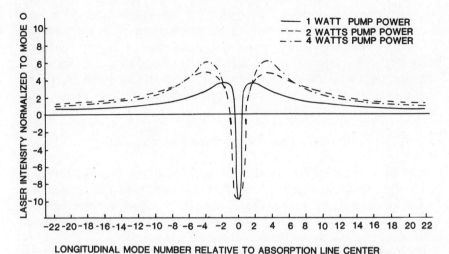

Fig. 12.4. Theoretical spectral shape of an intracavity-enhanced absorption peak with one of the model parameters (the mode coupling parameter) chosen to enhance the shape of the wings; reproduced from ref. (17) with permission.

The spectral line shape of the multiple-mode laser beam when an intracavity absorber is present is similar to the conventional absorption profile, but has enhanced reversed wings.

3. INSTRUMENTATION

The basic components for a laser intracavity absorption spectrophotometer, indicated in Fig. 12.1, include the active lasing medium and the sample cell, both located between the end mirrors that bound the laser cavity, the pump for the active lasing medium, and a laser beam detector. In addition, when the bandwidth of the absorbing sample is narrow relative to a multimode lasing bandwidth, the laser beam detector must have high spectral resolution in order to observe the intensity of only the absorbed laser mode(s).

When additional enhancement is desired, it may be possible to take advantage of the increase with time of the intracavity absorbance. In this case the laser is pulsed and the laser beam detector must be gated or have the ability to time resolve the signal in order to observe the signal near the end of the laser pulse.

Pulsed dye lasers are attractive because they are usually less difficult to operate and maintain than cw dye lasers. Important studies of intracavity absorption have been made with pulsed lasers. Unfortunately present pulsed lasers are relatively unstable compared to cw lasers or chopped cw lasers, and the best quantitative determinations of sample concentrations have been obtained with cw lasers. Therefore, we will concentrate on cw laser systems, keeping in mind that when stable pulsed lasers are available, the same principles will apply.

Factors that are important to obtaining optimum results will now be discussed along with several examples of typical intracavity absorption spectrometers. The examples are not meant to be exhaustive, but they include representative examples of some of the most successful systems for quantitative determinations of concentrations in samples.

3.1. Narrow-band Laser/Broad-band Absorber

One of the best systems for determining concentration of a broad-band absorber is described by Shirk, Harris, and Mitchell (18), who used a thoughtfully designed system with a carefully adjusted argon-ion-pumped jet-dye laser. Their description of how the laser is stabilized and aligned is worth reading by anyone who wants to obtain the best results from such a laser system.

Their advice includes mounting all optical components directly to the Invar laser cavity support, minimizing the number of optical surfaces and components, and avoiding secondary optical cavities by avoiding parallel surfaces and using windows with at least a 3° wedge angle between their two surfaces. Optical components should be as defect free as possible and have the highest-quality surface finish. TEM_{00} mode operation is important to reduce noise.

A minimum absorbance of 2×10^{-5} could be detected with their broad-band absorber system. This is about three orders of magnitude worse than can be achieved by a multimode wide-band laser with a narrow-band absorber, but is better than conventional absorption methods. Enhancement factors approached 600. For the case of a broad-band absorber, enhancement is an indication of the magnitude of multiple-pass and near-threshold effects. Enhancement helps to achieve excellent performance but is not as important as the minimum detectable absorbance because if noise is enhanced equally with the signal, there is no advantage. Therefore, noise must also be minimized.

3.1.1. Absorbance Null

A schematic diagram of the laser intracavity absorption spectrophotometer of Shirk et al. (18) is shown in Fig. 12.5. Instead of measuring the change in laser intensity that occurs when a sample is introduced into the cavity, a calibrated variable absorbance reference (variable loss) is added to the laser cavity. While observing the laser beam intensity, the absorbance of the calibrated variable reference is decreased (from an initially preset value) to just compensate for the absorbance increase caused by the sample. When the intensity of the beam is returned to its original value, the absorbance change of the variable reference equals the absorbance caused by the sample.

Because all wavelengths of the laser beam are equally absorbed by the broad-

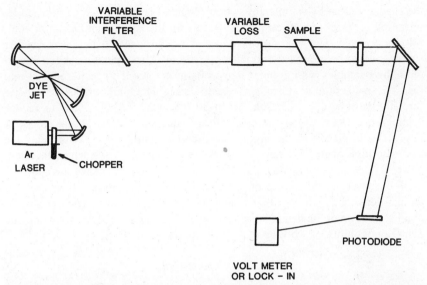

Fig. 12.5. Schematic diagram of a narrow-band laser intracavity absorption spectrophotometer with a calibrated variable loss. With the sample in the sample cell, the calibrated variable loss is adjusted to return the beam to the value it had when the blank was in the cell. The absorbance of the calibrated loss is then equal to the sample absorbance. The chopper reduces thermal lens effects by chopping the laser beam at a low-duty cycle; reproduced from ref. (18) with permission.

band absorbing sample, a broad-band variable absorbance reference can be used. In this case the variable absorbance reference is a Pockels cell with antireflection coatings on the outside windows. Voltages applied to the Pockels cell are less than 100 V because of the small absorbance changes needed. Good optical surface quality is necessary to avoid transverse mode hopping and amplitude instability. Careful alignment of the laser beam along the optical axis of the potassium dihydrogen phosphate crystal in the Pockels cell is necessary to avoid the problems caused by birefringence when the laser beam propagates at an angle to the crystal's optical axis. A simple test for proper alignment is to ensure that the light is symmetrically modulated when an ac signal is applied to the Pockels cell.

A similar null method was used earlier by Cresenzi and Shirk (19) with a flashlamp-pumped dye laser and a Kerr cell but suffered from the reproducibility problems of a pulsed dye laser.

3.1.2. Thermal Lens Problems

Heating by the dye laser beam causes thermal lens formation (Chapter 13) in the sample and the Pockels cell. A thermal lens distorts the laser beam and causes a change in the laser power that is often orders of magnitude greater than the change caused by sample absorption. Thermal heating could be reduced by operating the laser near threshold. However, for the laser system of Shirk et al. (18) thermal lens effects are greatest near threshold, and it is not clear whether stability would be improved or not near threshold. Also, operating above threshold allows samples with a wider range of concentrations to be introduced into the cavity without extinguishing the beam during the nulling procedure. Therefore, the laser is operated from 10 to 30% above threshold, and, to reduce heating effects, an optical chopper with a frequency of 130 Hz and an on/off ratio of $\frac{1}{40}$ is placed in the argon ion laser pumping beam. Fortunately, the detection limit of their system varies less than a factor of two when the pump power is increased from about 1% above threshold to 50% above threshold.

With respect to the choice of sample solvent, hydrocarbon solvents form a thermal lens much stronger than water. Water should form essentially no lens at all at 4°C where its density reaches a maximum, and the authors recommend that the possibility of working at 4°C be investigated.

3.1.3. Sample Cell

The sample cell is a flow-through cell designed with a Teflon spacer between quartz windows, which are placed at the Brewster angle to reduce reflection losses. A flow-through cell avoids the small but very significant changes in cavity configuration that occur when a cell is repeatedly removed and replaced in the cavity. The windows have an 8° wedge angle so that the laser beam is incident at the Brewster angle for the solution–window interface as well as for the air–window interface.

The precision of repetitive absorbance measurements on a single sample is $\pm 2 \times 10^{-5}$, and is $\pm 5 \times 10^{-5}$ for different samples of the same solution. Therefore, the detection limit is not determined by the stability of the laser but perhaps by heterogeneity of the sample solution or by changes in the sample cell caused by the small pressures that occur when changing the sample solution.

3.2. Broad-band Laser/Narrow-band Absorber

Considerations discussed in the previous section for the case of the broad-band absorber also apply to this case of the narrow-band absorber. However, because of the additional enhancement caused by mode competition, it is necessary for the detector to distinguish between modes that are absorbed and those that are not.

When the sample absorption peak is narrow relative to a broad-band laser beam, the greatest change in intensity occurs in those laser modes (wavelengths) that overlap the absorption peak; those are the modes to be observed. Therefore, unlike the case of a narrow-band laser, the laser detection system must have sufficient spectral resolution to isolate those modes or the one central mode that is absorbed the most by the sample. Spectral resolution is obtained by high-resolution spectrometers (with narrow slits, long focal lengths, and high grating orders) (15, 20), by interferometers (21), or by observing the fluorescence, optogalvanic, or photoacoustic signal from a cell external to the laser cavity that contains the same species that is being determined in the sample.

3.2.1. Spectrographs and Spectrometers

Early studies used photographic recording of the spectra. Photoelectric recording is more convenient for quantitative determinations. A photomultiplier tube or photodiode may be placed behind a single exit slit in the focal plane of a spectrometer to observe the intensity of the narrow band of wavelengths passing through the exit slit. The spectral irradiance produced by a laser beam in the plane of a spectrometer is relatively high, making possible the use of a photodiode array (15, 22) or vidicon (2, 23) to observe many wavelengths simultaneously.

To achieve the maximum resolving power of a spectrometer, care must be taken to focus the collimated laser beam, perhaps with a cylindrical lens, in order to cover all lines of the grating of the spectrometer with light from the entering laser beam.

3.2.2. Spectrally Matched Detectors

A relatively inexpensive high-resolution detection system (24) that essentially is spectrally matched to the absorbing sample is shown in Fig. 12.6. The sample

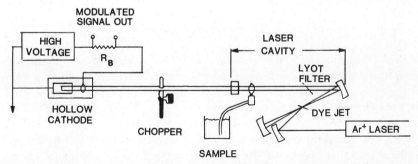

Fig. 12.6. Schematic diagram of a multimode laser intracavity absorption spectrophotometer with a flame sample cell. The hollow cathode lamp optogalvanic detector has narrow absorption lines that essentially are spectrally matched to the absorption lines in the sample, and therefore selectively responds to the intensities of the laser modes that are absorbed by the sample. The signal is modulated by the chopper for easier processing; reproduced from ref. (24) with permission.

cell is a flame that converts a nebulized solution sample into free atoms. The range of the many laser modes that are allowed to lase is selected by the Lyot filter. The free atoms absorb the few modes of the laser that overlap their narrow absorption peaks. The same type of atoms in the hollow cathode lamp are essentially spectrally matched to the absorption lines in the sample and therefore absorb these same few modes of the laser, causing a change in the electrical impedance of the lamp. The impedance change is detected as a change in voltage drop across resistor R_B. The optical chopper modulates the signal so that the modulated signal can be easily separated from the relatively high dc voltage across the resistor with the help of a lock-in amplifier. This modulated opto-galvanic signal from the hollow cathode lamp responds to the intensity of the laser beam only at the modes that overlap the atomic spectral lines of the elements in the lamp; other laser wavelengths cause at most a minor background signal in the lamp. The spectral resolution of the observation is in part determined by the spectral linewidths in the hollow cathode lamp, which are narrower than in the flame.

In similar wavelength-selective detectors, the laser beam enters a cell containing the atoms or molecules of interest, and the resulting fluorescence or photoacoustic signal (Chapter 5) is used as a measure of the intensity of the absorbed laser mode(s). Fluorescence has been used to detect different isotopes in I_2 vapor (10, 24, 25), and photoacoustic signals have been used to detect I_2 and Br_2 vapor (6).

To obtain linear response for these wavelength-selective detectors, it is important to keep the laser irradiance low enough to avoid saturating the absorber in the detector cell. It is also necessary to keep the concentration of the absorber low enough to avoid prefilter effects, and in fluorescence observations to avoid postfilter effects. Postfilter effects may also be avoided in fluorescence by using nonresonance fluorescence.

3.3. Time-Resolved Systems

Time-resolved intracavity spectroscopy has not yet been used for quantitative determinations, but the increasing enhancement that occurs throughout a laser pulse suggests that improved detection limits might be possible if the noise level does not change or if it improves during the pulse. Stoekel et al. (15) passed an argon ion laser pumping beam through an electro-optic switch in order to pulse a dye jet stream laser for up to 1 ms with a rise time of less than 1 μs. At the chosen time during the laser pulse, an acousto-optical grating (a diffraction grating formed by a standing acoustic wave in a Bragg cell) in the laser cavity deflects a 10-μs slice of the laser pulse to a spectrograph where the spectrum is recorded with a 1024-photodiode array. Figure 12.3 shows an example of their results.

Other time gating systems include a photomultiplier tube with a fast response time circuit or a gated photomultiplier tube that is turned on only during the last part of the laser pulse, when the enhancement factor is greatest. Either of these are followed by a boxcar integrator to average the results over many pulses.

4. APPLICATIONS

Applications of intracavity spectroscopy cover the determinations of concentrations of absorbing species in gases, solutions, and solids—from the visible to the infrared. Applications are limited to spectral regions of the active gain medium in a laser because frequency-doubled laser lines do not take part in the laser generation process that is affected by the intracavity absorber. Multiple-photon absorption can be used, however, to reach higher energy absorption transitions, and the enhancement of the intracavity method should make such absorption measurements practical.

4.1. Solutions

Except for the rare earths, there are few reports of the use of intracavity absorption to determine the concentrations of absorbing species in solutions. The most notable are Harris and Mitchell (27) and Harris and Williams (28) who used an optimized cw dye laser (18) to improve the detection limit for the determination of iron in solution by a factor of five compared to that for a conventional spectrophotometer. Although not a dramatic improvement (in part because this is the case of a broad-band absorber), it was helpful in their studies of ultrapurification methods for chemical standards.

Their detection limit was not limited by the carefully adjusted laser but by variations that occurred when the same sample was repeatedly introduced into the cavity via the flow cell. Similar sample introduction problems were the source of limitations in the conventional spectrophotometer method.

An early comparison of the cases of broad-band and narrow-band lasers for the determination of Eu^{3+} in methanol was made by Keller et al. in 1972 (6). Advantage was taken of the relatively narrow absorption bands of the rare earths in solution. The enhancement factor for the broadband laser system was about 400, and for the narrow-band laser system was much less (no numerical value was given).

A flashlamp-pumped laser system showed an enhancement factor of 100 for solutions of Ho^{3+} and Pr^{3+} (29). The spectral resolution was adequate to reveal the line profile. Data plotted as $\Delta P/P$ versus concentration were linear up to a value of 0.5.

A Q-switched, frequency-doubled ruby-laser-pumped dye laser system showed an intracavity absorbance that was linear up to 1.0 for solutions of Pr^{3+} and Eu^{3+} (22). A photodiode array was used in the focal plane of a spectrometer to obtain a spectral resolution of 0.25 nm, which was sufficient to resolve the absorption peaks.

More recently, rare-earth determinations (30) have shown enhancement factors of 10^3–10^4 for Eu^{3+}, producing a detection limit of 0.1 mg/mL in the presence of large amounts of Gd^{3+}. EDTA was added to chelate the Eu^{3+} and remove the effect of the anions. The enhanced absorption spectrum was independent of the Gd^{3+} concentration. Using a pulsed dye laser, intracavity spectral scans of Nd^{3+} in the range of 10^{-2}–$10^{-3}M$ showed a Stark splitting that was not present in the ordinary spectrum (31).

4.2. Gas-Phase Sample Cells

Intracavity atomic spectroscopy using flames, inductively coupled plasmas (ICP), laser microprobe plumes, and electrothermal atomizers as intracavity cells have been used for the determination of trace element concentrations in solutions. An advantage of the intracavity method is the ability to make absorption measurements in the presence of the intense background emission of some of these cells.

An enhancement factor of 2.5×10^4 was found for the determination of Na with an ICP (32) in the cavity of a cw dye laser. The detection limit was 0.6 ng/mL for Na and ranged from 1 to 10 μg/mL for Ba, Cu, Eu, U, and Zr lines in the wavelength range covered by the dye laser from 578 to 620 nm. Optogalvanic detection in a hollow cathode lamp containing the element of interest was used to obtain the spectral resolution required to observe the intensities of the modes absorbed by the atomic lines. The limiting source of noise was found to be wavelength fluctuations in the dye laser caused by ICP turbulence.

Several studies have been made using Na in a flame [e.g., (33–35)]. A detection limit of less than 1 ng/mL (in the original solution prior to atomization) was found for Na in a flame by Zalewski et al. (24). In this case the limit of detection was determined by the difficulty in preparing uncontaminated Na solutions.

A brief note in 1971 described the detection of Ba^+ and Sr in a flame (36). An air–acetylene flame was used to determine Na, Li, Sr, Ba, and Cs in the wavelength range from 450 to 680 nm (35). Detection limits for Na and Li were 0.2 and 5 ng mL^{-1}, respectively.

A Ta–Re alloy furnace atomizer (3100 K) placed in the cavity of a 40-ns pulse dye laser gave detection limits of 6×10^{-13} g for Fe and 2×10^{-12} g for Cr in 10-μL samples (37). The spectra were recorded photographically, and the film was calibrated with the pulsed laser in order to obtain an emulsion calibration consistent with the working conditions. As part of this same study, intracavity absorption from excited-state Fe lines was used to determine the excitation temperatures in the atomic vapor. A graphite furnace atomizer and a 5-μs flashlamp-pumped dye laser has been used to study rare-earth atoms (38).

Using a plume of a 7-J Nd laser microprobe atomizer (Chapter 19) in a dye laser cavity provided a detection limit of 3×10^{-6} g for Mo in a graphite target. The enhancement factor was estimated to be greater than 30.

The highest enhancement factors have been obtained with gas-phase samples that have excellent homogeneity and optical stability. The first report of using a cw laser (10) gave an enhancement factor of 10^5 for I_2 in an evacuated cell. Fluorescence from an external cell containing I_2 was used as the detector. A concentration of I_2 as low as 5×10^{11} molecules cm^{-3} was observed by (26) with a cw dye laser and fluorescence detector. A similar scheme using a flash-lamp-pumped dye laser showed a dynamic range of three orders of magnitude (39). Photoacoustic detection gave detection limits of 3 ng cm^{-3} for I_2 and 50 ng cm^{-3} for Br_2 with a cw dye laser (6). Na in a 15-cm intracavity cell was detected at a concentration of 5×10^5 atoms cm^{-3} (40).

Excited states of He with a lifetime of about 98 ns in a low-pressure microwave discharge have been studied as well as spectra of He_2 and C_2 in a hydrocarbon–helium discharge (2). The same group also studied short-lived excited states of Ne with a lifetime of 19×10^{-9} s in a microwave discharge (41). The number density was between 10^5 and 10^7 cm^{-3}.

An atomic beam of Ba was studied with a wavelength-scanned cw dye laser (23). A detection limit of 10^7 atoms/cm^3 was estimated. The beam was obtained by electron bombardment heating a small steel oven containing the sample. An optical multichannel analyzer was used with a 600-mm monochromator using a 1200-line per millimeter grating in the second order to obtain spectral line profiles.

A cw dye laser operated in a chamber at 50 torr was used to study the spectrum of O_2 from 580 to 630 nm. Absorption coefficients of 10^{-8} cm^{-1} were detectable (1). Weak transitions of HCl and O_2 have also been studied (20).

Very weak absorption lines of H_2O were studied with a dye laser whose cw argon ion laser pump was chopped in order to obtain 1-ms pulses. A spectrograph/photodiode array system with a resolving power of 140,000 was used to obtain the spectra. Absorption coefficients of 10^{-8} cm^{-1} were detectable.

Molecules of dimethyl-s-tetrazine in a supersonic beam were observed with a cw dye laser with an enhancement factor of 580. The detection limit appears to be about 10^{10} molecules cm^{-3} judging from the published spectral scans (42).

Molecules of C_3H_8 were observed by the C–H stretching transition near 3.39 μm. The mode competition between the 3.39-μm and the visible 0.63-μm modes of the He–Ne laser produced a change in the visible laser beam power, which was used as a measure of the infrared absorption (43). A detection limit of 9×10^{-11} g cm^{-3} was obtained, about one-fifth of the detection limit for a chromatographic conductivity detector.

Trace amounts of CH_4 in Ar at 1 atm were studied (3) with 17 lines of a deuterium fluoride laser in the spectral range of 3.6–4.0 μm. An absorption coefficient of 7×10^{-8} cm^{-1} gave a signal-to-noise ratio of 10.

A variety of small molecules and radicals have been detected, including HCO, NH_2, NO_2, NCO, C_2, CH, CN, HNO, CH_2, TiN, and TiO (13, 44–52). A Nd^{3+} laser has been used to observe the hydrogen stretch overtones of CH_4, C_2H_2, NH_3, HN_3, and HCN with a sensitivity thousands of times greater than with conventional infrared absorption at the fundamental frequencies. Harris (7) suggests that a Kr-pumped dye laser, tunable in the near infrared, would provide high sensitivity for the overtone transitions of almost any species with C–H, N–H, or O–H bonds. In addition to improved sensitivity, he indicates that there would be considerably fewer spectral interferences compared to conventional infrared absorption spectroscopy because of the relative simplicity of the overtone spectra.

5. FUTURE

Intracavity-enhanced spectroscopy is still in the early stages of development for the quantitative determination of analyte concentrations. With the present understanding of theory and practice it could begin to rapidly develop in applications where conventional absorption spectrophotometry is inadequate. For example, Harris (7) points out that the technique has the potential for making important measurements in combustion environments, especially using vibrational overtone spectroscopy. The use of two-photon intracavity spectroscopy could extend the useful spectral range of the method to other spectral regions. He also suggests that intracavity absorption may be a very sensitive way to study species absorbed on a reflecting surface that is a mirror in the laser cavity.

Making the intracavity measurements near the end of a laser pulse to take advantage of the increasing enhancement factor during the laser pulse may improve detection limits even further than has now been achieved if noise can also be reduced or at least does not increase. Reproducible long pulses for this technique are now made by chopping the cw pump beam of a stable cw dye laser (15).

Chopping the pump of a stable cw laser appears to generate more reproducible mode pulse lengths than the apparently random durations of modes in conventional cw laser operation (7). If this proves to be so, then an additional improvement in enhancement factor over cw laser operation would occur because only reproducible long-duration modes (with their larger enhancement factors) would contribute to the signals. It might be possible to improve the enhancement factor in this way over cw laser operation without time resolving the observations.

Clearly, the need is for a very stable laser system to achieve the advantages of intracavity-enhanced spectroscopy for the determination of concentration. Now that most of the requirements for improving intracavity measurements with a cw dye laser have been demonstrated, it may be possible to use this information to carefully design pulsed lasers and adjust them to obtain similar improvements. This would help to extend the useful spectral range of intracavity-enhanced measurements to regions now only accessible to pulsed lasers.

REFERENCES

1. W. T. Hill III, R. A. Abreu, T. W. Hansch, and A. L. Schawlow, *Opt. Commun.*, **32**, 96 (1980).
2. G. O. Brink, *J. Mol. Spectrosc.*, **90**, 353 (1981).
3. D. H. Leslie and G. L. Trusty, *Appl. Opt.*, **20**, 1941 (1981).
4. E. N. Antonov, P. S. Antsyferov, A. A. Kachanov, and V. G. Koloshnikov, *Opt. Commun.*, **41**, 131 (1982).
5. K. R. German, *Proc. Int. Conf. Lasers*, 128–132 (1981).
6. R. A. Keller, E. F. Zalewski, and N. C. Peterson, *J. Opt. Soc. Am.*, **62**, 319 (1972).
7. S. J. Harris, *Appl. Opt.*, **23**, 1311 (1984).
8. A. Yariv, *Introduction to Optical Electronics*, Holt, Rinehart and Winston, New York, 1976, Chapter 6.
9. E. J. Meehan, "Fundamentals of Spectrophotometry," Chapter 54, in Part 1, Vol. 5, I. M. Kolthoff and P. J. Elving, Eds., *Treatise on Analytical Chemistry*, Interscience, New York, 1964, p. 2757.
10. T. W. Hansch, A. L. Schawlow, and P. E. Toscheck, *IEEE J. Quantum Electron.*, **6**, 802 (1972).
11. C. L. Tang, H. Statz, and G. deMars, *J. Appl. Phys.*, **34**, 2289 (1963).
12. S. J. Harris, *Opt. Lett.*, **7**, 497 (1981).
13. T. P. Belikova, E. A. Sviridenkov, and A. F. Suchkov, *Opt. Spectrosc.*, **37**, 372 (1974).
14. V. M. Beav, T. P. Belikova, E. A. Sviridenkov, and A. F. Suchkov, *Sov. Phys. JEPT*, **47**, 21 (1978).
15. F. Stoeckel, M.-A. Melieres, and M. Chenevier, *J. Chem. Phys.*, **76**, 2191 (1982).
16. S. J. Harris and A. M. Weiner, *J. Chem. Phys.*, **74**, 3673 (1981).
17. G. O. Brink, *Opt. Commun.*, **32**, 123 (1980).
18. J. S. Shirk, T. D. Harris, and J. W. Mitchell, *Anal. Chem.*, **52**, 1701 (1980).
19. F. Crescenzi and J. S. Shirk, *Opt. Commun.*, **29**, 311 (1979).
20. R. G. Bray, W. Henke, S. K. Liu, K. V. Reddy, and M. J. Berry, *Chem. Phys. Lett.*, **47**, 213 (1977).
21. R. A. Keller, J. D. Simmons, and D. A. Jennings, *J. Opt. Soc. Am.*, **63**, 1552 (1973).
22. G. Horlick and E. G. Codding, *Anal. Chem.*, **46**, 133 (1974).
23. P. Kumar, G. O. Brink, S. Spence, and H. S. Lakkaraju, *Opt. Commun.*, **32**, 129 (1980).

24. E. F. Zalewski, R. A. Keller, and C. T. Apel, *Appl. Opt.*, **20**, 1584 (1981).
25. S. J. Harris, *J. Chem. Phys.*, **71**, 4001 (1979).
26. J. P. Hohimer and P. J. Hargis, Jr., *Anal. Chem.*, **51**, 930 (1979).
27. T. D. Harris and J. W. Mitchell, *Anal. Chem.*, **52**, 1706 (1980).
28. T. D. Harris and A. M. Williams, *Anal. Chem.*, **53**, 1727 (1981).
29. R. C. Spiker, Jr., and J. S. Shirk, *Anal. Chem.*, **46**, 572 (1974).
30. A. M. Udartsev, S. M. Mashakova, and A. V. Desyatkov, *Khim. Fiz.*, 1443 (1983).
31. M. V. Akhamanova, L. A. Gribov, S. G. Ivanov, and N. S. Stroganova, *Zh. Prikl. Spektrosk.*, **34**, 866 (1981).
32. S. W. Downey and N. S. Nogar, *Anal. Chem.*, **57**, 13 (1985).
33. R. B. Green and H. W. Latz, *Spectrosc. Lett.*, **7**, 419 (1974).
34. H. von Weyssenhoff and U. Rehling, *Z. Naturforsh.*, *A*, **29A**, 256 (1974).
35. M. Maeda, F. Ishitsuka, and Y. Miyazoe, *Opt. Commun.*, **13**, 314 (1975).
36. R. J. Thrash, H. von Weyssenhoff, and J. S. Shirk, *J. Chem. Phys.*, **55**, 4659 (1971).
37. V. S. Burakov, V. N. Verenik, V. A. Malashonok, S. V. Nechaev, and R. A. Puko, *Zh. Anal. Khim.*, **38**, 90 (1983).
38. V. N. Lopatko, M. V. Belokon, and A. N. Rubinov, *Zh. Prikl. Spektrosk.*, **39**, 1017 (1983).
39. F. J. Morgan, C. H. Dugan, and A. G. Lee, *Opt. Commun.*, **27**, 451 (1978).
40. M. Maeda, F. Ishitsuka, M. Matsumoro, and Y. Miyazoe, *Appl. Opt.*, **16**, 403 (1977).
41. G. O. Brink and S. M. Heider, *Opt. Lett.*, **6**, 366 (1981).
42. W. R. Lambert, P. M. Felker, and A. H. Zewail, *J. Chem. Phys.*, **74**, 4732 (1981).
43. J. D. Parli and D. W. Paul, *Anal. Chem.*, **54**, 1969 (1982).
44. G. H. Atkinson, A. H. Laufer, and M. J. Kurylo, *J. Chem. Phys.*, **59**, 350 (1973).
45. G. H. Atkinson, T. N. Heimlich, and M. W. Schuyler, *J. Chem. Phys.*, **66**, 5005 (1977).
46. O. M. Sarkisov, E. A. Sviridenkov, A. M. Udartsev, V. Zh. Ushanov, M. P. Frolov, and S. G. Cheskis, *Dolk. Akad. Nauk. SSSR*, **233**, 341 (1977).
47. J. H. Clark, C. B. Moore, and J. P. Reilly, *Int. J. Chem. Kinet.*, **10**, 427 (1978).
48. S. J. Harris and A. M. Weiner, *Opt. Lett.*, **6**, 434 (1981).
49. W. R. Anderson, J. A. Vanderhoff, A. J. Kotlar, M. A. Dewilde, and R. A. Beyer, *J. Chem. Phys.*, **77**, 1677 (1982).
50. N. G. Fedotov, V. I. Kozintsev, V. A. Nadtochenko, and O. M. Sarkisov, *A. F. Sil'Nitskii, Khim. Fiz.*, 1011 (1982).
51. P. I. Stepanov, E. N. Moskvitina, Yu. Ya. Kuzyakov, E. A. Sviridenkov, and A. N. Savchenko, *Vestn. Mosk. Univ. Ser. 2: Khim.*, **24**, 442 (1983).
52. M. V. Danileiko, A. M. Fal, M. T. Shpak, and L. P. Yatsenko, *Ukr. Fiz. Zh. (Russ. Ed.)*, **29**, 1109 (1984).

CHAPTER

13

THERMAL LENS EFFECT

JOEL M. HARRIS

Department of Chemistry
University of Utah
Salt Lake City, Utah

1. INTRODUCTION

1.1. Thermo-optical Absorption Measurements

The introduction of lasers into analytical spectroscopic instrumentation has produced, in many cases, significant improvements in detection performance compared to using conventional light sources. While many of these improvements arise from the greater optical power of the laser, a number of new methods rely on the unique coherence properties of laser radiation. A new class of sensitive, spectrophotometric techniques has been developed in which the heat produced by nonradiative decay of excited species acts to modify the optical properties of the sample (1). These thermo-optical absorption techniques may be classified

451

according to the spatial temperature distribution and corresponding refractive index change produced in the sample, the detection of which utilizes the coherence of the laser beam. For example, if the sample is uniformly heated by a large-area excitation beam, the change in optical path can be detected interferometrically (2, 3), which relies on the temporal coherence or monochromaticity of the laser. The spatial coherence of laser radiation, which is manifest in the small divergence of the beam, can be used to detect a thermal lens in the sample (4, 5) produced by focused laser excitation. The small divergence of the laser beam can similarly be used to measure the deflection caused by a thermal prism produced by optical excitation at an interface (6, 7). By splitting a laser beam and recombining the two beams within a sample, one can exploit the spatial and temporal coherence of the beams to generate an excitation interference pattern producing a thermal transmission grating that may be probed by diffraction (8, 9).

While the experimental arrangements and applications of these thermo-optical methods differ considerably from one another, there is a common thread running among them in the relationship between their sensitivities and the thermo-optical properties of a sample matrix. Although this chapter presents only the thermal lens effect and its numerous applications in analytical chemistry, much of the material would serve as a paradigm for understanding other laser-based methods of thermo-optical detection.

1.2. Thermal Lens Effect

The thermal lens effect or thermal blooming was discovered by Gordon, Leite, Moore, Porto, and Whinnery (4) only a few years after the invention of the continuous-wave (cw) laser. While carrying out an intracavity excitation of a liquid sample for Raman spectroscopy, the authors observed time-dependent changes in the laser intensity which they could attribute to localized heating of the sample. They derived a theory for the thermal defocusing of the beam, which predicted the magnitude and time constant of the response. Another contribution of this seminal paper was a proposal that the effect could be used as a sensitive method for measuring small absorbances.

The thermal lens effect may be described, as in Fig. 13.1, as a change in the optical path of the sample induced by absorption of a laser beam having a Gaussian radial intensity distribution. The greater heat produced at the beam center raises the temperature and, for most liquids that expand upon heating, lowers the refractive index, n. This creates a shorter optical path, nl, along the center of the beam path, where l is the physical length of the sample. The resulting thermo-optical element has the shape of a negative lens which causes the laser beam to diverge. The increased divergence of the beam may be observed some distance beyond the sample as a larger spot size or a lower intensity at the beam center.

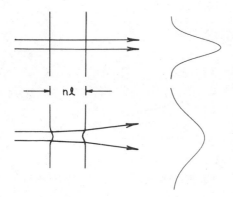

Fig. 13.1. A diagram of the thermal lens effect. A laser beam passes from left to right through the sample having an optical length bounded by the vertical lines. The initial unheated sample and far-field beam profile are shown at the top. A steady-state thermal lens and beam profile are shown below.

In the next section the theory of the thermal lens effect, which enables one to determine the absorbance of a sample by this technique, is reviewed. This theory also identifies the optical and sample-dependent factors that affect the measurement sensitivity. The various experimental approaches to the measurement are compared and applications to chemical analysis are described.

2. THERMAL LENS THEORY

2.1. Propagation of Gaussian Laser Beams

With the exception of superradiators (such as nitrogen lasers) and unstable resonators, most lasers are constrained to oscillate in a single, transverse mode. The intensity distribution of the resulting beam is everywhere diffraction limited, which makes it very sensitive to refractive index gradients such as a thermal lens. The intensity has a symmetric, Gaussian radial shape shown in Fig. 13.2 and given by

$$I(r) = \frac{2P}{\pi\omega^2} \exp\left(\frac{-2r^2}{\omega^2}\right) \tag{13.1}$$

where the beam radius or spot size, ω, converges and diverges symmetrically about a waist or focal point at $Z = 0$ according to

$$\omega^2 = \omega_0^2 \left[1 + \left(\frac{Z}{Z_c}\right)^2\right] \tag{13.2}$$

where confocal distance, $Z_c = \pi\omega_0^2/\lambda$, where ω_0 is the spot size at the waist. In the far-field region of the beam at large distances from the waist, $Z \gg Z_c$,

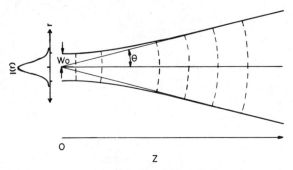

Fig. 13.2. Gaussian laser beam propagation. The solid lines indicate the spot size at any point, Z. The radial intensity distribution at the waist is shown at left. The dashed lines indicate the spherical phase fronts of the beam.

the spot size increases linearly with propagation, or equivalently, the divergence angle ϕ (θ in Fig. 13.2.) becomes constant

$$\phi = \tan^{-1} \frac{\omega}{Z} \cong \frac{\omega}{Z}$$

$$= \frac{\lambda}{\pi \omega_0} \tag{13.3}$$

The effect of an optical element such as a lens on the propagation of the laser beam is conveniently determined using the *ABCD* law (10), where A, B, C, and D are the elements of a ray transfer matrix for the optical element in question. For example, the ray transfer matrix of a free-space propagation of length Z is

$$\begin{bmatrix} A & B \\ C & D \end{bmatrix} = \begin{bmatrix} 1 & Z \\ 0 & 1 \end{bmatrix} \tag{13.4}$$

and that of a thin lens of focal length f is

$$\begin{bmatrix} A & B \\ C & D \end{bmatrix} = \begin{bmatrix} 1 & 0 \\ -1/f & 1 \end{bmatrix} \tag{13.5}$$

The ray transfer matrix of any system composed of N optical elements is the product, in reverse order of propagation, of the individual ray transfer matrices of each element:

$$\begin{bmatrix} A_s & B_s \\ C_s & D_s \end{bmatrix} = \begin{bmatrix} A_N & B_N \\ C_N & D_N \end{bmatrix} \cdot \cdot \cdot \begin{bmatrix} A_2 & B_2 \\ C_2 & D_2 \end{bmatrix} \begin{bmatrix} A_1 & B_1 \\ C_1 & D_1 \end{bmatrix} \tag{13.6}$$

When the origin of the optical system corresponds to the waist of a laser beam and when the refractive index at the end of the optical system is the same as the origin, the spot size of the emerging beam may be conveniently determined by (5)

$$\omega^2 = \omega_0^2 \frac{A^2 + B^2}{Z_c^2} \tag{13.7}$$

where ω_0 and Z_c are the spot size and confocal distance at the origin.

2.2. Predicting Thermal Lens Behavior

The ray transfer matrix method makes the analysis of a thermal lens experiment quite straightforward. To determine the influence of an absorbing sample on the beam, the induced temperature profile in the sample may be approximated as a parabola resulting in the formation of a lens of focal length, $f(t)$, which has a steady-state value (4, 11)

$$f(\infty) = \frac{\pi k \omega_1^2}{2.303 \, P(dn/dT) A} \tag{13.8}$$

where ω_1 and P are the spot size and power of the laser beam in the sample, and k, (dn/dT) and A are the thermal conductivity, change in refractive index with temperature, and decadic absorbance of the sample. A typical optical system to observe a thermal lens is shown in Fig. 13.3, which corresponds to a ray transfer matrix given by the product

$$\begin{bmatrix} A & B \\ C & D \end{bmatrix} = \begin{bmatrix} 1 & Z2 \\ 0 & 1 \end{bmatrix} \begin{bmatrix} 1 & 0 \\ -1/f(t) & 1 \end{bmatrix} \begin{bmatrix} 1 & Z1 \\ 0 & 1 \end{bmatrix} \tag{13.9}$$

Substituting A and B from the above system matrix into Eq. (13.7) and allowing $Z2$ to increase without bound so that the divergence of the beam becomes constant, yields the following expression for the far-field spot size:

$$\omega_2^2 = \omega_0^2 \, Z2^2 \left[\left(\frac{1}{f(t)} \right)^2 + \left(1 - \frac{Z1}{f(t)} \right)^2 \middle/ Z_c^2 \right] \tag{13.10}$$

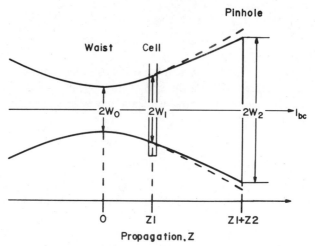

Fig. 13.3. Optical system for thermal lens measurements. The origin of the optical system is at the beam waist of spot size, ω_0, beyond which a sample cell is placed at a position, $Z1$. At a distance $Z2$ from the cell one infers the expansion of the beam spot size, ω_2, by a measurement of the beam center intensity.

Finally, the relative change to the far-field spot size is found by substitution, recalling that the sample focal length at the onset of illumination would be $f(0) = \infty$,

$$
\Delta\omega^2/\omega_2^2 = \frac{\omega_2^2(\infty) - \omega_2^2(0)}{\omega_2^2(0)}
$$

$$
= \frac{-2Z1}{f(\infty)} + \frac{Z1^2 + Z_c^2}{[f(\infty)]^2} \tag{13.11}
$$

Changes in the beam spot size are conveniently measured as changes in the intensity at the beam center since according to Eq. (13.1) these parameters are inversely proportional:

$$
I(r = 0, t) = I_{bc}(t) = \frac{2P}{\pi\omega^2(t)} \tag{13.12}
$$

Substituting Eqs. (13.12) and (13.8) into Eq. (13.11) with the assumption that the thermal lens is weak, so that the quadratic term may be neglected, predicts the sample position dependence of the thermal lens effect:

$$\frac{\Delta I_{bc}}{I_{bc}} = \frac{I_{bc}(0) - I_{bc}(\infty)}{I_{bc}(\infty)}$$

$$= \frac{2.303\,P(dn/dT)A}{\lambda k}\left[\frac{2Z1Z_c}{Z1^2 + Z_c^2}\right] \tag{13.13}$$

The position-dependent term in brackets is the product of a linear term in $Z1$, which indicates the effect of a lens of fixed focal length as it is moved through the waist, and a Lorentzian term, which accounts for the strength of the lens changing with the beam spot size. The product of these terms produces an antisymmetric curve as shown in Fig. 13.4, with peak response predicted to occur at $Z1 = \pm Z_c$. The antisymmetric shape has a physical explanation; a diverging thermal lens placed beyond the beam waist simply increases the divergence of the beam. The same lens placed before the waist decreases the convergence which, in turn, decreases the divergence after the beam propagates beyond its waist.

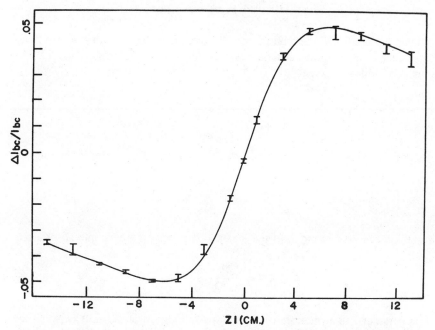

Fig. 13.4. Dependence of thermal lens response on sample cell position. Data shown with $\pm 1\,\sigma$ error bars are for a 1-cm pathlength of 2.0 ppm phenolthalein in aqueous 1.25mM NaOH. Argon ion laser excitation at $\lambda = 514.5$ nm, $P = 150$ mW. Smooth curve is a least-squares fit to Eq. (13.13).

The parabolic lens approximation that leads to Eq. (13.8) is extremely convenient for designing a thermal lens instrument using ray transfer matrices. It predicts the antisymmetric position dependence as shown in Fig. 13.4, which can be used to generate a differential response when sample and reference cells are located on opposite sides of the waist of a single laser beam (12). The parabolic lens model also provides a basis for understanding the sensitivity losses encountered with longer wavelength lasers in thermal lens measurements (13) and the departure from Beer's law for long pathlength sample cells (14). To achieve the most accurate predictions of thermal lens behavior, this simple model requires a few corrections that arise from the aberrant nature of the thermal lens (15, 16). The sample positions for maximum response are actually located at $Z1 = \pm\sqrt{3}\, Z_c$, and the sensitivity is a factor $\tan^{-1}(1/\sqrt{3}) \cong 0.52$ smaller than predicted by Eq. (13.13). The parabolic lens model, however, accurately predicts the relationship between the linear and quadratic terms, as in Eq. (13.11). In the next section the characteristics of thermal lens measurements will be described using a modified parabolic lens model (16) that accurately accounts for lens aberrations while retaining the intuitive attributes of the simpler theory.

2.3. Sensitivity and Temporal Characteristics

A sample placed $\sqrt{3}Z_c$ beyond the waist of a focused laser beam causes a time-dependent change in the far-field beam center intensity after the onset of illumination, which is given by

$$I_{bc}(t) = I_{bc}(0)\left[1 + \left(\frac{\theta}{1 + t_c/t}\right) + \frac{1}{2}\left(\frac{\theta}{1 + t_c/t}\right)^2\right]^{-1} \qquad (13.14)$$

where t_c, the time constant, depends on the spot size of the beam in the sample:

$$t_c = \frac{\omega_{1}^2\rho C_p}{4k} \qquad (13.15)$$

and θ indicates the strength of the lens and depends on the sample absorbance

$$\theta = \frac{2.303P(-dn/dT)A}{1.91\lambda k} \qquad (13.16)$$

This parameter is a measure of the thermally induced phase shift between the center and the edge of the beam at steady state (17).

The relative change in intensity induced by the thermal lens reaches a steady-state value after several time constants:

$$\frac{\Delta I_{bc}}{I_{bc}} = \theta + \frac{\theta^2}{2}$$

$$= 2.303EA + \frac{(2.303EA)^2}{2} \tag{13.17}$$

where $E = [-P(dn/dT)/1.91\lambda k]$ is the enhancement in sensitivity over a transmission measurement that would produce a relative intensity change

$$\frac{I_0 - I}{I_0} = 1 - 10^{-A} \cong 2.303A \tag{13.18}$$

for small absorbances.

The sensitivity of a thermal lens measurement, therefore, depends linearly on laser power. The choice of solvent can greatly influence the enhancement one observes for the laser power available, as shown in Table 13.1. Nonpolar solvents like carbon tetrachloride are particularly advantageous since they exhibit a large dn/dT and small thermal conductivity. The sensitivity of thermal lens measurement made in water under similar conditions would be 30 times smaller. The time constant of the thermal lens effect also varies with solvent to a lesser extent, as shown in Table 13.1. While the time constant depends quadratically on the beam spot size as indicated by Eq. (13.15), the sensitivity is independent of spot size. This is somewhat surprising since the strength of the thermal lens increases with decreasing spot size as shown by Eq. (13.8), however, the concomitant increase in the divergence angle of the beam, ϕ, with decreasing spot size, shown in Eq. (13.3), exactly cancels the effect of the stronger thermal lens.

It is instructive to examine the refractive index and temperature sensitivity of the thermal lens effect. Assuming a 1% change in the beam center intensity

Table 13.1. Thermo-optical Properties of Solvents for Thermal Lens Measurements[a]

Solvent	k (mW cm^{-1} K^{-1})	$10^4 \, dn/dT$ (K^{-1})	E/P^b (mW^{-1})	t_c^c (ms)
CCl$_4$	1.03	−5.9	4.7	33
Acetone	1.60	−5.4	2.8	27
Methanol	2.02	−4.2	1.7	25
Water	6.05	−1.0	0.14	18

[a] Data compiled in ref. 16.
[b] Enhancement per unit laser power in mW; $\lambda = 632.8$ nm.
[c] Time constant for $\omega_1 = 100 \, \mu$m, the beam spot size in the sample.

can be detected, the corresponding induced phase shift is $\theta = 0.01 = -2\pi \Delta n b/n\lambda_0$, where b is the sample pathlength and Δn is the refractive index change at the beam center compared with the edges. For a 1-cm pathlength and $\lambda_0 = 632.8$ nm, the relative change in refractive index is only $\Delta n/n = -1.0 \times 10^{-7}$. For a solvent having a large dn/dT such as carbon tetrachloride, this refractive index change corresponds to a temperature difference of $\Delta T = 2.5 \times 10^{-4}$ K. The sensitivity of a spatially coherent laser beam to such small refractive index gradients is the basis of the detection capabilities of the thermal lens effect.

3. MEASUREMENT APPROACHES

3.1. Single-Beam Measurements

The simplest optical configuration for a thermal lens measurement uses only a single laser beam both to produce the thermal lens in the sample and to probe its presence. This configuration, which included intracavity excitation (4), dominated the earliest studies of the thermal lens effect. Its chief attribute, as illustrated in Fig. 13.5, is both simplicity in the optics required and the ease of alignment. The laser beam is focused through the sample cell, which should be mounted on some form of translation stage to locate the position of optimum response. The expanding laser beam is directed at a pinhole or small area detector to sample the intensity at the beam center.

As the laser beam is slowly chopped, the initial intensity reflects the laser beam profile unaffected by a thermal gradient. As the sample continues to be illuminated, a thermal gradient develops that enlarges the far-field spot size and reduces the beam center intensity, as shown in Fig. 13.6. Two approaches may be used to gather data from this experiment. The simplest method (18) is to use a sample-and-hold circuit to capture the initial, $I_{bc}(0)$, and steady-state, $I_{bc}(\infty)$,

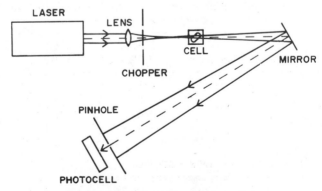

Fig. 13.5. Single-beam thermal lens experiment.

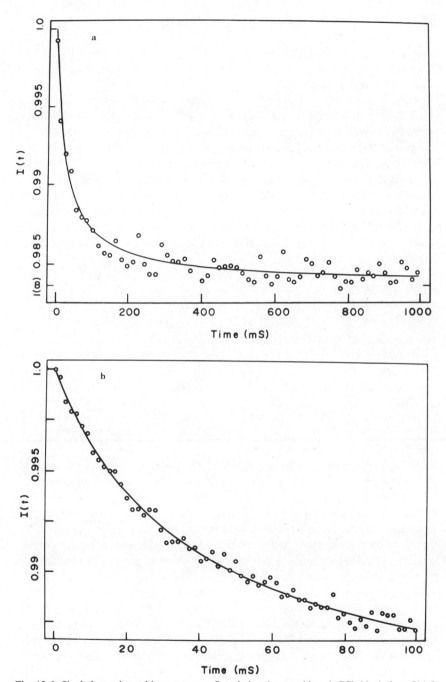

Fig. 13.6. Single-beam thermal lens response. Sample is a 1-cm pathlength CCl_4 blank; $\lambda = 514.5$ nm; $P = 160$ mW. Best fit to Eq. (13.14) is the solid line, with $t_c = 33$ ms. (a) Development of a steady-state response, $t_{max} \cong 30t_c$. (b) Optimal data range, $t_{max} \cong 3t_c$; reprinted from ref. (20) with permission.

461

intensities that may be interpreted using Eq. (13.17). As shown in Fig. 13.6, this method suffers from short-term intensity noise from the laser since only two points per experiment are detected.

A much more efficient procedure is to gather several hundred data points along a thermal lens transient and to fit Eq. (13.14) to the data (19, 20). This method of signal averaging has been shown (20) to reduce the measurement uncertainty by a statistical factor, $(n - 2)^{1/2}$, where n is the number of data points in the transient and two degrees of freedom are lost in fitting. This kinetic approach also reduces the experiment time, as shown in Fig. 13.6b, since one need not wait for a steady-state signal to develop. In addition, the dynamic range is greater since a stronger absorbing sample, producing a steady-state thermal lens governed by large aberrations and convection, can be measured as a weaker, well-behaved lens if kinetic data is gathered over a sufficiently short time period (20).

The single-beam method of thermal lens measurements can be used to produce a differential response by optical means (12) mentioned in the Section 2. Reference and unknown samples are located on opposite sides of the beam waist so that the thermal lens strengths of the two samples are subtracted by their opposite effect on the far-field beam spot size shown in Fig. 13.4. This differential configuration was found to exhibit excellent immunity from laser power fluctuations, but is limited to sufficiently weak thermal lenses, $\theta < 0.1$ [Eq. (13.16)] so that the divergence of the beam responds linearly to the absorbance in each cell.

3.2. Dual-Beam Configurations

In the first applications of the thermal lens effect for gas-phase dynamics (21) and for measuring overtone spectra (22), another optical configuration was introduced whereby one laser beam, absorbed by the sample, was used to form the lens, while a second laser beam probed the presence of the lens in the sample. A typical experimental arrangement is shown in Fig. 13.7. The pump-and-probe configuration is a necessary choice when using a pulsed laser to form the thermal lens (23, 24). This approach is powerful for generating absorption spectra over the broad wavelength range of a pulsed, tunable dye laser.

Using chopped, continuous-wave excitation, the pump-and-probe method allows the use of a lock-in amplifier to detect the modulation on the probe beam center intensity by the thermal lens (22). This configuration is convenient for real-time monitoring, as in chromatographic detection. While the signal processing with a lock-in amplifier is simpler than transient recording with a microcomputer used for single-beam experiments, the optical alignment constraints of the pump-and-probe configuration are much more severe (25), particularly when small-volume detection with tightly focused laser beams is attempted (26).

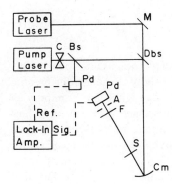

Fig. 13.7. Dual-beam thermal lens experiment. The pump laser beam is chopped at C, combined with the probe laser beam at a dichroic beam splitter, Dbs, and focused together by a concave mirror, Cm, through the sample, S. The pump beam is filtered out at F, and the probe beam center intensity is measured using an aperature, A, and photodiode Pd; reprinted from ref. (31) with permission.

Modeling the response of a dual-beam thermal lens experiment is more complicated than the single-beam theory presented in the previous section since one must take into account the confocal parameters and spot sizes of both pumping and probing laser beams (27). Using the simplifying assumption that the spot sizes of the two beams in the sample are approximately equal, which probably produces less than optimum sensitivity (28, 29), a numerical model for the dual-beam experiment was developed (30). An analytical solution to the frequency response was also published (31) and the results, expressed in terms of a reduced frequency, $v_0 = \pi v t_c$, where v is the chopping frequency and t_c is the thermal lens time constant, are shown in Fig. 13.8. While the response is found to decrease markedly for chopping at frequencies, $v \geq \frac{1}{10} t_c^{-1}$, the signal-to-noise ratio over a similar range of v_0 was found to be nearly constant (32).

An experimental comparison of the dual- and single-beam thermal lens methods was published (25). Under the constraint of equal spot sizes at the sample, the ratio of the sensitivities was found to agree with the theoretical prediction:

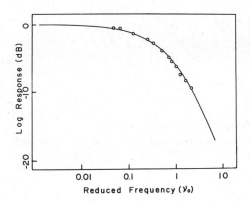

Fig. 13.8. Bode plot for the frequency response of a dual-beam thermal lens experiment. Reduced chopping frequency, $v_0 = \pi v t_c$, depends on the relative values of the time constant, t_c, and the chopping frequency, v. Data, circles, and theory, solid line, are plotted; reprinted from ref. (31) with permission.

$$\frac{E_p}{E_s} = \frac{(dn/dT)_p \lambda_s}{(dn/dT)_s \lambda_p} \tag{13.19}$$

where the s and p subscripts correspond to the single beam and probe beams, respectively. This result indicates a potentital sensitivity advantage for probing with a shorter wavelength laser beam, which would be significant for improving the sensitivity of infrared absorption measurements made with a thermal lens (13). The dynamic range of transient curve fitting of either single-beam or dual-beam data was found to be greater than lock-in amplifier detection, as shown in Fig. 13.9. The lock-in response saturates and then is reduced as the thermal lens becomes stronger, $2.303EA > 1.0$. The transient curve fitting, on the other hand, continues to be linear at five-fold greater absorbance levels. The reproducibilities and limits of detection for the single- and dual-beam configurations were found to be equivalent, probably affected by a common source of noise.

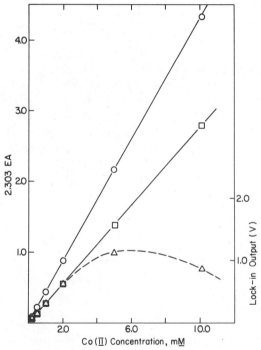

Fig. 13.9. Calibration curves for single- and dual-beam measurements. Excitation $\lambda_s = 514.5$ nm and the probe $\lambda_p = 632.8$ nm. Solid lines are from fitted transients from single-beam (circles) and dual-beam (squares) experiments. Dashed line and triangles correspond to lock-in detection of the probe beam; reprinted from ref. (25) with permission.

Since the intensity noise of the two lasers was quite different, the sample must be contributing to the uncertainty in the two measurements, either from convection (33) and/or small particles (34).

A new optical configuration for pump-and-probe thermal lens measurements is uniquely suited to small-volume detection, as shown in Fig. 13.10 (35, 36). In this arrangement the pumping and probing laser beams intersect at right angles in the sample, so that the probe beam only senses a refractive index gradient in one dimension equivalent to a cylindrical thermal lens. The induced modulation of the probe beam center intensity was shown to be linear with sample absorbance. The detection limits in terms of absorbance per unit length were 1×10^{-4} cm^{-1} in a probed volume of 25 pL using only 4 mW of pump power.

3.3. Methods of Detecting the Beam Expansion

The measurement of the change in beam spot size is conveniently accomplished by detecting the far-field beam intensity at its center as described above. This approach may not be optimal for two reasons. First, since the time-dependent noise on the laser beam is proportional to intensity, the best signal-to-noise ratio should be found where the greatest relative change in intensity is found for a small change in ω^2. While the absolute change is greatest at the beam center, the relative change maximizes as r approaches infinity. Second, monitoring at a single point is subject to spatial noise on the laser beam arising from imperfections in optical surfaces, dust, and optical inhomogeneities in the sample.

The first problem was addressed with little success by monitoring the laser beam intensity through an annulus having a radius $r \sim \omega$ (37). While the relative

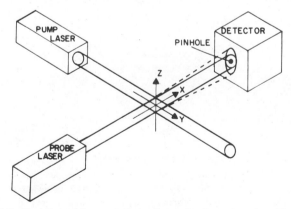

Fig. 13.10. Cylindrical thermal lens experiment for small-volume detection. The pump-and-probe laser beams intersect at right angles in the sample; reprinted from ref. (35) with permission.

intensity change was observed to be greater than at the beam center, the measurement appeared to be more sensitive to scattered light and spatial noise. Furthermore, the intensity measured no longer has the simple inverse relationship to ω^2 given by Eq. (13.12) but rather a mixed inverse and exponential response that changes with spot size.

An alternative approach that utilizes the intensity of the entire beam in one dimension is to capture the beam profile with a photodiode array detector followed by least squares fitting to a Gaussian shape to determine the change in beam spot size (38, 39). This method produces about a factor of four improvement in detection limits over comparable beam center measurements, indicating the value of spatial averaging as shown in Fig. 13.11. Only initial and steady-state profiles could be gathered, however, due to the data acquisition and processing requirements, which precluded the use of transient fitting or lock-in amplification for efficient averaging of the time-dependent noise.

To avoid this limitation, a recent study proposed the direct determination of the beam spot size using optical rather than electronic signal processing (40). The optical processor, which computes the second radial moment of the laser intensity distribution in two dimensions, $M_{2,2} = \omega^2(t)/2$, is shown in Fig. 13.12a. The laser beam is centered on a mask having a radially symmetric, parabolic transmission profile, Fig. 13.12b, which multiplies the beam by r^2. The intensity passed by the mask is collected by a lens and focused to obtain the integral of the product over the entire beam area. The time-dependent spot size of the beam, thus determined, has the expected inverse relationship with beam center detec-

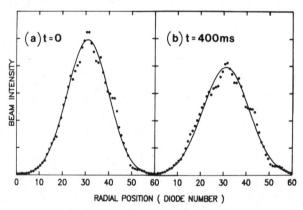

Fig. 13.11. Laser beam intensity profiles detected by a photodiode array. Solid line is a least-squares fit to Eq. (13.1). (a) Beam profile at the onset of illumination and (b) following 400 ms, which is nearly steady state. Sample is methylene blue in methanol, $A = 9 \times 10^{-3}$; reprinted from ref. (40) with permission.

a

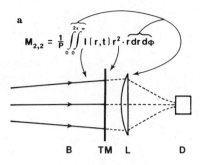

$$M_{2,2} = \frac{1}{P} \int_0^{2\pi}\int_0^{\infty} I(r,t) r^2 \cdot r\, dr\, d\varphi$$

    B   TM L   D

b

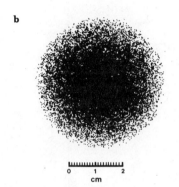

```
0    1    2
      cm
```

Fig. 13.12. (a) Optical processor for determining ω^2. B The laser beam in the far-field, TM is a parabolic transmission mask as shown in (b), L is a collecting lens and D is a detector; reprinted from ref. (40) with permission.

tion as in Fig. 13.13, which also shows the greater immunity to spatial noise. The gain in measurement precision over beam center detection provided by two-dimensional spatial averaging with the optical processor was a factor 16, corresponding to absorbance detection limits in CCl_4, $A_{min} = 1.1 \times 10^{-7}$, in a 1-cm pathlength cell.

4. ANALYTICAL APPLICATIONS

4.1. Overview

The thermal lens effect has begun to be developed as a new tool for trace-level determinations. With few exceptions most applications have utilized low to moderate power, discrete wavelength, cw gas lasers. In general, lasers of this type (e.g., He–Ne, CO_2, He–Cd, Ar^+, Kr^+) are sufficiently inexpensive and simple to operate that their use in routine determinations is feasible. In addition, the long time constant of the thermal lens effect, see Table 13.1, allows the integration of a large amount of energy from even a low-power, cw laser beam.

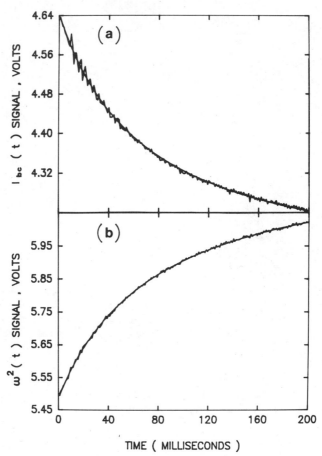

Fig. 13.13. Thermal lens transient signals. (a) Beam center measurements fit to Eq. (13.14). (b) Spot size detection fit to the inverse of Eq. (13.14). Sample is I_2 in CCl_4; reprinted from ref. (40) with permission.

By comparison, the sensitivity enhancement of pulsed laser-excited thermal lens measurements (24), corrected for lens aberrations (15, 16), is given by

$$E_p = -\frac{E_t(dn/dT)}{1.91\lambda kt_c} \tag{13.20}$$

where E_t is the total pulse energy and the laser pulse duration is much shorter than t_c. Comparing the cw enhancement, Eq. (13.17), reveals that a pulse having energy, $E_t = P \cdot t_c$, would produce equivalent enhancement as a cw laser of average power P. Given typical values of the time constant, t_c, a pulsed laser would require a large energy per pulse, ~ 1.5 mJ, to produce the equivalent thermal lens effect of a cw laser beam of only 50 mW.

Although discrete emission lines can be produced by cw gas lasers from the infrared through the ultraviolet, the lack of continuous tunability precludes the generation of spectra. The selectivity of an analytical method using such a light source must, therefore, be provided by chemical means such as a chromogenic reaction or chemical separation, while the thermal lens acts as a sensitive, quantitative but only semiselective detector. A second constraint on the design of an analytical method is the choice of solvent for the determination since differences of a factor 30 or more in sensitivity depend on the thermo-optical properties of the solvent, as shown in Table 13.1.

4.2. Spectrophotometric Determinations

The first analytical application of the thermal lens effect (18) illustrates both of the above concepts. The determination of copper at parts-per-million levels was carried out using a colorimetric reaction with EDTA monitored at 632.8 nm with a 4-mW He–Ne laser. To avoid the poor sensitivity of using water as the solvent, the determination was carried out in a $3:1$ acetone–water solution that produced a 10-fold increase in sensitivity over pure water and an absorbance detection limit of $A_{\min} = 1 \times 10^{-3}$. A subsequent study (24) improved considerably on these results by using a porphyrin ligand that had a 10^3 times greater molar absorbtivity and could be extracted into chloroform. The absorbance detection limits using pulsed laser excitation were lower by a factor of 2. The detection limits for the concentration of copper were ~ 30 ppb or 30 times lower than the earlier study.

Thermal lens measurements have also been carried out for the determination of iron. The studies have utilized 1,10-phenanthroline or its derivatives, which produce strongly absorbing, nonfluorescent complexes. Ion pair extraction into chloroform was shown to produce a 17 times greater sensitivity over aqueous samples and 10 times lower detection limits, $0.2nM$ iron, using a dual-beam experiment and transient recording of the signal (41). Similar detection limits were observed with photodiode array monitoring of the beam profile (39). While the cross beam, cylindrical thermal lens experiment produced larger concentration detection limits, $300nM$, in a mixed methanol–water solvent, the mass detection limit in the 25-pL probed volume corresponds to 0.4 fg of iron (36). A recent flow injection analysis using thermal lens detection produced superior concentration limits, $1.9nM$ iron in methanol–water solution, but mass limits in the larger probed volume of 35 fg (42).

Spectrophotometric determination of nonmetals has also been accomplished using the thermal lens effect. Nitrite (43) was detected at a $0.2nM$ concentration using a single-beam thermal lens experiment. In a similar experiment with a cw dye laser source, phosphorous was detected at a fixed wavelength as the blue, phosphomolybdate complex at a concentration of 0.2nM. Most recently, an intracavity thermal lens experiment utilizing a secondary resonant cavity for the sample was able to detect phosphorus at 0.4nM levels and arsenic at 1.3nM (44).

While the above studies utilized colorimetric reactions carried to completion for analytical selectivity with the analyte concentration being related to the amplitude of response, thermal lens detection has also been shown to be suitable for kinetic analysis where quantitation is inferred from the rate of change in the response (28). A dual-beam thermal lens experiment with lock-in detection provided convenient real-time monitoring of dopachrome from the enzyme-catalyzed oxidation of dopamine. Concentrations of dopamine as low as 1 μM could be detected in the initial slope of the kinetic curves.

Finally, for the functional group region of the infrared spectrum, single-wavelength thermal lens measurements may provide qualitative as well as quantitative information. The determination of hydrocarbons in condensed phase was accomplished using the 3.39-μm line from a He–Ne laser with a minimum detectable mass of 8 ng of trimethylpentane (13). The absorbance at this wavelength could be interpreted from functional group molar absorptivities of a series of hydrocarbons; the average specific absorption coefficient for saturated hydrocarbons at this wavelength was found to be 2.1 ($\pm$0.1) Lg^{-1} cm^{-1}.

4.3. Chromatographic Detection and Flow Injection Analysis

The sensitivity of the thermal lens technique, the small volume characteristics of a focused laser beam, and the more general applicability of thermal rather than fluorescence detection have made this method an attractive detector for liquid chromatography and flow injection analysis. Given the need for selectivity in thermal lens measurements, combining these methods would appear to be mutually beneficial. To use a thermal lens as a detector of a flowing process, the effect of flow on the measurement needed to be considered (45). The major disturbance of the thermal lens by the flow is caused by cross flow and mixing in the illuminated region of the sample. At excessive flow rates, a loss of sensitivity due to increased thermal conduction and an additional source of proportional noise from flow pulsations is observed. At lower flow velocities suitable for most applications, the performance is not severely degraded and a thermal diffusion time response can still be used to fit the data.

A second overall consideration for thermal lens monitoring of a flow process is detection volume. Based on the relationship between laser beam spot size and divergence, Eq. (13.2) and (13.3) above, a trade-off between the minimum sample volume and pathlength, b, is (46)

$$V_{min} \geq 48\lambda b^2 \qquad (13.21)$$

The quadratic dependence of minimum volume on pathlength assures that while concentration sensitivity of a smaller volume thermal lens measurement decreases due to reduced pathlength, the specific sensitivity increases at a faster rate.

The first application of thermal lens detection to liquid chromatography employed a single-beam experiment (47). A fast method of fitting thermal lens transient data during the 250-ms period between experiments was implemented, allowing the instrument to serve as a real-time monitor. Minimum detectable absorbance at 458 nm in methanol–water solution was $A_{min} = 1.5 \times 10^{-5}$ corresponding to 0.2 ng o-nitroaniline injected. Subsequent improvement of the transient data processing to account for error propagation (42) led to absorbance detection limits for flow injection analysis of $A_{min} = 8.5 \times 10^{-7}$ in 1-cm pathlength of CCl_4.

Further studies of thermal lens detection in liquid chromatography have been carried out. A dual-beam instrument was tested with conventional (32) and microbore (26) HPLC. The absorbance detection limits are around $A_{min} = 2 \times 10^{-6}$ for both experiments. A single-beam experiment whereby the thermal lens signal is measured as the second harmonic of the chopping waveform with a lock-in amplifier, in the presence of the large, fundamental modulation was used for conventional HPLC detection (48). The second-harmonic limits are about a factor of 2–3 worse than a comparable dual-beam experiment. Most recently, thermal lens detection was demonstrated for open tubular liquid chromatography, representing an enormous challenge for small-volume measurements (49). Absorbance detection limits are $A_{min} \sim 6 \times 10^{-5}$ in a 100-μm pathlength cell; the minimum detectable injection of o-nitroaniline contained 30 pg.

The combination of thermal lens detection and flow injection analysis is also mutually beneficial. The small volume of injection and the need for sensitivity in detection of flow injection analysis are well matched to the thermal lens capabilities. Sensitivity of detection in a spectrophotometric method does not always lead to lower detection limits in practical applications due to the uncertainties in the blank signal that are proportional to sensitivity (50). A significant contribution to the blank uncertainty arises from sample manipulation, where contamination, carryover, and sample cell realignment are likely sources of error when using conventional sample handling techniques.

Flow injection analysis has been advocated as a means of reducing contamination and carryover in ultratrace analysis with laser-based detectors (51). Advantages over conventional sample manipulation include sample processing in a closed, inert manifold, a high rinsing efficiency, and reduced consumption of ultrapure reagents. Thermal lens detection in flow injection analysis has been studied (42). Absorbance detection limits, $A_{min} = 8.5 \times 10^{-7}$, were observed for injections into CCl_4. Spectrophotometric detection of iron can be accomplished with a 100-μL injection of a 0.4-ppb solution containing less than 40 pg of analyte in a total analysis time of less than one minute.

4.4. Detection in Gases and Supercritical Fluids

Analytical applications of gas-phase thermal lens measurements have been comparatively fewer than for condensed-phase samples. The thermo-optical properties of gases are somewhat less favorable than liquids, particularly for cw excitation. The change in refractive index with temperature of a typical gas at 1 atm pressure, nitrogen for example, is only $0.9 \times 10^{-6} \, K^{-1}$, which is a factor of 600 smaller than liquid CCl_4 (51). The thermal conductivity of nitrogen is about four times smaller, leading to an enhancement per unit laser power, $E/P = 0.029 \, mW^{-1}$ at $\lambda = 632.8$ nm. This value is 160 times smaller than the enhancement for CCl_4 and 5 times smaller than for water. Due to the small density and heat capacity, time constants for the thermal lens effect in gases are very short. For 1 atm nitrogen and 100 μm beam spot size in the sample, the time constant predicted by Eq. (13.15) is $t_c = 0.11$ ms, indicating 250 times faster thermal relaxation than in liquids. The short time constants in gases point to a significant sensitivity advantage for pulsed excitation, since the breakeven pulse energy, $E_t = P \cdot t_c$, is proportional to t_c according to Eq. (13.20). Thus, for thermal lens measurements in nitrogen with a 100-μm spot size beam, a 10-W cw laser would be required to produce an enhancement, $E = 290$, while a pulsed laser would need to generate only 1.1 mJ pulses to achieve the same sensitivity.

The above concepts were described in detail in the first analytical paper on gas-phase thermal lens measurements (51). A 20-μJ pulsed dye laser was used to detect NO_2 at a 0.8-ppm level. Using much larger average power from a cw argon ion laser, 700 mW, and an improved optical configuration suitable for a long pathlength, 1 m, sample cell, improved detection limits of 5 ppb NO_2 were generated with chopped excitation and lock-in amplification of the modulation on an unfocused probe beam (52). The same optical configuration with a clever, windowless flow cell was used to generate infrared spectra of various organic vapors with a line tunable, cw CO_2 laser (53). Detection limits are 12 ppb for methanol, corresponding to an absorbance of $2.7 \times 10^{-7} \, cm^{-1}$. Pulsed CO_2 laser excitation has been used for thermal lens detection of dichlorodifluoromethane at 10 ppb in argon at 100 torr, which would extrapolate to sub-parts-per-billion levels at atmospheric pressure (54). The effects of transition saturation and comparisons with colinear photothermal deflection were also considered.

Thermal lens measurements in systems intermediate between gases and liquids, near the liquid–vapor critical point, do not behave as the average of either

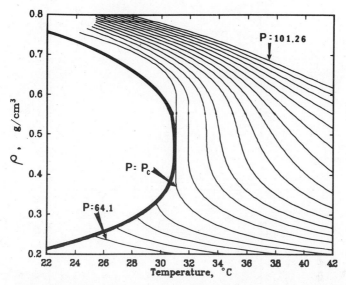

Fig. 13.14. Density versus temperature diagram for CO_2. Isobars separated by 2.2 atm are illustrated. The heavy line indicates the liquid–vapor coexistence region; reprinted from ref. (55) with permission.

case. Since the coefficient of thermal expansion diverges at the critical point, the resulting change in refractive index with temperature, dn/dT, becomes unbounded. While this results in excessive sensitivity giving rise to critical point opalescence, one can control the sample pressure and temperature to achieve large but finite sensitivity as shown in Fig. 13.14 where $(dn/dT)_p$ is proportional to the slope of the isobars. In the first study of thermal lens measurements in CO_2 near its critical point (55), enhancements per unit laser power of 100 times greater than CCl_4 were predicted and observed in a constant volume, high-pressure sample cell. No information about detection limits could be obtained, however, due to limitations in the batch-style sample handling.

In a subsequent study (56) flow injection of samples into a supercritical fluid stream provided a means of transferring samples from ambient laboratory conditions to the high-pressure, controlled temperature environment for detection. Temperature-dependent sensitivity studies were carried out along two isobars, as shown in Fig. 13.15a, where the maximum sensitivity observed is $E/P = 790$ mW^{-1} at a pressure and temperature of 4.5 atm and 3°C above the critical point, respectively. Absorbance detection limits scaled inversely with enhancement, as shown in Fig. 13.15b, with a minimum limit at the maximum sensitivity, $A_{min} = 2.4 \times 10^{-7}$, at a laser power of 50 mW. Flow injection proved to be a convenient means to exploit the thermo-optical properties of the solvent

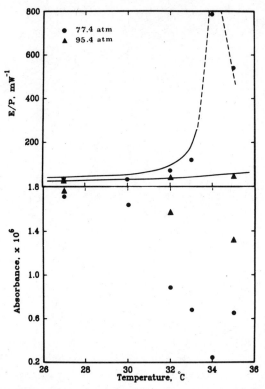

Fig. 13.15. (*a*) Thermal lens enhancement per milliwatt laser power in CO_2 for two isobars. Theoretical predictions are the lines, dashed in regions where the values are subject to numerical errors. (*b*) Absorbance detection limits in CO_2. Best case detection limit was $A_{min} = 2.4 \times 10^{-7}$; reprinted from ref. (56) with permisson.

for detection with no more difficulty or time required for sample manipulation than with normal liquid solvent.

ACKNOWLEDGMENT

This material is based upon work supported by the National Science Foundation under Grants CHE82-06898 and CHE85-06667.

REFERENCES

1. A. Hordvik, *Appl. Opt.*, **16,** 2827 (1973).
2. S. D. Woodruff and E. S. Yeung, *Anal. Chem.*, **54,** 1174 (1982).

3. D. A. Cremers and R. A. Keller, *Appl. Opt.*, **21,** 1654 (1982).

4. J. P. Gordon, R. C. C. Leite, R. S. Moore, S. P. S. Porto, and J. R. Whinnery, *J. Appl. Phys.*, **36,** 3 (1965).

5. J. M. Harris and N. J. Dovichi, *Anal. Chem.*, **52,** 695A (1980).

6. A. C. Boccara, D. Fournier, and J. Badoz, *Appl. Phys. Lett.*, **36,** 130 (1980).

7. T. I. Chen and M. D. Morris, *Anal. Chem.*, **56,** 19 (1984).

8. H. J. Eichler, *Opt. Acta.*, **24,** 631 (1977).

9. M. J. Pelletier, H. R. Thorsheim, and J. M. Harris, *Anal. Chem.*, **54,** 239 (1982).

10. H. Kogelnik and T. Li, *Proc. IEEE*, **54,** 1312 (1966).

11. J. R. Whinnery, *Accts. Chem. Res.*, **7,** 225 (1974).

12. N. J. Dovichi and J. M. Harris, *Anal. Chem.*, **52,** 2338 (1980).

13. C. A. Carter, J. M. Brady, and J. M. Harris, *Appl. Spectrosc.*, **36,** 309 (1982).

14. C. A. Carter and J. M. Harris, *Appl. Spectrosc.*, **37,** 166 (1983).

15. S. J. Sheldon, L. V. Knight, and J. M. Thorne, *Appl. Opt.*, **21,** 1663 (1982).

16. C. A. Carter and J. M. Harris, *Appl. Opt.*, **23,** 476 (1984).

17. C. Hu and J. R. Whinnery, *Appl. Opt.*, **12,** 72 (1973).

18. N. J. Dovichi and J. M. Harris, *Anal. Chem.*, **51,** 728 (1979).

19. J. H. Brannon and D. Magde, *J. Phys. Chem.*, **82,** 705 (1978).

20. N. J. Dovichi and J. M. Harris, *Anal. Chem.*, **53,** 106 (1981).

21. F. R. Grabiner, D. R. Siebart and G. W. Flynn, *Chem. Phys. Lett.*, **17,** 189 (1972).

22. M. E. Long, R. L. Swofford, and A. C. Albrecht, *Science*, **191,** 183 (1976).

23. A. J. Twarowski and D. S. Kliger, *Chem. Phys.*, **20,** 253 (1977).

24. K. Mori, T. Imasaka, and N. Ishibashi, *Anal. Chem.*, **54,** 2034 (1982).

25. C. A. Carter and J. M. Harris, *Anal. Chem.*, **55,** 1256 (1983).

26. C. E. Buffett and M. D. Morris, *Anal. Chem.*, **55,** 376 (1983).

27. H. L. Fang and R. L. Swofford in *Ultrasensitive Laser Spectroscopy*, D. S. Kliger, Ed., Academic Press, New York, 1983, Chapter 3.

28. J. P. Haushalter and M. D. Morris, *Appl. Spectrosc.*, **34,** 445 (1980).

29. J. W. Perry, E. A. Ryabov, and A. H. Zewail, *Laser Chem.*, **1,** 9 (1982).

30. H. L. Fang and R. L. Swofford, *J. Appl. Phys.*, **50,** 6609 (1979).

31. N. J. Dovichi and J. M. Harris, *Proc. Soc. Photoopt. Instr. Eng.*, **288,** 372 (1981)

32. C. E. Buffett and M. D. Morris, *Anal. Chem.*, **54,** 1824 (1982).

33. C. E. Buffett and M. D. Morris, *Appl. Spectrosc.*, **37,** 455 (1983).

34. R. Anthone, P. Flament, G. Gouesbet, M. Rhazi, and M. E. Weill, *Appl. Opt.*, **21,** 2 (1982).

35. N. J. Dovichi, T. G. Nolan, W. A. Weimer, *Anal. Chem.*, **56,** 1700 (1984).

36. T. G. Nolan, W. A. Weimer, and N. J. Dovichi, *Anal. Chem.*, **56,** 1704 (1984).

37. R. A. Leach and J. M. Harris, "Abstracts of Papers" 183rd National Meeting of the American Chemical Society, Las Vegas, April, 1982, American Chemical Society, Washington, D.C., Abstr. ANYL 124.

38. K. L. Jansen and J. M. Harris, *ibid.:* Abstr. ANYL 126.

39. K. Miyaishi, T. Imasaka, and N. Ishibashi, *Anal. Chem.*, **54,** 2039 (1982).

40. K. L. Jansen and J. M. Harris, *Anal. Chem.* **57,** 1698 (1985).

41. K. Miyaishi, T. Imasaka, and N. Ishibashi, *Anal. Chim. Acta*, **124,** 381 (1981).

42. R. A. Leach and J. M. Harris, *Anal. Chim. Acta*, **164,** 91 (1984).

43. K. Fujiwara, H. Uchiki, F. Shimokoshi, K. Tsunoda, K. Fuwa, and T. Kobayashi, *Appl. Spectrosc.*, **36,** 157 (1982).
44. V. I. Grishko, I. G. Yudelevich, and V. P. Grishko, *Anal. Chim. Acta*, **160,** 159 (1984).
45. N. J. Dovichi and J. M. Harris, *Anal. Chem.*, **53,** 689 (1981).
46. C. A. Carter and J. M. Harris, *Anal. Chem.*, **56,** 922 (1984).
47. R. A. Leach and J. M. Harris, *J. Chromatogr.*, **115,** 218 (1981).
48. T. J. Pang and M. D. Morris, *Anal. Chem.*, **56,** 1467 (1984).
49. M. J. Sepaniak, J. D. Vargo, C. N. Kettler, and M. P. Maskarinec, *Anal. Chem.*, **56,** 1252 (1984).
50. T. D. Harris, *Anal. Chem.*, **54,** 741A (1982).
51. K. Mori, T. Imasaka, and N. Ishibashi, *Anal. Chem.*, **55,** 1075 (1983).
52. T. Higashi, T. Imasaka, and N. Ishibashi, *Anal. Chem.*, **55,** 1907 (1983).
53. T. Higashi, T. Imasaka, and N. Ishibashi, *Anal. Chem.*, **56,** 2010 (1984).
54. G. R. Long and S. E. Bialkowski, *Anal. Chem.*, **56,** 1481 (1984).
55. R. A. Leach and J. M. Harris, *Anal. Chem.*, **56,** 1481 (1984).
56. R. A. Leach and J. M. Harris, *Anal. Chem.*, **56,** 2801 (1984).

CHAPTER

14

PICOSECOND SPECTROSCOPY
IN ANALYTICAL CHEMISTRY

M. J. WIRTH

Lawrence Livermore
National Laboratory
Livermore, California

G. J. BLANCHARD

Bell Communications Research, Inc.
Red Bank, New Jersey

1. INTRODUCTION

The utility of most spectroscopy methods in analytical chemistry originates from the uniqueness of the molecular and atomic spectra. The narrower the peaks in a given spectral range, the greater the amount of information. Thus, atomic electronic spectroscopy and molecular Raman spectroscopy are especially valuable. The parameter time does not gain the same type of advantage as wavelength because intensities in the time domain decay exponentially from a common time zero, whereas wavelength-resolved data appear as peaks of greatly varying positions. Time decays are therefore not as generally useful as spectra for quantitating components in mixtures. However, the wide use of time-resolved fluorescence in characterizing samples testifies to the fact that there are other types of valuable analytical information in spectroscopic time decays.

For molecular systems in liquid solution, much of the dynamical information is on the picosecond time scale. Nanosecond time decays are observed for the case of fluorescence lifetimes of a restricted set of molecules; many organic molecules have picosecond fluorescence lifetimes. Because the same fundamen-

477

tals of lasers and chemical systems apply to both the nanosecond and picosecond domains, no arbitrary division of the two time scales is made in this discussion. However, picosecond technology, but not nanosecond technology, is discussed extensively because an understanding of pulse generation is crucial to appreciating the potential of time-resolved spectroscopy in analytical chemistry.

There are three classes of applications of time-resolved spectroscopy to analytical chemistry. The first is the separation of interfering components based on different lifetimes. While there are numerous examples, a very illustrative one is the rejection of fluorescence from Raman scattering. This was first accomplished using nanosecond technology by which Raman spectra of highly fluorescent biological molecules were obtained (1). The use of time-resolved spectroscopy to separate signals of differing decay times has been adequately covered and will not be further discussed in this chapter. The second class of applications takes advantage of ultrahigh stability in the laser output when picosecond pulses are generated. The origin of this stability and its application to Raman spectroscopy are discussed in this chapter. The third class of applications is measurement of a dynamical quantity that characterizes the identity of a component. For example, fluorescence depolarization measurements are valuable for determining sizes and structures of macromolecules. This technique is useful in biochemistry and a discussion of it is included in this chapter. The basis for all three classes of applications can be better appreciated from a discussion of the essentials of laser pulse formation, as well as the nature of chemical phenomena on the picosecond time scale.

2. PRINCIPLES OF PULSE FORMATION

Lasers are used in picosecond spectroscopy because the formation of short pulses inherently requires coherent light, that is, well-defined phase relations among the light waves. A two-mirror laser cavity allows a wavelength to oscillate only if the wave constructively interferes with itself. This occurs when a cavity length can be integrably divided by one-half a wavelength. Figure 14.1 illustrates two allowed wavelengths of a cavity at one instant in time. Each of these wavelengths is termed a longitudinal mode. The allowed wavelengths of the mode satisfy the relationship $\lambda = 2L/n$, where n is an integer. The amplitude of the electric field must be zero at each mirror for the laser to oscillate. This boundary condition is characteristic of standing-wave behavior, arising from the fact that there is a 180° phase change upon reflection by a more refractive medium. Since only waves having zero amplitude at the mirrors can constructively interfere with themselves, the spatial phases are fixed. For a given mode the electric field amplitude varies with time, as determined by its optical frequency. However, the maximum amplitude at any point in space is determined by the spatial phase.

Fig. 14.1. Two allowed wavelengths in a laser cavity. Typically, the ratio of cavity length to wavelength is much greater than one. This is an illustrative example, however.

The space and time variation of the amplitude of a mode is described by Eq. (14.1).

$$E(x, t) = E_0\sin(kx)\sin(\omega t + \phi) \tag{14.1}$$

For a given cavity length, the number of modes oscillating is restricted by the wavelength dependence of the lasing transition and any optical components in the cavity. For the argon ion laser, the transition is relatively narrow and there are roughly 80 modes oscillating under typical operating conditions. The frequency spectrum of a multi-mode laser output is illustrated in Fig. 14.2. The laser emission is described by the envelope, but the spectrum of the laser output is a series of discrete spikes when emitted from a standing-wave cavity. Each spike corresponds to a longitudinal mode. The frequency width of each mode is determined by the mechanical stability of the cavity length. Since the frequency spacing of the modes is defined by the cavity length, the output of the laser can be made to be in a single mode by making the cavity sufficiently short that other modes are outside the bandwidth of the lasing transition. This can be accomplished more conveniently by inserting a pair of closely spaced plates, called an etalon, into a longer cavity. Since the time span of a lasing pulse and the allowed frequency width of the lasing transition are inversely related, for picosecond spectroscopy one desires many cavity modes to oscillate simulta-

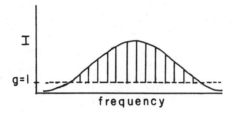

neously, and more widely spaced etalons are sometimes used to limit the large number of modes in the cavity.

The electric field amplitude in time and space is the sum of the amplitudes of all of the modes in the cavity.

$$E(x, t) = \sum_j E_{0j}\sin(k_j x)\sin(\omega_j t + \phi) \tag{14.2}$$

A pulse is formed in the cavity when the electric fields of the many modes add constructively at one point in time and destructively interfere at all other times. The square of the summation of the amplitudes of nine modes is illustrated in Fig. 14.3a, where the peak of the pulse is the maximum in the optical fringe pattern. The square of the amplitudes is the intensity. The square of the summation of 31 modes is shown in Fig. 14.3b, from which it is seen that more modes result in a narrower pulse. In generating Fig. 14.3, the distribution of E_0 values is a Gaussian having a width proportional to the number of modes. Since the amplitudes in the time and frequency domains are Fourier transforms of one another, the time profile of the pulse envelopes in Fig. 14.3 are also Gaussians.

The Fourier transform relation quantitatively predicts that the larger the number of modes in a given cavity (i.e., the wider the spectral distribution), the shorter the pulse in time. This is a very important concept in the design of picosecond lasers. It is the wide spectral bandwidth required that dictates the use of organic dyes for most picosecond pulse generation. The Nd:YAG and argon ion lasers generate only subnanosecond pulses due to their narrower bandwidths.

A crucial feature in picosecond pulse formation is that in order for the wave amplitudes to add constructively (as in Fig. 14.3), the phase factors of the waves, indicated as ϕ_j in Eq. (14.2), must be constant in time. The waves in Fig. 14.3 all have zero phase factors. Adding in a constant greater than zero for all modes has only the effect of moving the pulse in time. However, if a different phase for each mode is chosen, the pulse is destroyed. The effect of random phase factors for each mode is illustrated in Fig. 14.4. The resulting amplitude is randomly spread over all space and has the frequency components

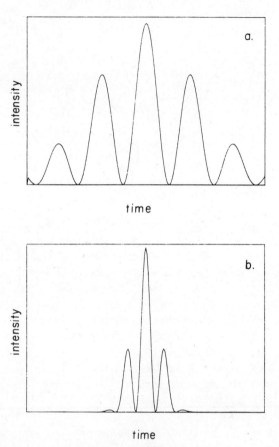

Fig. 14.3. (a) The square of the summation of the amplitudes of nine modes in a laser cavity. A Gaussian distribution of E_0 values was used, so the time profile of the laser pulse will also be Gaussian. (b) The square of the summation of the amplitudes of 31 modes in a laser cavity. The destructive interference away from the center of the pulse envelope is more complete than for (a) and the pulse is narrower in time.

of the modes. A dye laser having modes of random phase would clearly be unsuitable for time-resolved experiments, and additionally, it would have high-frequency noise on its output. By contrast, the phase-locked dye laser allows transform-limited time resolution and is quantum noise limited at high frequencies. A laser is commonly called "mode locked" when the phases of all modes are related, and it is important to achieve this state experimentally in order to accrue the advantages of picosecond time resolution and low noise detection.

When a laser is mode locked, the energy stored in the lasing transition is placed in one short pulse rather than continuously being used. Thus, the power

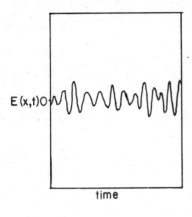

Fig. 14.4. The summation of the amplitudes of several modes with random phase relationships. This is representative of the very high frequency noise found on cw lasers.

in each pulse is much greater than that observed under non-mode-locked conditions. Since stimulated emission increases nonlinearly with increasing electric field amplitude, mode locking is a natural state for a laser. However, it is actually difficult in practice to achieve self-mode-locking because noise randomly shifts the phases of each mode in time. The chief source of noise is spontaneous emission, which generates an electric field amplitude outside of the pulse. This is equivalent to changing the mode phases to produce amplitude outside of the pulse. The key to mode locking a laser is to prevent lasing at all times outside of the window of the desired pulse.

The laser is mode locked by forcing all wavelengths to constructively interfere in only one specific region in time. By default, all phases must be equal at that point when this is achieved. Physically, this is accomplished by choosing a point in the cavity and forcing the intracavity light intensity to be above lasing threshold at only one instant in time. As a crude example, a fast shutter could be instantaneously opened and closed. Since the light in the cavity returns to the same point a certain time later, determined by the speed of light and the cavity length, the opening of the shutter must be repetitive at the cavity round trip rate. A mode-locked laser thus gives a train of identical pulses having a repetition rate determined by the cavity length.

There are three basic ways of mode locking a laser, and each involves preventing lasing at all but one short segment of time.

1. In active mode locking an external signal is applied to an element in the laser cavity that makes the loss high at all but the desired instant in time. Since the pulse repeats for any point in the cavity at the round trip transit time of light, or some multiple thereof, the loss applied to the cavity must be repetitive. In practice, active mode locking is achieved by placing an acousto-optic element in the cavity, such as a quartz prism or flat, and exciting acoustic waves in the quartz with between 0.5 and 1.0 W of rf power, typically at approximately 40 MHz. The acoustic waves deflect light out of the cavity, and only at the zero-

crossing points of the acoustic wave does the laser have minimal loss. The zero-crossing points thus define the small segment of time for the mode-locked pulse. Because the slope of the 40-MHz wave is not very steep near its zero-crossing point, acousto-optic mode locking is most compatible with long pulses, such as those of the argon ion or Nd:YAG lasers, which are 100 and 70 ps, respectively.

2. Passive mode locking is accomplished by placing an absorbing dye into a focal point in the laser cavity. The absorbance provides a high loss at all times except when a large pulse passes through the absorbing dye. The large pulse saturates the absorption, making the cavity loss low only for the mode-locked pulse itself. The dye recovers from saturation soon enough to prevent free running of the laser between pulses. Absorption by the absorber shortens the leading edge of the pulse, and the trailing edge is attenuated because the gain is depleted by the peak of the pulse (2). While passive mode locking generates pulses of less than a picosecond in duration (3), its chemical applications are limited because independently tunable passively mode-locked lasers cannot readily be synchronized to one another. In addition, the mode-locked tuning range is limited by the small number of known saturable absorber dyes.

3. Synchronous pumping is a method of generating picosecond pulses while retaining the ability to synchronize two or more dye lasers. With this technique the gain of the dye laser is modulated rather than the loss. The dye laser is pumped by a mode-locked laser such as an argon ion laser, and the cavity lengths of each are matched so that every time a dye laser pulse returns to the gain medium, an argon ion pulse arrives to amplify it. Due to the broad gain profile of the dye, the pulses are typically only 3–5 ps in duration. Since the pulse repetition rate of the dye laser is precisely determined by that of the pump laser, two dye lasers can be pumped by the same laser, resulting in excellent synchronization.

Recent advances in laser technology have improved the pulse quality of both passively mode-locked and synchronously pumped dye lasers. One improvement is to change the laser cavity into a ring configuration, which causes two pulses to circulate in the cavity in opposite directions. For passively mode-locked lasers the pulses collide in the absorber. By separating the pulse into two equal halves and sending them separately through the absorber, the leading edges are more strongly attenuated compared to a single pulse of twice the power. Pulses of 100 fs (0.1 ps) can be obtained from the passively mode-locked ring laser (3). The synchronously pumped dye laser is also improved using the ring configuration. Because there is no saturable absorber in the synchronously pumped dye laser, the advantage of the ring configuration has a different physical origin. Note that in a standing-wave laser cavity, excited states at cavity nodes are not used while excited states at wave crests are significantly depleted. In a ring laser there are no standing waves, thus there is a more efficient use of all excited molecules. Due to the lower overall depletion of the excited states, the ring configuration allows the center of the pulse to grow larger relative to the wings than does the standing-wave cavity. Synchronously pumped ring lasers provide pulses of about 0.8-ps duration (4).

A second advance in the generation of ultrashort pusles is the use of optical fibers to create a "soliton" laser. A solition is a nonlinear wave whose characteristic feature is that it maintains a constant time profile. An optical fiber can be constructed such that an infrared pulse must maintain a transform-limited width of several tens of femtoseconds. When this fiber is placed into the cavity of an infrared laser, the output is guaranteed to be this ultrashort width. The soliton laser has been successfully demonstrated for a color center laser in the near infrared, with pulses as short as 30 fs (5).

3. ULTRASENSITIVE RAMAN DETECTION USING PICOSECOND PULSES

A mode-locked laser is extremely stable on the nanosecond and picosecond time scales because the electric field amplitude is precisely determined by the phases, frequencies, and amplitudes of the modes. A cw laser, as illustrated earlier by Fig. 14.4, has abundant high-frequency noise. For the mode-locked laser, sources of noise such as optical or electronic instabilities occur on longer time scales, thus their effects can be minimized by detection on a short time scale. This high stability has been used to advantage in ultrasensitive Raman spectroscopy, where Raman spectra of surface monolayers have been obtained without the need for surface enhancement (6, 7).

In order to appreciate how high stability can be used for Raman detection, consider the following. The Raman transition is excited by two visible lasers having frequencies ω_L and ω_S, where the difference frequency, $\omega_L - \omega_S$, corresponds to the vibrational frequency of interest. The energy level diagram for this process is shown in Fig. 14.5. For every vibrational quantum excited, a photon of frequency ω_L is converted to a photon of frequency ω_S. The Raman intensity can thus be measured either by the loss on one beam or the gain on the other. This experiment is called "stimulated Raman gain–loss" and falls into the category of nonlinear Raman spectroscopy.

The Raman gain–loss signal is obtainable from the pump-probe experiment illustrated in Fig. 14.6. When the pump beam, ω_L, is modulated at 10 MHz, the signal of interest is the small fraction of the probe beam, ω_S, and has picked up the 10-MHz modulation. A high-frequency lock-in amplifier isolates the 10-MHz signal from the probe oeam. The amount of 10-MHz noise on the probe beam limits the size of the signal that can be detected. For a monolayer the gain on the probe beam is so small that the detection of intensity changes in the probe beam of only a few parts in 10^9 is required. This is many orders of magnitude more sensitivity than that achievable by the best absorbance mea-

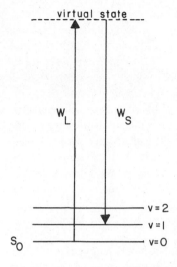

Fig. 14.5. The energy level diagram of the stimulated Raman gain–loss process. The lifetime of the virtual state is typically less than 1 ps.

surements; it corresponds to nearly the shot noise limit for a 20-mW probe beam. Accomplishment of such extraordinary detectability was first reported by two groups at Bell Laboratories (8, 9), where it was found that a 1-s integration time sufficiently averaged the low-frequency laser noise that the detectability was limited by the shot noise on the probe beam.

The achievement of shot-noise-limited detection was at first very difficult to apply on a routine basis because extreme care must be taken to minimize the

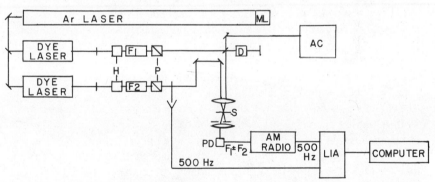

Fig. 14.6. Experimental apparatus for the stimulated Raman experiment. This figure shows the triple modulation scheme. The single modulation scheme is the same as making $F_1 = 0$ (dc) and eliminating the 500-Hz chopper from the lower beam. The detector is then just the lock-in amplifier set to detect frequency F_2. ML = mode locker, F_1 = frequency of modulation of ω_s, F_2 = frequency of modulation of ω_L, H = half-wave plate, P = Glan-Thompson prism polarizer, D = variable delay to optimize pulse overlap, AC = autocorrelator for pulse diagnosis, S = sample, PD = photodiode, LIA = lock-in amplifier. The AM radio must be capable of reception in the 10–20-MHz region.

effects of rf pickup from the 10-MHz modulation source and scatter of the pump beam into the diode detector. Recently, a valuable improvement on the detection scheme has been made to avoid these problems, allowing the sensitive measurements to be made more routinely (10, 11). This scheme involves triple modulation of the lasers, where both the pump and probe beams are modulated at MHz frequencies, and the pump beam is chopped at an audio frequency. The arrangement is illustrated in Fig. 14.6. The signal of interest is the megahertz sum or difference frequency on the probe, chopped at the audio frequency. The output of the diode detecting the probe beam is sent to an AM or single sideband radio tuned to the megahertz sum or difference frequency. The audio output of the radio is related to the Raman intensity and is measured by a low-frequency lock-in amplifier. Since there is no beam or electronic source at the sum or difference frequency, background interference is minimal. While the triple modulation scheme is slightly more complicated electronically, it is far more convenient experimentally.

The ultrasensitive Raman measurement stimulates the imagination to find ways of using the high stability of the mode-locked laser to improve other experiments. In general, the ultrahigh stability can be used to advantage in any measurement where the signal, but not the background, can be modulated at a high frequency. The reasons that Raman sensitivity is vastly improved by pump-probe detection are twofold. (1) The signal of interest can be modulated selectively, thus higher throughput is gained because a double monochromator is not needed. (2) The Raman signal is directed spatially toward the detector rather then being scattered uniformly in all directions. Thus, the Raman sensitivity is improved both by increasing the signal size and decreasing the background.

Fluorescence would appear to be a good candidate for pump-probe detection because of its similarity to Raman spectroscopy. The probe beam stimulates fluorescence emission for the same fundamental reason that Raman emission is stimulated. Raman scattering and fluorescence are the same spectroscopic phenomenon except for a phase relation. Emission that occurs in phase with the excitation light is Raman, while emission that is randomly phased with respect to the excitation is termed fluorescence (12). Since electronic intermediate states lose their phase coherence in less than a picosecond, Raman scattering occurs "instantaneously," and most of the light is emitted later in the form of fluorescence. This is why fluorescence is so much stronger than Raman scattering for systems with a reasonably large quantum yield. A stimulated fluorescence gain experiment is performed the same way as a corresponding Raman experiment. The wavelength of the modulated pump beam is tuned to the absorption band, and the probe is tuned to the emission band. As with the Raman experiment the induced modulation on the probe beam is the signal of interest.

Unlike Raman spectroscopy, conventional fluorescence is generally not limited by the strength of the signal but rather by background fluorescence from impurities. The pump-probe experiment does not reduce this limit because the fluorescence of the background is modulated along with that of the fluorophor. Further, fluorescence spectroscopy is already a sensitive technique; fluorescence from submonolayers can be detected. The advantage of using stimulated fluorescence gain lies not in the signal-to-noise ratio but in time resolution. The pump-probe experiment circumvents the use of relatively slow photomultiplier tubes. It has been shown that fluorescence lifetimes can be determined by monitoring the probe intensity as a function of probe delay (13), and the time resolution is thereby limited by the widths of the laser pulses rather than the photomultiplier response. Measuring fluorescence on the picosecond time scale is very useful because fluorescence depolarization can be detected, allowing a study of rotational diffusion for many systems of interest. These studies are discussed in the next section.

Since the already sensitive technique of fluorescence does not greatly benefit from pump-probe technology, a less sensitive spectroscopic measurement worth examining is conventional absorption spectroscopy. The detectability of this method is usually limited by the stability of the source. Pumping and probing at the same wavelength is analogous to the absorption measurement, and the modulated probe beam contains the absorbance information. Note that even though absorbance involves only one photon, it is amenable to a pump-probe experiment because the probe beam senses the modulated concentration of ground states. In principle, this would provide a significant improvement in detectability because the absorbance measurement is made to be equivalent to a fluorescence measurement. The probe beam interacts equally well whether it is causing excitation or emission. Also, the pump-probe absorption experiment ought to be more sensitive than the corresponding Raman experiment because electronic cross sections are larger than vibrational cross sections. However, one factor that makes this detectability more difficult to accomplish is that the absolute level of the signal is smaller. The transition in the absorption can saturate readily, while the virtual state involved in the Raman transition would never reasonably saturate. Pump levels 10–100 times smaller than those in the Raman experiment must be used, thus requiring a correspondingly weaker probe beam. The amplifiers needed before the lock-in amplifier or radio thus introduce more noise. The pump-probe absorption experiment appears promising but is in need of experimental evaluation.

Pump-probe absorption spectroscopy could be made into an even more powerful technique by employing two different wavelengths to improve the selectivity. For example, pentacene has strong absorption bands in both the visible and ultraviolet. By pumping the visible band, the absorbance of the UV band is modulated. Interferents that absorb only in the visible or only in the UV would be eliminated. This experiment is presently being developed.

One of the most challenging aspects of any sensitive pump-probe experiment, be it Raman, fluorescence or absorption, is to accomplish careful mode locking of the laser to minimize mode noise. The argon ion or other pump laser must be mode locked with the assistance of a fast photodiode and sampling oscilloscope, typically with an overall risetime of 50 ps. A frequency synthesizer is needed to provide stable acousto-optic modulation of the light in the cavity. Mode locking of the dye laser is more difficult because there are so many more possible modes. Photodiodes are not fast enough to diagnose dye laser pulses, thus autocorrelations are required (14). The shape of the autocorrelation trace reveals whether or not the pulses are well mode locked, and usually they are not. It has been found that the placement of an etalon in the cavity to reduce the number of modes makes shot-noise-limited behavior of the laser obtainable. The slow development of applications of ultrastable lasers is due to the induction time required in gaining expertise with control of the laser properties.

4. APPLICATION OF PICOSECOND SPECTROSCOPY TO ROTATIONAL DIFFUSION STUDIES

A problem that is outside the reach of conventional analytical spectroscopy is the characterization of macromolecules. Conventional spectra of macromolecules, particularly IR, are useful for determining functional group identities, but provide no information about molecular weight, conformation, or branching structure. For synthetic polymers the physical properties of materials are profoundly affected by molecular size as well as branching. Characterization of size by gel permeation chromatography is widely used in the chemical industry, while characterization of branching is more difficult. Motions of branches depend on their sizes, therefore, measurement of polymer dynamics over a range of time scales provides the branching information (15). For biological polymers, such as proteins, measurement problems have been different because nature has already selected workable sizes and structures, and the role of biochemists has been to understand these choices. However, with the growing use of genetic engineering to prepare unnaturally large quantities of biological materials, the need for determining the integrity of conformations has become industrially important. The use of time-resolved spectroscopic methods for probing molecular dynamics has thus become increasingly important for both synthetic and biological macromolecules. Some of the most valuable information, such as motions of polymer branches or protein residues, occur on the picosecond time scale, and their study will benefit from the use of pump-probe spectroscopy.

Debye showed that the rotational diffusion constant of a spherical particle is proportional to the volume of the particle (16):

$$D = \frac{kT}{\eta V} \qquad (14.3)$$

This shows the origin of interest in rotational diffusion for determining the size of macromolecules. Because molecules are irregular in shape, Perrin (17) extended the work of Debye to include rotation of general ellipsoids. The way that rotational diffusion is commonly examined is by fluorescence depolarization. The excitation photons select molecules of a given set of orientation, and the emitted photons are polarized as long as the molecules remain in the same orientation. After the molecules have become randomly oriented, the polarization of the emission is isotropic. Thus, the decay of the fluorescence anisotropy, defined as $R(t)$, is determined experimentally from

$$R(t) = \frac{I_{\parallel}(t) - I_{\perp}(t)}{I_{\parallel}(t) + 2I_{\perp}(t)} \tag{14.4}$$

For example, a fluorescence depolarization experiment involves measuring the anisotropy as a function of time, typically by measuring $I_{\parallel}(t)$ and $I_{\perp}(t)$ individually as parallel and perpendicularly polarized intensities at time t. The goal of measuring the anisotropy is to determine the rotational diffusion constant, D, to obtain information about the molecular volume. The rotational diffusion constant has components along the three cartesian molecular axes:

$$D = \tfrac{1}{3}(D_x + D_y + D_z) \tag{14.5}$$

The values of the components of D are determined by the shape of the molecule. The anisotropy is related to the rotational diffusion constant as the sum of five exponential terms. Chuang and Eisenthal have related the fluorescence anisotropy to the rotational diffusion constants (18):

$$R(t) = \tfrac{6}{5} q_x q_y \gamma_x \gamma_y\, e^{-3(D_z + D)t} \;+\; \tfrac{6}{5} q_y q_z \gamma_y \gamma_z\, e^{-3(D_x + D)t}$$

$$+ \tfrac{6}{5} q_z q_x \gamma_z \gamma_x\, e^{-3(D_y + D)t} \;+\; \tfrac{3}{10}(\beta + \alpha)e^{-(6D + 2\Delta)t}$$

$$+ \tfrac{3}{10}(\beta - \alpha)e^{-(6D - 2\Delta)t} \tag{14.6}$$

The symmetries of the absorbing and emitting transitions are contained in the form of the γ and q terms. The γ terms represent the component of the absorption, or pumped, transition dipole along the specified molecular axis, and the q terms are defined analogously for the emitting, or probed, transition. The value of each exponential is determined by the diffusion behavior, and the amount that each exponential contributes to the anisotropy is determined by the values of the q and γ components. α, β, and Δ are functions of γ, q and D_i. The complex nature of the experimentally measured $R(t)$ hampers quantitation of D.

Accurate quantitation of rotational diffusion behavior is difficult because many variables enter into $R(t)$. Spectroscopic measurements of rotational diffusion have frequently been published without knowledge of critical parameters only to be proven wrong later. It is generally necessary to make simplifications in interpreting $R(t)$ because it is extremely difficult to determine the individual components of a five-component exponential decay. It is the inability to validate simplifications that hampers the accuracy of interpretations. The anisotropy is the normalized difference between $I_{\parallel}(t)$ and $I_{\perp}(t)$. $R(t)$ is most commonly simplified by choosing spectroscopic excitation to be along a single molecular axis. For example, if the wavelength is chosen to coincide with a transition having x polarization, and if the emitting transition also has x-axis polarization, then all exponential terms having γ_y, γ_z, q_y, or q_z preexponential factors disappear. The resulting $R(t)$ function decays as a single exponential:

$$R(t) = -\tfrac{1}{5} e^{-6D_z t} \tag{14.7}$$

It is far easier, of course, to determine the time constant of a single exponential decay than to characterize a multiple-component decay. However, only molecules of high point group symmetry have a transition dipole exactly along a molecular axis, and this high symmetry is frequently perturbed in solution. Also, many molecules studied by rotational diffusion are of low symmetry. Nonetheless, because multiple exponentials are difficult, most experimental studies of rotational diffusion invoke the assumption of a single exponential decay.

If one desires accurate values of D from $R(t)$, three types of parameters must be known. First, the relation between the ground-state and excited-state rotational diffusion constants must be known because fluorescence is a measure of the excited-state behavior. In practice, such a relation is not yet generally known, and it is assumed that the two states behave identically. Second, the relative values of D_x, D_y, and D_z must be known because a single decay can only determine uniquely one parameter, which is D. Typically, these relative values are approximated from the shape of the molecule, without direct determination. Third, the exact values of the q and γ terms must be known. Transitions of species in the liquid state are not completely polarized along one axis, thus exact simplification of $R(t)$ to a single exponential does not occur. The extent to which other exponential terms contribute must be characterized by knowing the polarization of the transitions quantitatively. Such a determination requires polarization measurements to be made at time zero, before the molecule has reoriented. This is typically impossible with photomultiplier technology, and ideal values of the transition polarizations are assumed. Thus, for the three types of parameters that must be determined for interpreting anisotropy measurements, untested assumptions are made in fluorescence depolarization studies.

Picosecond measurement on model compounds can be used to evaluate all three of these common assumptions in rotational diffusion studies. First, pump-probe measurements can be made identically for both fluorescence and absorption processes, thereby allowing a comparison of ground- and excited-state behavior. Second, the decay curves can be monitored without significant convolution with the instrument function, allowing the fit of the relative values of D_x, D_y, and D_z to be evaluated. Third, the transition polarizations can be measured at times sufficiently early that the molecule has not had time to reorient.

Cresyl violet was studied recently using the pump-probe technique with triple modulation detection (19). By orienting the polarization of the probe beam parallel and perpendicular to that of the pump, it was possible to separately collect $I_{\parallel}(t)$ and $I_{\perp}(t)$, respectively. From these data $R(t)$ could be calculated, according to Eq. (14.4), at a variety of different probe wavelengths. From the $R(t)$ function it was possible to determine D for each state probed. It was found that the polarization ratio of the transition and the decay constant of $R(t)$ behaved as predicted from Eq. (14.6). By measuring the polarization ratio to obtain values for q_x, q_y, q_z and γ_x, γ_y, γ_z, it was possible to evaluate excited-state versus ground-state behavior. Even though the experimental anisotropy varied greatly as a function of wavelength, the data are consistent with the same rotational diffusion constant for both probed states. Thus, it was shown that the wavelength-dependent transition polarization was responsible for the anomalous behavior of $R(t)$. In the past, where determination of transition polarizations were not made, such variation has been interpreted as a change in the rotational diffusion constant (20).

The cresyl violet studies illustrate that important information is gained from picosecond measurements with independently tunable pulse trains. One factor that limits the routine application of such measurements is that the positions of the pulse trains in time move with respect to one another as the wavelength of one laser is changed. Spectral scans at one delay time are thus impractical. Further, for each time decay, zero delay must be located. While this is straightforward, it adds considerable time to the experiment. The problem of the relative temporal position of the pulses changing with wavelength is an inherent problem for dye lasers and originates from the fact that the amplification process is dynamic. This wavelength dependence can be circumvented only by using extracavity methods of wavelength scanning, such as Raman shifting (20) or continuum generation (21). These methods require high laser peak powers and therefore are not presently compatible with high repetition rates. Further advances may alleviate this problem, allowing valuable chemical information to be obtained without such significant experimental development time.

REFERENCES

1. J. M. Harris, R. W. Chrisman, F. E. Lytle, and R. S. Tobias, *Anal. Chem.*, **48**(13), 1937–1943 (1976).

2. J. Wiedmann and A. Penzkofer, *Opt. Commun.,* **30**(1), 107–112 (1979).
3. R. L. Fork, B. I. Greene, and C. V. Shank, *Appl. Phys. Lett.,* **21**(8), 348–350 (1972).
4. Coherent, Inc. Pico-Ring Brochure, 3/81.
5. C. V. Shank, R. L. Fork, R. Yen, R. H. Stolen, and W. J. Tomlinson, *Appl. Phys. Lett.,* **40**(9), 761–763 (1982).
6. J. P. Heritage and D. L. Allara, *Chem. Phys. Lett.,* **74,** 507 (1980).
7. P. Esherick, A. Owyoung, and C. W. Patterson, *J. Phys. Chem.,* **87,** 602 (1983).
8. B. F. Levine and C. G. Bethea, *IEEE J.Q.E.,* **16**(1), 85–88 (1980).
9. J. P. Heritage, *Appl. Phys. Lett.,* **34,** 470 (1979).
10. P. Bado, S. B. Wilson, and K. R. Wilson, *Rev. Sci. Instrum.,* **53,** 706 (1982).
11. L. Andor, A. Lorincz, J. Siemion, D. D. Smith, and S. A. Rice, *Rev. Sci. Instrum.,* **55,** 64–67 (1984).
12. R. M. Hochstrasser and F. A. Novak, *Chem. Phys. Lett.,* **48**(1), 1–6 (1977).
13. D. P. Millar, R. Shah and A. H. Zewail, *Chem. Phys. Lett.,* **66**(3), 435–440 (1979).
14. K. L. Sala, G. A Kenney-Wallace, and G. E. Hall, *IEEE J.Q.E.,* **16**(9), 990–996 (1980).
15. J. R. Lakowicz, B. P. Maliwal, H. Cherek, and A. Butler, *Biochemistry,* **22**(8), 1741–1752 (1983).
16. P. Debye, *Polar Molecules,* Chemical Catalog Co., New York, 1929, p. 84.
17. F. Perrin, *J. Phys. Radium,* **7,** 1 (1936).
18. T. J. Chuang and K. B. Eisenthal, *J. Chem. Phys.,* **57**(12), 5094–5097 (1972).
19. G. J. Blanchard and M. J. Wirth, *J. Chem. Phys.,* **82,** 39 (1984).
20. D. Reiser and A. Laubereau, *Ber. Bunsenges. Phys. Chem.,* **86,** 1106–1114 (1982).
21. R. R. Alfano and S. L. Shapiro, *Phys. Rev. Lett.,* **24,** 584 (1970).

CHAPTER

15

ANALYTICAL LIMITS OF ELECTROPHORETIC LIGHT SCATTERING

BENNIE R. WARE

Department of Chemistry
Syracuse University
Syracuse, New York

1. INTRODUCTION

The coherence and power of lasers as sources for light-scattering experiments have made it feasible to utilize the dynamic character of the scattered light to determine the properties of the motion of the scattering particles (1, 2). This principle has been used primarily for the study of translational diffusion, and

493

the published applications now number in the thousands. A major drawback of this approach, however, is the complication introduced when the scattering sample contains more than one species, as is often the case. The resulting data form is a sum of exponentials with undetermined amplitudes and time constants. In an attempt to deal with this problem, Ware and Flygare (3) announced in 1971 an experimental variation in which an electric field is applied to the sample as the dynamic light scattering is characterized. The resulting measurement, which has come to be called electrophoretic light scattering (ELS), permits the resolution of electrophoretically distinct species and at the same time provides an additional experimental parameter, the electrophoretic mobility. Essentially a hybrid of electrophoresis and dynamic light scattering, the ELS measurement is technically challenging. During the 1970s an extended effort in several laboratories led to an improved methodology that has made possible a variety of successful applications. Extensive reviews of the ELS technique have appeared elsewhere (4–7). Our purpose in this chapter is to examine the ELS technique from the analytical perspective in order to provide the prospective user with a set of criteria for the design of successful experiments. We begin with brief summaries of the principles and methodology of electrophoretic light scattering.

2. THEORY AND PRINCIPLES

Electrophoresis, the motion of charged particles in an applied electric field, is the basis of a variety of analytical techniques. In all of the classical methods the detection of electrophoretic motion relies on measuring the macroscopic displacement of the particles as a function of time. The hallmark advantage of ELS is the direct detection of electrophoretic velocity via the Doppler effect, without the requirement for any macroscopic displacement. We review here briefly the principles of electrophoretic motion and of Doppler detection that are required for our considerations of the analytical limits of the technique.

2.1. Electrophoresis

Application of an electric field to a solution of charged particles causes them to accelerate toward the electrode of opposite polarity until the frictional force of the viscous drag due to their translation through the stationary solvent matches the attractive force of the applied electric field. Within a short time [nanoseconds to microseconds (8)] the particles achieve a terminal velocity $\mathbf{V}$ which is proportional to the applied field strength $\mathbf{E}$. The constant of proportionality is called the electrophoretic mobility u.

$$\mathbf{V} = u\mathbf{E} \qquad (15.1)$$

The electrophoretic mobility of a particle is expected to be proportional to its electric charge Z and inversely proportional to its friction constant f. This intuitive relationship must be modified in magnitude to account for the shielding effects of the other ions that are present in solution. An exact relationship between u, Z, and f is not known, even for the simplest cases. A useful approximation, usually called Henry's law, can be written as (9, 10)

$$u = \frac{Ze}{f} \frac{\chi(\kappa R)}{(1 + \kappa R)} \tag{15.2}$$

where e is the electronic charge, R is the particle radius, and κ is the Debye–Huckel parameter given by

$$\kappa = \left(\frac{8\pi N_A e^2 \Gamma}{1000\epsilon kT}\right)^{1/2} \tag{15.3}$$

where N_A is the Avogadro number, ϵ is the dielectric constant, k is the Boltzmann constant, T is absolute temperature, and Γ, the ionic strength, is given by

$$\Gamma = \frac{1}{2} \sum_i C_i Z_i^2 \tag{15.4}$$

with C_i and Z_i, respectively, the molar concentration and electrovalence of the ith mobile ion species. In Eq. (15.2) the function $\chi(\kappa R)$, Henry's function (9, 10), is a sigmoid curve whose value monotonically increases from unity for κR less than 0.1 to a maximum of 1.5 for κR greater than 1000. Limiting forms of Eq. (15.2) can thus be deduced in the limits of large and small particles. To simplify further we will assume spherical particles, which permits us to write (9)

$$f = 6\pi\eta R \tag{15.5}$$

where η is the solvent viscosity. Then in the limit of small particles

$$u = \frac{Ze}{6\pi\eta\{R(1 + \kappa R)\}} \quad (\kappa R \ll 1) \tag{15.6}$$

In the limit of large particles

$$u = \frac{Ze}{4\pi\eta R^2 \kappa} \quad (\kappa R \gg 1) \tag{15.7}$$

Since the charge of a large particle is expected to reside on the particle's surface, the surface charge density s may be written as

$$s = \frac{Ze}{4\pi R^2} \qquad (15.8)$$

Substituting for Ze in Eq. (15.7) leads to the relation

$$u = \frac{s}{\eta \kappa} \qquad (\kappa R \gg 1) \qquad (15.9)$$

which was first derived by Smoluchowski (11). The important concept from Eq. (15.9) is that for large particles, the electrophoretic mobility is independent of particle size and determined directly by the surface charge density. In practice, the Debye–Huckel length $(1/\kappa)$ is generally on the order of 1 nm, so the limit of small particles is rarely achieved and the limit of large particles is quite common.

The typical magnitude of electrophoretic mobility is 1 μm cm/V s. The typical sustainable field strength is on the order of 10 V/cm. Thus, the typical electrophoretic velocity is 10 μm/s, or roughly 1 in./hr. Hence the measurement of electrophoretic mobilities requires either a considerable duration under stable conditions to achieve accurately measurable displacement, or, alternatively, a very accurate method for measuring small velocities.

2.2. The Doppler Effect

The fundamental principle of ELS is the measurement of electrophoretic drift velocities through the Doppler shifts of scattered laser light. The Doppler shift $\Delta\nu$ for detection of a wave of frequency ν_0 and velocity c when source and detector have a relative velocity $\mathbf{V}$ is given by

$$\Delta\nu = \frac{\mathbf{V}}{c}\nu_0 \qquad (15.10)$$

In a light-scattering experiment one has in general two Doppler shifts to consider, one between moving particles and the laser and the second between the particles and the detector. It is easy to show (12) that the former is given by $-\mathbf{k}_0 \cdot \mathbf{V}/2\pi$ and the latter is given by $\mathbf{k}_s \cdot \mathbf{V}/2\pi$, where $\mathbf{V}$ is the velocity vector of the particle, $\mathbf{k}_0$ is the wave vector of the incident light, and $\mathbf{k}_s$ is the wave vector of the scattered light. (Each wave vector has magnitude $2\pi/\lambda$). The net Doppler shift in hertz is thus given by

$$\Delta\nu = \frac{(\mathbf{k}_0 \cdot \mathbf{V} - \mathbf{k}_s \cdot \mathbf{V})}{2\pi} = \frac{\mathbf{K} \cdot \mathbf{V}}{2\pi} \qquad (15.11)$$

where $\mathbf{K} = \mathbf{k}_0 - \mathbf{k}_s$ is the experimental scattering vector. The magnitude of $\mathbf{K}$ is given by

$$\mathbf{K} = \frac{4\pi}{\lambda} \sin \frac{\theta}{2}$$

where λ is the wavelength of the laser light in the medium and θ, the scattering angle, is the angle between the incident and scattered wave vectors. Typical magnitudes of Doppler shifts in an ELS experiment are between 10 and 100 Hz. Although these shifts are very small in comparison to the frequency of the carrier wave ($\sim 10^{15}$ Hz), they can be measured quite accurately through the principle of optical beating, as will be discussed in Sections 3.5 and 3.6.

2.3. Effects of Diffusion and Heterogeneity

If all particles were in motion with a single velocity $\mathbf{V}$, then the ELS signal would be a single-frequency shift with magnitude $\mathbf{K} \cdot \mathbf{V}/2\pi$. In practice, there are two sources of dispersion that are of particular importance in considering the analytical limits of the technique. The first of these is the random thermal motion (diffusion) of each particle that is superimposed on the electrophoretic drift velocity. The second is the fact that individual particles may have different electrophoretic velocities, so that a population of particles will produce a distribution of Doppler shift frequencies.

The theory of the contribution of diffusion to the dynamic properties of scattered light has been treated and reviewed extensively (1, 2). It is important to note that the individual Brownian "jumps" of a particle in a small-molecule solvent such as water are very short in comparison to the wavelength of optical light and thus to the characteristic dimension of a light-scattering experiment $(2\pi/K)$. If one assumes that diffusion and electrophoresis are independent, then it can be shown that an ELS peak from a species with a single uniform electrophoretic mobility u and a single translational diffusion coefficient D is a Lorentzian peak at a Doppler shift of $\mathbf{K} \cdot \mathbf{V}/2\pi$ and with half-width at half-height given by $DK^2/2\pi$. Interesting exceptions to this result may be obtained under certain experimental conditions (13); but in general this result should hold for most cases of interest, and it has been verified quite precisely in careful experiments (14).

If there exists a distribution of electrophoretic mobilities in the sample, $f(u)$, then that distribution will be translated to a distribution of Doppler shifts $f(\Delta\nu)$

$= \mathbf{K} \cdot \mathbf{E} f(u)$. In the absence of diffusion, the ELS spectrum becomes a histogram of electrophoretic mobilities in the sample with each component weighted by its light-scattering cross section at the experimental scattering vector. If diffusion and electrophoretic heterogeneity are both significant, then the ELS spectrum is best represented as a sum of Lorentzian peaks, with each component having its own characteristic electrophoretic mobility (peak shift) and diffusion coefficient (peak width). When it is necessary to determine the relative contributions of diffusion and electrophoretic heterogeneity to the broadening of an ELS spectrum , two experimental manipulations may be used. First, increasing the electric field strength will cause a proportional increase in the heterogeneity broadening, whereas it should cause no increase in the diffusion broadening. Second, manipulation of the scattering angle has a relatively greater effect on diffusion broadening, which is proportional to K^2, than on heterogeneity broadening, which is proportional to K. At sufficiently low scattering angle, diffusion broadening should become negligible, but for technical reasons to be discussed later the limitations of low-angle detection may preclude this limit, particularly for relatively small scatterers which have correspondingly higher diffusion coefficients.

3. EXPERIMENTAL METHODOLOGY

Detailed descriptions of the experimental methodology of electrophoretic light scattering have been published elsewhere (5–7, 15). This section includes a brief description of the most important considerations, with particular emphasis on those aspects that are related most closely to the analytical limits of the technique.

3.1. Sample Preparation

Electrophoresis and light scattering are both techniques that require special considerations in sample preparation. ELS samples must be prepared with both sets of considerations in mind. Light scattering is not a very selective technique. All particles scatter light to some degree, and large particles scatter much more intensely than small particles. Hence, contamination by aggregates, unwanted particles, and "dust" must be minimized. These contaminants are not necessarily resolved as such in the ELS spectrum since the particles of interest often adsorb onto them, giving them a similar surface charge density. The result may be an ELS peak at or near the proper mobility but with a relative intensity that is too great and a width that is too narrow in comparison to the intended sample of pure species. Particulate contamination is obviously a more serious problem when the particles of interest are realtively small. The problem of particulate contamination may be dealt with by filtration, centrifugation, or column chromatography. This problem defines the concentration limit for many samples.

Whereas most light-scattering experiments have little restriction for the choice of solvent (only refractive index and absorbance), the simultaneous electrophoretic requirements of ELS are considerable. The electrophoretic mobility of any species depends on the solvent. Generally, electrophoretic mobility increases, and nuisance joule heating decreases, at lower ionic strength. However, there are a variety of considerations that mandate against lowering the ionic strength beyond certain limits. Counterions, buffer ions, and stabilizing added salt are required for a meaningful experiment. The concentration of ions that may bind or adsorb to the species of interest must be known, and in some cases that requires special ion buffers. In virtually all ELS experiments the pH must be buffered, and this requirement places a lower limit on the ionic strength. As a general rule experiments conducted under conditions of ionic strength less than $10^{-2}M$ must be interpreted with extreme caution.

3.2. Laser

A schematic diagram of an ELS apparatus is shown in Fig. 15.1. In all cases the light source must be a stable laser. Fortunately, the requirements for the laser are minimal by today's standards. Relatively low power (15 mW) He–Ne lasers, which are reliable and inexpensive, suffice for the great majority of applications, provided they have little modulation noise ($<1\%$ rms). More intense Ar^+ or Kr^+ lasers are sometimes employed to increase the scattering signal from weakly scattering samples. There is an advantage both in scattering cross section and in photomultiplier tube efficiency in using the blue wavelengths of Ar^+ lasers, but these advantages are usually not worth the increased purchase price and maintenance costs.

3.3. Sample Chamber

The design of the ELS sample chamber is probably the most critical component for optimization of the analytical power of the instrument. An ELS chamber must serve simultaneously the objectives of laser light scattering and electrophoresis. The optics must permit entry and exit of the incident laser beam and must allow collection of the scattered light over the desired range of scattering angles. Electrodes must be placed in electrical contact with the sample in a manner that allows for application of a maximum electric field of accurately known field strength and with a minimum of artifacts that may result from electro-osmosis, electrode reactions, or the joule heat generated by the passage of the current. Sample volume should be minimized. Other considerations include the ease of cleaning, assembly, alignment, and sample exchanges.

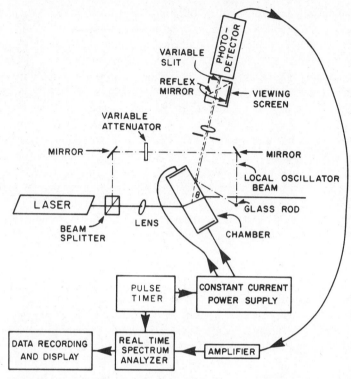

Fig. 15.1. Schematic diagram of an electrophoretic light-scattering apparatus.

The two major options in chamber design are distinguished by the placement of the electrodes. The more conventional approach separates the electrodes by a channel, in which the laser Doppler measurement is made (3, 4, 5, 8). This approach avoids effects of electrode reactions and permits stabilization of the observed region against convection. In this design, electro-osmosis, the movement of the solvent as a result of the applied electric field, must be calibrated or minimized by reducing the electrical charge of the chamber walls. An alternative approach is to pass the laser between two closely spaced parallel-plate electrodes immersed in the sample (6, 16). This design obviates the problem of electro-osmosis but introduces possible artifacts due to electrode reactions, and the optical geometry requires the measurement to include a portion of the sample to which an electric field is not applied. The two approaches have been contrasted in detail elsewhere (4–8).

We have constructed and tested at least 30 different ELS chamber designs over the past 15 years, and we have found the design first described by Haas and Ware (17) and subsequently modified by Smith and Ware (5) to be the best choice for extending the analytical limits of the technique. A simplified diagram of this design is shown in Fig. 15.2. This chamber design reduces the sample volume and the thermal time constant by placing the two electrodes just outside a narrow channel formed by dielectric spacers. The electrodes are hemicylindrical in order to obtain a uniform current density, which can be roughly 15 times lower than the current density in the channel. This design permits the application of a much higher electric field in the scattering volume before achieving the critical electrode current density above which unwanted side reactions may occur. The flat, narrowly spaced surfaces of this gap inhibit convection, while the curved walls of the electrode region encourage convection. Thus, heat is conducted out of the gap without convection and then carried by convection to the electrodes, which are thermally conducting and in thermal contact with the large metal chamber halves, whose temperature is maintained by circulating coolant. Sample is injected and subsequently removed from the chamber by insertion of a hypodermic syringe needle. The chamber must be disassembled, cleaned, and (optionally) coated with methylcellulose to reduce electro-osmosis after 10–20 hr of use. Detailed procedures are available elsewhere (5), and a simplified design has been published by Schmitz (18).

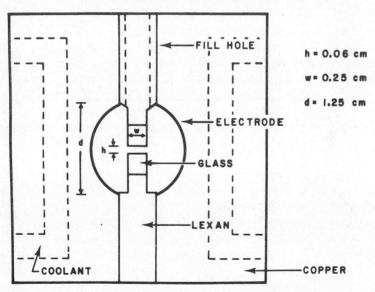

Fig. 15.2. Electrophoretic light-scattering chamber. This modified channel design was first developed by Haas and Ware (17) and then adopted by Smith and Ware (5) in the form shown above. The two large halves are fabricated from copper or brass, then silver plated. The electrode material is cemented to the two halves by electrically conducting epoxy. Dielectric Lexan spacers with (optional) glass tips form the conducting channel in which the ELS measurement is made. Glass windows are mounted above and below the plane of the drawing to form the sample region, which has a total volume of about 0.25 mL.

3.4. Application of the Electric Field

It is imperative that the scattering particles are drifting in a constant electric field strength over the duration of the measurement. However, considerations of joule heating, net displacement of the scatterers, and possible electrode reactions generally dictate an intermittent pattern of field application. The optimal selection of waveform of the electrophoretic field depends on chamber design. Parallel-plate electrodes are closely spaced and small in area, so the dissipation of heat is rapid and efficient. On the other hand this design is critically sensitive to electrode reactions and particle displacement. Hence, the general practice is to apply a square-wave field of frequency on the order of hertz. The frequent reversals of electrode polarity minimize electrode reactions and net particle displacement, but they also introduce a complication into the ELS spectrum. With each polarity reversal the phase change of the Doppler signal is reversed. The result is a set of modulation sidebands on the ELS peak (8, 19). These sidebands carry no useful information, but they are not a particular problem if the ELS spectrum consists of a single symmetric peak. If there are distinct species in the sample that would produce closely spaced ELS peaks, then the sidebands from each will overlap and reduce the resolution of the technique.

For the channel-type chambers such as the design employed in our laboratory, electrode reactions and net displacements of scattering particles are not major problems, but there is an increased thermal time constant associated with the longer pathlength of the current. At sufficiently low field strengths, the applied field can be held constant for the duration of the measurement (1–2 min). However, it is generally preferable to achieve higher field strengths through the use of shorter-duration electric field pulses. These pulses are generally of a duration somewhat greater than the minimum time necessary for the transient capture of one real-time spectrum. For example, a resolution of 0.5 Hz in the ELS spectrum would require a pulse duration of at least 2 s. The interval between pulses is selected, to permit heat dissipation, in consideration of the thermal time constant of the chamber (8). The polarity of the pulses is alternated to obviate any problems associated with net displacement of the particles.

Regardless of the pattern of the electric field application, it is important that the mode of application assures constant current through each pulse rather than constant voltage. In a region of solution with conductivity σ and constant cross-sectional area A, the passage of a current of magnitude i produces an electric field strength E given by

$$E = \frac{i}{A\sigma} \qquad (15.13)$$

Hence, the maintenance of constant current assures a constant electric field, provided σ remains constant. In practice, σ increases during a measurement because the joule heating raises the temperature and lowers the viscosity of the solvent. However, the electrophoretic mobility should be increased by the same factor [Eq. (15.9)]. Thus, the rise in viscosity, which may be as great as 10%, is compensated to first order by this fortuitous cancellation, and the measured electrophoretic mobility is the appropriate value for the initial condition of the experiment, provided the rise in temperature has not caused any significant change in the conformation or charge of the scattering particles. This fortuitous cancellation does not obtain if a constant voltage is applied to the electrodes. In addition, overvoltage effects and the linear inhomogeneity of the solution caused by the migration of all ions will generally introduce substantial errors into the determination of the electric field strength in constant-voltage operations.

3.5. Detection Optics

The heart of the detection process is the principle of optical heterodyning (1–7). The Doppler shifts caused by electrophoretic motion are far too small (10–200 Hz) to be observed directly, so it is necessary to mix the scattered light with a reference beam in order to measure the beat frequency. The reference beam, or local oscillator, is usually just a portion of the incident beam. We have found it optimal to direct the local oscillator around the scattering chamber with a set of mirrors, passing also through a variable attenuator as shown in Fig. 15.1. The final deflection utilizes a curved surface, such as a small glass rod, the reflections from which are reflected again by the window of the scattering chamber along the same axis as the scattered light. The collection optics are designed to form a real image of the scattering volume and the local oscillator on the photomultiplier tube (5). This real image may be viewed on a screen by rotating a reflex mirror into position. The respective images of the scattering volume and the local oscillator are superimposed by manipulation of the glass rod while observing the viewing screen. The appropriate portion of the image is selected by adjusting the slit width using a micrometer driving pin. The reflex mirror is then rotated out of the optical path to permit the aligned scattered and local oscillator light to fall upon the photocathode.

Though it is not necessary to match the pathlengths of the scattered and reference beams exactly, it is important that the pathlengths not vary greatly (i.e., by more than a few centimeters). The ability to measure a Doppler shift on a laser line whose time-averaged linewidth is much greater than the magnitude of the shift is dependent on the beating principle. Comparison of scattered and reference beams over roughly equivalent pathlengths effectively eliminates the time-dependent phase drifts of the incident laser. The only phase-shift differences between scattered light and local oscillator are the Doppler shifts imposed on the scattered photons by the electrophoretic drift of the scattering particles. In general, there are many different Doppler shift magnitudes in the scattered signal. In principle, these signals may all beat with each other to produce a number of spurious signals. For this reason the local oscillator intensity must be greater than the scattered signal. On the other hand a local oscillator intensity that is too great introduces other sources of spurious signals, such as laser power–supply modulation sidebands and mode noise on the laser line. In practice, a local oscillator ratio of about 10 is optimal. The optical arrangement shown in Fig. 15.1 permits easy selection of the local oscillator ratio using the variable attenuator while reading the magnitudes of the photocurrent.

3.6. Signal Processing

The ELS signal is an oscillating photocurrent, the frequencies of which are the Doppler shifts caused by the motion of the scatterers in the electric field. The predetection signal-to-noise ratio is often quite high (>1), and the frequencies of interest are in the range of 1–200 Hz. The clearest form for viewing these data is the frequency spectrum, so the instrument of choice is a spectrum analyzer. Commerical spectrum analyzers are available at moderate cost that have all of the necessary characteristics. We use a Hewlett-Packard 3582A. An alternative approach would be to program a digitizing computer to Fourier transform the signal as it is collected. Another alternative is to use an analog autocorrelator and to Fourier transform the data for inspection. Because of the low frequencies and high signal levels, photon counting is not an attractive option. The photocurrent is passed through a dropping resistor to ground, and the voltages above the resistor are amplified by an audio frequency amplifier with variable high- and low-frequency cutoffs. The amplification is set to match the input range of the spectrum analyzer. The spectrum analyzer must be real-time efficient, must have a good baseline flatness specification ($\sim .1\%$), and must be capable of signal averaging. It is desirable for the spectrum analyzer to be operated in the transient capture mode with an external trigger.

   In our apparatus a specially designed timing circuit applies the constant-current impulse for electrophoresis and triggers the spectrum analyzer for transient capture later, terminating the electrophoretic current shortly after the data collection is complete. This process is repeated for a selectable number of captures, and the successive spectra are signal averaged. There are two distinct motivations for signal averaging. The first is to improve the ratio of signal to random noise. The ratio improves by the factor $N^{1/2}$, where N is the number of independent spectra averaged. This motivation is particularly important for dilute solutions of relatively small scatterers. The second motivation for signal averaging, important for very large particles in dilute suspension, is simply to improve the statistics of the measurement by sampling a larger number of particles. In this case it is important to be certain that successive spectra are collected from different sets of particles, which may require stirring between pulses or some other physical manipulation of the sample.

3.7. Data Analysis

The mathematical rigor that is essential for interpretation of other types of dynamic light-scattering data is not generally useful for ELS. The fundamental difference is that the dispersion of spectral intensity in an ELS spectrum is a manifestation, in part or in total, of the electrophoretic heterogeneity of the sample. Since the distribution of mobilities in a given sample cannot be assumed or expected to follow any particular functional form, elaborate fitting procedures are not in order. The best way to represent the subtleties of the data is simply to display the spectrum itself.

The correspondence between Doppler shift and electrophoretic mobility for a given spectrum is worked out using Eq. (15.11) to obtain the velocity, then dividing by the electric field, presumably from Eq. (15.13). The mobility of any peak can then be assigned. For interpreting the shape and width of the distribution, it is essential to have established, by varying the electric field strength or scattering angle or both, whether diffusion is a significant source of broadening in the spectra. If diffusion is negligible, there is a one-to-one correspondence between Doppler shift frequency and electrophoretic mobility. If diffusion broadening is significant, the positions of the peak centers can be assigned to electrophoretic mobilities, but obviously the peak is broader than the true mobility distribution, and the abscissa of the spectrum cannot be labeled in terms of velocity or electrophoretic mobility.

If there is more than one peak in the spectrum, the relative intensities of the peaks present may be of considerable interest. Each species is represented in the spectrum in proportion to the product of its concentration and its light-scattering cross section. The latter may be a complicated function of particle size, particle shape, and magnitude of the experimental scattering vector (20, 21).

Interpretation of electrophoretic mobilities is sometimes carried further to calculate the zeta potential, the absolute surface charge density, and even the absolute number of charges. Unfortunately, the level of approximation inherent in these calculations is much greater than experimental error, and the result is often more misleading than instructive. We caution against placing undue significance on the absolute magnitude of the electrophoretic mobility or any quantity calculated from it; it simply is not a very revealing number. The shape of the mobility distribution, the relative mobilities of related species, and the change in the mobility in response to chemical or biological stimuli are observations that, in our experience, are more likely to lead to meaningful interpretations of fundamental significance.

4. ANALYTICAL LIMITS

In assessing the capabilities of ELS and any alternative techniques for electrophoretic investigations, there are several questions to be addressed. How much material is required? How much time is required to make a measurement? What are the respective accuracies and precisions of the measured quantities? What is the ability of the technique to resolve distinct species? In each case these questions may be addressed theoretically or from the perspective of practical experience. The emphasis in this chapter will be on the latter, using the relevant equations to explain the practical observations where appropriate. One cautionary note may be in order. The practical limits quoted apply to the capabilities of our laboratory at the present stage of development. Several other laboratories have demonstrated comparable capabilities in most cases. The ELS technique is experimentally challenging, however, and the new investigator may find that a great deal of effort is required to achieve the limits cited. However, commercial ELS apparatus is now coming to market, and there is hope that advanced, even improved, instrumental capabilities may be available to new investigators at modest cost.

4.1. Concentration Limit

The concentration limit of ELS is set effectively by the considerations of performing a low-frequency dynamic light-scattering experiment. Hence, the concentration of each scatterer must be weighted by its scattering cross section at the experimental scattering vector. The scattering cross section depends on the mass, shape, and refractive index increment of the scatterer (9, 20, 21). For an optically dense species in the size range of microns, it may be relatively easy to make an ELS measurement on a single particle.

Concentration considerations become quite severe for smaller particles for two reasons. First, the scattering cross section depends on the mass of the particle. For small particles the scattering cross section per particle depends on the square of the mass of each particle. Second, the smaller the particle, the greater the diffusion coefficient, so the effects of diffusion broadening (to be discussed later) degrade the sensitivity of the measurement. Although there can be no absolute lower limit specified for molecular weight, it is safe to say that ELS experiments on species of molecular weight below 10^5 are challenging and below 10^4 are unknown. In this range the lowest concentration reported in a published ELS study was for tetramers and dimers of hemoglobin (MW, 64,000 and 32,000, respectively) which were observed in a range of concentrations down to about 0.005% by weight (22). As a practical index for the low-concentration limit for small particles in an ELS measurement, take the product of the molecular weight (g/mol) times the weight concentration (g/L) times the refractive index increment (L/g, typically on the order of 10^{-4}). If this product does not exceed unity, the measurement is probably impractical. If the product exceeds 100, the measurement should be straightforward. Between these two values lies temptation and difficulty.

The foregoing discussion has incorporated our experience in the removal of dust and other particles from ELS samples. In practice, the removal of particulate contamination establishes the concentration limit in a great majority of samples. For large species purity is essentially the only criterion for the concentration limit. Particles in the micron range can be observed at virtually any concentration, but at low concentration they may be significantly contaminated by the dust particles always present in solvents and on the surfaces of vessels used to contain the sample. In our studies on blood cells, which are generally between 5 and 20 μm in diameter, we chose to work at concentrations around 10^6 cells/mL, which was adequate to make particulate contamination negligible.

4.2. Accuracy and Precision of Electrophoretic Mobility

A variety of factors enter into the calculation of the absolute accuracy of electrophoretic mobility in an ELS measurement. Most of the sources of error contribute a few tenths of 1%, and an overall accuracy limit of better than 1% is difficult to attain. Precision estimates from experience on optimal samples suggests a precision of between 0.1 and 1%, with about 0.5% as a typical number, indicating that roughly half the error is systematic. The optimal accuracy and precision are not easy to achieve, so a careful evaluation of the factors that enter into the accuracy of the measurement is most worthwhile.

The accuracy of the Doppler shift determination is set by the method of signal processing, by the signal-to-noise ratio, and by the various contributions to line broadening. Typical Doppler shift magnitudes are in the range of 30–80 Hz, so the frequency resolution must be in the range of 0.3–0.8 Hz to achieve a nominal 1% accuracy. A real-time spectrum analyzer with about 250 regularly spaced channels should be adequate to provide a spectral determination that is not the limiting parameter of overall accuracy, provided it is used properly. There may be a temptation to operate the spectrum analyzer at a higher-than-necessary frequency scale because the determination on higher scales can be accomplished faster. The ELS spectral peak then appears in the first few channels of the spectrum analyzer, appears narrow, and has a good signal-to-noise ratio. This procedure is obviously the equivalent of using a spectrum analyzer with many fewer channels, and it can only lead to a reduced accuracy of determination. The equivalent operation is quite common among workers who process the data with autocorrelators. An attractive autocorrelation function can be obtained in a short time by measuring only the first two or three oscillations, but the accuracy of the determination of the Doppler shift may be no greater than $\pm 25\%$. An exception to this rule can be made if it can be assumed that the ELS signal is a single symmetric peak, for which very few data points are required for an accurate determination. Experience has shown that this assumption is rarely justified. The most reliable procedure is to determine the complete range of frequencies from 0 Hz to the maximum frequency of interest. The sampling theorem tells us that in order to accomplish this we must sample $2n$ data points to calculate an n-point spectrum and that this procedure will require a time t, where t^{-1} is the frequency resolution of this measurement and n/t is the maximum frequency determined.

Spurious sources of line broadening also reduce the accuracy of determination of the Doppler shift. Thermal convection must be limited by chamber design and maintenance of a sufficiently low current. In channel-type chambers convection is a threshold effect, and thus it causes very little line broadening below the critical current density. Field heterogeneity can be another source of spurious line broadening, particularly in narrow parallel-plate chambers (6). In channel-type chambers an electro-osmotic flow profile can be an additional source of line broadening. The combined sources of spurious broadening are difficult to quantify, but experience suggests that they contribute to the linewidth of most ELS spectra to an extent of the order of a few percent of the line shift. The error introduced into the determination of the peak position would normally be on the order of a few tenths of 1%.

A fundamental source of line broadening in ELS measurements is introduced by the effects of thermal diffusion. In the very first paper in this field, Ware and Flygare (3) defined the analytical resolution r of this new technique to be the ratio of the Doppler shift to the diffusion-broadened half-width:

$$r = \frac{\mathbf{K} \cdot \mathbf{V}}{DK^2} \tag{15.14}$$

Normally $\mathbf{V}$ is aligned to be perpendicular to $\mathbf{K}$, so Eq. (15.14) simplifies to

$$r = \frac{V}{DK} = \frac{uE\lambda_0}{D4\pi n \, \sin(\theta/2)} \tag{15.15}$$

In practical terms Eq. (15.15) tells us that the analytical resolution of a diffusion-broadened ELS spectrum can be improved either by increasing the electric field strength or by decreasing the scattering angle. Note that decreasing the scattering angle actually decreases the Doppler shift, rendering it even more difficult to measure with desired precision, but decreases the diffusion width as K^2, so that the line becomes sharper. Electrophoretic mobilities do not depend much on particle size, in fact do not vary greatly at all, but diffusion coefficients vary over several orders of magnitude, roughly as the linear characteristic dimension of the particle. Hence, diffusion broadening is a serious analytical limitation for ELS on particles in the 1–5-nm size domain and is essentially negligible for very large particles. In the latter case the measurement is generally performed at a conveniently high scattering angle in order to maximize the magnitude of the Doppler shift. If it can be assumed that other instrumental broadening factors are independent of scattering angle, then the optimal angle for electrophoretic resolution of a diffusion-broadened ELS peak is the angle at which the diffusion width is equal to the instrumental width (6). It should be emphasized, however, that the minimization of diffusion broadening is desirable only if one's primary goal is to optimize electrophoretic resolution. The ability to measure simultaneously the translational diffusion coefficient and the electrophoretic mobility is a prime advantage of ELS and a major motivation for its invention.

Once the magnitude of the Doppler shift ($\Delta \nu$) has been determined, there are four other parameters required for the calculation of electrophoretic mobility:

$$u = \frac{\Delta \nu \lambda_0}{2nE \, \sin(\theta/2)} \tag{15.16}$$

The incident vacuum wavelength λ_0 is generally known to at least four significant figures and thus is not a source of error. The refraction index n enters directly into the calculation of electrophoretic mobility and is also required for the determination of the scattering angle since θ is defined in the scattering medium and refraction must consequently be taken into account. Refractive indices are easily determined to an accuracy of 0.1% and thus are not a source of significant error unless the experimenter carelessly disregards this fact and inserts a standard value for water or buffer. The determination of the scattering angle is generally done either by measuring the distances of the vector triangle on the scattering table or by the use of a precision rotating stage or goniometer. If these measurements are performed with care, an accuracy of 0.1% in θ is achievable except at very low scattering angles. Note that the finite size of the acceptance angle of the detection optics assures a range of scattering angles that sets an absolute limit on the accuracy in this parameter. When measurements are performed at low angle ($< 10°$), it is generally necessary to restrict the acceptance solid angle, to a degree that may reduce the signal-to-noise ratio, in order to achieve satisfactory accuracy in θ. At low angles, $\sin(\theta/2) \cong \frac{1}{2} \sin \theta \cong \theta/2$, so a 1% error in θ would produce a 1% error in u.

The electric field is determined from Eq. (15.13), so there are three factors to consider. The area A is easily measured in channel-type chambers. For chambers such as the design in Fig. 15.2, A must be measured after each assembly, preferably with an optical comparator. The error in this measurement should be kept well below 1%. In parallel-plate chambers the area A is more difficult to measure, to keep constant, and to keep uniform, making this design more difficult to employ with equivalent accuracy. Fringe field effects must also be considered in using Eq. (15.13) for the calculation of the field strength between parallel plates (6). The electrophoretic current i can be held constant and measured to an accuracy of better than 0.1%, so it is not a source of error. The electrical conductivity of the sample can be measured precisely, but an absolute accuracy better than 1% requires considerable care. Most commercial conductivity meters are not much better than 1% in accuracy.

From the foregoing discussion it should be clear that many limiting factors prohibit an accuracy in the determination of the electrophoretic mobility of greater than about 1%. As examples of accurate determinations of the electrophoretic mobility of a uniform sample, we show in Fig. 15.3 two spectra of human red blood cells. Spectrum (a) was taken in physiological ionic strength (0.145M) and spectrum (b) in an isosmotic, reduced ionic strength buffer (0.025M). The shift-to-width ratio is about 11 in spectrum (a) and about 27 in spectrum (b). Most of this difference is due to the fact that the mobility in

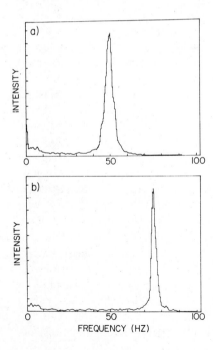

Fig. 15.3. Electrophoretic light-scattering spectra of fresh human red blood cells. Spectrum (a) was collected from cells suspended in Hanks' balanced salt solution (0.145M ionic strength). Spectrum (b) was collected from cells suspended in a medium containing 25mM NaCl 2.5mM HEPES buffer (pH 7.4) and 4.2% sorbitol. For both spectra the scattering angle was 37.0° and the temperature was 25.0°C. The electric field strength was 33.9 V/cm for spectrum (a) and 31.2 V/cm in spectrum (b). The walls of the chamber were coated with methyl cellulose to reduce electro-osmosis. The final correction for electro-osmosis was 2 Hz. The respective electrophoretic mobilities for the two media were 1.06 μm cm/V s for spectrum (a), and 1.94 μm cm/V for spectrum (b). Spectra were taken by Dr. Lindsay Plank.

spectrum (b) is nearly twice as great because of the reduced ionic strength. Experimental details are given in the figure caption. Both mobilities quoted should be accurate to about ±1% and are reproducible at about that level of precision.

The accuracy limit of about 1% is quite typical for transport coefficients. Indeed, the electrophoretic mobility is not a parameter for which determinations with an accuracy of better than 1% are likely to be very meaningful. The electrophoretic mobility depends on the surface charge density of the particle and the viscosity of the medium. The viscosity is usually not determined to an accuracy better than 1%, but such accurate determinations of viscosity, along with precise temperature control, would be necessary in order to make accurate values of the mobility meaningful. The surface charge density is a function of the pH, the ionic strength, and the concentration of any ions in solution that may bind at the particle's surface. In order to interpret the electrophoretic mobility at the 1% accuracy at which it is currently measurable, it is necessary to buffer the pH and to measure it accurately (0.01 pH unit) and to measure (and often to buffer) the free concentration of ions that may bind to the surface of the particle and are thus in equilibrium with it. Finally, as pointed out in Section 2.1, the theories that are used to interpret electrophoretic mobility in terms of more fundamental parameters are not accurate at the level of current experimental accuracy, so the motivation to improve experimental accuracy is not sufficiently great to justify the effort that would be required.

4.3. Accuracy and Precision of Diffusion Coefficients

The most accurate method of measuring translational diffusion coefficients is dynamic light scattering (1, 2), which has a precision and accuracy of about 1%. The complicating effects of introducing the electric field for an ELS measurement seriously compromise the accuracy of determining the diffusion coefficient. The most accurate published account of diffusion coefficients measured in an ELS experiment was the study of hemoglobin by Haas and Ware (14, 17). They were able to fit their ELS spectra to shifted Lorentzian functions with the correct half-width. However, the half-width of this measurement was not sufficiently precise to study the dimer–tetramer equilibrium at low hemoglobin concentration (down to 0.005% by weight). These studies also demonstrate the lowest concentration limits achieved and the lowest scattering angles reported (2°).

The primary utility of measuring ELS linewidths to determine diffusion coefficients is that it provides a simultaneous monitor of the state of conformation or aggregation of a system being studied for charge effects by ELS. Often it is of interest to measure electrophoretic mobilities in response to the addition of agents that may alter the surface charge of the particle. These alterations of surface charge may induce a change in conformation or state of aggregation of the system that must also be measured in order to interpret the electrophoretic changes. Properly performed ELS measurements provide both types of information from the same measurement. Two examples will illustrate this point. Siegel and Ware (23) studied the affinity of synaptic vesicles and synaptosomal membranes (from guinea pig brain) for divalent cations to test a hypothesis regarding synaptic transmission of nerve impulses. The vesicles were titrated with divalent cations and the electrophoretic mobility was measured by ELS after the addition of each increment of titrant. The ability to determine the diffusion coefficients via the ELS linewidths provided a simultaneous determination of the extent to which the surface charge reduction induced vesicle aggregation. As a second example, Rhee and Ware (24) and Yen et al. (25) studied the condensation of multivalent cations on DNA. Trivalent and tetravalent ions caused a collapse of the DNA conformation, from an extended to a more compact form, at the point of condensation saturation predicted by current theories. Divalent ions did not condense to a sufficient extent to induce the conformational transition. In these experiments the dual nature of the ELS measurement permitted the measurement of the charge reduction via the electrophoretic mobility and the conformational transition via the diffusion-broadened ELS linewidth.

The accuracy in the diffusion coefficient necessary for the detection of aggregation and major conformational transitions is not difficult to achieve. The primary consideration is to select a scattering angle at which diffusion broadening is substantial and the electrophoretic mobility is still measurable to an acceptable accuracy. It is important also to select a frequency scale on the spectrum analyzer that permits the resolution of a large number of frequencies, at least 10 points, within the two half-heights of the ELS peak. Aggregating and flocculating systems often exhibit time-dependent behavior, so precision in linewidth measurements may be quite poor, and a time-dependent trend in linewidth may be discernible. The potential of this application of ELS for study of the basic nature of flocculation and aggregation processes is an exciting prospect for future investigations.

4.4. Resolution of Electrophoretically Distinct Species

Even though ELS does not permit the physical separation of distinct species, the analytical distinction among electrophoretically inequivalent species is a major motivation of many experiments. The resolving power of ELS is certainly related to its accuracy, but several sources of error (such as errors in electric field strength and scattering angle) are equivalent for all species and therefore do not degrade resolution. The primary issue in the resolution of two species is the ratio of the difference of their Doppler shifts to the sum of the half-widths of their respective ELS peaks (each in the absence of the other). For peaks of equal intensity, if this ratio is greater than one, the two species are said to be resolved. In fact, for ratios as low as 0.1, a reproducible and meaningful distinction can often be made. In general, the resolution ratio must be greater to resolve peaks of unequal intensity, as with any spectroscopic technique.

If the linewidths of the individual species are both diffusion broadened, then the resolution may be increased by increasing the electric field and/or decreasing the scattering angle. Once the diffusive components of both peaks have been minimized as contributing factors to the linewidth, there is no other experimental manipulation that will improve the resolution. It may, of course, be possible to alter the solution conditions so that the species to be resolved have a greater difference in electrophoretic mobility.

There are numerous examples in the literature of ELS experiments in which electrophoretic resolution among distinct species was an important capability. An example is shown in Fig. 15.4. In this experiment rat serosal mast cells were stimulated to secrete by the addition of rabbit anit-rat F(ab')$_2$ antiserum. Figure 15.4 shows the dose response. A fraction of antiserum diluted 1:100 produced electrophoretic alterations in only a few cells, but these cells are readily discernible in the higher-mobility portion of the ELS spectrum. Serum diluted 1:50 altered about half the cells, and serum diluted 1:25 altered virtually all of the cells. The interpretation of these experiments (with associated controls) was that the interior surface of the secretory vesicle (granules) bears a high net negative charge, which adds to the negative surface charge density of the mast cell when the exocytic secretory process transforms the mast cell granule interior to become part of the mast cell exterior. These spectra were collected separately, and the superposition of the unstimulated mast cell peaks gives a good indication of the reproducibility of the technique. The individual sharp peaks in the stim-

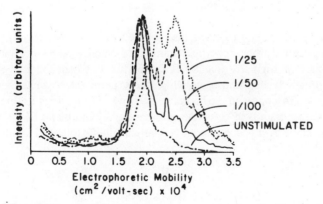

Fig. 15.4. Effect of immunological stimulation on rat serosal mast cells. The fractions represent dilution factors of rabbit anti-rat F(ab')$_2$ antiserum. Increasing concentration of this antiserum clearly increases the proportion of cells in the higher-mobility population. Reprinted from ref. (26) with permission.

ulated (higher-mobility) region probably correspond to individual cells. Details of this experiment are given in Petty et al. (26).

4.5. Measurement Time Scales

The duration of an ELS measurement is an important consideration in several contexts. The total length of time to make a measurement dictates the throughput for the study of multiple samples and establishes the period of time over which samples must be stable. The minimum time for the performance of independent determinations establishes the time resolution of kinetic studies for samples that are changing in time. The spectroscopic time scale sets the limits on the time scale of averaging of dynamic parameters during the collection of data. We will proceed from the shortest time scale to the longest.

The spectroscopic time scale of an ELS measurement is on the order of milliseconds to seconds. Fast processes such as charge fluctuations due to proton exchange are thus not resolved. The electrophoretic mobility measured corresponds to a time-averaged charge and a time-averaged conformation. There are some changes in conformation and perhaps in the electrical charge of scattering particles that may be expected to be on the time scale of the ELS measurement, and such charges can be shown theoretically to have a kinetic effect on the shape of the ELS spectrum (8, 27). Unfortunately, an experimental example has not yet been produced.

The minimum time for collection of independent ELS spectra is the time for the collection of one spectrum, roughly 1–4 s. For small scatterers it is usually necessary to signal average to achieve the desired signal-to-noise ratio; for large scatterers it is usually necessary to signal average (with stirring or mixing) to achieve adequate sampling statistics. The full time scale for most measurements is generally between 5 and 15 min. The time to change samples is only a few seconds.

The greatest advantage of ELS is the relatively short time necessary to construct a complete electrophoretic histogram for a heterogeneous population of particles. This is particularly true for larger particles, where the conventional technique is microelectrophoresis, in which the experimenter selects individual particles in a microscope and clocks them by hand. The savings in time, with equal or better accuracy and complete automation, should make ELS the clear technique of choice as it becomes commercially available to a wider community of users.

5. APPLICATIONS

ELS has been applied successfully to virtually all types of biological particles, including proteins, nucleic acids, viruses, organelles, membrane vesicles, and many types of living cells, and to synthetic polyelectrolytes and several types of synthetic and natural particles. A comprehensive survey of the literature through about 1981 was presented by Ware and Haas (7). Our purpose here is to present only a brief discussion of which types of applications have shown ELS superior to alternative technology, in view of the analytical capabilities of the technique already outlined in this chapter. In this context it is useful to distinguish on the basis of the size domain of the particles to be investigated.

In the lowest applicable molecular weight range (10^4–10^5), there have been few successful applications because of the low scattering cross sections and high diffusion coefficients and also because the measurement of electrophoretic mobilities in solution has ceased to be a typical characterization for such molecules. When this measurement is desired, ELS is probably superior to moving-boundary (Tiselius) electrophoresis, provided the ionic strength can be kept relatively low ($<0.1M$). In cases for which electrophoretic resolution is the desired result and a determination of the magnitude of the free-solution electrophoretic mobility is not essential, the combined electrophoretic-chromatographic separation techniques using gels and other supporting media are generally far superior to ELS or moving-boundary methods. An interesting and important exception to this rule is the self-associating system, such as the dimer–tetramer equilibrium of hemoglobin mentioned earlier, for which ELS has the analytical advantage over separative techniques in that it does not perturb the equilibrium in order to make the measurement.

For the electrophoretic characterization of large particles (>1 μm), ELS is clearly the technique of choice. Much of this work is in the area of industrial characterization of particles used for such diverse applications as pigment stabilization, coatings adherence, xerographic imaging, fiberglass fabrication, and other materials characterizations. The ELS technique is rapid, accurate, and sufficiently automatic to permit operation by a technician. Over the past few years a number of industrial laboratories have replaced conventional microelectrophoresis with ELS, and with the new availability of inexpensive commercial ELS apparatus, this trend can be expected to increase. Although there are very few publications in this area, industrial particle characterization laboratories are probably the single largest community of users. Two recent reports from academic laboratories will serve as good introductions for the interested reader (28, 29).

Most of the published applications of ELS have been in the area of the electrophoretic investigation of living cells, which are generally between 5 and 30 μm in diameter. Two recent reviews discuss this area comprehensively (7, 30). The ELS technique is ideal for the electrophoretic characterization of living cells, particularly because the rapidity of data collection improves the likelihood of biological stability over the period of the measurement. It was originally hoped that precise electrophoretic characterization would be a useful means of identifying and studying subpopulations of cells, and considerable progress was made toward that goal. However, contemporary with this work has been the development of highly specific chemical and genetic techniques, particularly monoclonal antibodies, that are clearly far superior for the same purpose. It now appears likely that future emphasis in the application of ELS to living cells will be placed on the study of those limited cases for which the surface charge density is a parameter of intrinsic interest. Such a case was illustrated in Fig. 15.4.

There is an intermediate size range (~ 0.1–1 mm) for which the application of either moving-boundary or microelectrophoresis techniques is difficult. Applications of ELS in this area have been quite successful. ELS spectra for particles in this size range are generally broadened both by diffusion and by electrophoretic heterogeneity, so that both charge density distributions and size effects (aggregation, changes in conformation) can be studied simultaneously. Common specimens in this size range include viruses, vesicles, and certain polyelectrolytes. Two recent examples of ELS applications are the study by Yen et al. of the electrokinetic behavior of inside-out vesicles from human red cell membranes (31) and the study by Kitano et al. of the Ca^{2+} affinity of vesicles prepared from rod outer segments (32).

A large proportion of recent ELS applications have dealt with solutions of linear polyelectrolytes, both synthetic and biological. The physical behavior of polyelectrolytes has received renewed attention in recent years, both from theorists and from experimentalists employing a variety of techniques. The data from ELS reveal both electrical and conformational properties of these macromolecules, and ELS studies have proved to be incisive elements of progress in this field. Drifford et al. (33, 34) have studied the electrophoretic mobilities of polyelectrolytes as a function of added salt and have interpreted their data in terms of an apparent charge. Schmitz and co-workers have used ELS to study charge and conformational aspects of mononucleosomes, polynucleosomes, and linear DNA; an interesting feature of their methodology is the use of applied sinusoidal electric fields to investigate conformational dynamics (35–37). Wilcoxon and Schurr (38) and Zero and Ware (39) have used ELS to study the electrophoretic mobilities and ELS linewidths of poly-L-lysine as a function of added salt through the so-called ordinary-extraordinary transition. Below the transition salt concentration, the dynamic light-scattering data reveal a greatly reduced apparent diffusion coefficient. The ELS studies showed that electrophoretic mobilities were not much reduced by the transition and that the ELS linewidths underwent a change at the transition that was much less severe than the zero-field linewidths.

Much of the recent ELS work in our laboratory has focused on the question of counterion condensation onto linear polyelectrolytes. Using ELS as an indicator of reduced charge and macromolecular conformation, we have titrated DNA solutions with divalent, trivalent, and tetravalent cations to observe the condensation of the counterions (24, 25). The results agreed qualitatively with the counterion condensation theory of Manning (40), but several quantitative

discrepancies were revealed. The most fundamental test of the concept of counterion condensation theory required the synthesis of a special polyelectrolyte.
The theory states that condensation occurs when a parameter ξ is greater than
one (in univalent salt). This parameter ξ is given by

$$\xi = \frac{q^2}{\epsilon kTb} \tag{15.17}$$

where q is the proton charge, ϵ is the bulk dielectric constant of the solvent, k
is Boltzmann's constant, T is the absolute temperature, and b is the linear charge
spacing along the polyelectrolyte. For most polyions ξ is greater than one (for
DNA $\xi = 4.2$) so the theory had been tested only by measuring the extent of
counterion condensation. Klein and Ware (41) synthesized poly[(dimethylimino)
hexylene bromide] for which $\xi = 0.82$ in water and for which ξ can be raised
continuously to about 1.85 by addition of methanol (to lower ϵ). Viscosity
measurements were performed to demonstrate that the polymer maintains an
extended configuration throughout this range. Then ELS measurements were

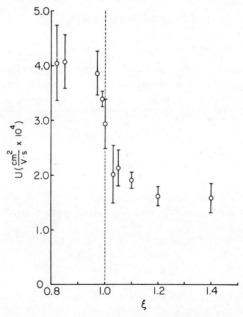

Fig. 15.5. Plot of the mean electrophoretic mobility ($\pm$ standard deviation) of 6,6-ionene in 4.0mM
KBr as a function of the reduced charge density parameter ξ, which was increased from the aqueous
value of 0.82 by addition of methanol. The polymer concentration was 9.95×10^{-5} g/L. All
measurements performed at 20°C; mobilities were corrected to the viscosity of water at 20°C. The
sharp reduction at $\xi = 1$ is a direct confirmation of the concept of counterion condensation. Reprinted
from ref. (41) with permission.

performed to measure the electrophoretic mobility as a function of ξ for a direct test of counterion condensation. The dramatic results are shown in Fig. 15.5. At precisely $\xi = 1$, the electrophoretic mobility dropped by more than a factor of two. The reduction in mobility was even greater than the theory predicts, but the intuitive notion of counterion condensation is clearly upheld. Klein and Ware suggested that the physical basis for the departure from theory is the fact that near $\xi = 1$, the spacing of condensed counterions predicted by counterion condensation theory is greater than the Debye screening length.

There is every reason to believe that ELS will continue to be a valuable technique with many exciting applications. One must always remember to select physical techniques judiciously and to use them in combination. We have found that a combination of ELS with photon correlation spectroscopy and fluorescence photobleaching recovery forms a powerful array of complementary techniques for characterizing the hydrodynamics of complex systems (39, 42, 43). It is also our experience that the continuous application of new physical techniques to the solution of important problems creates the finest context for the extension of analytical capabilities.

ACKNOWLEDGMENTS

Electrophoretic light scattering has been supported in our laboratory in recent years by the National Institutes of Health (GM-27633 and GM-33786). I am grateful to my many co-workers whose insights have added much to the development of this technique. Dr. Lindsay Plank took the spectra in Fig. 15.3. Dr. Plank and Ms. Terry Dowd contributed helpful suggestions on the manuscript.

REFERENCES

1. B. Chu, *Laser Light Scattering*, Academic Press, New York, 1974.
2. B. J. Berne and R. Pecora, *Dynamic Light Scattering*, Wiley-Interscience, New York, 1976.
3. B. R. Ware and W. H. Flygare, *Chem. Phys. Lett*, **12,** 81 (1971).
4. B. R. Ware, *Adv. Colloid Interface Sci.*, **4,** 1 (1974).
5. B. A. Smith and B. R. Ware, in D. M. Hercules, G. M. Hieftje, L. R. Snyder, and M. A. Evenson, Eds., *Contemporary Topics in Analytical and Clinical Chemistry*, vol. 2, Plenum, New York, 1978, pp. 29–54.
6. E. E. Uzgiris, *Progr. Surface Sci.*, **10,** 53 (1981).
7. B. R. Ware and D. D. Haas, in R. I. Sha'afi and S. M. Fernandez, Eds., *Fast Methods in Physical Biochemistry and Cell Biology*, Elsevier, New York, 1983, pp. 173–220.
8. D. D. Haas, "Electrophoretic Light Scattering of Dilute Protein Solutions," Ph.D. dissertation, Harvard University, Cambridge, MA, 1978.

9. C. Tanford, *Physical Chemistry of Macromolecules*, Wiley, New York, 1961.
10. D. C. Henry, *Proc. Roy. Soc.*, **A133**, 106 (1931).
11. M. Smoluchowski, *Z. Physik. Chem.*, **92**, 129 (1981), through ref. 9.
12. B. R. Ware, in C. B. Moore, Ed., *Chemical and Biochemical Applications of Lasers*, Academic Press, New York, 1977, pp. 199–239.
13. M. B. Weissman and B. R. Ware, *J. Chem. Phys.*, **68**, 5069 (1978).
14. D. D. Haas and B. R. Ware, *Biochemistry*, **17**, 4946 (1978).
15. B. R. Ware, in J. C. Earnshaw and M. W. Steer, Eds., *The Application of Laser Light Scattering to the Study of Biological Motion*, NATO Advanced Study Institute, Life Sci, Vol. 59, Plenum, New York, 1983, pp. 89–122.
16. E. E. Uzgiris, *Opt. Commun.*, **6**, 55 (1972).
17. D. D. Haas and B. R. Ware, *Anal. Biochem.*, **74**, 175 (1976).
18. K. S. Schmitz, *Chem. Phys. Lett.*, **63**, 259 (1979).
19. A. J. Bennett and E. E. Uzgiris, *Phys. Rev. A*, **8**, 2662 (1973).
20. M. Kerker, *The Scattering of Light and Other Electromagnetic Radiation*, Academic Press, New York, 1969.
21. M. B. Huglin, *Light Scattering from Polymer Solutions*, Academic Press, New York, 1972.
22. D. D. Haas and B. R. Ware, *Biochemistry*, **17**, 4946 (1978).
23. D. P. Siegel and B. R. Ware, *Biophys. J.*, **30**, 159 (1980).
24. K. W. Rhee and B. R. Ware, *J. Chem. Phys.*, **78**, 3349 (1983).
25. W. S. Yen, K. W. Rhee, and B. R. Ware, *J. Phys. Chem.*, **87**, 2148 (1983).
26. H. R. Petty, B. R. Ware, and S. I. Wasserman, *Biophys. J.*, **30**, 41 (1980).
27. B. J. Berne and R. Gininger, *Biopolymers*, **12**, 1161 (1973).
28. R. C. Fairey and B. R. Jennings, *J. Colloid Interface Sci.*, **85**, 205 (1982).
29. J. R. Goff and P. Luner, *J. Colloid Interface Sci.*, **99**, 468 (1984).
30. E. E. Uzgiris, *Adv. Colloid Interface Sci.*, **14**, 75 (1981).
31. W. S. Yen, R. W. Mercer, B. R. Ware, and P. B. Dunham, *Biochim. Biophys. Acta*, **689**, 290 (1982).
32. T. Kitano, T. Chang, G. B. Caflisch, D. M. Piatte, and H. Yu, *Biochemistry*, **22**, 4019 (1983).
33. M. Drifford, A. Mener, P. Tivant, P. Nectoux, and J. P. Dalbiez, *Rev. Phys. Appl.*, **16**, 19 (1981).
34. M. Drifford, P. Tivant, F. Bencheikh-Larbi, K. Tabti, C. Rochas, and M. Rinaudo, *J. Phys. Chem.*, **88**, 1414 (1984).
35. K. S. Schmitz, *Biopolymers*, **21**, 1383 (1982).
36. K. S. Schmitz, N. Parthasarathy, and E. Vottler, *Chem. Phys.*, **66**, 187 (1982).
37. K. Schmitz, *Chem. Phys.*, **79**, 297 (1983).
38. J. P. Wilcoxon and J. M. Schurr, *J. Chem. Phys.*, **78**, 3354 (1983).
39. K. Zero and B. R. Ware, *J. Chem. Phys.*, **80**, 1616 (1984).
40. G. S. Manning, *Q. Rev. Biophys.*, **11**, 179 (1978).
41. J. W. Klein and B. R. Ware, *J. Chem. Phys.*, **80**, 1334 (1984).
42. B. R. Ware, D. Cyr, S. Gorti, and F. Lanni, in *Measurement of Suspended Particles by Quasi-Elastic Light Scattering*, B. Dahneke, Ed., Wiley, New York, 1983, pp. 255–289.
43. B. R. Ware, in *Spectroscopy and the Dynamics of Biological Systems*, P. M. Bayley and R. E. Dale, Eds., Academic Press, New York, 1984.

CHAPTER

16

LASER FLOW CYTOMETRY

SCOTT J. HEIN

Department of Chemistry
Oregon State University
Corvallis, Oregon

LAWRENCE C. THOMAS

Department of Chemistry
Seattle University
Seattle, Washington

1. INTRODUCTION

The direct measurement of properties of isolated biological cells and particles
in a flowing stream and their subsequent separation from a large population of

521

cells has become an important analytical method for biological, biochemical, and medical research. This procedure, referred to as flow cytometry, exploits unique characteristics offered by lasers in spectroscopic applications.

In flow cytometry cells in a single-cell suspension pass single file through a flow chamber. In this chamber one or more signals associated with biochemical and physical properties of the cell such as fluorescence, absorbance, or light scattering are measured. Subsequent separation of analyte cells from nonanalyte cells may then be based on the results of the measurements.

Many methods have been developed over the past several years for analysis and separation of specific biological cells or groups of cells. Typical analytical methods used by the cell biologist are electrophoresis, surface-affinity-based methods, centrifugation, light and electron microscopy, as well as biological methods employing selective survival of analyte cells. Using such techniques, however, it is difficult or impossible to obtain information about single cells and to separate cells with special characteristics from a bulk group of cells within a reasonable period of time. The development of flow cytometry and related cell sorting techniques such as fluorescence-activated cell sorting (FACS) has provided an alternative to conventional cytometric methods. These procedures are both sensitive and selective, while allowing for rapid analysis and separation of single cells and particles.

The earliest cellular flow analyzer was described by Coulter in 1956 (1). This analyzer monitored the change in conductance of an electrolyte solution caused by cells passing between a pair of electrodes in a small flow chamber. The magnitude of the conductance change could be related to cell volume. Coulter's instrument could only be used to count and size cells, however, not to physically separate them. It was not until 1965 that Fulwyler incorporated Coulter's analyzer into the first true cell sorter (2). This instrument sorted cells on the basis of cell volume with the separation accomplished using an electrostatic deflection system originally developed for ink-jet printers.

For many studies dealing with complex intracellular processes, cell volume is not a sufficiently selective characteristic for isolation of target cells from a population. Consequently, instruments have been developed that probe and isolate single cells based on other, more selective physical parameters. For example, in 1965 a cell analyzer was developed by Kamentsky (3) that measured the absorbance of ultraviolet radiation as cells passed through a flow chamber illuminated by a conventional mercury source.

To avoid problems associated with the use of conventional sources, such as spatial instability, wide excitation bandwidths, stray light, and low power, laser sources were adapted to flow cytometry. One such cytometer was described in 1969 that analyzed up to 10^5 cells per minute based on cellular fluorescence excited by an argon ion laser (4). During that same year, Mullaney reported an analyzer in which a helium–neon laser was used as a source for low-angle light-scattering measurements (5). The light-scattering data were used to obtain cell size information (6). Again, these instruments could analyze, but not sort, the cells of interest. To overcome this limitation, a cytometer that both analyzed and sorted stained cells based on a laser-induced fluorescence signal was developed by Herzenberg in 1972 (7). This instrument also used an argon ion laser as its excitation source but, in addition, utilized a sorting system similar to Fulwyler's (2), making it one of the first fluorescence-activated cell sorters.

These instruments are single-parameter instruments, analyzing or sorting cell samples based on a single parameter such as fluorescence or light scattering. In 1973 a multiparameter instrument was described that could measure Coulter volume, two separate wavebands of fluorescence emission, and light scattering on individual cells passing through a specially designed flow chamber (8). Multiparameter cell measurements improve the ability to discriminate between different types of cells by increasing the number of physical and biochemical properties that can be used to characterize them. This instrument, which used an argon ion laser as its light source, was the forerunner of today's multiparameter flow cytometers and cell sorters.

As is the case for many spectroscopic applications, the laser is an excellent source for inducing optical signals such as fluorescence and light scattering for analyses via flow cytometry. The laser offers a stable, intense, coherent, and nearly monochromatic source of light that can be focused into a very small volume and directed over distances much more easily than can conventional sources. The increased radiance associated with laser light enhances the sensitivity of optical methods in flow cytometry, allowing measurements of weak intrinsic cellular fluorescence (autofluorescence). Moreover, the coherent nature of laser light permits the beam to be focused to a cross section smaller than normal cell cross sections. This allows surface or internal morphological information to be obtained. In addition, the laser beam can be shaped and polarized, which is advantageous in certain applications. Thus, lasers can be utilized in flow cytometry to increase both sensitivity and selectivity, thereby allowing measurements to be made that are not feasible with conventional light sources. Consequently, many flow cytometers employ one or more lasers as excitation sources for cell measurements.

2. PRINCIPLES

There have been several excellent reviews written on flow cytometry and FACS, including reviews in the general scientific literature (9–11), the biological and medical literature (12–14), the biochemical and biophysical literature (15, 16), the immunological literature (17), and the chemical literature (18). Also, a comprehensive reference text has been written exclusively on flow cytometry and sorting (19). Many of these reviews include comprehensive discussions of the principles of these methods. Consequently, only an outline of basic principles is discussed herein.

Since flow cytometry is a method for the analysis of individual cells in a flowing stream, the first step in a flow cytometric analysis is to obtain cells in a single-cell suspension. Body fluids such as blood and semen are ideal samples for this type of analysis since they naturally contain single, isolated cells, and hence dispersal induced damage is avoided. If the sample is solid tissue, then the cells must be dispersed into suspension through grinding, mincing, sonication, or chemical dispersal. The cells must then be fixed to allow dye permeation and to prevent autolysis and bacterial growth. Finally, in order to obtain useful information from the analysis, the cells must have properties that can be probed and detected by the instrumentation.

The cells can be analyzed directly through measurement of an intrinsic physical property. However, the signal obtained from an intrinsic property is often too weak to be detected and must be enhanced through modification or treatment of the cell. Flow cytometry often utilizes fluorescence measurements for analysis and sorting, thus a typical modification involves staining the cell or subcellular structure with a fluorescent dye. The cells can either be stained in solution or in situ (20), and the specific cellular process under examination will largely determine which dye is used in an application. The dispersed cells are forced single file under moderate pressure through a flow chamber. Measurements are then made on the cells as they flow through observation volumes within the chamber.

Many cellular properties are associated with the various signals measured in flow cytometry. The most commonly measured signals in flow cytometry are fluorescence and light scattering. Cells exhibiting autofluorescence or that have been stoichiometrically stained with a fluorescent dye should exhibit fluorescence intensities that are proportional to the amount of fluorophore present. Consequently, fluorescence intensity measurements yield information about the content of target substances in the cell. In addition, with high spatial resolution measurements, information is obtained about the distribution of stained target substances along the cell axis.

The light scattered by a cell traversing a laser beam is due to a complex combination of reflection, refraction, and diffraction. Scattered light collected at low forward angles (0.5°–2.5°) to the excitation axis is primarily due to diffraction and is related to the gross size of the cell (5, 6). As the collection angle increases from 2.5°, refractive index and reflection off internal structures become more important. Thus, scattered light collected at larger angles can be related to internal structures such as the nucleus (21). It should be stressed that the relationship between scattered light and these cellular properties is complex and still not well characterized.

Some of the more common signal–property relationships are compiled in Table 16.1. A more complete list is found in a review by Jovin (22). A large number of different signals may be chosen for measurement in flow cytometric analyses, providing the method with a great deal of flexibility. Furthermore, with a properly designed flow cytometer several combinations of these signals can be measured concurrently on an individual cell. Consequently, cell classification can be based on several criteria, thereby increasing the selectivity of the method.

Table 16.1. Signals Commonly Measured in Flow Cytometry and the Cellular Properties Associated with Each Signal

Signal Measured	Cellular Property (Reference)
Fluorescence	
Intensity	Cellular content (i.e., DNA) (29, 74, 106)
	Spatial distribution (37–39)
	Viability (96, 107, 108)
	Cell-surface receptor binding and antigen expression (109, 110)
Spectra	Effect of cellular (especially time dependent) process on fluorophore (42)
Pulse width	Cell and nuclear diameter (111, 112)
	Chromosomal arm length (113)
	Size (114)
Polarization	Environment (34, 115)
	Mobility (113)
	Intracellular microviscosity (116)
Resonance energy transfer	Relative location on cell surface (117)
Absorbance	
Intensity	Length (47)
	Viability (118)
	Cross-sectional area (48, 118)
Light Scattering	
Low angle	Size (5, 6, 112, 113)
	Viability (118, 119)
	Cell-surface receptor binding (120)
Wide angle	Internal structure: granularity, reflectivity, nuclear shape (21, 48, 121)
Multiangle	Collectively: shape, internal structure, refractive index, size (43–45)
Wide versus low	Particle asymmetry (122)
Scattering ratios	Size (123)
Pulse width	Size (48)

Finally, after measurements have been made on a cell in the flow chamber, analyte cells can be separated from cells that do not exhibit the property or properties of interest. A typical cell sorter includes a flow chamber attached to a piezoelectric transducer, which oscillates at 20,000–40,000 Hz. With proper choice of oscillation frequency and proper design of the flow chamber, the stream of cells is broken into uniform, individual droplets as they exit from the flow chamber nozzle. The instrument is adjusted so that each droplet contains either one or no cells, and droplets containing cells are bracketed by several empty droplets. The measurements that were made on the cells are synchronized with the droplets containing the cells. If a cell meets all the sorting criteria, the droplet is charged by applying an electrical potential, typically ± 50 V, to the saline stream. The charged droplets are then deflected into a collection tube or other collection apparatus by applying a high-voltage, constant potential field across the region of falling drops. Potentials up to 3000 V are typical, and the degree and direction of deflection depends on both the magnitude of the potential field and the mass-to-charge ratio of the droplet. Uncharged droplets containing nonanalyte cells pass undeflected into a waste container.

In this manner a large number of cells from a sampled population are analyzed in a short period of time. This allows for a dramatic increase in the statistical significance of the measurement over previous nonflow methods of cell analysis and sorting. Commercially available systems can sort about 5000 cells per second and can analyze without sorting 10,000 cells per second. Thus, a small subpopulation of cells can be isolated based on measurements made on the cells during flow. This makes flow cytometry a powerful tool for cellular analysis. Therefore, flow cytometry is an important technique for both fundamental and applied studies in fields such as cell biology, medicine, cancer biology and therapy, and immunobiology. The variety of procedures to which flow cytometry can be applied to single-cell analyses is expanding rapidly along with researchers' abilities to adapt new techniques and procedures to existing methods.

3. INSTRUMENTATION

Many different kinds of instrumentation can be utilized in flow cytometry, providing the method with a great deal of flexibility. Consequently, it is difficult to describe available instrumentation in terms of a single configuration. Therefore, aspects of the instrumentation will be discussed separately, with reference to the simple single-laser dual-parameter instrument illustrated in Fig. 16.1.

3.1. Flow Chamber

The measurements made on cells in flow cytometry are greatly dependent on the ability of the flow chamber to present isolated single cells reproducibly to

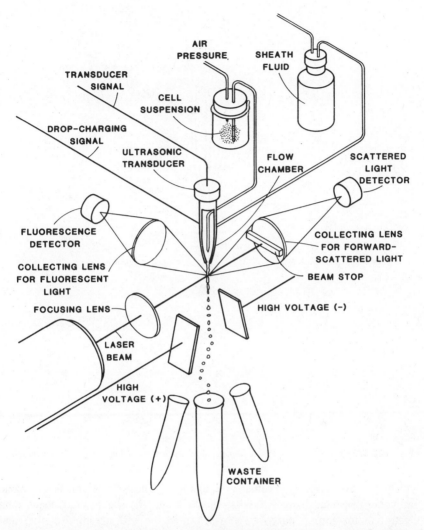

Fig. 16.1. Diagram of a single-laser, dual-paramter flow cytometer. This instrument is configured to measure fluorescence intensity and low-angle light scattering.

each of the measurement volumes. The flow chamber must also be able to isolate the single cells in individual droplets of reproducible mass and charge for separation in an electric field. Thus, the design of the flow chamber is crucial to the success of flow cytometry. The flow chamber (see Fig. 16.2) typically consists of an inner sample tube and an outer sheath tube used to direct the flow of a sheath fluid around the inner sample tube. The isolated cells flow under moderate pressures from a sample reservoir through the inner sample nozzle,

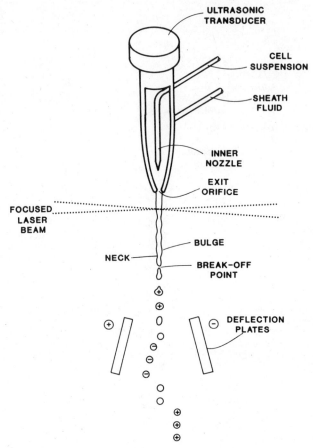

Fig. 16.2. Diagram of a typical flow cytometer flow chamber including sorting instrumentation.

which is usually several micrometers in diameter. The cells exit the inner nozzle and come in contact with a flowing sheath fluid. The sheath fluid, usually an electrolyte for sorting purposes, flows around the cell stream, isolating and hydrodynamically orienting the cells in the center of the flow chamber. Flow rates of the sheath fluid and cell stream are adjusted so that nonturbulent, laminar flow conditions exist. In this manner the cells, enclosed in the sheath fluid, are made to follow the same reproducible path through the flow chamber and, with careful design of the flow chamber, with the same relative orientation (23). This ensures that each cell will be illuminated equally by the light source, allowing reproducible measurements to be made. Kay and Wheelis have studied different nozzle geometries with respect to the orientation of gynecologic cells and suggest that a nozzle that encloses the cell in an elliptical sheath, as opposed to a circular or slit-shaped sheath, gives the best overall performance (24). Optical measurements can be made in the flow chamber itself if the sheath tube is constructed of materials that do not interfere with measurements in the wavelength region of study. For instance, for measurements in the ultraviolet region, the nozzle is made of high-quality quartz or fused silica. The stream of cells exits the flow chamber through an orifice that is usually 50–200 μm in diameter. Several instruments make measurements on the stream of cells in free space after exiting the flow chamber rather than in the flow chamber itself.

The nozzle assembly is normally attached to a piezoelectric crystal that vibrates at frequencies of 20–40 kHz. These periodic vibrations cause perturbations in the cylindrical stream exiting the nozzle. The perturbations, in the form of velocity and pressure fluctuations, cause the stream to form into alternate necks and bulges (see Fig. 16.2), which eventually break into droplets downstream from the nozzle (25). To maintain the highest sorting precision, these droplets must have extremely uniform sizes, shapes, and masses. However, even the presence of the cell in the stream during droplet formation can cause irregularities in the size and mass of the droplet. For example, if the cell is positioned in the neck of a forming droplet, the resultant drop is much more likely to be nonuniform than if the cell is located in the center of the bulging region. Therefore, care must be taken in instrumental design if very high sorting accuracy is required (25, 26). If the optical measurements are made on cells in free space after exiting the nozzle, rather than in the flow chamber itself, the measurements are made prior to the section of the stream where the formation of necks and bulges would make measurements difficult or irreproducible.

3.2. Light Sources

Optical measurements in flow cytometry require a stable light source to induce the optical phenomena. Requirements for the source vary, depending on the nature of the individual measurement. However, high powers are generally required since the fluorescence and light-scattering signals are proportional to the intensity of the excitation source. In addition, the source radiation should be able to be focused to at least cellular dimensions or even much smaller volumes if morphology is to be probed. Finally, the source must be adaptable to the design of the flow chamber, especially in multiparameter instruments where several different measurements requiring different types of instrumentation are made in close proximity to each other, often using the same light source.

Cellular fluorescence is a commonly measured optical signal in flow cytometry. The fluorescence can be weak autofluorescence characteristic of the cell or it can be a stronger fluorescence that usually arises from fluorescent dyes with which the cells have been stained. The choice of excitation source will depend on the type of fluorescence under examination: weak autofluorescence requires excitation by a high-power source, while measurements on cells stained with highly fluorescent dyes may be made with a low-power source. In addition, the wavelengths available from the source must be compatible with the excitation characteristics of the fluorescent component of the cell. Light-scattering measurements, which often accompany fluorescence measurements in flow cytometry, require a source that exhibits low divergence so that the scattered light can be measured at small forward angles from the excitation beam.

The laser meets these requirements and consequently is utilized as the optical source in many flow cytometers. Peters has compared the use of a conventional mercury source with the use of a laser source in flow cytometry (27). He suggests that for applications using dyes that are not efficiently excited by available laser wavelengths a conventional source gives better results at a lower cost. However, for studying weak intrinsic fluorescence or fluorescence from a weakly bound dye, the higher power available from lasers is necessary. In addition, optics used in laser applications and the directional, coherent nature of laser radiation provide more flexibility in design of the flow chamber. This allows for multi-parameter measurements and sorting to be accomplished more easily than with nonlaser sources. Modern continuous-wave lasers are also much more stable than conventional sources. Finally, if the internal morphology of the cell is to be probed, the source must be focused into a volume smaller than cellular dimensions. The coherent nature of laser light allows this to be accomplished. Thus, if common fixed-wavelength lasers do not excite cellular fluorescence effectively, the use of tunable dye lasers as the excitation source may be advantageous, although more expensive and complex. Commercially produced flow cytometers utilizing tunable laser sources are becoming more available.

The most commonly used lasers for flow cytometry are the argon ion laser with lines in the blue and green, the krypton ion laser with emission in the red and violet, the helium–neon laser with red output, and the helium–cadmium laser with outputs in the blue and ultraviolet. For optimum efficiency, the laser must be operated in the TEM_{00} mode, which has a Gaussian irradiance distribution. This mode exhibits the lowest diffraction losses, lowest divergence, and allows the beam to be focused into the smallest volume, ensuring optimal illumination of the flowing stream. If the laser is operating in another mode, such as the bimodal TEM_{01}, it should be tuned back to TEM_{00} (17). The lasers are usually operated at powers ranging from milliwatts to several watts. Lasers used for flow cytometry and their respective emission lines are summarized in Table 16.2.

In addition to being effective sources for fluorescence measurements, lasers are also good sources for light-scattering measurements. Laser beams exhibit low divergence and high spatial coherence: this allows the laser beam to be focused into a very small volume with an associated increase in power density. Similarly, because of the low divergence and high directionality of the beam,

Table 16.2. Lasers Used in Flow Cytometry and Their Respective Emission Wavelengths

Laser		Wavelengths Utilized (nm)
Argon ion	Primary	488, 515
	Secondary	458, 466, 473, 477, 497, 502
Helium–neon	Primary	633
Helium–cadmium	Primary	325, 441
Krypton ion	Primary	647, 568, 531, 676
	Secondary	521, 483, 476

light scattered when cells intercept the laser beam can be collected at low angles relative to the excitation axis using a suitable lens. A beam stop is often used to prevent laser radiation from interfering with the detection of scattered light (see Fig. 16.1). Information about cell size and cell viability is obtained from these low-angle light-scattering measurements.

In addition to instruments employing a single laser, dual-laser instruments have been developed (28–30) and are now commercially available. A block diagram of a dual-laser, multiparameter flow cytometer is shown in Fig. 16.3.

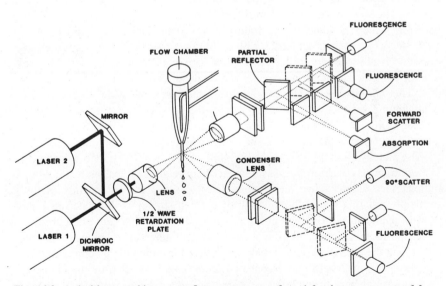

Fig. 16.3. A dual-laser, multiparameter flow cytometer configured for the measurement of four different wave bands of fluorescence emission, as well as low-angle forward light scattering, 90° or wide-angle light scattering, and cellular absorbance. Dashed optical elements indicate dichroic mirrors.

These instruments provide greater flexibility in analyses. The beams can be spatially adjusted to allow for simultaneous or sequential excitation of fluorescence (29). With sequential illumination of the cells there is a delay of about 20 μs between successive excitations causing fluorescence emissions to be temporally separated. This procedure increases the range of dyes that can be used and the substances that can be analyzed. One of these instruments (28) utilizes a dual-wavelength helium–cadmium laser along with an argon ion laser, providing three spatially separated beams. Up to five light-scattering measurements can be made at different angles, as well as fluorescence at three different excitation wavelengths. Another dual-laser instrument uses a helium–neon laser to supplement an argon ion laser (31). The signal from the helium–neon laser is used in conjunction with the fluorescence pulse from cells intercepting the argon ion laser to monitor the transit time of individual cells through the instrument. This information is then used to improve the timing of the charging pulses for cell sorting.

3.3. Optics

In order to successfully utilize lasers in flow cytometry, particularly in multiparameter instruments, special care must be taken in the design of the optics. Excitation optics include mirrors, polarizing optics, and beam-shaping lenses. The laser beam may be focused into a circular area, approximately 50 μm in diameter, which is equal to or slightly larger than normal cell dimensions. Alternately, two cylindrical lenses with mutually perpendicular axes are used to shape the beam to yield an elliptical cross section of about 10 μm $\times$ 90 μm with the elongated dimension perpendicular to the direction of the flowing stream (13). This elliptical beam-shaping improves the precision of the measurement by alleviating variances due to wandering of the flowing stream, thereby ensuring that each cell is illuminated equally. In addition, with proper focusing the narrow dimension of the ellipse can be made small enough to view small sections of cells, thereby allowing internal morphology or surface structure to be probed.

Fluorescence polarization measurements may require special optical configurations such as half-wave retardation plates to select the polarization of the excitation radiation or epi-illumination of the cell stream as illustrated in Fig. 16.4. In applications utilizing epi-illumination, dichroic filters are often used to reflect the laser radiation into the flow cell with fluorescence being viewed at 180° along the excitation optical path rather than the common 90° configuration (32). Longer wavelength fluorescence emission is not reflected by the dichroic filter and passes back through the same dichroic to a detector (Fig. 16.4). It has been suggested that this configuration compensates for the emission anisotropy of the optical system as well as misalignment of the electric vector of the excitation radiation relative to the detector polarization, thereby increasing the precision of the measurements (32).

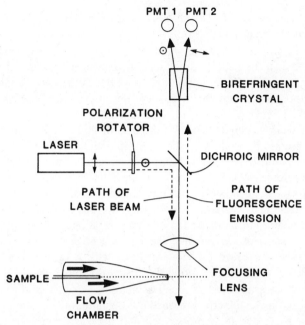

Fig. 16.4. Diagram of a flow cytometric optical system configured for epi-illumination of the cell stream and fluorescence depolarization measurements using a birefringent crystal.

Collection optics for fluorescence and scattered light include collimating and focusing lenses, mirrors, filters, and beam stops. Microscope objectives are often used in place of conventional single-element lenses to focus the fluorescence or scattered light onto a photodetector. These optics are arranged such that measurements may be made either concurrently or sequentially over a short distance within the flow chamber. A flow cytometer has been reported that uses an ellipsoidal mirror around the flow chamber, increasing the efficiency of fluorescence collection by about 58% (33). The detector is located at one focus of the ellipse while the sample–laser intersection point is located at the other. In fluorescence polarization measurements, additional optics such as calcite crystals exhibiting birefringence have been used to route the perpendicular and parallel polarization components of the fluorescence emission to different detectors (32, 34). If the excitation radiation is polarized, the signal at each of these detectors is used to obtain a depolarization ratio that can be related to the mobility or environment of a target substance. Similarly, dichroic filters have been used to separate fluorescence emission wave bands. The dichroic filter transmits a portion of the emission wave band to one photodetector but also acts as a mirror, reflecting the remaining wavelengths to a second photodetector (8). The choice of which type of emission filter to use depends on the type of measurement, the wavelength of the fluorescence emission, and the wavelength of the laser emission. Loken discusses the considerations necessary for determining which filters to use for flow cytometry (17, 35).

In the usual instrumental configuration the laser beam is focused into a volume slightly larger than the cellular cross section. However, this makes it difficult to obtain any morphological information about the cell surface or interior. The slit-scan method overcomes this difficulty. Because of the coherent nature of laser light, the laser beam can be focused to a small, typically $1 \times 5 \ \mu m$, slit-shaped cross section, perpendicular to the axis of flow. By monitoring the fluorescence pulse profile as the cells flow past the elliptical "slit" of laser radiation, low-resolution morphological information is obtained (36–38). The resulting measurements give time-dependent fluorescence intensities that correspond to the concentration, and consequently distribution, of fluorophore along a cell axis. Thus, variations in fluorescence due to intracellular structure or surface structure are monitored. For instance, slit-scan flow cytometry has been used to determine the number and position of centromeres along mammalian chromosomes based on fluorescence intensity variations arising from DNA-specific stains distributed along the chromosome (18, 39).

The incorrect determination of a cell as abnormal or misclassification of the cell by slit-scan flow cytometry is termed a false alarm (40). A large percentage of false alarms are due to improper cell and nucleus orientation in the flow chamber or to binucleate or coincident multiple cells. Several investigators have studied ways in which these false alarms can be minimized (37–40). One method (38, 41) uses multiple slits. Three spatially orthogonal detectors monitor the slit-scanned fluorescence from three separate excitation slits. A microcomputer analyzes the data from the three detectors to determine whether each cell should be rejected by the analysis. Viewing fluorescence from all three axes increases the frequency of identification of anomalies and allows for more confident cell classification.

3.4. Detectors

For the laser to be effectively utilized as an optical source in flow cytometry, care must be taken in the choice of detector: the detector must be sensitive enough to take advantage of the high powers available from lasers and must be fast enough to respond to analysis rates of up to 5000 cells per second. Most commonly, fluorescence intensity is the signal measured in flow cytometry. Consequently, photomultiplier tubes (PMTs) are often utilized as detectors. The PMT is excellent for such measurements due to its fast response and great sensitivity in the spectral region where detectable fluorescence from cells and stains most often occurs. Depending on the design of the cytometer and the number of measurements being made, several PMTs can be used concurrently.

In addition to the PMT, photodiode arrays and vidicons (36), microchannel plate intensifiers (38), and silicon intensified target (SIT) vidicons (37) have been used as detectors for fluorescence emission intensity. Fluorescence emission spectra have been obtained from cells via flow cytometry using a grating to disperse the emission radiation, and a vidicon detector (42). In this instrument an emission bandpass of 320 nm is dispersed over the 500-channel vidicon and spectra with adequate signal-to-noise ratios are obtained in a few seconds.

In most cases the fluorescence is viewed in the normal configuration of 90° to the excitation axis. However, in measurements using epi-illumination, the fluorescence is viewed at 180°, back along the excitation axis as described previously (32).

Detectors for light-scattering measurements are usually photodiodes with optics configured to collect the scattered light at various forward angles from the laser excitation beam axis. It has been shown that there is an angle dependence in the information obtained from scattering signals (43). With a single photodiode detector, as shown in Fig. 16.5a, the measured scattered light signal is integrated over a wide collection angle. This can lead to errors since cells that give the same integrated scattered light signal may actually be of different sizes, and have different refractive indices or other properties (44).

Several approaches are used to increase the amount of information obtained from the scatter signal, thus reducing errors associated with integrated measurements. One method utilizes two laser excitation sources and two separate scattered light detectors. One of the detectors is fixed, while the other is adjusted to various angles relative to the excitation axis and the fixed detector (44). This allows the angular dependence of the scattered light to be determined. Another configuration utilizes a circular photodiode array consisting of 32 concentric rings, each ring a separate photodiode (45, 46). The scattered light is detected at 32 different angles relative to the laser beam–sample stream intersection. The exit window of this detector limits the system to a maximum collection angle of 30°. This type of scatter detection is illustrated in Fig. 16.5b. A third configuration collects the multiangle scattered light signals using five concentric rings of fiber optics bundles, each leading to a different PMT (47). Thus, five different angles of scattered light are detected and analyzed simultaneously. These multiangle detectors allow for improved characterization of cells, or groups of cells, based on their light-scattering properties.

In addition to the fluorescence and light-scattering measurements, absorbance measurements have been made with flow cytometry. The detector for this type of measurement is usually a photodiode (28, 47), although silicon solar cells are also used (48).

Finally, electrical measurements are also made on cells in flow. An example is the measurement of conductivity using the Coulter method to obtain cell size information (1). The principles of this sizing method have been reviewed (49). In addition, instrumental designs have been suggested (8, 50) that allow elec-

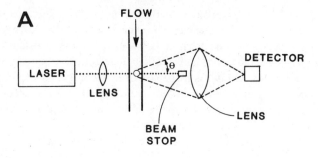

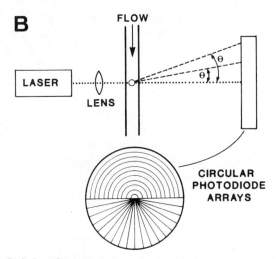

Fig. 16.5. (a) Optical configuration for the detection of the integrated light-scattering signal using a collecting lens and a single photodiode detector; (b) optical configuration for multiangle light-scattering measurements using a circular photodiode array consisting of 32 separate, concentric photodiode rings.

trical measurements to be made concurrently with optical measurements. However, there is some evidence that electronic cell sizing prior to optical measurements can cause degradation of fluorophores (51). Thus, if such multiparameter measurements are being made, care should be taken in both experimental procedure and instrumental design to minimize such interactive effects.

3.5. Cell Sorting

These various types of measurements are used to determine if a particular cell matches the characteristics of target cells under study. If it does, the individual cell can be separated from a large population with high precision. This sorting is accomplished by adjusting flow rates, original cellular concentration, and vibrational frequencies of the piezoelectric transducer so that each droplet formed in the postnozzle region is either empty or contains only one cell. The droplets containing cells are bracketed by several empty droplets to improve sorting efficiency. In this manner the cells are isolated in the electrolyte medium of the sheath fluid and can be sorted individually.

The nature and magnitude of signals obtained for an individual cell are analyzed to determine if the cell meets the sorting criteria. The sorting requirements can be as simple as setting a threshold fluorescence signal intensity (7) or as complex as monitoring 10 separate signals and sorting on the basis of 5 combinations of these 10 signals (52). If a cell meets the sorting requirements, a delay timer that is synchronized with the flowing cells is activated and, upon completion of the delay, a charge is applied to the stream. This charge can be varied in amplitude and polarity and is usually timed to be applied when the position of the target cell immediately precedes the breakoff point of the stream. The droplet breaks off and remains charged while the rest of the stream is returned to ground potential. The charged droplets fall between a pair of high-voltage plates that deflect the droplets into a collection vessel (see Figs. 16.1 and 16.2). The direction and degree of deflection depend on the mass of the droplet and the magnitude of the applied charge. Consequently, the sorting precision is highly dependent on the reproducible formation and charging of the droplets.

In the normal configuration the target cells are deflected into collection containers while the nonanalyte cells or fragments pass undeflected into a waste container. It has been suggested, however, that a better configuration for sorting a single subpopulation from a much larger population is to do the reverse; deflect the unwanted cells while allowing the analyte cells to pass through to the collection vessel (25, 52). This limits the method to sorting only one subpopulation at a time, but the sacrifice may be worthwhile if very high precision in sorting or in the placement of the sorted cells is required.

The sorted cells can be collected in a sample container, such as a culture tube, and saved in solution for further analyses by other methods. This requires the cells remain viable during the analysis. Consequently, instrumental and experimental design must ensure the sterility and viability of target cells. Alternatively, the cells can be collected so that a single cell's identity and measured properties are maintained. This is accomplished by deflecting the cells onto a microscope slide mounted on a computer-controlled xy translator (53). The cells are deposited in a matrix, with each matrix position correlated by computer with the signals obtained during the analysis. The isolated cells are then observed by conventional light microscopy to obtain additional information about individual cells (53). Similarly, another instrument collects cells on 16-mm motion picture film (54). The sorted cells are directed onto the film and adhere to the gelatin surface, allowing 100,000 cells to be stored on 100 m of film.

3.6. Signal Processing

The typical electronic signal that arises from fluorescence or light-scattering measurements in flow cytometry is a current or voltage pulse. These pulses arise as each cell crosses and interacts with the laser beam as shown in Fig. 16.6a.

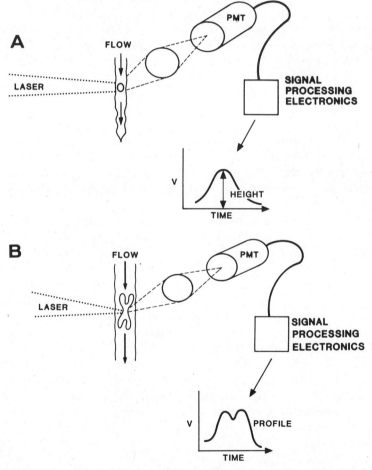

Fig. 16.6. (a) Voltage pulse arising as a cell flows past a moderately focused laser beam. The parameter most often recorded in this configuration is pulse height, which is related to the content of a fluorophore; (b) the voltage pulse profile arising from illumination of a cell or chromosome by a highly focused "slit" of laser radiation. Examination of the pulse profile is important in this configuration and is related to the axial distribution of the fluorophore along the cell.

During the period that the cell is within the laser field, light is absorbed, scattered, or emitted as fluorescence. The resulting signal pulse has a characteristic width related to the length of time the cell or stained portion of the cell interacts with the laser beam, and consequently to such parameters as flow rate, cell size, and laser beam size. The height of each pulse is a function of the intensity of the fluorescence or scatter signal, and is thereby related to the content of chromophores in the cell and the size and shape of the cell. The pulse area is a measure of the overall content or concentration of the chromophores present in the cell. Finally, the pulse profile is related to the axial distribution of chromophores along the length of the cell. Each of these pulse characteristics can be measured to obtain information about the cell.

Pulse height (Fig. 16.6a) is the most commonly measured parameter, especially in experiments where the laser beam diameter is greater than or equal to the cell diameter. Consequently, commercially available pulse height analyzers used extensively in gamma-ray spectroscopy are often used to process flow cytometry signals. In applications such as slit-scan flow cytometry, where the laser beam diameter is smaller than the cell or particle diameter, the pulse profile (Fig. 16.6b) rather than pulse height is used to determine such morphological information as the number of centromeres present in each chromosome or the presence of binucleate or coincident cells.

These signals are processed in a variety of ways. In a simple flow cytometer without cell sorting capabilities, the signals are digitized and stored in a computer for postrun analysis of the data. In complex multiparameter cell sorters, a high-speed computer is an integral part of the system. The computer must monitor the signal from each different channel at cellular flow rates of up to $5000 \ s^{-1}$. In addition, the computer must quickly determine if a cell meets preset sorting criteria based on each individual measurement or on some combination of several measurements. If the cell meets the sorting criteria, then the computer must determine into which collection container the cell is to be sorted and trigger the sorting electronics to apply the correct charge at the correct time. Thus, the timing and synchronization of the cells through the system is important if high precision is to be maintained at high sorting and analysis rates. This requires that flow rates be very stable and reproducible. Also, the electronics must be fast enough to process the signals and be designed to minimize signal pulse distortion. After the sample has been analyzed and the various signals obtained during the run have been stored in memory, the computer processes the data according to algorithms selected by the operator.

3.7. Data Evaluation

The output from a flow cytometry data system is different than the output obtained from a typical spectroscopic experiment. Spectroscopists generally deal with spectra that are plots of a dependent signal versus an independent experi-

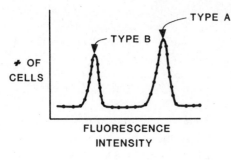

Fig. 16.7. Histogram plot of number of cells-versus-fluorescence intensity for a hypothetical sample containing two types of cells with different DNA content. Type B cells have lower DNA content and consequently correspond to the lower-intensity mode.

mental variable such as wavelength or frequency. However, the information generated from a flow cytometry experiment typically includes histograms relating the frequency of occurrence or number of cells versus the magnitude of a measured parameter. This is illustrated in Fig. 16.7, which might result from an experiment carried out for the purpose of determining the number of cells from a sampled population that fall into each of two different subpopulations, discriminated via fluorescence intensity. Cell type A might contain twice as much DNA as cell type B.

To effect such an experiment an entire sample population is stained with a DNA-specific fluorescent dye that very reproducibly binds to the DNA in each cell type: if cell type A has twice as much DNA as cell type B, it should also bind twice the amount of dye. Consequently, when the entire population is analyzed using flow cytometry, type A cells yield greater fluorescence intensities. Thus, if only type A and B cells are present, a histogram plot of number of cells versus the fluorescence intensity should give two peaks as shown in Fig. 16.7. In this case the independent variable is the fluorescence intensity that is a characteristic of each cell type, while the dependent variable is the number of cells of each type.

Other plots include correlation plots, known as dot plots, where two experimentally observed parameters are plotted against each other. An extension of the correlation plot is the three-dimensional isometric plot where two different parameters are plotted on the x and y axes, and the number of cells or occurrences is plotted along the z axis. Similarly, contour plots are also frequently used to evaluate flow cytometry data (17).

4. EXPERIMENTAL CONSIDERATIONS AND SAMPLE PREPARATION

The most time-consuming and critical part of a flow cytometry analysis is usually the experimental preparation of the cell sample rather than actual instrumental operation. In order to obtain useful results and effectively utilize the instrumentation, the experiment must be carefully designed with consideration given to factors such as stains and staining techniques and methods of fixation and dispersal. The most well-designed, finely tuned flow cytometer is useless if the sample is damaged, incorrectly stained, or the cells are present in undispersed clumps that can clog the nozzle or give spurious results.

4.1. Cell Dispersal

Flow cytometry analyses require single-cell suspensions. If the cells to be analyzed are bound in a matrix such as tissue, the cells must be released from the matrix through a dispersal technique. Dispersal methods are chosen so as to minimize unwanted disturbances to the analyte cells. Monodisperse samples such as blood or semen are ideal for flow cytometry as they can be prepared without dispersal. Monolayer cell samples can usually be dispersed through gentle agitation (55). Cells in solid tissue samples are dispersed and then filtered or centrifuged to remove fragments and particles. Solid tissue dispersal techniques include mechanical dispersion through grinding, mincing (55), or forcing the sample through a mesh (56). In addition, sonication (57) and syringing (56) are used as mechanical dispersal methods. Because it is often difficult to totally disperse the sample mechanically without disrupting cellular integrity, these mechanical methods are often used in conjunction with other methods in a preparative capacity. An example of a less traumatic dispersal method is the enzymatic dissociation of the cell matrix using a proteolytic enzyme, commonly trypsin (56, 58–60). Trypsin hydrolyzes the membranes of dead or dying cells and cellular debris, while at the same time disrupting cell surface membrane adhesion properties. In addition, trypsin cannot diffuse through a living cell membrane, and thus living cells are unaffected and cell viability is maintained (14). Chemical methods of dispersal using chelating agents such as EDTA (61, 62) have also met with some success. The method chosen to disperse the tissue sample should yield a monodisperse liquid suspension of cells that is free of small clumps and larger aggregates. The method should also minimize alteration of the physical structure and biological function of the cell.

4.2. Cell Fixation

After the cells have been dispersed, the next step in the cell preparation procedure is usually cell fixation. Fixation involves treating the cells in such a manner that they retain their form and important cellular and biochemical characteristics during storage or at least until staining and analysis. In addition, fixation can be used to render the cells permeable to the dyes used in staining for fluorescence measurements. Finally, fixation helps to prevent autolysis, reaggregation of the dispersed cells, and bacterial growth in the sample. The fixative should accomplish this without disrupting the morphology of the cell or changing characteristics such as staining specificity.

Many different fixatives are in use, and the choice of fixative depends on the type of sample and analysis being conducted. Examples of common fixing solutions are 50, 70, and 90% solutions of aqueous ethanol (20, 56, 58, 63); 70% ethanol acidified with concentrated hydrochloric acid (64); 25% saline ethanol (59, 65); 50% methanol (66); as well as formalin (56), paraformaldehyde (67), and mixtures such as 85% methanol/10% formalin/5% acetic acid (20). The cells are suspended in the fixative for as little as 30 min or as long as 12–18 hr to allow diffusion of the fixative throughout the sample. The fixation is often carried out at low temperatures, typically 4°C, to prevent reaggregation of the dispersed sample. In addition, stained cells can often be successfully stored in the fixative at these low temperatures for 1–2 weeks without loss of fluorescence or introduction of artifacts until the analysis can be completed (68). Fixation can be carried out before staining, concurrently with staining, or even after staining, depending on the procedure.

4.3. Staining for Fluorescence Analysis

Once the cells have been dispersed and fixed, they must be stained for fluorescence analysis. The fluorescent dyes utilized for cell staining in flow cytometry must meet several requirements. Obviously, the dyes must exhibit strong fluorescence in a wavelength region that can be selectively excited and detected by the instrumentation. In order for the dye to fluoresce, it must necessarily absorb radiation. Thus, the excitation maxima for the dyes must coincide with available laser lines. Also, the dye should bind strongly and specifically to the particular cellular component that is to be studied. This ensures that both the fluorescence emission intensity is proportional to the amount of the stained component present in the cell and that the dye will not diffuse into the cellular system causing nonspecific fluorescence of nontarget structures. Finally, the staining procedure, as with all preparative procedures, should not disrupt the cell sample integrity.

There are several classes of dyes available for fluorescence staining in flow cytometry. The choice of which stain to use depends not only on the nature of the analysis but also on the sample requirements. In most cases each dye will require a different staining procedure. Thus, the reagents involved in the procedure, as well as the number of steps and overall staining time, should be considered with respect to the cell sample. Examples of dyes used in various staining applications are compiled in Table 16.3. This list is not comprehensive but gives examples of several more commonly used stains in their typical applications.

DNA stains generally fall into three classes: Feulgen, intercalating, and nonintercalating base-pair binding stains. One of the most widely used staining procedures is the Feulgen procedure. The Feulgen procedure involves acid hydrolysis of the DNA purine bases. Acid hydrolysis exposes aldehyde functional groups, which can be selectively and quantitatively attacked by an aldehyde-

Table 16.3. Common Dyes Used for Fluorescence Staining in Flow Cytometry

Type	Name	Class	Reference
DNA			
	Acriflavine	Feulgen	20, 124
	Proflavine	Feulgen	124
	Ethidium bromide	Intercalating	69, 77, 80, 81
	Propidium iodide	Intercalating	58, 66, 69, 125
	Acridine orange	Intercalating	69, 70, 72, 126
	Hoechst 33258	A–T binding	65, 80, 94, 127
	Hoechst 33342	A–T binding	55, 65, 89, 128
	DAPI (4′-6-Diamidino-2-phenylindole)	A–T binding	55, 89, 128
	DIPI (4′-6-bis[2′-imidazolinyl-4H-5H]-2-phenylindole)	A–T binding	78, 129
	Mithramycin	G–C binding	56, 59, 60, 89, 130
	Chromomycin A_3	G–C binding	56, 78, 80, 94
Protein			
	Fluorescein isothiocyanate		56, 69, 89, 109, 110
	Fluorescamine		89, 115, 131
Viability			
	Trypan blue		69, 108, 118
	Ethidium bromide		108

specific staining reagent. Examples of Feulgen stains are acriflavine, and a similar compound, proflavine.

Another common class of DNA stains are the intercalating dyes. These dyes bind between (intercalate) the strands of double-stranded nucleic acids in positions that are sterically favored. Common examples of intercalating dyes are propidium iodide (PI), ethidium bromide (EB), and acridine orange (AO). These intercalating dyes bind to both double-stranded DNA and to RNA. Consequently, if DNA specificity is required in the analysis, any double-stranded RNA present in the cell must be denatured using, for example, the enzyme RNase. Acridine orange is a particularly interesting dye that emits green fluorescence when the molecule is isolated in between the strands of nucleic acids. However, when acridine orange is bound to surfaces such as cytoplasmic structures, the dye molecules stack polymerically. The interaction between neighboring dye molecules in the stack causes a shift of the fluorescence emission to the orange (69). It is thereby feasible to obtain information simultaneously about both DNA and cytoplasmic content. Similarly, acridine orange bound to single-stranded RNA exhibits red fluorescence emission. Thus, simultaneous measurement of RNA and DNA content is possible (70).

Finally, there is a class of DNA stains that bind nonintercalatively to specific

DNA base pairs. These stains show a preference for either the adenine–thymine (A–T) base pairs, or guanine–cytosine (G–C) base pairs of DNA. Examples of G–C specific stains are the antitumor antibiotics mithramycin, and chromonycin A_3. Examples of stains which exhibit A–T binding preference are the bis-benzimidazole dyes, Hoechst 33258, and Hoechst 33342. These dyes have the advantage that they are specific for particular base pairs on DNA, and thus no RNA denaturation step is required. In addition, no acid hydrolysis step is required as in the Feulgen procedure.

Examples of two protein or cytoplasmic stains are fluorescein isothiocyanate (FITC) and fluorescamine. Such stains are used alone or in combination with DNA stains mentioned above for concurrent protein-DNA measurements (56, 58, 71). These types of measurements give information about which biochemical processes occur during different parts of the cell cycle. The requirements for dual staining of a sample are that the two stains have overlapping excitation spectra that also coincide with one of the laser lines and that their emission spectra are sufficiently resolved to allow independent detection of each emission wave band. An example of a dye combination that has been successfully used in such an application is propidium iodide–fluorescein isothiocyanate (56, 58).

Another useful class of dyes are the viability stains. Many dyes such as trypan blue and ethidium bromide are excluded from living cells and consequently cannot stain them significantly. When the cell membrane has been damaged or ruptured, these dyes may then permeate the membrane and stain the cell. Consequently, cell viability can be probed on the basis of fluorescence intensity. This procedure is useful in applications such as evaluating new fixing or staining procedures with respect to cell viability. Viability probes are also used to purify a cell sample via FACS. This ensures that the sample, which is to be analyzed further by flow cytometry or other techniques, consists largely of viable cells.

The methods chosen to disperse, fix, and stain the cells should result in a monodisperse suspension, free of clumps and large amounts of debris. The resulting sample should be relatively stable, not subject to autolysis, bacterial growth, or changes in fluorescence intensity. The stains should be specific for the substances intended and should stain stoichiometrically to ensure that fluorescence intensity is proportional to the amount of target substance present. In order to assure that these criteria are met and that the sample is ready for introduction into the flow cytometer, a representative sample from the cell population is often visually examined under a fluorescence microscope to determine that no clumps are present and that staining has been uniform.

5. APPLICATIONS

The power of flow cytometry as an analytical technique is demonstrated by the wide range of applications for which it is used. Flow cytometry is applied not only to cell biology studies but also to disciplines such as enzymology, immunology, virology, hematology, oncology, and cytogenetics. Journals such as *Cytometry* and *The Journal of Histochemistry and Cytochemistry* publish current investigations using flow cytometry. In addition, applications of flow cytometry and cell sorting have been reviewed by several authors (13–17).

5.1. Cell Cycle Analysis

One important application of flow cytometry is cell cycle analysis (59, 66, 72–75). Cell reproduction occurs in an ordered sequence of events known as the cell cycle, during which cellular DNA is replicated. This is required for each daughter cell to contain a complete copy of the genetic information originally found in the parent cell after cell division. The phase of the cell cycle when DNA synthesis occurs is known as the S (synthesis) phase. Prior to the S phase there is a period of cellular growth known as the G_1 (growth$_1$) phase, during which the various organelles and cytoplasm grow in preparation for DNA synthesis. After the S phase there is another period of growth, G_2, in which the cells prepare for mitosis. During mitosis, the genetic material of the cell, DNA in the form of chromosomes, separates into two identical sets of new chromosomes. The cell then divides to form two cells, with each daughter cell carrying an identical set of the parent cell's DNA. Once cell division has occurred, the newly formed cells can begin the G_1 phase.

The DNA content of cells varies as the cell cycle progresses from one phase to another. In G_1 the cell contains the normal amount of DNA. As the cell proceeds through S phase, the DNA content increases until G_2 is reached, at which time the cell has twice the DNA content as in G_1. Finally, after mitosis and cell division, the DNA content of each daughter cell corresponds to the normal level. Thus, by measuring the DNA content of a population of cells, information is obtained about the number of cells in each phase of the cell cycle. Flow cytometry is very effective for this purpose since selective staining of cellular DNA with fluorescent dyes is possible.

Using flow cytometry, a sample of several thousand cells can be rapidly analyzed, and the population distribution of cell cycle development can be estimated via fluorescence intensity measurements which correspond to DNA content. An unsynchronized, randomly distributed population of cells might result in a DNA histogram similar to that shown in Fig. 16.8a. Note the bimodal distribution of DNA content. The lower intensity mode in Fig. 16.8a represents cells in the G_1 phase. These cells have not begun to synthesize DNA and have the least amount of DNA for normal cells. Synthesis (S) phase cells are represented by the moderate-intensity region between the two peaks. The DNA content of the cells in the synthesis phase increases until they contain as much DNA as cells in the G_2 phase or mitosis before cell division. These cells have twice

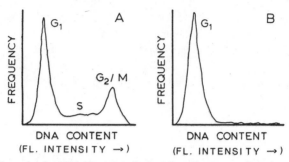

Fig. 16.8. (a) Example of a DNA histogram obtained from fluorescence intensity measurements of a sample taken from an unsynchronized cell population distributed randomly through the cell cycle; (b) Histogram obtained after application of a blocking agent that halts S phase DNA synthesis, causing the cells to arrest in the G_1 phase.

as much DNA as the cells in the G_1 phase and consequently correspond to the higher intensity mode in Fig. 16.8a.

This type of cellular analysis is important in medical studies involving chemicals that affect the cell cycle of target cell populations. For these purposes flow cytometry allows for rapid and precise descriptions of cell populations at various times after the administration of a drug. Thus, chemicals can be investigated that selectively interfere with the cellular development of specific groups of target cells, such as cancer cells (76). For example, chemotherapeutic drugs that block the growth of cancer cells by preventing the cells from synthesizing DNA or preventing mitosis may be examined. A drug that blocks a population of cells from the synthesis phase yields a low-intensity unimodel histogram as shown in Fig. 16.8b. In addition, FACS allows sorting of cells in mitosis from cells in interphase (G_1 + S + G_2), which is important for such procedures as chromosome karyotyping (72).

5.2. Chromosome Karyotyping

Another important application of flow cytometry is chromosome karyotyping (60, 77–79). Chromosome typing, or karyotyping, is a method by which individual chromosomes are sorted and identified. Karyotyping is usually carried out on chromosomes obtained from cells in the metaphase stage of mitosis. Metaphase chromosomes consist of pairs of chromatids connected together at a central point, called the kinetochore or centromere, forming the familiar X shape. Each type of chromosome has a distinctive shape, size, and centromere position; thus they can be distinguished from one another. The chromosomal characteristics from a single cell, including number of chromosomes and the individual characteristics described above, are used to assess its karyotype.

Classical karyotyping involves several procedures. First, The cell cycle must be blocked in the metaphase stage of mitosis by a suitable blocking agent. The cell and nuclear membranes are then disrupted to allow access to the chromosomes. The chromosomes are then stained and photographed under a microscope. Finally, the photographs are enlarged, and each individual chromosome is characterized, typically being cut from the picture by hand. The chromosomes are then classified and categorized according to parameters such as size and position of the centromere to obtain the karyotype. These karyotypes are used both to determine an individual organism's identity, much like a fingerprint, and also to determine if any chromosomal genetic anomalies are present, such as the presence of too many chromosomes.

This classical method of karyotyping has the disadvantage that it is tedious, labor intensive, and slow. In addition, since the classification is visual, human subjectivity can perturb the classification. On the other hand flow cytometry allows rapid, automated karyotyping of large numbers of chromosomes. This is possible through stoichiometric staining of chromosomal DNA (80). Since each chromosome contains a characteristic amount of DNA, each chromosome is identified and differentiated on the basis of fluorescence intensity.

One major effort in developing flow cytometric karyotyping has been toward obtaining the fluorescence intensity resolution needed to discriminate between chromosomes having only slight differences in DNA content. Consequently, experimental procedures have been developed that focus on reproducible preparation and staining of chromosomes (81). Also, great care must be taken in instrumental design to ensure that each chromosome is illuminated equally by the laser, with special considerations regarding flow paths and chromosome orientation. These special considerations are of particular importance for karyotyping of human chromosomes. Human cells normally have 22 pairs of chromosomes plus 2 sex chromosomes. Many of these chromosomes have nearly identical DNA content, requiring the highest possible fluorescence intensity resolution to discriminate them. Early attempts at the karyotyping of human chromosomes yielded fluorescence intensity distributions which separated the chromosomes into only 7 resolved fractions (77). More recently, however, workers using a dual-laser system and double staining with the DNA-specific dyes DIPI and chromomycin were able to sort human chromosomes into 22 fractions and could identify 21 unique chromosomal types; only 3 chromosomes remained unresolved (78).

Chromosome karyotyping is particularly useful for identifying chromosome defects associated with special syndromes. The presence of an extra chromosome in humans often causes severe abnormalities including mental deficiency and physical deformities. An example is Down's syndrome which has been linked to the existence of an extra chromosome 21 either alone or attached to another chromosome. If a child is born with Down's syndrome, both the child's and parent's chromosomes can be karyotyped to determine the probability of their having other children with the syndrome. High-resolution flow cytometric karyotyping has also been used to assign another syndrome, McArdle's syndrome, to chromosome 11 (78). The assignment of the syndrome to the previously unkown chromosome location should help future diagnoses and study of the syndrome.

Chromosome karyotyping is also applied in the area of amniocentesis and genetic screening (82). In this type of analysis fetal cells are often obtained from the amniotic fluid. A karyotype study is then performed on the cells to determine the sex of the fetus or the presence of abnormalities such as Down's syndrome. Flow cytometry offers a rapid, accurate method by which to perform such karyotypes. Similarly, to avoid the dangers associated with current methods of sampling fetal blood for such disorders as hemophilia, it has been suggested that flow cytometry could be used to detect and separate fetal blood cells directly from the maternal circulatory system (82). This method of obtaining fetal blood cells would eliminate the 4–5% fetal mortality rate associated with current sampling methods.

5.3. Dosimetry

Flow cytometry is applied in the area of biological dosimetry for the assessment of exposure to radiation or chemical mutagens. Radiation hazards and environmental exposure to toxic chemicals are great international concerns. However, even in cases where exposures have been identified and quantified, it is difficult to assess the extent of damage to a human population. It has been suggested that an appropriate indicator of radiation and chemical exposure may be measurements of the effects on the human reproductive system (83). In particular, studies of chromosome aberrations, sister chromatid exchanges, and sperm abnormalities seem to offer viable methods for such dosimetry. One major disadvantage of chromosome and sperm studies, however, is obtaining statistically significant results for low dosages (83). Analyses of very large samples are needed to obtain required levels of significance, a tedious job using conventional methods. High analysis rates, accuracy, and resultant statistical significance permitted by flow cytometry would seem to be an answer to this problem.

One biological indicator for damage due to chemical exposure is the presence of chromosomes with more than one centromere. These aberrations, often in the form of chromosomes with two centromeres, dicentrics, can be successfully identified using slit-scan flow cytometry (18, 39). There is less DNA present at the centromere than along each of the chromatids; consequently, a plot of fluorescence intensity versus length along the chromosome yields profiles with minima corresponding to each centromere location. Thus, profiles with two or more minima can be classified as aberrations. This type of analysis is rapid, allowing the analysis of 1000 chromosomes s^{-1}, and may greatly increase the ability to identify individuals with an incidence of aberrations above normal levels.

In addition, variability in both sperm DNA content and morphology can be used as biological dosimeters. It has been suggested that variability in both sperm head shape (84) and sperm DNA content (85) can be linked to exposure to chemical mutagens. Flow cytometry has been used successfully to rapidly and accurately identify these variations (86–88).

5.4. Miscellaneous Applications

Other applications of flow cytometry and FACS include limnological applications (89, 90), blood cell counting and classification (91), evaluation of myocardial injury (92), as well as viral (93) and bacterial (94) classification. Flow cytometry has been used to perform analyses of tumor cells (95–98), for the screening of gynecologic smears for abnormal cells (99–102), and for the evaluation of lung cancer (103–105). In addition, many new applications should evolve as experimental procedures continue to develop and improve and as flow cytometers become more available to investigators who need them.

6. CONCLUSION

A great deal of time and energy has gone into the development of flow cytometry. This research and development has yielded powerful instruments that are unsurpassed for use in applications requiring rapid, single-cell analyses. Advantages of flow cytometry include the ability to make simultaneous multiparameter measurements on individual cells at very high analysis rates. High analysis rates allow rapid measurements of many cells, offering substanial improvements in statistical significance as compared to other methods. In addition, analysis criteria are objective, not subject to human mistakes or individual bias. Finally, laser-based flow cytometry is sensitive, selective, and can be applied to a wide variety of sample types including nonbiological samples. Further, flow cytometry with cell sorting allows a large population of cells to be analyzed and target cells to be separated from the population with high precision.

The major limitations in flow cytometry at this time seem to involve experimental factors such as uniformity of dispersal, selectivity of stains, and reproducibility of staining techniques. Highly reproducible cell preparation is required to optimally utilize flow cytometry. However, methods of cell preparation have not developed as quickly as has the instrumentation. Therefore, recent work in the field has emphasized the development of better experimental procedures.

Flow cytometry will continue to be a very attractive technique for the biological, biochemical, and medical sciences as new procedures are developed that effectively utilize existing instrumentation. In addition, flow cytometry should be easily adapted to fields outside the biological sciences that require rapid analysis of single particles. For example, in the area of colloid and surface chemistry, rapid analyses of the properties of colloidal particles might be possible.

REFERENCES

1. W. H. Coulter, *Natl. Electron. Conf. Proc.*, **12**, 1034 (1956).
2. M. J. Fulwyler, *Science*, **150**, 910 (1965).
3. L. A. Kamentski, M. R. Melamed, and H. Derman, *Science*, **150**, 630 (1965).
4. M. A. Van Dilla, T. T. Trujillo, P. F. Mullaney, and J. R. Coulter, *Science*, **163**, 1213 (1969).
5. P. F. Mullaney, M. A. Van Dilla, J. R. Coulter, and P. N. Dean, *Rev. Sci. Inst.*, **40**, 1029 (1969).
6. P. F. Mullaney and P. N. Dean, *Appl. Opt.*, **8**, 2361 (1969).
7. W. A. Bonner, H. R. Hulett, R. G. Sweet, and L. A. Herzenberg, *Rev. Sci. Inst.*, **43**, 404 (1972).
8. J. A. Steinkamp, M. J. Fulwyler, J. R. Coulter, R. D. Hiebert, J. L. Horney, and P. F. Mullaney, *Rev. Sci. Inst.*, **44**, 1301 (1973).
9. J. L. Marx, *Science*, **188**, 821 (1975).
10. L. A. Herzenberg, R. G. Sweet, and L. A. Herzenberg, *Scientific American*, **234**, 108 (1976).
11. P. K. Horan and L. L. Wheeless, Jr., *Science*, **198**, 149 (1977).
12. P. F. Mullaney and J. H. Jett, "Flow Cytometry: An Overview," in *Lasers in Medicine and Biology*, NATO Advanced Study Institute Series, Vol. 34A, F. Hillenkamp, R. Pratesi, and C. A. Sacchi, Eds., Plenum, New York, 1980, p. 179.
13. P. F. Mullaney, J. A. Steinkamp, H. A. Crissman, L. S. Cram, and D. M. Holm, "Laser Flow Microphotometers for Rapid Analysis of Individual Mammalian Cells," in *Laser Applications in Medicine and Biology*, Vol. 2, M. L. Wolbarsht, Ed., Plenum, New York, 1974, p. 151.
14. H. A. Crissman, P. F. Mullaney, J. A. Steinkamp, "Methods and Applications of Flow Systems for Analysis and Sorting of Mammalian Cells," in *Methods in Cell Biology*, Vol. 9, D. M. Prescott Ed., Academic Press, New York, 1975, p. 179.
15. H. S. Kruth, *Anal. Biochem.*, **125**, 225 (1982).
16. D. J. Arndt-Jovin and T. M. Jovin, "Automated Cell Sorting with Flow Systems," in *Annual Review of Biophysics and Bioengineering*, Vol. 7, L. J. Mullins, Ed., Annual Reviews, Palo Alto, CA, 1978, pp. 527–558.
17. M. R. Loken and A. M. Stall, *J. Immunol. Methods*, **50**, R85 (1982).
18. D. Pinkel, *Anal. Chem.*, **54**, 503A (1982).
19. M. R. Melamed, P. F. Mullaney, and M. L. Mendelsohn, Eds., *Flow Cytometry and Sorting*, Wiley, New York, 1979.

20. J. W. Levinson, R. G. Langlois, V. M. Maher, and J. J. McCormick, *J. Histochem. Cytochem.*, **26**, 680 (1978).
21. A. Brunsting and P. F. Mullaney, *Biophys. J.*, **14**, 439 (1974).
22. T. M. Jovin and D. J. Arndt-Jovin, *TIBS*, August, 214 (1980).
23. V. Kachel, E. Kordwig, and E. Glossner, *J. Histochem. Cytochem.*, **25**, 774 (1977).
24. D. B. Kay and L. L. Wheeless, Jr., *J. Histochem. Cytochem.*, **25**, 870 (1977).
25. R. T. Stovel, *J. Histochem. Cytochem.*, **25**, 813 (1977).
26. J. T. Merrill, P. N. Dean, and J. W. Gray, *J. Histochem. Cytochem.*, **27**, 280 (1979).
27. D. C. Peters, *J. Histochem. Cytochem.*, **27**, 241 (1979).
28. H. M. Shapiro, E. R. Schildkraut, R. Curbelo, R. B. Turner, R. H. Webb, D. C. Brown, and M. J. Block, *J. Histochem. Cytochem.*, **25**, 836 (1977).
29. J. A. Steinkamp, D. A. Orlicky, and H. A. Crissman, *J. Histochem. Cytochem.*, **27**, 273 (1979).
30. P. N. Dean and D. Pinkel, *J. Histochem. Cytochem.*, **26**, 622 (1978).
31. J. C. Martin, S. R. McLaughlin, and R. D. Hiebert, *J. Histochem. Cytochem.*, **27**, 277 (1979).
32. W. G. Eisert and W. Beisker, *Biophys. J.*, **31**, 97 (1980).
33. M. J. Skogen-Hagenson, G. C. Salzman, P. F. Mullaney, and W. H. Brockman, *J. Histochem. Cytochem.*, **25**, 784 (1977).
34. M. Epstein, A. Norman, D. Pinkel, and R. Udkoff, *J. Histochem. Cytochem.*, **25**, 821 (1977).
35. M. R. Loken, *J. Histochem. Cytochem.*, **28**, 1136 (1980).
36. D. B. Kay, J. L. Cambier, and L. L. Wheeless, Jr., *J. Histochem. Cytochem.*, **27**, 329 (1979).
37. L. L. Wheeless, Jr., D. B. Kay, M. A. Cambier, J. L. Cambier, and S. F. Patten, Jr., *J. Histochem. Cytochem.*, **25**, 864 (1977).
38. J. L. Cambier, D. B. Kay, and L. L. Wheeless, Jr., *J. Histochem. Cytochem.*, **27**, 321 (1979).
39. J. W. Gray, D. Peters, J. T. Merrill, R. Martin, and M. A. Van Dilla, *J. Histochem. Cytochem.*, **27**, 441 (1979).
40. L. L. Wheeless, Jr., J. L. Cambier, M. A. Cambier, D. B. Kay, L. L. Wightman, and S. F. Patten, Jr., *J. Histochem. Cytochem.*, **27**, 596 (1979).
41. J. L. Cambier and L. L. Wheeless, Jr., *J. Histochem. Cytochem.*, **27**, 325 (1979).
42. C. G. Wade, R. H. Rhyne, Jr., W. H. Woodruff, D. P. Bloch, and J. C. Bartholomew, *J. Histochem. Cytochem.*, **27**, 1049 (1979).
43. M. Kerker, H. Chew, P. J. McNulty, J. P. Kratohvil, D. D. Cooke, M. Sculley, and M.-P. Lee, *J. Histochem. Cytochem.*, **27**, 250 (1979).
44. M. R. Loken, R. G. Sweet, and L. A. Herzenberg, *J. Histochem. Cytochem.*, **24**, 284 (1976).
45. P. F. Mullaney, J. M. Crowell, G. C. Salzman, J. C. Martin, R. D. Hiebert, and C. A. Goad, *J. Histochem. Cytochem.*, **24**, 298 (1976).
46. G. C. Salzman, J. M. Crowell, C. A. Goad, K. M. Hansen, R. D. Hiebert, P. M. LaBauve, J. C. Martin, M. L. Ingram, and P. F. Mullaney, *Clin. Chem.*, **21**, 1297 (1975).
47. W. G. Eisert, *J. Histochem. Cytochem.*, **27**, 404 (1979).

48. T. K. Sharpless, M. Bartholdi, and M. R. Melamed, *J. Histochem. Cytochem.*, **25**, 845 (1977).

49. V. Kachel, *J. Histochem. Cytochem.*, **24**, 211 (1976).

50. R. A. Thomas, T. A. Yopp, B. D. Watson, D. H. K. Hindman, B. F. Cameron, S. B. Leif, R. C. Leif, L. Roque, and W. Britt, *J. Histochem. Cytochem.*, **25**, 827 (1977).

51. O. Alabaster, D. L. Glaubiger, V. T. Hamilton, S. A. Bentley, S. E. Shackney, K. S. Skramstad, and R. F. Chen, *J. Histochem. Cytochem.*, **28**, 330 (1980).

52. D. J. Arndt-Jovin and T. M. Jovin, *J. Histochem. Cytochem.*, **22**, 622 (1974).

53. R. T. Stovel and R. G. Sweet, *J. Histochem. Cytochem.*, **27**, 284 (1979).

54. E. R. Schildkraut, M. Hercher, H. M. Shapiro, R. E. Young, N. Matsu, D. C. Brown, and R. H. Webb, *J. Histochem. Cytochem.*, **27**, 289 (1979).

55. I. W. Taylor, *J. Histochem. Cytochem.*, **28**, 1021 (1980).

56. H. A. Crissman, M. S. Oka, and J. A. Steinkamp, *J. Histochem. Cytochem.*, **24**, 64 (1976).

57. U. Moller and J. K. Larsen, *Cell Tissue Kinet.*, **12**, 203 (1979).

58. H. A. Crissman and J. A. Steinkamp, *J. Cell Biol.*, **59**, 766 (1973).

59. H. A. Crissman and R. A. Tobey, *Science*, **184**, 1297 (1974).

60. M. L. Meistrich, W. Gohde, R. A. White, and J. Schumann, *Nature*, **274**, 821 (1978).

61. A. L. Kisch, R. O. Kelley, H. Crissman, and L. Paxton, *J. Cell Biol.*, **57**, 38 (1973).

62. P. M. Kraemer, R. A. Tobey, and M. A. Van Dilla, *J. Cell Physiol.*, **81**, 305 (1973).

63. E. Tannenbaum, M. Cassidy, O. Alabaster, and C. Herman, *J. Histochem. Cytochem.*, **26**, 145 (1978).

64. P. B. Coulson and R. Tyndall, *J. Histochem. Cytochem.*, **26**, 713 (1978).

65. D. J. Arndt-Jovin and T. M. Jovin, *J. Histochem. Cytochem.*, **25**, 585 (1977).

66. A. Krishan, *J. Cell Biol.*, **66**, 188 (1975).

67. L. L. Lanier and N. L. Warner, *J. Immunol. Methods*, **47**, 25 (1981).

68. A. Krishan, *Stain Tech.*, **52**, 339 (1977).

69. L. R. Adams, *J. Histochem. Cytochem.*, **25**, 965 (1977).

70. F. Traganos, Z. Darzynkiewicz, T. Sharpless, and M. R. Melamed, *J. Histochem. Cytochem.*, **25**, 46 (1977).

71. M. Stohr, M. Vogt-Schaden, M. Knobloch, R. Vogel, and G. Futterman, *Stain Tech.*, **53**, 205 (1978).

72. Z. Darzynkiewicz, F. Traganos, T. Sharpless, and M. R. Melamed, *J. Histochem. Cytochem.*, **25**, 875 (1977).

73. D. W. Galbraith, K. R. Harkins, J. M. Maddox, N. M. Ayres, D. P. Sharma, and E. Firoozabady, *Science*, **220**, 1049 (1983).

74. F. Dolbeare, H. Gratzner, M. G. Pallavicini, and J. W. Gray, *Proc. Natl. Acad. Science USA*, **80**, 5573 (1983).

75. M. Kubbies and G. Pierron, *Exp. Cell Res.*, **149**, 57 (1983).

76. S. W. Dean and M. Fox, *J. Cell Sci.*, **64**, 265 (1983).

77. J. W. Gray, A. V. Carrano, D. H. Moore II, L. L. Steinmetz, J. Minkler, B. H. Mayall, M. L. Mendelsohn, and M. A. Van Dilla, *Clin. Chem.*, **21**, 1258 (1975).

78. R. V. Lebo, F. Gorin, R. J. Fletterick, F.-T. Kao, M.-C. Cheung, B. D. Bruce, and Y. W. Kan, *Science*, **225**, 57 (1984).

79. J. W. Gray, A. V. Carrano, L. L. Steinmetz, M. A. Van Dilla, D. H. Moore II, B. H. Mayall, and M. L. Mendelsohn, *Proc. Natl. Acad. Sci. USA*, **72**, 1231 (1975).

80. R. H. Jensen, R. G. Langlois, and B. H. Mayall, *J. Histochem. Cytochem.*, **25**, 954 (1977).

81. R. Sillar and B. D. Young, *J. Histochem. Cytochem.*, **29**, 74 (1981).

82. P. T. Rowley, *Science*, **225**, 138 (1984).

83. T. H. Maugh II, *Science*, **215**, 643 (1982).

84. D. A. Benaron, J. W. Gray, B. L. Gledhill, S. Lake, A. J. Wyrobek, and A. T. Young, *Cytometry*, **2**, 344 (1982).

85. B. Gledhill, S. Lake, and P. N. Dean, in *Flow Cytometry and Sorting*, M. R. Melamed, P. F. Mullaney, and M. L. Mendelsohn, Eds., Wiley, New York, 1979, p. 481.

86. D. Pinkel, P. Dean, S. Lake, D. Peters, M. Mendelsohn, J. Gray, M. Van Dilla, and B. Gledhill, *J. Histochem. Cytochem.*, **27**, 353 (1979).

87. M. A. Van Dilla, B. L. Gledhill, S. Lake, P. N. Dean, J. W. Gray, V. Kachel, B. Barlogie, and W. Gohde, *J. Histochem. Cytochem.*, **25**, 763 (1977).

88. D. P. Evenson and M. R. Melamed, *J. Histochem. Cytochem.*, **31**, 248 (1983).

89. C. M. Yentsch, P. K. Horan, K. Muirhead, Q. Dortch, E. Haugen, L. Legendre, L. S. Murphy, M. J. Perry, D. A. Phinney, S. A. Pomponi, R. W. Spinrad, M. Wood, C. S. Yentsch, and B. J. Zahuranec, *Limnol. Oceanog.*, **28**, 1275 (1983).

90. L. J. Ong, A. N. Glazer, and J. B. Waterbury, *Science*, **224**, 80 (1984).

91. H. M. Shapiro, E. R. Schildkraut, R. Curbelo, C. W. Laird, R. B. Turner, and T. Hirschfeld, *J. Histochem. Cytochem.*, **24**, 396 (1976).

92. B. A. Khaw, J. Scott, J. T. Fallon, S. L. Cahill, E. Haber, and C. Homcy, *Science*, **217**, 1050 (1982).

93. M. Hercher, W. Mueller, and H. M. Shapiro, *J. Histochem. Cytochem.*, **27**, 350 (1979).

94. M. A. Van Dilla, R. G. Langlois, D. Pinkel, D. Yajko, and W. K. Hadley, *Science*, **220**, 620 (1983).

95. N. T. Van, M. Raber, G. H. Barrows, B. Barlogie, *Science*, **224**, 876 (1984).

96. M. M. Zatz, B. J. Mathieson, C. Kanellopoulos-Langevin, and S. O. Sharrow, *J. Immunol.*, **126**, 608 (1981).

97. M. N. Raber, B. Barlogie, and M. Luna, *Cancer*, **53**, 1705 (1984).

98. M. Anniko, B. Tribukait, and J. Wersall, *Cancer*, **53**, 1708 (1984).

99. M. C. Habbersett, M. Shapiro, B. Bunnag, I. Nishiya, and C. Herman, *J. Histochem. Cytochem.*, **27**, 536 (1979).

100. W. A. Linden, K. Ochlich, H. Baisch, K.-U. Scholz, H.-J. Mauss, H.-E. Stegner, D. S. Joshi, C. T. Wu, I. Koprowska, and C. Nicolini, *J. Histochem. Cytochem.*, **27**, 529 (1979).

101. J. E. Gill, L. L. Wheeless, Jr., C. Hanna-Madden, and R. J. Marisa, *J. Histochem. Cytochem.*, **27**, 591 (1979).

102. D. L. Barrett, R. H. Jensen, E. B. King, P. N. Dean, and B. H. Mayall, *J. Histochem. Cytochem.*, **27**, 573 (1979).

103. J. K. Frost, H. W. Tyrer, N. J. Pressman, C. D. Albright, M. H. Vansickel, and G. W. Gill, *J. Histochem. Cytochem.*, **27**, 545 (1979).

104. H. W. Tyrer, J. F. Golden, M. H. Vansickel, C. K. Echols, J. K. Frost, S. S. West, N. J. Pressman, C. D. Albright, L. A. Adams, and G. W. Gill, *J. Histochem. Cytochem.*, **27**, 552 (1979).

105. J. K. Frost, H. W. Tyrer, N. J. Pressman, L. A. Adams, M. H. Vansickel, C. D. Albright, G. W. Gill, and S. M. Tiffany, *J. Histochem. Cytochem.*, **27**, 557 (1979).

106. M. J. McCutcheon and R. G. Miller, *J. Histochem. Cytochem.*, **27**, 246 (1979).

107. B. Rotman and B. W. Papermaster, *Proc. Natl. Acad. Sci. USA*, **55**, 134 (1966).

108. P. K. Horan and J. W. Kappler, *J. Immunol. Methods*, **18**, 309 (1977).

109. L. L. Lanier and M. R. Liken, *J. Immunol.* **132**, 151 (1984).

110. B. Bohn, *Exp. Cell Res.*, **103**, 39 (1976).

111. J. A. Steinkamp, K. M. Hansen, and H. A. Crissman, *J. Histochem. Cytochem.*, **24**, 292 (1976).

112. T. K. Sharpless and M. R. Melamed, *J. Histochem. Cytochem.*, **24**, 257 (1976).

113. L. S. Cram, D. J. Arndt-Jovin, B. G. Grimwade, and T. M. Jovin, *J. Histochem. Cytochem.*, **27**, 445 (1979).

114. J. F. Leary, P. Todd, J. C. S. Wood, and J. H. Jett, *J. Histochem. Cytochem.*, **27**, 315 (1979).

115. D. J. Arndt-Jovin, W. Ostertag, H. Eisen, F. Klimek, and T. M. Jovin, *J. Histochem. Cytochem.*, **24**, 332 (1976).

116. T. Lindmo and H. B. Steen, *Biophys. J.*, **18**, 173 (1977).

117. P. E. Wanda and J. D. Smith, *J. Histochem. Cytochem.*, **30**, 1297 (1982).

118. A. Penttila, E. M. McDowell, and B. J. Trump, *J. Histochem. Cytochem.*, **25**, 9 (1977).

119. M. R. Loken and L. A. Herzenberg, *Annals N.Y. Acad. Sci.*, **254**, 163 (1975).

120. B. Bohn, *Molec. Cell. Endocrinol.*, **20**, 1 (1980).

121. S. M. Watt, A. W. Burgess, D. Metcalf, and F. L. Battye, *J. Histochem. Cytochem.*, **28**, 934 (1980).

122. M. R. Loken, D. R. Parks, and L. A. Herzenberg, *J. Histochem. Cytochem.*, **25**, 790 (1977).

123. T. M. Jovin, S. J. Morris, G. Striker, H. A. Schultens, M. Digweed, and D. J. Arndt-Jovin, *J. Histochem. Cytochem.*, **24**, 269 (1976).

124. J. E. Gill and M. M. Jotz, *J. Histochem. Cytochem.*, **22**, 470 (1974).

125. L. S. Cram, E. R. Gomez, C. O. Thoen, J. C. Forslund, and J. H. Jett, *J. Histochem. Cytochem.*, **24**, 383 (1976).

126. F. Traganos, A. J. Gorski, Z. Darzynkiewicz, T. Sharpless, and M. R. Melamed, *J. Histochem. Cytochem.*, **25**, 881 (1977).

127. S. A. Latt and G. Stetten, *J. Histochem. Cytochem.*, **24**, 24 (1976).

128. K. J. Puite and W. R. R. Ten Broeke, *Plant Sci. Lett.*, **32**, 79 (1983).

129. W. Schnedl, O. Dann, and D. Schweizer, *Eur. J. Cell Biol.*, **20**, 290 (1980).

130. V. T. Hamilton, M. C. Habbersett, and C. J. Herman, *J. Histochem. Cytochem.*, **28**, 1125 (1980).

131. S. P. Hawkes and J. C. Bartholomew, *Proc. Natl. Acad. Sci. USA*, **74**, 1626 (1977).

PART

VI

LASERS WITH OTHER METHODS

CHAPTER

17

LASER SPECTROSCOPY FOR DETECTION IN CHROMATOGRAPHY

EDWARD S. YEUNG

Department of Chemistry and Ames Laboratory
Iowa state University
Ames, Iowa

1. INTRODUCTION

Separations is an important branch of analytical chemistry. The development of hyphenated techniques in chemical analysis makes it natural to combine lasers

557

and chromatography. With increasing interest in complex samples in clinical (1), environmental (2), and energy-related (3) areas, analytical measurements almost always require some degree of prior separation to sort out the individual components or to remove interfering species. Laser spectroscopy is no exception. Despite the emergence of highly selective spectroscopy techniques, it is unlikely that spectroscopy alone can provide much insight into the components in a complex mixture. So, there are two ways to look at the use of laser-based detectors in chromatography. The skeptic will suggest that it is unreasonable to interface any chromatograph to a detector that costs as much as a factor of 10 more, that may still be plagued with problems of reliability and convenience, and that may only be advantageous in limited situations (in contrast to mass spectrometry). On the other hand, the proponent can counter by suggesting that laser spectroscopy, in its various forms as outlined in the rest of this book, does indeed provide valuable information to the analytical chemist, and that such measurements are generally greatly enhanced if one incorporates some form of chromatographic separation prior to the spectroscopic measurement, with almost no increase in cost or inconvenience. It is with this latter point of view that the following discussions are put forward.

The two main types of separations in the analytical laboratory are gas chromatography (GC) and liquid chromatography (LC). Even though only 10–15% of the 6 million compounds in the chemical registry are suitable for GC analysis because of stability and volatility, roughly one-half of current routine problems are studied with GC. The reason is that currently both the separatory power and the detection methods in GC are substantially better than those in LC. The development of better detection methods in LC is therefore of greater importance. The remainder of this chapter is then heavily weighted toward LC detectors, even though the concepts are typically also applicable to GC detectors.

Optical detectors are already widely accepted in liquid chromatography. A recent survey (4) shows that more than 91% of all LC work presently relies on an optical detector. In GC, the only optical method that has been used routinely is Fourier transform infrared (FTIR) spectroscopy. It is necessary for us to first examine the requirements for interfacing laser spectroscopy to chromatography. Then, we can consider the various properties of the laser to see how one can gain an advantage over using conventional light sources. Finally, some specific examples of working systems will be described, so that the start-of-the-art capabilities can be appreciated, and that the likelihood of further advances can be assessed.

2. REQUIREMENTS OF CHROMATOGRAPHIC DETECTORS

2.1. Detectability

The scope of application of any analytical technique is usually inversely related to its limit of detection (LOD). So, there is always a need for a better LOD in chromatographic detectors. For GC, the LOD is already quite good (picogram range) for the case of the flame ionization detector, the electron capture detector, and the mass spectrometer. Unless the detector also gives unique information, it is not likely to be competitive. For LC, the presence of the large amounts of solvent makes it difficult to interface to the above GC detectors, although some combinations have been demonstrated recently (5, 6). Commercial LC detectors provide a LOD in the range of micrograms for refractive index, nanograms for absorption, and picograms for fluorescence. Any improvements will have to be compared with these values.

One must separately consider the concentration LOD and the mass LOD. When the sample size is essentially unlimited, only the former is relevant. When the amount of sample is limited, the latter will be the primary concern. To maximize the concentration detectability, one can inject larger quantities, up to the point where the chromatographic resolution is degraded, either because of column saturation or because of band broadening. The detector should be designed to contain as much of the chromatographic peak as possible. For the absorption detector, this can mean longer absorption paths to enhance the LOD. Generally, optimization is achieved with a detector volume $\frac{1}{5}$ to $\frac{1}{10}$ of the typical peak volume for the chromatographic system. In GC, for example, this precludes the use of extemely long-path, multiple-reflection gas cells for improving the concentration LOD. To maximize mass detectability, the smaller microcolumns for LC are advantageous. The corresponding detectors must maintain a reasonable concentration LOD despite the smaller volume. For example, the concentration LOD of the deflection type of refractive index detector does not depend on pathlength, so that its mass LOD can potentially be very good.

The signal-to-noise (S/N) ratio that determines detectability often depends on the time constant used. For the new generation of high-speed columns in LC, time constants must be below 1 s. To maximize the advantage of speed, one must then improve the LOD of LC detectors for the shorter time constants.

2.2. Selectivity

For the analysis of complex samples, detectability without selectivity is not very useful because of interference among unresolved components in a chromatogram. In GC, chromatographic resolution is already quite good, so that detector selectivity is not a serious concern. In LC, chromatographic resolution is relatively poor. If a detector responds to only a fraction of the components present, well-resolved chromatograms can be obtained even though the components are not physically separated. So, detector selectivity greatly enhances the amount of information that can be obtained from the sample. On the other hand, many components will go undetected, unless the selectivity of the detector can be "tuned" to first respond to a certain group of components and then respond to other groups of components in the sample. An example is the absorption detector, provided of course that the solvent is transparent in the spectral regions of interest. There will always be a need for a totally nonselective detector, so that in the initial survey of an unknown sample all information is preserved. After the initial survey, one can then use detectors with tailored selectivity to obtain specific information. This is where new optical methods, particularly those based on lasers, are highly attractive.

2.3. Volume

Detector volumes are usually not a problem in GC since gas densities are typically 1000 times lower than liquids. Naturally, make-up gas or solvent can always be used, but detectability will be compromised. Table 17.1 shows a comparison of operating conditions for various LC systems. Tabulated are typical values rather than upper or lower limits for these separations. A conventional column is a 4.6-mm diameter, 25-cm long column packed with 10-μm particles. A microbore column is taken to be 300 μm in diameter, 45 cm long and packed with 3-μm particles (7). A packed microcapillary is defined as a 75-μm diameter, 29-m long column packed with 30-μm particles (8). An open microtubular column is unpacked and is 10 μm in diameter and 10 m long. The peak volumes are estimated by 4σ, where σ is the half-width at half-height of a peak with a capacity factor, $k = 1$, following standard procedures (9):

$$\sigma = \frac{\pi d^2 L(1 + k)}{\sqrt{N}} \tag{17.1}$$

where N is the number of theoretical plates, d is the diameter, and L is the length of the column. It is useful to consider a physical length of the chromatographic peak, which is calculated by dividing the peak volume by the cross-sectional area of the column. Another parameter of interest is the linear flow velocity, which is determined by dividing the solvent flow rate by the cross-sectional area of the column. To preserve the available resolving power of the

Table 17.1. Operating Parameters for LC

	Conventional	Microbore	Packed Microcapillary	Open Microtubular
Column diameter (μm)	4600	300	75	10
Peak volume (μL)	500	5	2	0.01
Peak length (mm)	30	70	450	127
Flow rate (μL/min)	1000	10	3	0.05
Linear flow rate (mm/min)	60	140	680	600

column, the injected volume and the volume of the detector must both be substantially smaller than the peak volume. This requires that the detector volume be at least a factor of 5, and preferably a factor of 10, smaller than the peak volume. Clearly, standard detectors in the several microliter range cannot be used except for conventional LC. In scaling down the volumes of conventional detectors for microcolumns, one must also consider the desirable optical pathlength of the detector. In Table 17.1 the pathlengths are derived for detectors with diameters identical to those of columns. For conventional and for microbore LC, it is possible to reduce this diameter immediately after the column to increase the peak length. For the other two, the diameters are already small and further reduction is technically difficult. The length of the detector should again be $\frac{1}{5}$ to $\frac{1}{10}$ the peak length to preserve the resolution. The reason that length is an important parameter is because some detectors show length-dependent sensitivities. So, the shape of the detector is as important as its volume. An important point is that the relevant volume contains the optical region but may be much larger than that. So, the optical system must be designed to allow measurement as close to the outlet of the column as possible.

2.4. Dynamic Range

There are two types of dynamic range to consider. The first has to do with the difference between the maximum and the minimum concentrations of analyte that can be used. To handle variations in sample concentrations, a large working dynamic range is desirable. Linearity over the working range is preferred but not necessary. The low concentration end has already been discussed in Section 2.1. In GC, the upper limit in concentration is generally dictated by the capacity of the column, and not by the detector. The same is true for the refractive index detector in LC. However, the maximum concentration range for the absorption and the fluorescence detector in LC is substantially below that of column saturation. For example, absorption detectors at best can be used up to an absorbance of 2.5. For a species with a molar absorptivity of 10^4 L/mol cm and a molecular weight of 200, this translates to a maximum injected quantity of 25 μg for that component if the peak volume is 0.5 mL and the optical pathlength is 1 cm. This is considerably below the loading capacities of conventional columns. For microbore columns (1-mm i.d.), peak volumes are typically 20 μL so that the maximum injected quantity is 1 μg. The fluorescence detector has an even lower limit on the maximum concentration range because the excitation light must not be depleted substantially on passing through the sample. Naturally, dilution of the sample or shortening the optical pathlength can extend the maximum concentration range, but the dynamic range is unchanged because the minimum concentration range is increased proportionately.

The other type of dynamic range reflects the ability to detect a small change on top of a large background signal, for example, due to the solvent. As one goes toward shorter wavelengths for absorption measurements, solvent absorption becomes a severe problem. Even though a reference flow cell can be used to compensate for solvent absorption, the lower overall light intensity degrades the S/N ratio, with a corresponding degradation in the LOD. Most absorption detectors can measure a change of 0.001 absorbance with a background of 1.0 absorbance for a dynamic range of 10^3. In contrast, the S/N ratio in the refractive index (RI) detector is preserved regardless of the solvent refractive index. So, a LOD of 10^{-7} RI units for a change in solvent RI of 0.1 units becomes a dynamic range of 10^6, which is better than the absorption detector. This type of dynamic range is the relevant figure of merit when applying quantitation schemes without standards (10–12).

2.5. Sample Preservation

To take advantage of selective detectors and not lose information, it is desirable to place several of these detectors (different selectivities) in series. Most optical detectors are inherently nondestructive, except for possible photochemistry. What is typically more problematic is the extracolumn broadening introduced after each detector. It is thus desirable to perform as many different measurements as possible in the same optical region. The detectors must also be arranged so that the integrity of each is not affected by the previous detector. For example, the RI detector is particularly sensitive to temperature changes, so that it should be placed before any absorption detector.

2.6. Cost and Convenience

The research instrument is generally more complex and more costly than what eventually may appear as a routine instrument. Without the sophistication, one may not be able to learn enough about the method to refine it during the developmental stage. Typically, when the need increases for a particular type of instrumentation, additional effort in engineering can simplify the design and make it less costly. Most laser-based chromatographic detectors are still at a development stage. On the other hand, some of these are based on highly reliable lasers and optics and are ready for the routine analytical laboratory.

3. RELEVANT LASER PROPERTIES

Lasers have many special characteristics that make them good tools for spectroscopic measurements, as discussed in the various other chapters in this book. We shall only be concerned with those that are relevant to the design of chromatographic detectors, and indicate why they are important.

3.1. Power

Even though power is the most obvious of the properties of lasers, it does not lead to improved measurements in every case. Table 17.2 shows a comparison of lasers and conventional light sources in terms of power output. Two types of power should be separately considered. The average power is the energy available over a 1-s interval, which is a typical integration time in chromatography. The peak power is the maximum photon flux during a laser pulse, which may sound much more impressive. Dye lasers are the only realistic tunable source and can be derived from most other lasers at an efficiency of 1–10%. Frequency doubling can convert visible light into UV light at an efficiency of 1–10%. The important point about Table 17.2 is that laser light with these powers can be adapted to illuminate very small volumes, so that it can be used more effectively in chromatographic detectors.

A high average power is advantageous in absorption measurements, if the system can be pushed to the shot noise limit. Fluorescence measurements can be improved if the background is independent of laser power. The same is true for Raman scattering, light scattering, photoacoustic spectroscopy, thermal lens calorimetry, photoionization, and photoconductivity. However, for most liquid-phase systems, the limiting factor is the signal from the solvent, so that improvements in the S/N ratio must involve other design factors as well. High peak powers can be used to saturate the excitation in fluorescence, so that

Table 17.2. Properties of Light Sources

Type	Wavelength (nm)	Power (W)[a]	
Conventional[b]			
D_2 (50 W)	230	1.6×10^{-3}	
Hg arc (5 W)	254	1.5×10^{-2}	
Halogen–tungsten (100 W)	600	2.7×10^{-2}	
Lasers			
ArF	193	0.5	(5×10^6)
KrF	248	1.5	(1×10^7)
Nd:YAG	266	0.5	(5×10^6)
XeCl	308	0.6	(6×10^6)
HeCd	325	5×10^{-3}	
N_2	337	0.2	(1×10^6)
Ruby	347	1.0	(1×10^8)
Nd:YAG	355	1.0	(1×10^7)
Ar ion	514	4.0	
Nd:YAG	532	2.0	(2×10^7)
Sync-pumped dye	590	0.2	(6×10^2)

[a] For pulsed lasers the peak powers follow in parentheses.
[b] $f/1.0$ collection efficiency assumed.

fluctuations in the source intensity do not affect the signal (13). A unique feature of high peak powers is that other optical effects can be induced due to the simultaneous interaction with more than one photon. These include two-photon absorption and nonlinear Raman effects, which provide new types of selectivity in chromatographic detectors.

3.2. Collimation

Presently, detector volume is limited by the required pathlength for a given sensitivity and the area of the light beam. Ideally, one should scale down the area and not the pathlength to preserve detectability. However, conventional light sources cannot be molded so that a large fraction of their power fits into a small volume because the sources are inherently divergent.

The optical characteristics of a laser cavity usually result in a well-defined spatial distribution of the output intensity. The simplest of these is Gaussian in shape and is known as a TEM_{00} mode (transverse electric and magnetic). This type of light beam is best suited for focusing on a small spot. The change in radius of a Gaussian beam at the diffraction limit as it propagates through the focal point, O, is shown in Fig. 17.1. The radius (half-intensity points), ω, at any location, r, is related to the radius at the focal point, ω_0, such that,

$$\omega^2 = \omega_0^2 \left(1 + r^2/z^2\right) \tag{17.2}$$

where z is a characteristic distance where the beam radius is $\sqrt{2}$ times that at the focal point. The region where the laser remains reasonably small is then $\pm z$ from the focal point. For a given wavelength of light, λ, z can be determined as

$$z = \frac{\pi \omega_0^2}{\lambda} \tag{17.3}$$

To construct a flow cell for chromatography, one can consider a cylindrical cavity enclosing the optical region in Fig. 17.1. The radius of the cell has to be twice the beam radius to pass most of the light, and this gives a value of $2\sqrt{2}\omega_0$ for the cell radius. The length of the cell is then $2z$. So, the cell volume is

$$V = \frac{16\pi^2 \omega_0^4}{\lambda} \tag{17.4}$$

or

$$V = 16z^2\lambda \tag{17.5}$$

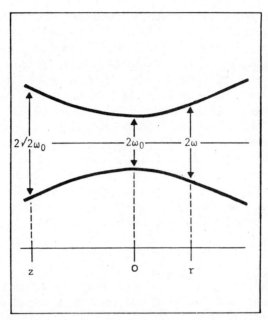

Fig. 17.1. Radius of a Gaussian beam near the beam waist according to Eq. (17.2)

Using the proper optics, one can mold the laser beam to any value of ω_0. Equation (17.4) shows that if the beam can be focused to smaller and smaller radii, the detector volume decreases rapidly. In most cases, however, some interaction length is needed, so that the cell volume is dependent on the square of the desired pathlength as given by Eq. (17.5). If the pathlength is 1 cm and the wavelength of light is 300 nm, Eq. (17.5) gives a cell volume of 120 nL. If one can sacrifice some of the intensity, and can tolerate scattered light from the cell walls, smaller cell volumes than that predicted by Eq. (17.4) can be used. For the cases shown in Table 17.1, a pathlength of 1 mm is satisfactory even in the most demanding case of open microtubular columns. So, a cell volume of 1 nL is all that is required. In addition to providing small volumes, the highly collimated beam allows spatial filtering, that is, separation of stray light from the signal, so that detectability can be improved. Superior collimation is also the basis for specialized optical methods such as interferometry and thermal lens calorimetry.

3.3. Monochromaticity

Molecular spectral widths in solutions are on the order of tens of nanometers in the visible and UV regions. So, there is no direct benefit from using highly monochromatic lasers. The indirect benefit is due to discrimination between fluorescence (spectrally broad) and Raman and Rayleigh scattering (spectrally narrow). Figure 17.2 shows the relative widths of background scattering when the excitation is broad (17.2a) and when the excitation is narrow (17.2b). The available windows for observing fluorescence are thus much wider in the second case, and interference is reduced. If Raman scattering from the analyte is of interest, the second case also minimizes spectral overlap. In GC, molecular absorption lines can be very narrow. So, the use of lasers can preserve selectivity in the measurements.

Monochromaticity can also lead to improvements in measurements not involving absorption. In interferometry, the stability of the interference pattern is dependent on the frequency stability of the light source. In polarimetry (14) and in refractive index measurements, frequency stability eliminates the dispersion effects so that baseline drifts can be avoided.

3.4. Temporal Resolution

Chromatographic events are usually substantially longer than 1 ms, even for high-speed columns in LC so that there is no direct advantage in using pulsed lasers. A possible benefit is that more photons are available within one time constant of the chromatographic detector. Measurements that are limited by photon shot noise can be extended in their useful ranges. Indirectly, the available temporal resolution can be used in various ways to avoid interference. In fluorescence detection the limit is typically Raman and Rayleigh scattering from the

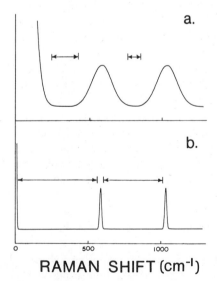

RAMAN SHIFT (cm⁻¹)

Fig. 17.2. Scattering background due to solvent for (a) 10-nm wide excitation source and (b) monochromatic excitation source. Arrows define usable spectral windows for fluorescence.

solvent. These have different time dependences compared to fluorescence (15). So, electronic gating (16) can be used to favor fluorescence. In Raman detection a major contributor to the background is fluorescence, so the reverse concept can be used to favor Raman scattering. The fluorescence lifetimes of condensed-phase systems are around 2–50 ns. It is difficult for conventional light sources to deposit enough energy into the system at these short time intervals. Lasers are therefore necessary. The extent of discrimination can be as high as a factor of 300. This is based on a lifetime of 50 ns and a detector response time of 150 ps. It is interesting to note that the present limit is due to the response time of phototubes and not the pulse duration of the laser.

The temporal properties of lasers can also be used to distinguish between the rapidly decaying signal and the much slower temperature drift in the environment in thermal-optical methods (17). In photoacoustic spectroscopy the time dependence of the pressure waves can be used to sort out the absorption from the sample and the absorption from the cell windows (18).

3.5. Polarization

Some lasers are inherently linearly polarized due to the optics in the cavity or the crystal structure of the lasing medium. This, however, only increases the efficiency for producing polarized light by a factor of two over conventional light sources. More important is the fact that the highly collimated laser beam can enhance the rejection ability of the polarizing optics (19) so that higher purity can be obtained, for example, for optical rotation detectors. Other than this advantage, an inherently polarized laser beam is no better than polarized conventional light sources in discriminating between Raman emission and fluorescence, in circular dichroism measurements, or in determining depolarization ratios in light-scattering instruments.

4. LASER-BASED CHROMATOGRAPHIC DETECTORS

4.1. Refractive Index (RI)

Despite the lack of sensitivity in commercial RI detectors, the universal nature of the response makes them a valuable tool in LC. New versions of an RI detector, or modifications of old versions, that provide improved detectability will always be needed. Also, existing instruments all have volumes larger than 5 μL and are thus incompatible with microbore columns. Miniaturization is an important design criterion. For GC, RI detection is also feasible, but so far the lack of special information and the poor detectability have discouraged the development of an RI detector.

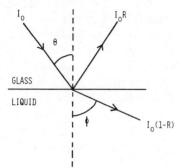

Fig. 17.3. Ray diagram for reflectance type of RI detector. θ, angle of incidence; ϕ, angle of refraction; I_o, incident intensity; and R, reflectance according to Eqs. (17.6) and (17.7).

The reflection type of RI detector is shown in Fig. 17.3. Near the critical angle, the amount of light reflected at the glass–liquid interface is given by the Fresnel equations:

$$R_s = \left(\frac{\cos\theta - \sqrt{n^2 - \sin^2\theta}}{\cos\theta + \sqrt{n^2 - \sin^2\theta}} \right)^2 \tag{17.6}$$

$$R_p = \left(\frac{-n^2\cos\theta + \sqrt{n^2 - \sin^2\theta}}{n^2\cos\theta + \sqrt{n^2 - \sin^2\theta}} \right)^2 \tag{17.7}$$

where the subscripts s and p refer to the polarization direction of the light beam being perpendicular and parallel to the plane of incidence, respectively, θ is the angle of incidence, and $n = n_1/n_2$ is the ratio of RIs of the two media. It can be shown that the change in reflected intensity has a much sharper dependence on the angle for the p polarization. This means that the standard reflectance type of RI detector (20) will have improved detectability if p-polarized light is used. The transmitted beam is monitored because of the reduced influence of flicker noise. A roughened surface is placed on the opposite side of the liquid so that the backscattered transmitted beam is viewed with a photodiode, with appropriate apertures to isolate the reflected beam. The shot noise limit can be improved if instead the actual transmitted beam is monitored, by coupling it out through another glass prism on the opposite side of the liquid. Then, a laser source will be desirable. The collimated nature of the laser beam should also be beneficial since θ in Eq. (17.7) will then have less spread, which tends to degrade the sharp angular dependence. The real potential of the reflectance type of RI detector is the very small volumes that can be achieved in principle. Only the interface takes part in the signal, as long as the liquid is thicker than a few wavelengths. A focused laser beam also reduces the width of the interaction region. So, future improvements in this type of detector should be possible.

The deflection type of RI detector (21) also depends only on refraction at the interface of the liquid and the cell. In principle, the volume can be very small. Unfortunately, the prism-shaped cell is difficult to miniaturize to the dimensions of a focused laser beam. So, the only advantage is derived from the better degree of collimation, which enhances the function of the position-sensitive detector.

The commercial interferometric RI detector (22) uses white light interference in a two-beam arrangement. The detectability depends on the contrast in the interference fringes. For that reason, a Fabry–Perot interferometer is inherently superior (23). The Fabry-Perot interferometer consists of simply two end mirrors that can be translated relative to each other by a piezoelectric crystal. As the mirror separation d is changed, constructive and destructive interferences occur for monochromatic light as shown in Fig. 17.4, and the peaks are given by the expression

$$m\lambda = 2\,dn \tag{17.8}$$

where λ is the wavelength of light, n is the RI, and m is any integer. So, the idea is to monitor the change in location of the interference peak rather than the transmitted intensity. Flicker noise in the light source thus only contributes indirectly. To compensate for temperature changes and thus baseline drifts, a double-beam arrangement can be used (24). This is shown in Fig. 17.5. There, the reference flow cell can be used to follow even the frequency instabilities of the laser, and a routine He–Ne laser can be used. Figure 17.6 shows a chromatogram from such an instrument, providing a LOD of 4×10^{-9} RI units (S/N = 3).

Three additional improvements can be made in interferometric RI detectors. The original concept (23, 24) relies on a computer to locate the interference peak during the scanning of the interferometer. This limits the scanning rate to

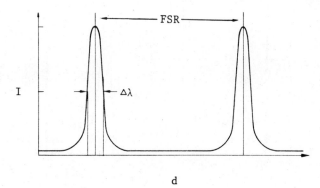

Fig. 17.4. Transmission properties of a Fabry–Perot interferometer as a function of mirror separation, d. $\Delta\lambda$, width of interference peak; and FSR, free spectral range.

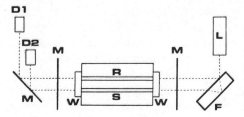

Fig. 17.5. RI detector based on a dual-beam Fabry–Perot geometry. M, mirrors; W, antireflection-coated cell windows; R, reference flow cell; S, sample flow cell; L, laser; F, optical flat; and D1, D2, photodetectors.

3 Hz. We have recently built an analog system to perform the same function, increasing the scanning rate to 100 Hz (25). The S/N ratio is thus increased. It is also possible to use the confocal geometry for the interferometer, that is, concave end mirrors rather than plane-parallel mirrors. A natural beam waist of very small volume exists in that arrangement so that coupling to microbore LC is easy. These confocal interferometers can achieve a resolution (FSR/$\Delta\lambda$ in

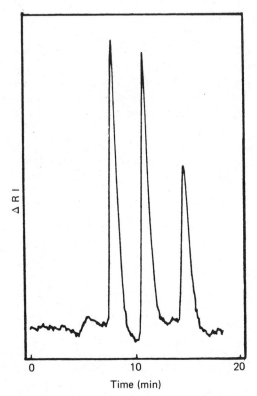

Fig. 17.6. RI chromatogram of glucose, sucrose, and raffinose for 0.72 μg of each injected. Column: C_{18} 10 μm, 4.6-mm i.d., 25-cm long; eluent: water at 0.5 mL/min.

Fig. 17.4) of 200, so that a 1-cm cavity can measure a RI change of 8×10^{-8} units. The order of 20 ng can thus be detected. A third possibility is to stabilize the interferometer at the half-intensity point on the constructive interference peak in Fig. 17.4. Then, RI changes are converted to intensity changes at the phototube. This is essentially the same principle as in the commercial instrument (22), but the LOD is improved because of the sharper peak.

4.2. Absorption

The ultimate limit in detectability in absorption is shot noise. So, in principle, the higher photon flux available in lasers is advantageous, particularly considering the small volumes that can be used. However, flicker noise in lasers usually dominates the background, unless special modulation techniques are used. At the high concentration end, the larger photon flux can extend the dynamic range of absorption measurements. To benefit from this, thermal problems due to the amount of absorption must be dealt with so that false absorption signals can be avoided. To go from 2 to 4 in absorbance units, one only gains a factor of two in dynamic range. The light intensity, however, decreases by a factor of 100. The trade-off is seldom worthwhile.

A unique feature of laser cavities is the presence of a threshold in the lasing process. At the threshold any additional absorption in the cavity will stop lasing completely. So, one has essentially infinite sensitivity for absorption. In practice, one works slightly above the threshold to assure stability, but the output intensity is still roughly exponentially dependent on the absorbance inside the laser cavity. There is also the multipass advantage inside the laser cavity. To actually measure small changes in absorption, one must be able to control the other parameters for the laser very well. An approach is to introduce a variable loss in the laser cavity by a Pockels cell (26). When absorption occurs, the loss is reduced by an applied potential on the Pockels cell to regain the identical output intensity. Identical lasing conditions are then guaranteed for the laser, and an absorbance LOD of 5×10^{-5} is achieved.

In the infrared region one can make use of some special conditions for mode competition in a laser to indirectly determine absorption (27). A He–Ne laser can be made to operate simultaneously at two wavelengths, 0.633 and 3.39 μm. These two transitions compete for the same group of excited states in the electrical discharge. Now if one of these wavelengths falls in the absorption band of a molecular species inside the laser cavity, the particular transition will show less gain. The competition for excited molecules is then decreased, and the other transition can then achieve a higher power level. So, absorption at one of the He–Ne laser wavelengths is reflected in the increase in output intensity at the other wavelength. A working system is shown in Fig. 17.7 (27). About 1.5 mW of output is obtained at each wavelength without any absorbing species.

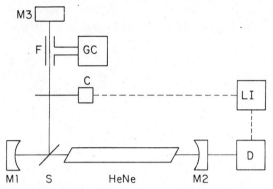

Fig. 17.7. Intracavity mode competition absorbance detector for GC. F, flow cell; M1, M2, M3, cavity mirrors for He–Ne laser; C, mechanical chopper; S, beam splitter; D, photodetector; and LI, lock-in amplifier.

When $10\mu L$ of methane is injected into the gas chromatograph, the absorption produced at the 3.39 μm laser line is enhanced by a factor of 1.2 for the 5-cm flow cell inside the cavity compared to the absorption outside the cavity. In contrast, the change in the laser intensity at 0.633 μm due to the presence of the eluted methane is a factor of 46.7. This can be explained since a small change in the upper state population induced by absorption at the high-gain infrared line results in a large change in the output of the low-gain visible line. The linear dynamic range of this detector is about 8 orders of magnitude in concentration, with detectability of 9×10^{-11} g/cm^3. The fixed wavelength of 3.39 μm limits the utility to the various hydrocarbons, however.

4.3. Photoacoustic Spectroscopy

The heat produced after absorption can be monitored as a pressure wave and thus an acoustic signal. This has been discussed in more detail in Chapter 5. This way of measuring absorption directly avoids the shot noise limit when detecting small changes on top of a large intensity in conventional measurements. Contributions from stray light, losses from scattering, and beam distortion from thermal lensing can all be eliminated from the background. Lasers can deposit a much larger amount of energy per pulse, so the acoustic signals are enhanced.

In GC, absorption is not a widely used mode of detection because of the availability of other fine detectors. GC–FTIR provides structural information, and has become a useful tool. Incorporation of the photoacoustic mode of detection in FTIR instruments can enhance the LOD for absorption, but in the open cell with flowing gas, the major problem is environment acoustic noise, and the LOD does not show much improvement. The utility of lasers in such situations has been limited by the lack of tunability of IR lasers. In closed-cell geometries, the major problem is window absorption and thus a background signal. This contribution cannot be reduced even with a higher intensity source.

In LC, the interface between the transducer and the liquid is more difficult to construct, even though the actual interaction volume can be quite small. There are also problems due to flow, convection, and absorption of the scattered light in the cell walls. Coupling has been tried using direct contact (28) through the housing of the transducer (29) and through a metallic foil (30). The last arrangement is shown in Fig. 17.8 and gives a detectability of 1.2×10^{-5} absorbance

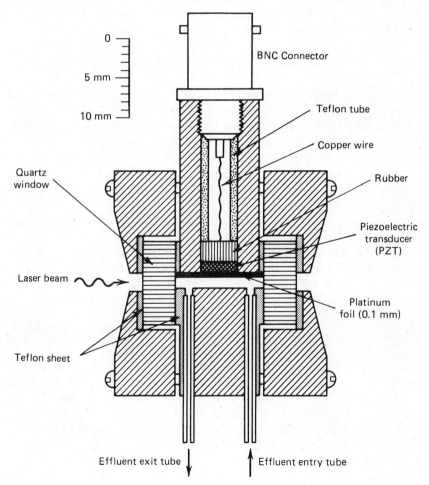

Fig. 17.8. Photoacoustic detector for LC. Reprinted with permission from ref. (30).

units when 500 mW of 488-nm radiation is used. Presently, all transducers for interfacing with LC have fairly large sensitive areas, and detectability will degrade with miniaturization. The 20-μL cell in Fig. 17.8 is already approaching the practical limit. In the IR region LOD in LC is mainly determined by solvent absorption, and laser photoacoustic detection offers no advantage over FTIR methods.

Photoacoustic spectroscopy has been very useful for studying solid surfaces because the heat that is produced can be coupled efficiently through the gas above the surface to the microphones. Lasers can take the place of the 1-mW average power in conventional sources (31) to produce a larger signal when scanning the surface of thin-layer chromatographic plates, provided that the substrate does not absorb. The short pulses lasers are capable of producing can be used to gate the proper acoustic signal to isolate environmental noise.

4.4. Thermal-optical Methods

When heat is produced from absorption, expansion of the medium occurs. This gives rise to a host of thermal-optical methods. In LC, the heating causes a change in the refractive index that can be predicted from the change in number density. So, all of the RI detectors discussed in Section 4.1 can be made into indirect detectors for absorption, as long as light producing the absorption can be properly introduced into the optical region. A subtle point is that the exciting radiation must be coupled through the longest dimension in the cell for maximum sensitivity.

An LC absorption detector based on the Fabry–Perot interferometer is shown in Fig. 17.9 (23). This is essentially the system in Fig. 17.5 allowing for an additional excitation laser. Only a single beam is used because the experiment is double beam in time by switching the excitation laser on and off. The temperature increase in the cell for small absorption is

$$\Delta T = \frac{2.303AI}{C_pDa} \tag{17.9}$$

where A is the absorbance, I is the total amount of light energy, C_p is the heat capacity of the solvent, D is the density of the solvent, and a is the area of the cell. Equation (17.9) is valid if all light absorbed does become heat in a time scale that is short compared to thermal conduction out of the optical region. Using typical values for common LC solvents, a RI detectability of 4×10^{-9} (Section 4.1) will mean an absorption LOD of 8×10^{-7}. The arrangement in Fig. 17.9 allows an LOD of 2.6×10^{-6} absorbance units (S/N = 3) for 60 mW of excitation at 514 nm. Noise is due to pump fluctuations, a slow data acquisition time (15 s), and solvent absorption. Since then, we have been able

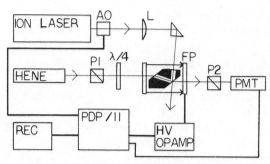

Fig. 17.9. Indirect absorption LC detector based on the Fabry–Perot interferometer, FP. AO, optical shutter; P1, P2, λ/4, polarizing optics for stray light reduction; L, lens; PMT, photomultiplier tube; PDP/11, minicomputer; and REC, recorder.

to use a 1-cm long, 8-μL flow cell with a time constant of 1 s by introducing the excitation laser via a rotating prism inside the laser cavity. A still better arrangement is to rely on the transmission properties of the interferometer mirrors to introduce the excitation laser collinearly through the mirrors. Also, the confocal Fabry–Perot interferometer should be superior because of the smaller area [Eq. (17.9)] and the smaller volume.

Other interferometers can be used. For example, a GC detector has been demonstrated using a Mach–Zender interferometer and a CO_2 excitation laser (32). This concept has been explained in more detail in Chapter 6. A LOD of 0.08 ppb of injected SF_6, or a total injected quantity of 3 pg, was demonstrated. For liquids, a Jamin interferometer with uncoated quartz flats has been used to provide one strong beam and one weak beam in two cells (33). The difference reflects the increased amount of light absorbed from the strong beam. With stabilization, 100 mW of excitation allowed a LOD of 5×10^{-6} absorbance units (34). However, the 1-min data acquisition time is impractical for LC applications. In general, these two-beam interferometers are less sensitive than multiple-beam interferometers (e.g., Fabry–Perot) because of the lower contrast associated with the former.

Instead of probing the change in bulk RI after absorption, one can also probe the RI gradient that is generated. Since laser beams can assume a Gaussian distribution in intensity across the diameter, the central region, for the same absorbance, will be heated up more than the edges of the beam. This is the basis for thermal lens calorimetry (35). The RI gradient generated is equivalent to a negative lens, which changes the spot size of the beam at far fields. If the detector is limited by an appropriate aperture, a change in intensity will result. The thermal lens technique is discussed in more detail in Chapter 13. It shares the advantage of all indirect absorption methods in that one does not measure the small difference between the incident and the transmitted intensities. However, the detector does monitor a decrease in the far-field intensity of the probe laser, and will be subjected to noise from that laser. The thermal lens is also flow sensitive, since transverse flow equilibrates the thermal gradient and longitudinal flow depletes the heated liquid in the optical region. At flow rates typical of LC, usable results are obtained despite a decrease in sensitivity (36). Using 190 mW of a chopped cw laser at 458 nm, nitroaniline isomers can be detected at 1.5×10^{-5} absorbance units in a 1-cm cell with internal volume of 18 μL. The time constant there was 5 s due to the need to deconvolute the time-dependent signal to average out short-term noise in the laser. A more recent report (37) gives a LOD of 10^{-6} absorbance units in a 8-μL cell. It appears that the slower flow rates in microcolumn LC are in fact beneficial to thermal lens measurements. The smaller cell volume also allows more efficient heating and better heat dissipation between data points. Thermal lens calorimetry is probably the easiest of the indirect absorption techniques to set up. On the other hand, the dependence on a well-defined laser beam profile prevents the use of high-power lasers with poor mode structures.

When two laser beams derived from the same laser at wavelength λ intersect each other at an angle θ, interference occurs so that planes of constructive and destructive interference appear, separated by a spacing Λ, such that

$$\Lambda = \frac{\lambda}{2 \sin (\theta/2)} \qquad (17.10)$$

If absorption occurs, the resulting RI change will cause the equivalent of a diffraction grating to be formed in the liquid (38). This can then be probed by a second laser in the diffraction mode, with the diffracted intensity being related to the amount of absorption. This concept has been demonstrated in an optical region of 0.17 μL with a 50-μs pulse of 15 μJ. The LOD was 7×10^{-4} absorbance units. The advantage over thermal lensing is that detectability is not limited by the presence of a background intensity.

A laser beam can also be used to probe the RI gradient by deflection (39). The maximum gradient occurs about one beam radius away from the center of the excitation beam. A second laser directed toward this region will be deflected according to Snell's law. Using about 100 mW of excitation, the LOD is 6×10^{-7} absorbance units.

There is an important difference between monitoring the bulk RI change versus RI gradients. In the former one should use a cell with volume comparable to that of the actual optical region, so that temperature equilibration can be rapid, and no extra solvent needs to be heated up. In the latter one must have a cell somewhat larger than the optical region so that temperature equilibration does not set in in the time scale of the data acquisition. So, the ultimate cell volume for interferometry can be smaller than those for other thermal-optical methods. For all of these, solvent absorption seems to be a limiting factor for detectability.

4.5. Fluorescence

Since fluorescence intensity increases with excitation intensity, lasers are natural choices for the excitation light. However, commercial fluorescence detectors for LC provide very respectable LODs already. The limiting factors are stray light, fluorescence from cell walls or windows, and fluorescence or Raman scattering from the solvent. To improve detection, these other factors must be dealt with in addition to increasing the excitation intensity. We have already discussed the advantage of using monochromatic radiation for excitation to discriminate against solvent Raman scattering. The contributions from stray light can be reduced by using better collimated laser beams and appropriate apertures. Fluorescence from the solvent can only be avoided through purification. So, the development of laser-based fluorescence detectors has been tied to the cell design to avoid contributions from the cell walls.

It is possible to do away with the cell wall completely by suspending a droplet of solution between the outlet capillary and a rod immediately below it (40). Figure 17.10 shows a recent version of this arrangement (41) that has a volume of 3 μL, although the optical region is much smaller. Since fluorescence is collected at 90°, any change in the shape of the droplet due to temperature, degassing, thermal lensing, eluent flow rate, eluent composition, and vibrations can modify the collection efficiency and thus the signal. Still, detectabilities in the picogram level have been demonstrated. Another approach to eliminate the cell wall is by a free-falling jet from the column outlet (42). When the flow rates are higher than 1.2 mL/min, droplets will not be formed and a well-shaped column of solvent will be maintained under the narrow-bore syringe needle that is connected to the column. Even gradients can be used in conjunction with the

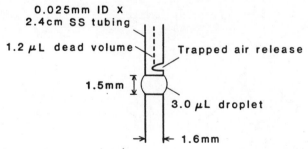

Fig. 17.10. Flowing droplet for fluorescence detection in LC. Reprinted from ref. (41) with permission.

jet. Using 1 W of laser excitation at the UV lines of a Kr ion laser, 20 fg of fluoranthene has been detected. The optical region is only 1 nL, but the dead volumn is substantially larger. The high flow rates are also not compatible with microbore columns.

The cell wall can be effectively isolated from the optical region by injecting the chromatographic effluent into the center of an ensheathing solvent stream (43). No mixing will occur if laminar flow is maintained. A 53-nL optical region is possible, but some mixing at the entrance and extra connections make the actual dead volume larger. A detectability of 53 pg is reported for 8 mW of excitation at 488 nm. The flow rate again makes it difficult to interface with microbore LC.

If a glass or quartz capillary can be connected directly to the outlet of the column, a laser beam can be used to irradiate a region very close to the outlet, and the dead volume is minimized. For packed microcapillary or open micro-tubular systems, the end of the column can serve directly as the cell. By using selected fused silica, one can reduce fluorescence from the cell material (42). To avoid collecting light from the cell walls, an optical fiber can be used (44). As shown in Fig. 17.11, transmission of the optical fiber is due to internal reflection at angles larger than the critical angle i_c. So there is a limited acceptance angle defined by θ such that

$$\sin \theta \leq \frac{(n_f^2 - n_c^2)^{1/2}}{n_e} \tag{17.11}$$

where n_f, n_c, and n_e are the RI of the fiber, cladding, and eluent, respectively. So, if the distance from the optical region, d, is properly chosen, light originating

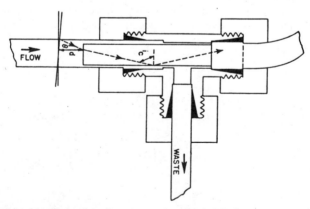

Fig. 17.11. Fiber optics flow cell for fluorescence detection in LC. θ, acceptance angle; d, distance from excitation region; and i_c, critical angle for optical fiber.

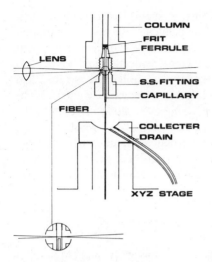

Fig. 17.12. Fiber optics flow cell for microbore LC. Lower left is an enlargement of the optical region.

from the cell walls will not be transmitted by the optical fiber. Typical acceptance angles for optical fibers allow an f/number of $f/0.5$, which is excellent for collection. The dimensions of the optical fiber also make it ideal for interfacing with a monochromator to further isolate the fluorescence signal. A fiber-optics-based flow cell suitable for microbore LC is shown in Fig. 17.12. The internal diameter of the quartz capillary is 300 μm, and the outer diameter of the optical fiber is 150 μm. The volume of the optical region is 10 nL and the total dead volume is 0.3 μL. It should be possible to have the optical region even closer to the frit of the column to reduce the dead volume even more. The performance of such a detector is shown in Fig. 17.13, where a LOD of 1 pg is determined. Since some packed microbore columns are actually 300 μm in internal diameter, this arrangement can be used for essentially on-column detection. In fact, optical fibers are available down to tens of microns in diameter for use with even smaller

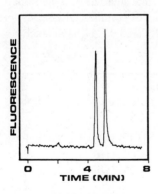

Fig. 17.13. Chromatogram from fluorescence detector shown in Fig. 17.12 for amino derivatives of NBD excited by 200 mW of 488-nm radiation. Column: 3 μm C_{18}, 2-mm i.d., 15-cm long; eluent: 35:65 H_2O–acetonitrile; flow: 0.17 mL/min.

capillaries. In designing the proper interfacing, it should be remembered that the fluorescence intensity is proportional to the pathlength so that concentration sensitivity does decrease in going to smaller capillaries. The optical fiber should also have a diameter as large as possible to increase collection efficiency.

Irradiating the optical region perpendicular to the capillary is not the ideal approach because it does not maximize the length relative to the volume. Technical difficulties aside, it is better to introduce the laser along the bore of the capillary tube, which only needs to be as large as the laser beam waist according to Eq. (17.3). An aperture can then define the optical region beginning from the end of the column to a distance z, which is a fraction of the peak length in Table 17.1. The windows passing the laser can be placed slightly further away to isolate scattered light.

4.6. Two-photon Excited Fluorescence

When the molecular energy levels are separated by an energy equal to the sum of the energies of two photons, simultaneous absorption can occur. This normally weak process is enhanced by high-power lasers, and useful LOD can be obtained by monitoring the fluorescence after two-photon absorption by photon counting (45). Using 1 W of a cw laser at 514 nm, excitation of electronic levels is possible with the equivalent energy provided by 257-nm light, and 1 ng of oxadiazoles can be detected. The spectroscopic information available is complementary to that from normal fluorescence. Two-photon excitation is more selective because a third energy level must be involved and because the polarizations and the wavelengths of the individual photons can be controlled (46). Interestingly, the two-photon excited fluorescence can be observed from the same optical region as the normal fluorescence (47) and is essentially extra information that does not lead to additional band broadening in the chromatography.

The extent of two-photon excitation is dependent on the power (W/cm^2) of the laser. Assuming all lasers can be focused to the same small area in the same interaction length, one can compare the various types of commercial laser systems in their suitability for two-photon excitation. Table 17.3 shows such a

Table 17.3 Two-photon Excitation Sources

Laser Type	Peak Power (W)	Pulse Width (ns)	Repetition Rate (Hz)	Duty Cycle	Relative Efficiency
Ar ion	1	10^9	1	1	1
Cu vapor	7×10^4	30	5000	1.5×10^{-4}	7.4×10^5
YAG-dye	5×10^6	4	10	4×10^{-8}	1×10^6
Sync-pumped dye	2000	0.15	4×10^6	6×10^{-4}	2.4×10^3
Excimer-dye	10^6	5	200	10^{-6}	1×10^6

comparison, with the efficiencies normalized to that of the first entry. It is desirable to have very high peak powers, but it is also necessary to have a reasonable duty cycle to collect enough photons in a 1-s interval. The average power should also be low to avoid thermal lensing, which defocuses the laser beam. A final consideration is that the pulse-to-pulse reproducibility must be good since the signal varies as the square of the power. So, a high repetition rate allows signal averaging to improve the S/N ratio.

4.7. Raman and Light-scattering Methods

Raman scattering provides functional group information. The various kinds of Raman methods have been discussed in Chapter 10. As chromatographic detectors, the LOD is quite poor. The low number densities in GC have prevented any realistic applications of Raman detectors. In LC, the solvent background is the major problem. All successful applications have to rely on the resonance enhancement that is available when the excitation wavelength approaches an actual absorption level. Raman detection has been demonstrated using a 14-μL flow cell and 700 mW of radiation at 458 nm (48). A multichannel detector gives the Raman spectrum in a 1000 cm^{-1} region in 5 s for concentrations as low as 1 μg. Similar detectabilities are obtained using coherent anti-Stokes Raman spectroscopy in a sophisticated optical arrangement (49). In principle, the latter concept can be used even below the 1-μL cell volume that is reported.

Nephelometry has been demonstrated as an LC detector in conjunction with postcolumn precipitation for the detection of lipids (50). A flow cell of 17 μL is irradiated with 0.5 mW of laser light at 633 nm, and the scattered light is collected by fiber optics. About 0.5 μg can be detected. An interesting concept based on vaporizing the solvent and then observing the scattered light from the solutes contributes only about 0.2 μL to band broadening (51). A 1-mW He–Ne laser is sufficient for detecting solutes in the 0.5-μg range. Quasielectric light scattering can probe particles in the 50-Å–2-μm range. Particle size, shape, molecular weight, and molecular rotations can be studied. A commercial version of this instrument is already available (52). The LOD is only in the 5–100-μg/mL range at the detector, but the information is complementary to other types of detectors.

4.8. Photoconductivity and Photoionization

Increased excitation intensity should lead to increased signals in photoionization and photoconductivity. Temporal discrimination against the background is possible with pulsed lasers. Even two-photon ionization is possible at high peak powers (53). In the gas phase, either for GC or for LC effluents vaporized after the column, photoionization in general gives useful sensitivities (54). Even for the less probable two-photon photoionization event, resonance enhancement allows the order of 10 pg to be detected when a pulsed N_2 laser at 337 nm is used in a 5-μL flow cell (55). It is clear that high-intensity pulsed sources can bring improvements in LOD over conventional photoionization (56) and photoconductivity (57) schemes, for example, if discrete-line UV lasers can be used. The additional advantage is that photoionization and photoconductivity can be monitored in the same optical region as fluorescence with minimal modifications.

4.9. Optical Rotation

Optical activity is an interesting property of molecules. The rotation of the polarization direction of light by chiral molecules is usually an indication of biological activity. In complex samples of clinical, geological, or biological importance, this is a useful type of selectivity. Commercial instruments do not have the sensitivity or the small volumes required of LC. A laser-based polarimeter has been demonstrated for LC applications in the study of sugars in urine (58), cholesterol in serum (59), and chiral components in coal extracts (60). The key to success is the better rejection of stray light and the reduction of birefringence in the optical components when a small, collimated laser beam is used (61). For analytical-scale LC a cell of 5-cm length and 100-μL volume seems to be the best compromise. Our most recent version has been interfaced to microbore LC with a detector volume of 1 μL and an optical pathlength of 1 cm (62). The cell is shown in Fig. 17.14. The actual optical volume is even smaller, and 1 μL is limited by the machining process. Since the flow rates are substantially lower in microbore LC, it is possible to use exotic solvents. Figure 17.15a shows a chromatogram of l-2-octanol and several normal alkanes eluted by an optically active eluent. Even though the normal alkanes are not optically

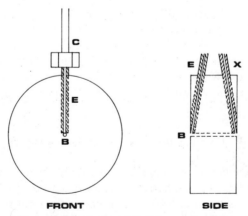

Fig. 17.14. Polarimetric flow cell for microbore LC, 2.5-cm o.d. and 1-cm long. C, column; E, entrance capillary; X, exit capillary; and B, 1-mm i.d. bore for laser beam.

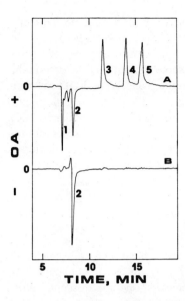

Fig. 17.15. Optical activity chromatogram of a mixture of *l*-2-octanol (2), *n*-decane (3), *n*-tetradecane (4), and *n*-hexadecane (5). The void peak is 1. A, optically active eluent; and B, optically inactive eluent. Column: 5 μm C_{18}, 1-mm i.d., 25-cm long; and flow: 20 μL/min.

active, they displace an equivalent amount of the eluent when they pass through the detector. So, a signal is recorded. In contrast, Fig. 17.15b shows the same test mixture eluted by the racemic mixture of the same eluent. Only *l*-2-octanol, which is itself optically active, shows up in the chromatogram. So, the utility of the polarimetric detector for LC is extended to include *optically inactive* materials. This is then a universal detector, and everything shows up unless it has exactly the same specific rotation as the eluent. Considering the LOD of 10 ng (S/N = 3 for a *difference* of specific rotation between the analyte and the eluent of 100°) and the small volume, the polarimetric detector has the potential of replacing the RI detector as the workhorse in the analytical laboratory. When the same sample is eluted first by an optically active eluent and then by its racemic mixture, as in Fig. 17.15, the two chromatograms allow quantitation without standards and without identification (62). The specific rotation of the analyte can also be obtained from these chromatograms, thus allowing the determination of the optical purity of materials without requiring the separation of the enantiomers. We have calculated the cost of using an optically active eluent with microbore columns and found that it is in fact cheaper than using UV-grade LC eluents with standard analytical-scale columns. This is because the eluent does not have to be optically pure, if one can sacrifice a little sensitivity.

It is interesting to note that the polarimetric detector, when an optically active eluent is used, can function as an indirect absorption detector for LC. This is because heat produced by absorption will cause a net expansion in the eluent and thus a reduced number density in the optical region. So, a decrease in optical rotation should be observed. Since the polarimetric detector has a LOD better than the corresponding RI detector, the absorption LOD monitored indirectly should also be better compared to interferometry (Section 4.4). It should also be noted that so far only convenient, visible lasers have been used for polarimetry in LC. If UV lasers are used, there should be a significant increase in the specific rotations of molecules near real absorption bands. The LOD can then be improved further.

The improved polarization purity of lasers does not directly influence LOD in circular dichroism measurements since that is essentially an absorption process. However, if one can approach the shot noise limit in circular dichroism measurements, the higher intensity from lasers will become advantageous.

5. SUMMARY

It is apparent that optical methods based on lasers have many advantages to offer when properly combined with chromatography. In fact, chromatographic information can be enhanced through laser detectors, and spectroscopic information can be enhanced through "sample modulation" that results from chromatography. The popularity of these methods depends very much on having available reliable lasers. Systems based on continuous, low-power lasers such as He–Ne or He–Cd lasers are ready for routine use. On the other hand, reliable UV lasers, preferably tunable ones, are needed before the scope of application of absorption methods can be broadened. At least in the area of microcolumn LC, lasers may be the only suitable light source due to the small volume that is required. There, even complicated instrumentation can be justified. Currently, it may be desirable to design the chromatographic system (e.g., precolumn derivatization) to fit available laser wavelengths, such as the 325-nm output of the He–Cd laser. As laser technology continues to advance, these technical limitations are likely to disappear, and laser-based chromatographic detectors will be able to reach their full potential.

ACKNOWLEDGMENTS

The author thanks the many co-workers in his laboratory who have contributed to the work described here, particularly S. D. Woodruff, M. J. Sepaniak, J. C. Kuo, D. R. Bobbitt, R. E. Synovec, and S. A. Wilson, and the U.S. Department of Energy, Office of Basic Energy Sciences, Division of Chemical Sciences, for partial research support through the Ames Laboratory, Iowa State University, under contract No. W-7405-eng-82.

REFERENCES

1. G. D. Mack and R. B. Ashworth, *J. Chromatogr.*, **16,** 93 (1978).
2. W. A. Saner, G. E. Fitzgerald, and J. P. Welsh, *Anal. Chem.*, **48,** 1747 (1976).
3. D. E. Anders and W. E. Robinson, *Geochim. Cosmochim. Acta*, **35,** 661 (1977).
4. S. A. Borman, *Anal. Chem.*, **54,** 327A (1982).
5. V. L. McGuffin and M. Novotny, *Anal. Chem.*, **53,** 946 (1981).
6. T. Covey and J. Henion, *Anal. Chem.*, **55,** 2275 (1983).
7. F. J. Yang, *J. Chromatogr.*, **236,** 265 (1982).
8. T. Tsuda and M. Novotny, *Anal. Chem.*, **50,** 271 (1978).
9. J. H. Knox and M. T. Gilbert, *J. Chromatogr.*, **186,** 405 (1979).
10. R. E. Synovec and E. S. Yeung, *Anal. Chem.*, **54,** 1599 (1983).
11. R. E. Synovec and E. S. Yeung, *J. Chromatogr.*, **283,** 183 (1984).
12. S. A. Wilson and E. S. Yeung, *Anal. Chim Acta*, **157,** 53 (1984).
13. G. D. Boutilier, J. D. Winefordner, and N. Omenetto, *Appl. Opt.*, **17,** 3482 (1978).
14. A. L. Cummings, H. P. Layer, and R. J. Hocken, in *Lasers in Chemical Analysis*, G. M. Hieftje, J. C. Travis, and F. E. Lytle, Eds., Humana Press, Clifton, NJ, 1981, p. 291.
15. J. M. Friedman and R. M. Hochstrasser, *Chem. Phys.*, **5,** 155 (1974).
16. H. Merkelo, S. R. Hartman, T. Mar, and G. S. S. Grovindjee, *Science*, **164,** 301 (1969).
17. N. J. Dovichi and J. M. Harris, *Anal. Chem.*, **53,** 689 (1981).
18. C. K. N. Patel and A. C. Tam, *Chem. Phys. Lett.*, **62,** 511 (1979).
19. C. E. Moeller and D. R. Grieser, *Appl. Opt.*, **8,** 206 (1969).
20. R. D. Conlon, *Rev. Sci. Instrum.*, **34,** 1418 (1961).
21. D. Zaukelies and A. A. Frost, *Anal. Chem.*, **21,** 743 (1949).
22. Optilab 902 refractometer, Vallingby, Sweden.
23. S. D. Woodruff and E. S. Yeung, *Anal. Chem.*, **54,** 1175 (1982).
24. S. D. Woodruff and E. S. Yeung, *Anal. Chem.*, **54,** 2124 (1982).
25. G. Chen and E. S. Yeung, unpublished results.
26. J. S. Shirk, T. D. Harris, and J. W. Mitchell, *Anal. Chem.*, **52,** 1701 (1980).
27. J. D. Paril, D. W. Paul, and R. B. Green, *Anal. Chem.*, **54,** 1969 (1982).
28. W. Lahmann, H. J. Ludewig, and H. Welling, *Anal. Chem.*, **49,** 549 (1977).
29. A. C. Tam and C. K. N. Patel, *Nature*, **280,** 302 (1979).
30. S. Oda and T. Sawada, *Anal. Chem.*, **53,** 471 (1981).
31. V. A. Fishman and A. J. Bard, *Anal. Chem.*, **53,** 102 (1981).
32. H. B. Lin, J. S. Gaffney, and A. J. Campillo, *J. Chromatogr.*, **206,** 205 (1981).
33. J. Stone, *J. Opt. Soc. Am.*, **62,** 327 (1972).
34. D. A. Cremers and R. A. Keller, *Appl. Opt.*, **21,** 1654 (1982).
35. J. M. Harris and N. J. Dovichi, *Anal. Chem.*, **52,** 695A (1980).
36. R. A. Leach and J. M. Harris, *J. Chromatogr.*, **218,** 15 (1981).
37. C. E. Buffett and M. D. Morris, *Anal. Chem.*, **54,** 1824 (1982).
38. M. J. Pelletier, H. R. Thorshelm, and J. M. Harris, *Anal. Chem.*, **54,** 239 (1982).
39. A. C. Boccara, D. Fournier, W. Jackson, and N. M. Amer, *Opt. Lett.*, **5,** 377 (1980).

40. G. J. Diebold and R. N. Zare, *Science*, **196**, 1439 (1977).
41. R. N. Zare, private communication.
42. S. Folestad, L. Johnson, B. Josefsson, and B. Galle, *Anal. Chem.*, **54**, 925 (1982).
43. L. W. Hershberger, J. B. Callis, and G. D. Christian, *Anal. Chem.*, **51**, 1444 (1979).
44. M. J. Sepaniak and E. S. Yeung, *J. Chromatogr.*, **190**, 377 (1980).
45. M. J. Sepaniak and E. S. Yeung, *Anal. Chem.*, **49**, 1554 (1977).
46. E. S. Yeung and M. J. Sepaniak, *Anal. Chem.*, **52**, 1465A (1980).
47. M. J. Sepaniak and E. S. Yeung, *J. Chromatogr.*, **211**, 95 (1981).
48. M. D'Orazio and V. Schimpf, *Anal. Chem.*, **53**, 809 (1981).
49. L. A. Carreira, L. B. Rogers, L. P. Goss, G. W. Martin, R. M. Irwin, R. Von Wandruszka, and D. A. Berkowitz, *Chem. Biomed. Environ. Instrum.*, **10**, 249 (1980).
50. J. W. Jorgenson, S. L. Smith, and M. Novotny, *J. Chromatogr.*, **142**, 233 (1977).
51. A. Stolyhwo, H. Colin, and G. Guiochon, *J. Chromatogr.*, **265**, 1 (1983).
52. M. L. McConnell, *Anal. Chem.*, **53**, 1007A (1981).
53. E. Voightman, A. Jurgensen, and J. D. Winefordner, *Anal. Chem.*, **53**, 1921 (1981).
54. J. T. Schmermund and D. C. Locke, *Anal. Lett.*, **8**, 611 (1975).
55. S. Yamada, A. Hino, and T. Ogawa, *Anal. Chim Acta*, **156**, 273 (1984).
56. D. C. Locke, B. S. Dhingra, and A. D. Baker, *Anal. Chem.*, **54**, 447 (1982).
57. D. J. Popovich, J. B. Dixon, and B. J. Ehrlich, *J. Chromatogr. Sci.*, **17**, 643 (1979).
58. J. C. Kuo and E. S. Yeung, *J. Chromatogr.*, 223, 321 (1981).
59. J. C. Kuo and E. S. Yeung, *J. Chromatogr.*, **229**, 293 (1982).
60. D. R. Bobbitt, B. H. Reitsma, A. Rougvie, E. S. Yeung, T. Aida, Y. Chen, B. F. Smith, T. G. Squires, and C. G. Venier, *Fuel*, **64**, 114 (1985).
61. E. S. Yeung, L. E. Steenhoek, S. D. Woodruff, and J. C. Kuo, *Anal. Chem.*, **52**, 1399 (1980).
62. D. R. Bobbitt and E. S. Yeung, *Anal. Chem.*, **56**, 1577 (1984).

CHAPTER

18

LASER IONIZATION TECHNIQUES FOR ANALYTICAL MASS SPECTROMETRY

ROBERT S. HOUK

Ames Laboratory-USDOE and Department of Chemistry
Iowa State University
Ames, Iowa

1. INTRODUCTION AND SCOPE

Lasers are efficient ways to add energy to atoms or molecules so as to generate signals of use to an analytical chemist. One such interaction between a laser and a sample is the generation of gas-phase ions as the analytical signal (e.g., Chapter 3). Mass spectrometry (MS) is the most sensitive and selective way to measure ions. Thus, it is natural that lasers should be investigated and used in ionization techniques for analytical MS. The first such investigations took place in the early 1960s (1–4). Since that time there have been over 900 publications in the field (5, 6). Of course, there have also been numerous fundamental or diagnostic studies of the interaction of lasers with gases or solids. To some extent this work by physicists and physical chemists has shown the way for

587

subsequent use by analytical chemists, a characteristic that laser MS has in common with many other emerging analytical technologies. Analytical laser MS has progressed to the point where such instruments are commercially available, which is a sign that at least some people feel this concept to be of analytical value. Leybold-Heraeus is now marketing its third generation of laser MS called LAMMA, an acronym for laser microprobe mass analyzer. Cambridge Consultants in the United Kingdom sells a laser MS instrument called LIMA, and Nicolet also offers laser desorption capability with its Fourier transform (FT)–MS.

This chapter emphasizes techniques and analytical applications in which the laser energy contributes to ionization. Also, lasers have been used extensively to vaporize or pyrolyze samples for subsequent ionization by a separate energy source, for example, electron impact (EI) (7, 8). This chapter does not attempt to review all the developments in laser ionization in detail. Rather, it is intended to survey the major areas and experimental techniques of current interest as indicated by the section headings. At the time this is written these studies represent present and possible future contributions of laser MS to analytical science.

2. ELEMENTAL ANALYSIS OF SOLIDS

When a laser beam of sufficient power density strikes a solid surface, material is ejected into the gas phase. Lasers are thus effective ways to vaporize localized sections of solids. The ejected material contains significant populations of positive atomic ions, which suggests the use of a laser ion source for elemental analysis of solids. This concept represents the first application of lasers to analytical MS over 20 years ago (1, 2). Lasers also possess several other unique features that make them attractive as elemental ion sources. The rate of energy transfer to the specimen (i.e., the power density) can be controlled via the power supplied by the laser or by varying the beam area impacting the sample surface (9, 10). The laser beam transfers energy to a small discrete region of the sample. Thus, either spatial profiling of elemental concentrations or rastering of the laser to sample a representative section of the whole sample is possible. The laser beam does not add charge to the sample surface so that analysis of either insulators or conductors is possible. The properties of the microplasma formed as the laser strikes the sample will largely determine the extent to which these potential attributes will be observed in reality. Therefore, a brief description of the laser interaction with the sample is in order.

2.1. Generation of Atomic Ions from a Solid

Introduction of a solid into the ion source is a straightforward task. The sample is merely inserted into the vacuum chamber in a position such that the laser can irradiate it. Two basic geometries for the interaction between laser and sample have been used to date. Most instruments designed primarily for elemental analysis use the front-side irradiation geometry shown in Fig. 18.1a. Various incident angles between the laser and the sample surface have been used, as have various collection angles between the surface and the ion optical axis of the mass spectrometer (5). The microplasma formed propagates through a solid angle that is essentially symmetrical about the normal to the specimen surface for various angles of irradiation (11, 12). Previous versions of the LAMMA employed the back-side irradiation geometry shown in Fig. 18.1b. Here the sample is thin enough for the laser to drill a hole completely through it. Ion formation and collection occur on the opposite side of the sample relative to the incident laser beam. The newest generation of the LAMMA offers either front- or back-side irradiation. Q-switched lasers with pulse lengths of 10–100 ns are typical. A frequency-quadrupled Nd:YAG laser ($\lambda = 265$ nm) is generally used with the LAMMA; the LIMA typically employs either the fundamental (1.06 μm) or doubled (530 mm) output of a Nd:YAG laser (13–15). Eloy has used lasers in the visible and ultraviolet, some with very short pulses (~ 3 ns) (16–18). Other workers employ infrared lasers of the Nd:YAG, ruby, or CO_2 variety (9, 11, 19–26). External optics are used to focus the laser. The spot position is controlled either optically or by translating the specimen. Microscopic viewing of the specimen during analysis is also valuable.

As the laser initially strikes the sample surface some of the laser energy is absorbed by the sample and some is reflected. The relative fractions of absorbed

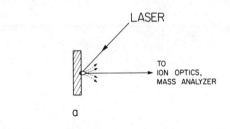

a

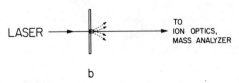

b

Fig. 18.1. Spatial relationships between laser beam and sample: (a) front-side irradiation geometry, (b) back-side irradiation geometry.

versus reflected light depend on the absorption characteristics of the sample for the laser wavelength used. At low power density ($\lesssim 10^8$ W cm^{-2}) positive ions are readily observed from volatile elements of low ionization energy, for example, Na, K, and Pb. These ions are likely formed by ordinary thermal ionization in a relatively cool microplasma. Cluster ions of inorganic constituents (e.g., Na$_n$SO$_m^+$) (27–32) and intact organic ions are also observed at these low power densities (see Section 3). The efficiency of atomization and ionization under these conditions varies widely between different elements and will also depend on the particular sample matrix investigated. Thus, low power densities are in general undesirable for quantitative elemental analysis. However, at low power density there is some indication that short-range order in the solid specimen is preserved to some extent in the resulting vapor-phase species. If so, the mass spectra observed may be correlated with the chemical states of various inorganic constituents in the solid specimen (particularly anions) (33). Thus, laser mass spectra at low power densities may reflect elemental speciation (e.g., coordination state) in solid samples.

At higher power density ($\gtrsim 10^9$ W cm^{-2}) the laser transfers considerably more energy to the sample surface. The vaporization process typically leaves a crater in the specimen surface as the vaporized material is ejected. A dense microplasma forms from vaporized material above the sample surface. The laser pulse can transfer energy directly to free electrons in the microplasma by processes such as inverse bremsstrahlung:

$$e^- + X + h\nu_L \rightarrow e^-_{\text{fast}} + X \tag{18.1}$$

Here $h\nu_L$ represents a photon of laser radiation and X represents a third body (most likely a positive ion). Direct absorption of laser energy by plasma electrons is most probable if ν_L matches a natural frequency for an oscillation or a wave in the microplasma. The likelihood of absorption can be evaluated qualitatively by comparing ν_L to the plasma frequency ν_P (11, 12, 34–37):

$$\nu_P = \left(\frac{4\pi n_e e^2}{m_e}\right)^{1/2} \tag{18.2}$$

$$\nu_P \text{ (Hz)} = 8.9 \times 10^3 (n_e)^{1/2} \tag{18.3}$$

Here n_e is the electron number density (cm^{-3}) in the microplasma, e the electron charge, and m_e the electron mass. Some references list a different numerical factor for Eq. (18.3) based on different units for n_e or expression of plasma frequency in angular terms. A plot of plasma reflection and transmission characteristics as functions of the ratio ν_L/ν_P is shown in Fig. 18.2. For $\nu_L/\nu_P > 1$ the plasma transmits the laser radiation efficiently. For $\nu_L/\nu_P \ll 1$ the plasma reflects the laser radiation. Absorption is feasible for $\nu_L/\nu_P \lesssim 1$.

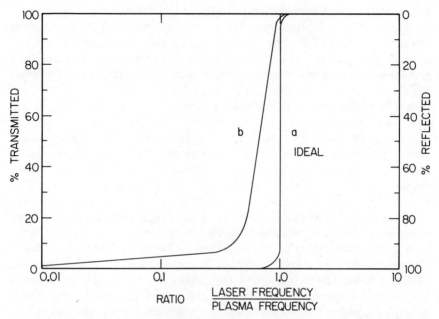

Fig. 18.2. Plasma transmission and reflection characteristics as functions of ν_L/ν_P: (a) ideal behavior (sharp resonance at $\nu_L = \nu_P$), (b) real behavior (resonance frequency broadened to $\nu_L \lesssim \nu_P$).

It may seem unlikely that n_e in the microplasma would be precisely the appropriate value for absorption of laser radiation. Actually, n_e in the microplasma varies with both spatial position and time, which facilitates a match between ν_L and ν_P somewhere in the plasma. Also, absorption can occur for a range of ν_L less than ν_P (Fig. 18.2b) depending on other plasma parameters such as electron collision frequency (12, 34, 36–38). Other mechanisms such as multiphoton absorption may also contribute to energy transfer from the laser to the species vaporized by the specimen. At any rate, conditions in the microplasma are sufficiently energetic for atomization and ionization to be quite efficient. For example, there is considerable empirical evidence that virtually all the sputtered material is present as atomic ions for front-side irradiation of a metal sample with a Nd:YAG laser at power densities $\gtrsim 10^9$ W cm^{-2} (5, 9, 22, 39, 40). On the debit side these ionization conditions for elemental analysis yield some multiply charged ions (10). They also generate ions with a considerable spread of kinetic energy (up to hundreds of electron volts), perhaps due to shock wave formation as the plasma expands away from the sample surface (9, 11, 12, 34–37, 41–44).

For irradiation at a given wavelength the number of multiply charged ions, the electron temperature, and the ion kinetic energies generally increase with power density, indicating that a sensitive compromise in power density is necessary. Ion–electron recombination is likely (particularly for the more highly charged ions) after the laser pulse is off. Thus, the plasma cools as it expands into the vacuum chamber. The surviving ions are collected, mass resolved, and detected. It is generally possible to identify a compromise power density for a given specimen such that these surviving ions are predominantly singly charged (9, 10, 39, 40).

It is likely that laser power density, pulse duration, and irradiation geometry may all affect ion production. Values of electron number density corresponding to $\nu_P = \nu_L$ are listed for various lasers in Table 18.1. The electron number density likely for a laser microplasma is comparable to the total density of heavy particles (atoms and ions) and is likely to be somewhere within the n_e range indicated in Table 18.1. Higher electron densities are required for absorption of shorter wavelength laser radiation by the plasma. Thus, laser wavelength may be an important parameter in determining the energy content of the plasma, which influences the energies and identities of the ionic species detected by the mass spectrometer. Work in Ames (10) and the Soviet Union (9) indicates that the focal position of the laser relative to the sample surface is also an important variable. Because these parameters vary widely among the various analytical laser mass spectrometers, it is difficult to make global statements about what is or is not happening in laser microplasmas, which geometry is best, and so on. Interested readers can consult several thorough reviews of laser interactions with solids and plasmas for more information about this area (11, 12, 34–38).

2.2. MS Instrumentation

Time-of-flight (TOF) and magnetic sector mass spectrometers have been used to mass resolve and detect laser-produced ions for analytical purposes. Two

Table 18.1. Laser Frequency (ν_L) and Electron Number Density Corresponding to Equivalent Plasma Frequency for Various Lasers

Laser Type	λ	ν_L (s^{-1})	n_e (cm^{-3}) for $\nu_L = \nu_P^a$
Frequency-quadrupled Nd:YAG	265 nm	1.1×10^{15}	1.5×10^{22}
Frequency-doubled, ruby	347 nm	8.6×10^{14}	9.4×10^{21}
Ruby	694 nm	4.3×10^{14}	2.4×10^{21}
Nd:YAG	1.06 μm	2.8×10^{14}	1.0×10^{21}
CO_2	10.6 μm	2.8×10^{13}	1.0×10^{19}

[a]Calculated from Eq. (18.3).

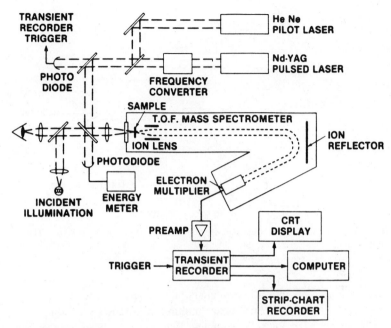

Fig. 18.3. Schematic diagram of LAMMA 500. Note ion reflecting mirror and back-side irradiation geometry. Reproduced from ref. (48) with permission.

such devices are illustrated in Figs. 18.3 and 18.4. The TOF device used with the present version of the LAMMA incorporates an ion reflecting mirror (45) to help compensate for the kinetic energy spread of the extracted ions. Of the sector instruments both single- (16–18) and double-focusing devices have been used. The instrument constructed in Ames and most of the Russian instruments use the Mattauch–Herzog double-focusing geometry, which provides for energy analysis of the extracted ion beam and also facilitates either photographic or electrical detection of the mass-resolved ion beams. Intense ion currents can be generated (particularly at power densities $\gtrsim 10^9$ W cm^{-2}) because of the efficiency with which ions can be formed and extracted, as illustrated by the following example. Suppose a single laser shot into a metallic Cr sample erodes a crater that is 0.5 μm deep $\times$ 15 μm diameter. Approximately 2.4×10^{-10} cm^3 or 1.8 ng of Cr is removed by this shot. Suppose atomization and ionization are complete in the resulting microplasma. The single shot yields approximately 2×10^{13} ions. An impurity element M (atomic weight 100 g mol^{-1}) present at 1 ppmw (parts per million by weight) would yield 10^7 ions per shot. There have been complaints of space charge effects in the extracted ion beam (46) and discharges in the ion source because the ion density was too high!

The relation of intensity data to the corresponding concentrations in the sam-

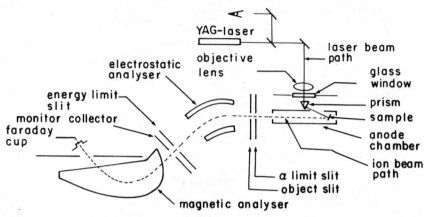

Fig. 18.4. Schematic diagram of Mattauch–Herzog laser MS constructed in Ames. The monitor collector is actually a grid that intercepts a fraction of the total ion beam. Note front-side irradiation geometry. Reproduced from ref. (40) with permission.

ple is an important operation in any quantitative analytical technique. In particular, it is difficult or costly to prepare multiple calibration standards of known elemental composition for solid samples. Solids are likely to be spatially heterogeneous as well. Therefore, techniques that can analyze solids without need for external calibration standards are especially desirable. This goal is accomplished with the Mattauch–Herzog geometry by intercepting approximately 15% of the ions at an ion collecting grid located between the two sectors (Fig. 18.4). The ion beam at this total ion collector has not yet been mass dispersed. Thus, the signal at the total ion collector is proportional to the total number of ions generated in the ionization process. The detection efficiency of the total ion collector is readily measured by comparing the total ion signal to the mass-resolved signal for a major peak of a pure element, for example, $^{56}Fe^+$ for an Fe sample. If ionization in the laser-induced plasma is complete and multiply charged ions are few enough to be neglected, the ratio of signal at the MS collector for a particular m/z (mass-to-charge) value to the signal at the total ion collector yields a direct measure of the atomic concentration of the element of interest. Possible discrimination in the MS detector can be identified and corrected for (22, 47). Integration of these two signals helps average out shot-to-shot variations in laser power density. For this quantitation method the laser microplasma used should be of sufficient power density for atomization and ionization to be complete and uniform for various elements.

At any particular instant in time a magnetic sector mass spectrometer with electrical ion detection only observes ions at those m/z values for which there are detectors. Ions at other m/z values are lost. Indeed, the ion transmission efficiency of these devices is low, for example, only 10^{-6} for the Ames instrument. Fortunately, the laser produces plenty of ions to partially compensate for this poor detection efficiency. Compared to a magnetic sector, a TOF mass spectrometer should be able to collect, mass resolve, and detect a larger fraction of the total ions produced by a transient ionization event such as a laser pulse. The TOF device is capable of obtaining a complete mass spectrum from a single laser shot, which is the reason this mass analysis method is used in the LAMMA and LIMA devices. The difficulty with TOF is that it requires an ion beam that is homogeneous in direction and energy and of a limited temporal duration. A TOF measurement also places stringent requirements on the dynamic range and response of the ion detector and recording electronics. Some of these problems have been described by Simons (48) and by Surkyn and Adams (49) in their efforts to use the LAMMA for isotopic analysis. Research in several laboratories is addressing the question of whether these requirements can be satisfied for the energetic plasmas and widely varying concentration ranges encountered in elemental analysis.

2.3. Analytical Characteristics and Selected Applications

As stated above, if the laser power density is appropriate and the analyte signal is referred to the total ion signal, then the ratio of analyte ion (corrected for isotopic abundance) to total ion signal is proportional to the analyte concentration. The extent to which this claim is valid has been evaluated by several workers. Typically, they analyzed standard reference materials (SRMs) of certified elemental composition and compared the determined and certified values. A relative sensitivity coefficient (RSC) is sometimes used to correct the determined values into agreement with the certified values

$$\text{Certified value} = \frac{\text{determined value}}{\text{RSC}} \tag{18.4}$$

The closeness of these RSCs to unity determines the validity of the internal calibration scheme. Bykovskii et al. have reported RSCs of 0.99 ± 0.03 (one standard deviation) for determination of major constituents of various semiconducting, ionic, and metallic substrates (47). Other Russian workers also claim that RSCs for various elements are essentially unity for laser power densities of 10^9–10^{10} W cm^{-2}. They indicated this by correctly determining 37 out of the 43 certified elements in a particular geological SRM (9). Representative data for elemental analysis of a steel SRM by the Ames group is presented in Table 18.2. The internal calibration scheme yields concentration values that agree with the certified values within an accuracy of approximately 20% for metals present at 0.1–1 ppma (parts per million atomic). The RSC for V does lie outside this range for reasons that are not clear at this time. Several Th standards doped

Table 18.2. Elemental Analysis of NBS Steel SRM 462 by Laser MS

| Element | Concentration (ppma) | | RSC |
	Determined	Certified	
V	0.044	0.064	0.69
Cr	0.975	0.803	1.09
Mn	1.16	0.965	1.20
Co	0.086	0.105	0.82
Ni	0.532	0.673	0.79
Cu	0.142	0.178	0.80

with varying amounts of Mo were also analysed as solids by laser MS. These samples were then dissolved, and the Mo content determined spectrophotometrically. The Mo concentrations determined by the two methods agreed well, as illustrated in Fig. 18.5 (50). These data indicate the validity of the internal calibration approach for at least semiquantitative analysis as well as the ability of the laser source to generate ions from a refractory metal like Th.

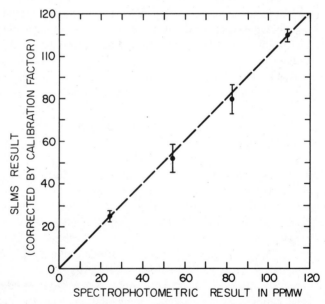

Fig. 18.5. Determination of Mo doped into various Th standards. Correlation of Mo concentration by scanning laser MS with results by solution spectrophotometry. The "calibration factor" corrects for the isotopic abundance of the Mo peak and the relative detection efficiencies of the monitor collector and the Faraday cup (Fig. 18.4).

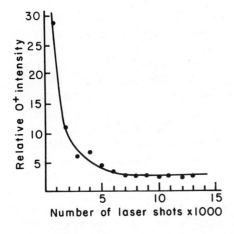

Fig. 18.6. Cleaning of O from sample by repetitive laser shots. Reproduced from ref. (40) with permission.

Laser MS is also capable of determining C, N, and O in metals. Initial bombardment of the specimen yielded an O^+ signal that was anomalously high (Fig. 18.6). Therefore, surface oxygen was cleaned from the sample by spacing successive laser shots side by side through a $600 \times 600\ \mu$m area of the sample. This rastering process was done via computer control of the optical arrangement that directed the laser onto the specimen. After approximately 7000 shots the O^+ signal stabilized and analytical data representative of the bulk O concentration could be obtained. The cleaning process took only 20 s because of the rather high repetition rate (333 Hz) of the laser used. Analytical results for C (0.02–0.4 wt. %), N (0.003–0.014 wt. %), and O (0.002–0.106 wt. %) in NBS steel SRMs 1040–1047 agreed with the specified values as follows: for a total of 24 determinations (3 elements in 8 samples) 14 of the determined results were within 10% (relative deviation) of the NBS data, 6 agreed within 30%, and the remaining 4 were within 50%. Accuracy and precision data shown in Table 18.3 indicate that O can be readily determined at ppma levels. The detection limit for O was approximately 0.03 ppma (40).

Table 18.3. Determination of Trace O in Metals by Laser MS (40)

		O Concentration (ppma)		
Material	NBS Ref. No.	Certified	Determined[a]	Percent Relative Standard Deviation
Maraging iron	1094	16	17.6	5.0
Platinum	680a	50	48	11

[a] Determined value represents the mean of five replicate measurements on the same sample loading.

In the determination of bulk elemental composition, it is convenient to be able to raster the laser or translate the sample relative to the laser so that a representative fraction of the specimen is sampled. Alternatively, the ability of the laser to generate ions from a localized region of the specimen can be exploited to measure spatial concentration profiles for particular elements. The spatial resolution is limited by the dimensions of the crater formed as the sample material is vaporized. Most of the work to date has emphasized lateral resolution (along the sample surface) rather than depth resolution into the specimen. The instrument in Ames is used extensively and routinely to yield elemental concentration profiles in metals for studies of thermal or electrotransport and is capable of lateral resolution of approximately 20 μm (50–52). Typical laser MS data for such a spatially resolved study are shown in Fig. 18.7. Elemental analysis of individual fluid aliquots included into mineral samples has also been performed by laser MS (17, 18). The LAMMA device employs a very finely focused laser and back-side irradiation geometry to yield lateral resolution of approximately 1 μm. Its TOF mass analyzer is also equipped to obtain a complete mass spectrum from a single shot. These characteristics of LAMMA reflect the original intent of its manufacturer to market it for spatially resolved elemental analysis of biological specimens. Numerous applications in this area have been described in symposia and proceedings devoted solely to LAMMA work. Some examples include studies of Pb and U uptake by algae (53, 54), accumulation

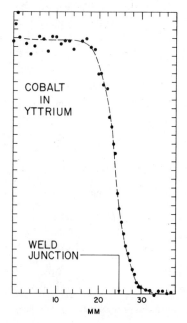

Fig. 18.7. Concentration profile for Co in Y. The intensity of Co$^+$ is plotted on the vertical axis in arbitrary units; lateral displacement along the sample rod is plotted on the horizontal axis. Reproduced from ref. (50) with permission.

of alkali metals at protein sites in muscles (55), and identification of the spatial distribution of metallic additives in the walls of wood cells (56). The laser punches a clean round hole through thin samples such as these. Microscopic examination of the sample between shots permits correlation of the mass spectrum with the hole position in the sample.

These biological applications also illustrate the ability of the laser to generate ions from electrically insulating samples. Bingham and Salter (25, 26) emphasize this capability and also point out that it is not necessary to risk contaminating the sample by mixing it with a conducting matrix as is often done in emission spectrometry or spark source MS. Bulk minerals have been analyzed directly with spatial resolution by laser MS (19). The LAMMA device has been used for localized analysis of particulates and mineral fibers such as asbestos (27, 49, 57–60). In this LAMMA work the laser is generally directed to graze the side of the sample particle from the back side. Possible matrix effects and their influence on the interpretation of LAMMA spectra have been described (33, 49). Nevertheless, localized analysis of individual particles and insulating matrices are unique features of laser MS. Many of these sample types are also difficult or slow to dissolve for atomic spectrochemical analyses so that a technique suitable for direct elemental analyses would be welcome.

This section concludes with the author's impression of the eventual impact laser MS is likely to have on methodology for elemental analysis. Elemental analysis of solutions and many conducting solids is done reasonably well by existing techniques such as atomic absorption spectrometry, plasma emission and plasma mass spectrometry, and arc–spark emission spectrometry. Direct elemental analysis of dielectric materials is routinely done by X-ray fluorescence spectrometry, which often lacks sufficient sensitivity for determination of trace impurities. Laser MS could contribute to analytical science as a general technique for trace analysis of solids (or determination of minor and major elements, too) with either spatial resolution or raster sampling for determining bulk elemental concentrations. If these objectives can be accomplished routinely for conducting or insulating solids without calibration standards or sample preparation, then laser MS could assume an important place among the arsenal of elemental analytical techniques even if the instruments continue to be expensive ($\geq$ \$500,000).

2.4. Reaction Chemistry of Atomic Metal Ions

Freiser's group at Purdue University has used laser ionization of a metallic sample as a way to generate singly charged atomic metal ions in a fashion similar to that described above. The metal ions are mass selected and stored via ion cyclotron resonance (ICR) in a FT–MS cell (61–63). Gaseous compounds of interest are added to the cell. The products and kinetics of reactions between metal ions and added gas can then be followed by FT–MS (64–71). Collisionally induced dissociation (CID) can even be used to provide structural information for the reaction products. Furthermore, a particular reaction product (e.g., an Fe^+–olefin complex) can itself be stored selectively and its reaction chemistry investigated. These concepts have been used extensively to study gas-phase metal ion chemistry. For analytical purposes, they may also permit use of metal ions as selective reagents for chemical ionization (CI) of particular classes of organic compounds (65). At any rate this work represents a clever and interesting use for atomic metal ions generated by laser impact.

3. LASER DESORPTION OF ORGANIC COMPOUNDS

Atomization in a laser microplasma is not complete if the laser power density is sufficiently low. In the course of an investigation into laser pyrolysis MS (7, 8) Vastola and co-workers at Pennsylvania State University observed intact parent ions from organic compounds such as polynuclear aromatic hydrocarbons (PAHs) and amino acids without an auxiliary ionization source. This work showed that molecular ions could be formed directly by laser bombardment of a solid organic matrix. As expected, the degree of fragmentation induced by the laser increased with power density (72, 73). These concepts attracted wider attention in 1978 when Kistemaker and co-workers described laser desorption (LD) of intact parent ions and fragments from such nonvolatile compounds as oligosaccharides, peptides, and nucleosides (74). At that time most mass spectra of nonvolatile compounds were being obtained by field desorption (FD). The classical problem with mass spectrometry of these molecules had been that the higher the molecular weight and the more polar groups on the molecule, the greater the difficulty encountered in volatilizing it without thermal degradation. The work in LD as well as other bombardment methods such as fission fragment desorption (FFD), secondary ion MS (SIMS), and fast atom bombardment (FAB) has accomplished much more than simply augmenting the burgeoning array of acronyms imposed upon the modern analytical chemist. It is now almost routine to obtain useful mass spectra (i.e., parent ions for molecular weight determination and structurally significant fragment ions) from compounds that were previously considered too large and/or polar to get into the gas phase unless derivatized. FAB has been the most successful of the bombardment techniques that make this possible. The acceptance of FAB stems partly from its ready and inexpensive adaptability to existing mass spectrometers. LD is less frequently used than FAB, and is not a substitute for FAB by any means. LD does possess some unique features such as spatial resolution. Furthermore, the extent of fragmentation in LD can be controlled to some extent through appropriate choice of power density supplied to the specimen. The analytical potential of LD has been judged sufficient to generate over 50 research papers and several review articles (75–78) since Kistemaker's 1978 paper. Most of these research papers either report LD spectra of particular compounds or deal with fundamental aspects of desorption, ionization, and fragmentation. These topics will be described below, as will initial applications of LD to actual analytical problems.

3.1. Ionization and Mass Analysis in LD

As is the case for elemental analysis, the solid sample is inserted into the vacuum system and irradiated by the laser from either the front or back side (Fig. 18.1). Pulsed lasers much like those described in Section 2.1 are typically used. Frontside irradiation is generally employed on home-built LD mass spectrometers, into which the sample is introduced by coating it onto a direct insertion probe. The older versions of the LAMMA employed back-side irradiation and were thus limited to thin specimens. The present LAMMA 1000 employs front or back-side irradiation. The sample layer is generally kept relatively thin; its morphology and that of the supporting substrate may influence the coupling of laser energy into the sample and the resulting mass spectra. In most experiments the laser is the only energy source for ionization, even if the actual ionization event takes place in a region physically distinct from the laser. Examples of techniques in which an additional ionization process also contributes to the mass spectra are described separately in Section 3.4.

In common with the other bombardment techniques, the laser adds energy to the sample matrix and/or support material. This added energy somehow causes ejection of molecular species and fragments, which are eventually detected as ions by the MS. The relative merits of TOF versus scanning mass analyzers as described in Section 2.2 are also applicable for LD. The dynamic range limitations of TOF are probably less of a problem in LD than in elemental analysis; TOF remains the technique of choice if a complete spectrum is desired from a single shot. Magnetic sectors of normal geometry, reversed geometry (79–81), and with multichannel detection (74, 82) have been also used, as have single-quadrupole (80, 83–86), triple-quadrupole (87), and FT (88) mass analyzers. Some of this work has demonstrated the value of the extra information and selectivity provided by CID of a selected ion into a known daughter ion for detection.

The precise physical nature of ionization in LD is uncertain at this time. This is another characteristic that LD shares with other bombardment ionization techniques. Despite the wide variations in lasers, operating conditions, and mass spectrometers, there are enough common features in the many published LD spectra for some likely processes to be identified. Cooks and others have suggested that the common characteristics of mass spectra obtained by various desorption techniques are a natural consequence of fundamental similarities in ion production processes (75, 79, 89–92). Hercules and co-workers (76, 77) have suggested that the general features of LD mass spectra correlate with the simple, "common-sense" ionization model depicted in Fig. 18.8. A small area of the sample and its support experience the energetic laser beam directly (region 1). The laser rapidly heats the surrounding regions of the sample and support, causing thermal desorption and ionization (region 2). The importance of thermal ionization is indicated by the fact that the intact parent ions for many molecules often contain attached Na^+ or K^+, a phenomenon referred to as cationization. For some ionic compounds (e.g., quaternary ammonium cations) desorption of an intact, "preionized" species can occur. Ion–molecule reactions also likely contribute either in the relatively dense region adjacent to the laser (region 2) or as the desorbed species expand into the vacuum system (region 3). Fragment ions may be formed by either:

1. Direct action of the laser (region 1), which would likely favor small fragments;
2. thermal degradation in region 2, which is supported by the observation that fragment ions as well as parents are commonly cationized; or
3. conventional unimolecular decomposition of an excited or unstable ion in the gas phase (79).

These concepts are intended more as a preliminary qualitative description than as a comprehensive model. Doubtless there are particular combinations of

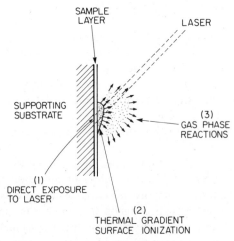

Fig. 18.8. Schematic diagram of ionization processes in LD with front-side irradiation geometry. Adapted from ref. (76).

analyte compounds, irradiation geometry, and laser parameters that favor a particular process over others.

3.2. Some LD Mass Spectra

A convenient means to evaluate the likelihood of the processes described above is to correlate them with the observed mass spectra. Thus, several groups have reported extensively on the mass spectral characteristics of known compounds ionized by LD. This is an obvious first step because it is necessary to have at least an empirical understanding of the mass spectra before any ionization technique can be used for analysis or structural identification.

It is not the purpose of this section to dissect small differences in spectra observed on various instruments by various groups. However, the field is sufficiently in hand to permit some general observations that are more or less consistent with the LD processes suggested above. Laser wavelength apparently has little influence on either the absolute or relative intensities of various ions even if the laser wavelength corresponds to an absorption band of the analyte. This is consistent with the concept that direct transfer of laser energy occurs largely to the support rather than via absorption by the sample (76–78, 92–94). The power densities used for LD ($\leq 10^8$ W cm^{-2}) are generally less than would be used for producing energetic microplasmas and elemental ions ($\gtrsim 10^9$ W cm^{-2}). LD mass spectra have even been generated using a continuous-wave (cw) CO_2 laser at as little as 20 W cm^{-2} (84, 95, 96). Usually the extent of fragmentation increases with power density (84, 92, 97–100). An example is shown in Fig. 18.9. Fragmentation for some compounds has been reported to be insensitive to power density for back-side irradiation, however (101). Most compounds exhibit a power density threshold below which LD ions are not observed (96, 97, 99, 100). For species that are cationized by alkali metal ions, there is some evidence that this threshold corresponds roughly to that for production of Na^+ or K^+ by simple thermal evaporation and ionization (78, 96, 99, 102–104). Ionization and fragmentation characteristics of some compounds desorbed thermally from rapidly heated filaments are similar to those obtained by LD (105). Kistemaker and co-workers have reported results of LD experiments designed to identify the importance of thermal processes in LD relative to the thermal conductivity and temperature of the support (95, 96, 99, 104). The morphology (i.e., roughness) of the support and sample matrix may also influence the rate of heat transfer to the sample at a given power density (104, 106, 107). Some workers have stated that it can be tricky to reproduce LD spectra from shot to shot or even from sample to sample. These observations are reasonable if the ionization and fragmentation efficiencies are sensitive to experimental factors such as the rates of energy transfer to the support, molecular desorption, and thermionic generation of Na^+ and K^+.

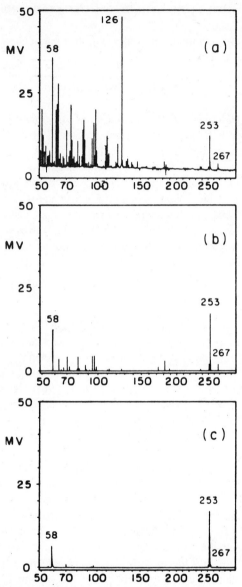

Fig. 18.9. LD mass spectra of the diquaternary ammonium salt below as a function of laser power. $(M\!\!-\!\!H)^+ = m/z$ 267, $(M\!\!-\!\!CH_3)^+ = m/z$ 253, and $(CH_3)_2N\!\!=\!\!CH_2^+ = m/z$ 58. The fragment of $m/z = 126$ is probably the structure II. Spectrum (a) is at 2.5×10^{10} W cm^{-2}, (b) at 8.0×10^9 W cm^{-2}, and (c) at 5.5×10^9 W cm^{-2}. The horizontal axis is calibrated in atomic mass units. Reproduced from ref. (100) with permission.

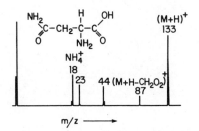

Fig. 18.10. Positive ion LD mass spectrum of aspar-agine. Note protonated molecular ion $(M + H)^+$ and fragment ions. Reproduced from ref. (76) with permission.

For polar species that readily protonate or cationize, even-electron ions (e.g., $(M+H)^+$, $(M+K)^+$ or $(M+Na)^+$, Figs. 18.10 and 18.11) are much more prominent than odd-electron ions (e.g., $M^{+\cdot}$). In conventional ion sources even-electron ions are generally favored by chemical ionization (CI) processes similar to those postulated for thermally excited and gas-phase reaction zones in LD (regions 2 and 3, Fig. 18.8). In contrast EI processes tend to favor odd-electron ions more than CI does. PAHs, which are thermally stable and susceptible to photoionization (see Section 4.2 below), are the most prominent class of compounds whose LD spectra show $M^{+\cdot}$ in greater abundance than protonated or cationized parent ions (76, 77, 98). Negative ions are readily detected as $(M-H)^-$ for many compounds, particularly those known to form such ions via acid-base chemistry. Some compounds exhibit amphoteric properties, that is, they can be detected as both $(M+H)^+$ and $(M-H)^-$, often under the same experimental conditions (Fig. 18.12). Many biological molecules of interest are composed of concatenated molecular units. Examples include oligosaccharides, oligopeptides, nucleosides, and nucleotides. Under LD these molecules generally yield intact parent ions (protonated and/or cationized). They also show some tendency to fragment at the links between units either under the same or slightly

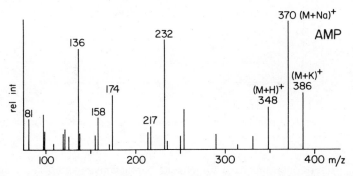

Fig. 18.11. Positive ion LD mass spectrum of adenosine-5'-monophosphoric acid (AMP) showing cationization of parent species by Na^+ and K^+. The fragment peaks at m/z 136, 158, and 174 correspond to $(B + H)^+$, $(B + Na)^+$, and $(B + K)^+$ where B represents the base part of AMP. The isotope peaks containing $^{41}K^+$ have been omitted for clarity (74).

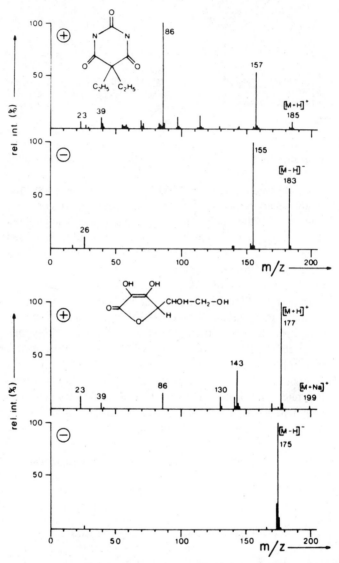

Fig. 18.12. Positive and negative ion LD mass spectra of barbital (upper) and ascorbic acid (lower). Reproduced from ref. (30) with permission.

"warmer" laser conditions used to generate parent ions. Examples are shown in Figs. 18.11 and 18.13. Fragment ions corresponding to elimination of small stable molecules (e.g., H_2O) are also common. LD spectra containing parent ions and fragments have also been obtained for organometallic compounds such as β-diketonates (80) and cobalamins (MW $\approx$ 1340–1580) (108). Irradiation of

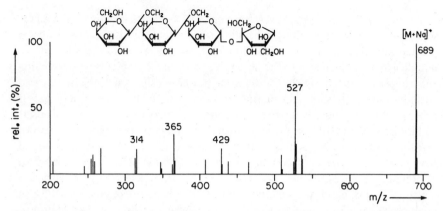

Fig. 18.13. LD mass spectrum of stachyose mixed with NaCl. The peaks at *m/z* 365 and 527 correspond to di- and trisaccharide moieties complexed with Na. Reproduced from ref. (30) with permission.

polymers yields $C_nH_m^{\pm}$ clusters and fragments characteristic of side-chain units (109–111). This behavior is reasonable if the polymer matrix is too large and intertwined to be desorbed intact so that fragment pieces are the only ions observed. Many molecules also yield small fragments such as CN^- and CNO^-, which perhaps emanate from the region directly exposed to the laser (76, 109). There is also some tendency for cluster ions [e.g., $(2M+H)^+$] to be formed (75, 98), which admittedly would be inconvenient for assigning structures of unknown compounds by LD.

3.3. Analytical Applications of LD

At the time this was written published accounts of the use of LD for identification and semiquantitative analysis of specific organic compounds were just beginning to appear in the literature. Most of this work sought to demonstrate the utility of LD to obtain a mass spectrum from a small amount of material or a localized sample area. To this end LD mass spectra have been shown for organic compounds adsorbed onto individual particulates and asbestos fibers (57, 58, 112–114). The latter studies showed that reportedly high levels of pthalate compounds on asbestos were likely to be artifacts from contamination by storage of bulk asbestos samples in plastic bags. Cooks and co-workers have identified naturally occurring quaternary compounds such as candicine by LD combined with CID (81). Hercules' group and others have also reported extensively on LD of quaternary ammonium salts (92, 100, 101, 115). Dutta and Talmi have used the LAMMA device for spatially resolved microprobe analysis of organic structures in coal macerals (116), and Vanderborgh and Jones report LD studies of coals and shales (117).

As was the case with LAMMA for elemental analysis, quantitation in LD is somewhat of a problem. It is the author's view that this is to some extent a problem inherent to spatially resolved techniques. Better spatial resolution implies less material exposed to the laser with consequent problems with sampling and sample heterogeneity. In addition, the power density is not precisely reproducible from shot to shot, and the fragmentation patterns are dependent on power density. Determining the relative concentrations of two similar compounds is within the scope of LD provided their mass spectral characteristics are known or can be verified (101). It is also the case that LD has not yet been tested strenuously for analysis of complex mixtures, for which CID may be required to provide the necessary selectivity. Of course, mass selection of a particular ion for CID dispenses with the other ions formed, so selectivity via CID and microprobe capability would seem to be somewhat incompatible. Another possible approach is to use LD to obtain mass spectra in conjunction with a chromatographic separation. For example, Vestal and Hardin (85, 86) have described LD of nonvolatile compounds sprayed onto a moving belt interface. Here again problems with sampling, ion collection (a scanning quadrupole was used), and ionization reproducibility surfaced. Also, Novak et al. (77) report the use of LD to identify individual spots from various dyes separated by thin-layer chromatography (TLC).

3.4. Combinations of LD with Other Ionization Techniques

Cotter and co-workers have demonstrated that laser-desorbed neutral species can be ionized by either EI or CI. This auxillary ionization greatly increased ion yields for the compounds and laser conditions they investigated. In their studies desorption of neutrals occurred before cationization, that is, cationization can occur in the gas phase after the neutral precursor desorbs (78, 102, 103, 105, 118, 119). Thus, these experiments have yielded very interesting information about the relative time scale of various steps in LD. Because neutral desorption continues for a considerable time (~ 500 μs to 1 ms) after the laser pulse, it is possible that LD coupled with EI or CI may be more readily adapted to scanning mass spectrometers than LD alone. Yost et al. have demonstrated that this concept with CI permits use of LD with a triple-quadrupole mass spectrometer for selective determination of several antiepileptic drugs (87). These workers also point out that LD coupled with CI helps reduce variations in fragment ion intensities with laser power density. Hercules and co-workers have demonstrated that aromatic nitro compounds or $NaNO_3$ can be mixed with PAHs to yield $(M-O+H)^+$ ions from the PAH under LD. In this fashion the additive serves as a CI reagent, with the CI reaction occurring either in the soild or gas phase (77, 120). By the same token, widespread occurrence of reactions like these could cause the spectra obtained from mixtures to differ from those of the corresponding pure compounds, which would make LD of limited value for direct analysis of complex organic samples. Ramaley et al. report use of a laser to desorb compounds from specific spots on a TLC plate. An auxiliary reagent gas was then used to sweep the desorbed material into a CI source. However, LD proved no better and in some ways worse than simple thermal desorption by heating with a flashlamp (121).

Schulten and colleagues have shown that a laser can be used to "assist" or augment ionization and fragment formation in field desorption (FD) (122–124). In these experiments the sample is deposited on the microneedles of a typical activated wire emitter for FD. As the high voltage is applied to induce FD, the sample is also irradiated with the laser. Low power densities enhance the parent ion peaks much like warming the emitter often does in conventional FD. The extent of fragmentation can be increased by increasing the laser power density. Organometallic compounds can be fragmented down to the bare metal ion for isotope ratio measurements with only nanogram quantities of compound. These capabilities of laser-assisted FD may make it valuable for the isotope ratio measurements necessary for use of enriched, stable metal isotopes as tracers of elemental fates and pathways in biomedical and enviromental systems.

4. PHOTOIONIZATION OF GASEOUS SPECIES

In many applications MS with an EI or CI source has become the technique of choice for analysis of samples that are either gases or can be vaporized readily. A case in point is the almost universal use of gas chromatography (GC)–MS with EI and/or CI for multicomponent analysis of organic mixtures. It has long been known that photons of appropriate energy can ionize gaseous atoms or molecules.

$$A + n(h\nu) \rightarrow A^+ + e^- \tag{18.5}$$

For most atoms or molecules ionization via single-photon absorption ($n = 1$) requires a light source in the vacuum ultraviolet (VUV). Photoionization by single-photon absorption in the VUV is a very valuable tool for probing energy levels of atoms and molecules (125, 126) and is even the basis for a selective GC detector (127). The advent of lasers makes multiphoton ionization (MPI, $n \geq 2$) feasible for UV and visible photons. However, photoionization must be able to do something that EI or CI cannot do if it is to compete with or complement these accepted ionization techniques. There is evidence that photoionization (particularly MPI) can be made to selectively ionize a particular analyte if the source wavelength and laser power density are selected appropriately. Ionization selectivity for MPI is certainly greater than for EI and possibly greater than for CI. It is also possible that molecular fragmentation might be more readily varied in MPI with less of a sensitivity sacrifice than is the case for EI. In this section the principles and applications of MPI for mass spectrometric analysis of atoms and molecules are described. For both atomic and molecular systems the energy level structure for the absorber influences the ionization process, hence the potential for selectivity. Ionization of atoms will be described first, they being simpler spectroscopic systems.

4.1. Resonance Ionization Mass Spectrometry of Atoms

The accomplishments of several workers in selective detection of minute amounts of particular atoms by resonance ionization (RI) spectroscopy have been described in Chapter 4. Hurst's group at Oak Ridge National Laboratory has been particularly active in this area. The discussion below is limited to the use of RI as an ionization technique with a mass spectrometric measurement (RI–MS). In many cases the ionization process alone is sufficiently selective that mass analysis of the ions produced is not necessary. However, it is natural that the two techniques be used in conjunction because each is both sensitive and selective. Because the fundamental principles of selective ionization of atoms by RI are similar to those used in MPI of molecules, their description here is a convenient introduction for the rest of this section. The theory, characteristics, and applications of RI and RI–MS have been described in detail elsewhere (128–139).

A simple depiction of some RI processes is shown in Fig. 18.14. The analytical objective depicted in this figure is selective ionization of either atom A in the presence of B and C or atom B in the presence of A and C. The three elements have similar ionization energies so that neither EI, thermal ionization (TI), nor single-photon ionization would be selective for one in the presence of

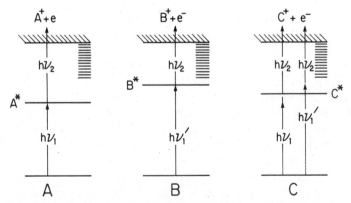

Fig. 18.14. Schematic diagram of atomic energy levels and transitions in RIS. Photon $h\nu_1$ populates A* but not B* or C*. Photon $h\nu_1'$ populates B* but not A* or C*.

the others. However, the energies of the excited electronic states in the three neutral atoms differ significantly. These energy level differences can be exploited as follows. The laser is tuned to $h\nu_1$. Absorption at this wavelength populates the A* state depicted in Fig. 18.14. Excited states of B and C are much less efficiently populated (if at all) than A* because the laser energy does not match a bound-bound electronic transition for B or C. Absorption of a second photon ($h\nu_2$) by A* takes A above its ionization limit, resulting in formation of A^+. Of course, the latter statement is valid for absorption of two identical photons only if the intermediate state A* lies more than half way "up the ladder" to the ionization limit. At an appropriate power density the photoionization yield of A^+ is much greater than for B or C because the intermediate A* state is resonant with the photon energy $h\nu_1$. If the laser frequency is changed to $h\nu_1'$ B will selectively absorb two photons and be ionized to B^+. C is not ionized because C* is not populated by excitation at either $h\nu_1$ or $h\nu_1'$. "Two-color" photoionization systems ($h\nu_1 \neq h\nu_2$) are feasible in which two photons of different energy may be absorbed. Absorption of three or more photons to yield ionization is also feasible. An appropriate juxtaposition of absorption events can be found that will ionize essentially any element in the periodic table (130, 131).

Use of a MS measurement to monitor the RI signal (e.g., Fig. 18.15) provides mass selective information for a particular element. Isobaric or neighboring peaks from different elements can be detected separately with a low-resolution mass analyzer with efficient ion transmission because the ionization process is selective for one element at a time. An example is shown for two rare-earth

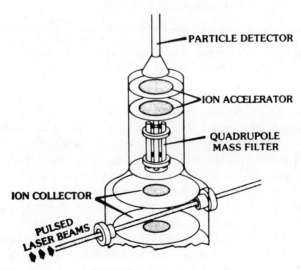

Fig. 8.15. Schematic diagram of quadrupole mass spectrometer with RI source for detection of ^{81}Kr. Reproduced from ref. (134) with permission.

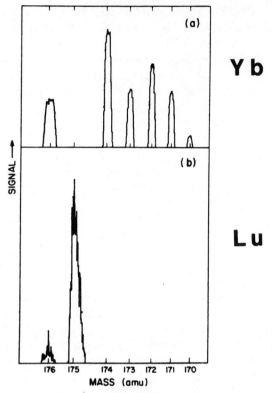

Fig. 18.16. (a) Thermal ionization of equimolar Lu/Yb mixture. Only Yb is observed because it is more volatile and easily ionized than Lu. (b) RI mass spectrum of same sample. Only Lu is observed because the laser is tuned to a wavelength absorbed by Lu but not by Yb. Signal-to-noise ratio is poorer than in (a) because of the low duty cycle ($\sim 10^{-5}$) in the RI experiment. Reproduced from ref. (139) with permission.

elements in Fig. 18.16. RI–MS can determine isotope ratios or monitor a selected isotopic peak for mixtures of elements such as these without prior chemical separation (136, 137, 139). Because the RI process is selective, the background gas in the source is not ionized extensively. Pulse counting ion detection is capable of counting individual ions provided they arrive at a rate greater than the fluctuation in dark current, which can be made less than 1 count per second. Thus RI–MS is capable of determining minute amounts of one element in the presence of a large excess of other elements. For example, as few as 300 atoms of ^{81}Kr can be detected in the presence of a total of 10^7 stable Kr isotopes and 10^{12} other species (135). Numerous other demonstrations of RI–MS have also been reported with either continuous (140) or pulsed (141) atom sources. The atom source is a crucial aspect of RI–MS in that real samples for which elemental analysis is desired only rarely come in the form of free, ground-state, neutral atoms. Transforming a sample into free atoms generally requires energy, which consequently leaves some of the atoms in excited states. These excited atoms could then be photoionized to the detriment of selectivity for one particular analyte in the presence of other elements. In light of the title of this book some particularly interesting examples use a laser for vaporizing and atomizing a solid sample and additional laser(s) for excitation and ionization (133, 142). If solid samples can be efficiently ablated or otherwise transformed into atoms without matrix effects or extensive population of excited states, then RI–MS could indeed become a valuable technique for general elemental analysis.

4.2. Multiphoton Ionization of Molecules for MS

Johnson (143, 144) and Dalby (145) were the first to use laser MPI as a tool to probe molecular energy levels. Some particular features of MPI that were desirable for these studies were (a) transitions forbidden for one photon are often allowed for two or more photons, (b) the generated ion current provides a sensitive measure of absorption, and (c) special VUV light sources and techniques are not required. Prior to 1977 these MPI experiments simply measured the total ionization rate by collecting ions at an electrode near the laser beam. Los (146), Bernstein (147, 148), Schlag (149), Letokhov (150), and Reilly (151) then showed that the ions produced by MPI could be detected by a mass spectrometer. Although this original work was primarily fundamental in nature, these authors also described potential analytical applications of MPI. A brief discussion of ionization processes in MPI (78, 152–159) will show the reasons for these claims.

In general, MPI can occur by one of the two processes shown in Fig. 18.17. Initial absorption promotes the molecule to an intermediate state. In the non-resonant mechanism (NRMPI) some or all of the intermediate states are virtual, that is, they are not real energy levels of the molecule and have very short lifetimes ($\sim 10^{-15}$ s). In general, NRMPI can selectively ionize a particular compound only if its ionization energy is lower than those of other compounds present in the ion source. If one of the initial absortion steps promotes the molecule to a real intermediate state, the overall efficiency of excitation and ionization is greatly enhanced, largely because the longer-lived ($\gtrsim 10^{-9}$ s) real intermediate state has a much higher probability of absorbing another photon. Thus, this latter process is called resonance-enhanced MPI (REMPI, Fig. 18.17) and can occur at power densities lower than those necessary for NRMPI. Consider a mixture of the hypothetical compounds ABC and (ABC)' (Fig. 18.18). As was the case in the schematic depiction of RI for atoms, these molecules have similar ionization energies. If they are structural or substitutional isomers, their EI spectra will be quite similar and it may also be difficult to separate them

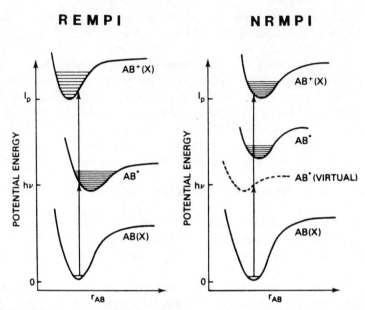

Fig. 18.17. Schematic energy level diagrams for resonance-enhanced multiphoton ionization (REMPI) and nonresonant multiphoton ionization (NRMPI). Reproduced from ref. (159) with permission.

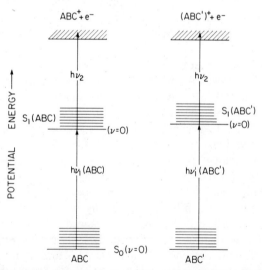

Fig. 18.18. REMPI for the selective ionization of the molecules ABC and ABC'. Their excited singlet states (S_1) differ in energy. Excitation at $h\nu_1$ enhances the yield of ABC$^+$ relative to (ABC')$^+$, vice versa for excitation at $h\nu_1'$.

614

chromatographically. Thus, it is difficult to determine ABC in the presence of ABC' by conventional GC–MS, particularly if one is present in large excess or if their relative concentrations vary widely from sample to sample. NRMPI will not discriminate between these molecules either unless their cross sections for absorption and ionization should happen to differ greatly. Compare this situation with a REMPI experiment. If the two neutral molecules have excited electronic states that differ significantly in energy, a tunable laser can be used to make the initial absorption resonant for ABC but not for ABC' (Fig. 18.18). Thus, ABC* is generated efficiently and can readily absorb more photons to yield ABC$^+$. The photoion yield for ABC' will be much lower for excitation at $h\nu_1$, however, because the less favorable NRMPI is the only mechanism for (ABC')$^+$ formation. Figure 18.19 shows a schematic diagram of a TOF device for MPI–MS. Again, with pulsed lasers it is desirable to collect and detect as large a fraction of the transient ionization signal as possible, hence the popularity of TOF measurements. Like LD, MPI is feasible in a FT–MS cell (160, 161), which contains ions effectively and also makes high-resolution, accurate mass measurement and CID possible (61, 161). Recent improvements in mass spectral performance of

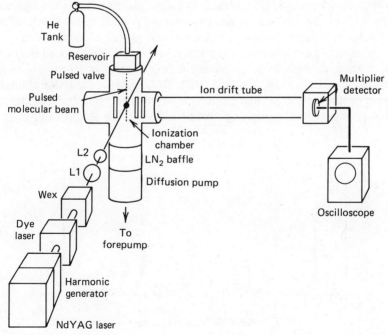

Fig. 18.19 TOF–MS for MPI. Note He reservoir and pulsed valve for supersonic jet generation. Reproduced from ref. (165) with permission.

the quadrupole ion trap (162) may also make this device useful as a sensitive, inexpensive mass analyzer for MPI.

Aromatic molecules are the classic subjects for MPI studies. Some examples of mass spectra for such compounds are shown in Figs. 18.20 and 18.21. As power density increases the relative ratios small fragments–large fragments–intact parent ions generally increase while the total ion yield often saturates (157, 158, 163). The fragmentation patterns observed in MPI of benzene and other molecules are somewhat different than EI fragmentation patterns. Fragmentation can be quite extensive as indicated by the observation of C^+ from benzene, which requires absorption of at least nine 3-eV photons. There are several possible mechanisms for fragment ion formation and some uncertainty about their relative importance. For this discussion it is sufficient to note that either the ionic or neutral fragments formed from dissociation of M^+ likely absorb more photons and themselves fragment or ionize to yield the observed mass spectrum (154–158). However, ions like C^+ and $C_nH_m^+$ (Fig. 18.20) are hardly of much value in assigning a structure or identifying a compound. It remains to be seen whether structurally significant fragmentation can be induced and adequately controlled for a variety of analyte molecules and if normal fluctuations in laser power ($\sim \pm 10\%$) make fragmentation difficult to reproduce.

It is possible to obtain MPI spectra with the sample gas added to the ionization cell by a simple flow system. In this case thermal population of vibrational and rotational levels hurts selectivity because initial photon absorption can occur from the manifold of vibrational and rotational sublevels in the ground electronic state. Adding the sample by entraining it in a supersonic jet expansion has numerous advantages (154, 158, 163–167). During the jet expansion collisions between sample molecules and carrier gas depopulate excited vibrational and rotational sublevels. The resulting "cold" molecules are almost completely in

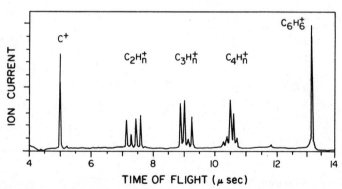

Fig. 18.20. TOF mass spectrum of benzene obtained by MPI. Note extensive fragmentation. Reproduced from ref. (158) with permission.

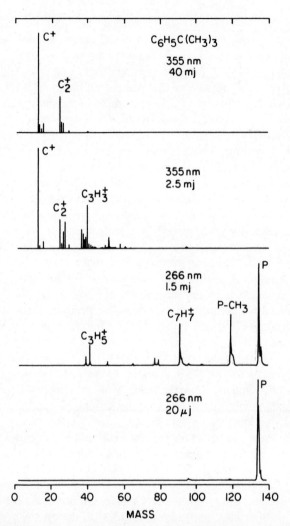

Fig. 18.21. NRMPI mass spectra of *t*-butylbenzene. Note fragmentation increases with power density. Reproduced from ref. (156) with permission.

their lowest rovibronic levels. With sufficient rotational and vibrational cooling it should even be possible to find a resonant wavelength suitable for ionizing ABC' (i.e., the molecule with the more energetic first excited state in Fig. 18.18) with some selectivity over ABC, provided the excited-state vibronic levels of the two molecules are sufficiently different in energy. The gas dynamic characteristics of the supersonic jet also help keep the "cold" sample molecules localized in a finite region and moving with a well-defined velocity. Molecules in the jet can thus be efficiently interrogated by the laser and resolved by the mass analyzer. Use of a continuous jet with GC sample introduction is feasible despite the low duty cycle in which ions are produced (typically $\sim 10^{-7}$ for a laser pulsed at 10 Hz). The fact that mass spectra can be detected at all at such a low duty cycle is a measure of the inherent efficiency of MPI (particularly REMPI) and ion detection by MS. Improvement of the duty cycle with a consequent increase in sensitivity should result if a pulsed valve can be used for sample introduction (165). Such a valve would form a repetitive, transient jet that could be synchronized with the laser so that a much larger fraction of the sample would be exposed to the laser shot. There is some question, however, as to whether a pulsed jet with minimal dead volume can be constructed that will operate reliably at GC temperatures (168). Nevertheless, a supersonic jet is clearly the technique of choice for introducing gaseous samples, particularly if ionization selectivity is necessary. A high-temperature pulsed valve would also facilitate a supersonic jet analog of the classical direct insertion probe for introduction of batch-type samples into a MPI source.

As far as analytical applications of molecular MPI are concerned, at this writing the technique is in the stage where the selectivity and fragmentation patterns observed for compounds of interest are being verified. Lubman (165, 169, 170), Decorpo (159), and Reilly (171, 172) have been active in these sorts of studies. As a portent of things to come, Reilly has even been able to demonstrate some ionization selectivity for NRMPI (172). Perhaps the most pertinent analytical objective for MPI–MS research is the selective detection of PAH isomers in a GC effluent. Isomeric forms of these compounds are difficult to resolve chromatographically and have quite similar EI mass spectra, yet their carcinogenicity or mutagenicity are thought to vary considerably. The first excited states for these molecules are at sufficiently different energies that it should be possible to selectively excite each of them individually by an appropriate REMPI scheme. Conventional EI or CI data could be obtained at the same time by splitting the column effluent into a separate MS such as one of the inexpensive ($\sim$ \$50,000) devices being marketed for capillary GC detection. If REMPI proves capable of "picking out" particular molecules of interest for a wide range of applications, it could also solve analytical problems with shorter chromatographic runs or less exhaustive optimization of the column conditions for separation. In addition, considerable progress has been made in techniques for introducing liquid chromatographic (LC) effluent into an ion source. Several of the LC–MS interface concepts such as thermospray (173) or direct liquid introduction (174) entrain the analyte in a molecular beam of sorts (although probably without efficient vibrational and rotational cooling). A selective ionization technique for LC–MS would be particularly valuable because LC separations are unlikely to be as good as for capillary GC, that is, coeluting compounds are likely to be more of a problem in LC. In the author's opinion these are some ways in which MPI–MS may contribute to future analytical science. This ionization technique could become particularly valuable if MPI proves sufficiently sensitive and selective for a wide range of compounds and if the lasers used are sufficiently powerful, tunable, and reliable.

ACKNOWLEDGMENT

Ames Laboratory is operated for the U.S. Department of Energy by Iowa State University under Contract No. W-7405-Eng-82. This research was supported by the Director for Energy Research, Office of Basic Energy Sciences.

REFERENCES

1. R. E. Honig, *Appl. Phys. Lett.*, **3**, 8 (1963).
2. R. E. Honig and J. R. Woolston, *Appl. Phys. Lett.*, **2**, 138 (1963).
3. K. A. Lincoln, *Anal. Chem.*, **37**, 541 (1965).
4. N. C. Fenner and N. R. Daly, *Rev. Sci. Instrum.*, **37**, 1068 (1966).
5. R. J. Conzemius and J. M. Capellen, *Int. J. Mass Spectrom. Ion Phys.*, **34**, 197 (1980).
6. R. J. Conzemius, D. S. Simons, Z. Shankai, and G. D. Byrd, *Microbeam Analysis—1983*, **18**, 301 (1983).
7. J. Haverkamp and P. G. Kistemaker, *Int. J. Mass Spectrom. Ion Phys.*, **45**, 275 (1982).
8. N. E. Vanderborgh, "Laser Induced Pyrolysis Techniques," in C. E. R. Jones and C. A. Cramers, Eds., *Analytical Pyrolysis*, Elsevier, Amsterdam, 1977, p. 235.
9. I. D. Kovalev, G. A. Maksimov, A. I. Suchkov, and N. V. Larin, *Int. J. Mass Spectrom. Ion Phys.*, **27**, 101 (1978).
10. R. J. Conzemius, Z. Shankai, R. S. Houk, and H. J. Svec, *Int. J. Mass Spectrom. Ion Processes*, **61**, 277 (1984).
11. B. E. Knox, *Dynamic Mass Spectrom.*, **2**, 61 (1971); "Laser Ion Source Analysis of Solids," in A. J. Ahearn, Ed., *Trace Analysis by Mass Spectometry*, Academic Press, New York, 1972, Chapter 14.
12. J. F. Ready, *Effects of High-Power Laser Radiation*, Academic Press, New York, 1971.
13. T. Dingle, B. W. Griffiths, and J. C. Ruckman, *Vacuum*, **31**, 571 (1981).
14. T. Dingle, B. W. Griffiths, J. C. Ruckman, and C. A. Evans, Jr., *Microbeam Analysis—1982*, **17**, 365 (1982).
15. B. K. Furman and C. A. Evans, Jr., *Microbeam Analysis—1982*, **17**, 222 (1982).
16. J. F. Eloy, *Microscopica Acta, Suppl. 2*, 307 (1978).
17. E. Deloule and J. F. Eloy, *Chem. Geol.*, **37**, 191 (1982).
18. J. F. Eloy, M. Leleu, and E. Unsold, *Int. J. Mass Spectrom. Ion Phys.*, **47**, 39 (1983).

19. R. J. Conzemius and H. J. Svec, *Anal. Chem.*, **50**, 1854 (1978).
20. R. J. Conzemius, *Microbeam Analysis—1982*, **17**, 369 (1982).
21. A. I. Busygin, *Sov. Tech. Phys. Lett.*, **3**, 459 (1977).
22. Yu. A. Bykovskii, G. I. Zhuravlev, V. I. Belousov, V. M. Gladskoi, V. G. Degtyarev, and V. N. Nevolin, *Ind. Lab. USSR* (Engl. trans.), **44**, 799 (1978).
23. Yu. A. Bykovskii, L. M. Bobenkov, T. A. Basova, V. I. Belousov, V. M. Gladskoi, V. V. Gorshkov, V. G. Degtyarev, I. D. Laptev, and V. N. Novelin, *Instrum. Exp. Tech. USSR*, **20**, 508 (1977).
24. J. A. J. Jansen and A. W. Witmer, *Spectrochim. Acta*, **37B**, 483 (1982).
25. R. A. Bingham and P. L. Salter, *Int. J. Mass Spectrom. Ion Phys.*, **21**, 133 (1976).
26. R. A. Bingham and P. L. Salter, *Anal. Chem.*, **48**, 1735 (1976).
27. F. J. Bruynseels and R. E. Van Grieken, *Anal. Chem.*, **56**, 871 (1984).
28. F. J. Bruynseels and R. E. Van Grieken, *Spectrochim. Acta*, **38B**, 853 (1983).
29. E. Michiels and R. Gijbels, *Anal. Chem.*, **56**, 1115 (1984).
30. H. J. Heinen, S. Meier, H. Vogt, and R. Wechsung, *Adv. Mass Spectrom.*, **8A**, 942 (1980).
31. N. Fürstenau and F. Hillenkamp, *Int. J. Mass Spectrom. Ion Phys.*, **37**, 135 (1981).
32. N. Fürstenau, F. Hillenkamp, and R. Nitsche, *Int. J. Mass Spectrom. Ion Phys.*, **31**, 85 (1979).
33. E. Denoyer, R. E. Van Grieken, F. Adams, and D. F. S. Natusch, *Anal. Chem.*, **54**, 26A (1982).
34. T. P. Hughes, *Plasmas and Laser Light*, Wiley, New York, 1975.
35. H. Hora, *Physics of Laser Driven Plasmas*, Wiley, New York, 1981.
36. P. K. Carroll and E. T. Kennedy, *Contemp. Phys.*, **22**, 61 (1981).
37. J. Dawson, P. Kaw, and B. Green, *Phys. Fluids*, **12**, 875 (1969).
38. I. Opauszky, *Pure Appl. Chem.*, **54**, 879 (1982).
39. Yu. A. Bykovskii, T. A. Basova, V. I. Belousov, V. M. Gladskoi, V. V. Gorshkov, V. G. Degtyarev, I. D. Laptev, and V. N. Nevolin, *Sov. Phys. Tech. Phys.*, **21**, 761 (1976).
40. Z. Shankai, R. J. Conzemius, and H. J. Svec, *Anal. Chem.*, **56**, 382 (1984).
41. R. Dinger, K. Rohr, and H. Weber, *J. Phys. D: Appl. Phys.*, **13**, 2301 (1980).
42. I. D. Kovalev, N. V. Larin, G. A. Maksimov, A. I. Suchkov, and I. Yu. Feoktisov, *Sov. Phys. Tech. Phys.*, **25**, 259 (1980).
43. C. L. S. Lewis, P. F. Cunningham, L. Pina, A. K. Roy, and J. M. Ward, *J. Phys. D: Appl. Phys.*, **15**, 69 (1982).
44. M. Newstein and N. Solimeme, *IEEE J. Quant. Electronics*, **QE-17**, 2085 (1981).
45. B. A. Mamyrin, V. I. Karataev, D. V. Shmikk, and V. A. Zagulin, *Sov. Phys.—JETP*, **37**, 45 (1973); B. A. Mamyrin and D. V. Shmikk, *Sov. Phys.—JETP*, **49**, 762 (1979).
46. A. I. Boriskin, A. S. Bryukhanov, Yu. A. Bykovskii, V. M. Eremenko, and I. D. Laptev, *Opt.-Mekh. Prom-st.*, **6**, 48 (1983).
47. Yu. A. Bykovskii, G. I. Zhuravlev, V. M. Gladskoi, V. G. Degtyarev, and V. N. Nevolin, *Sov. Phys. Tech. Phys.*, **23**, 225 (1978).
48. D. S. Simons, *Int. J. Mass Spectrom. Ion Processes*, **55**, 15 (1983).

49. P. Surkyn and F. Adams, *J. Trace Microprobe Tech.*, **1**, 79 (1982).
50. R. J. Conzemius, F. A. Schmidt, and H. J. Svec, *Anal. Chem.*, **53**, 1899 (1981).
51. R. J. Conzemius and H. J. Svec, "Analysis of Solids by Spark Source and Laser Mass Spectrometry," in D. R. Rossington, R. A. Condrate, and R. L. Snyder, Eds., *Advances in Materials Characterization*, Plenum, New York, 1983, p. 59.
52. M. S. Beck, F. A. Schmidt, and R. J. Conzemius, *J. Electrochem. Soc.*, **131**, 2169 (1984).
53. B. Sprey and H.-P. Bochem, *Fresenius Z. Anal. Chem.*, **308**, 239 (1981).
54. D. W. Lorch and H. Schäfer, *Fresenius Z. Anal. Chem.*, **308**, 246 (1981).
55. L. Edelmann, *Fresenius Z. Anal. Chem.*, **308**, 218 (1981).
56. P. Klein and J. Bauch, *Fresenius Z. Anal. Chem.*, **308**, 283 (1981).
57. P. K. Dutta, D. C. Rigano, R. A. Hofstader, E. Denoyer, D. F. S. Natusch, and F. C. Adams, *Anal. Chem.*, **56**, 302 (1984).
58. J. De Waele, P. Van Espen, E. Vansant, and F. Adams, *Microbeam Analysis— 1982*, **17**, 371 (1982).
59. K. R. Spurny, J. Schörmann, and R. Kaufmann, *Fresenius Z. Anal. Chem.*, **308**, 274 (1981).
60. P. Wieser, R. Wurster, and U. Haas, *Fresenius Z. Anal. Chem.*, **308**, 260 (1981).
61. M. B. Comisarow and A. G. Marshall, *Chem. Phys. Lett.*, **25**, 282 (1974); **26**, 489 (1974).
62. C. L. Wilkins and M. L. Gross, *Anal. Chem.*, **53**, 1661A (1981).
63. R. B. Cody, R. C. Burnier, W. D. Reents, Jr., T. J. Carlin, D. A. McCrery, R. K. Lengel, and B. S. Freiser, *Int. J. Mass Spectrom. Ion Phys.*, **33**, 37 (1980).
64. R. C. Burnier, T. J. Carlin, W. D. Reents, Jr., R. B. Cody, R. K. Lengel, and B. S. Freiser, *J. Am. Chem. Soc.*, **101**, 7127 (1979).
65. R. C. Burnier, G. D. Byrd, and B. S. Freiser, *Anal. Chem.*, **52**, 1641 (1980); *J. Am. Chem. Soc.*, **103**, 4360 (1981); *J. Am. Chem. Soc.*, **104**, 3565 (1982).
66. G. D. Byrd and B. S. Freiser, *J. Am. Chem. Soc.*, **104**, 5944 (1982).
67. T. J. Carlin, M. B. Wise, and B. S. Freiser, *Inorg. Chem.*, **20**, 2743 (1981).
68. D. B. Jacobson, G. D. Byrd, and B. S. Freiser, *J. Am. Chem. Soc.*, **104**, 2320 (1982).
69. D. B. Jacobson and B. S. Freiser, *J. Am. Chem. Soc.*, **105**, 736, 5197, 7484 (1983); *Organometallics*, **3**, 513 (1984).
70. T. J. Carlin, L. Sallans, C. J. Cassady, D. B. Jacobson, and B. S. Freiser, *J. Am. Chem. Soc.*, **105**, 6320 (1983).
71. T. J. Carlin, T. C. Jackson, D. B. Jacobson, and B. S. Freiser, *J. Am. Chem. Soc.*, **106**, 1252 (1984).
72. F. J. Vastola and A. J. Pirone, *Adv. Mass Spectrom.*, **4**, 107 (1968).
73. F. J. Vastola, R. O. Mumma, and A. J. Pirone, *Org. Mass Spectrom.*, **3**, 101 (1970).
74. M. A. Posthumus, P. G. Kistemaker, H. L. C. Meuzelaar, and M. C. Ten Noever de Brauw, *Anal. Chem.*, **50**, 985 (1978).
75. F. Hillenkamp, *Int. J. Mass Spectrom. Ion Phys.*, **45**, 305 (1982).
76. D. M. Hercules, R. J. Day, K. Balasanmugam, T. A. Dang, and C. P. Li, *Anal. Chem.*, **54**, 280A (1982).
77. F. P. Novak, K. Balasanmugam, K. Viswanadham, C. D. Parker, Z. A. Wilk,

D. Mattern, and D. M. Hercules, *Int. J. Mass Spectrom. Ion Phys.*, **53**, 135 (1983).

78. R. J. Cotter, *Anal. Chem.*, **56**, 485A (1984).
79. D. Zakett, A. E. Schoen, R. G. Cooks, and P. H. Hemberger, *J. Am. Chem. Soc.*, **103**, 1295 (1981).
80. J. L. Pierce, K. L. Busch, R. G. Cooks, and R. A. Walton, *Inorg. Chem.*, **21**, 2597 (1982).
81. D. V. Davis, R. G. Cooks, B. N. Meyer, and J. L. McLaughlin, *Anal. Chem.*, **55**, 1302 (1983).
82. G. J. Q. van der Peyl, W. J. van der Zande, K. Bederski, A. J. H. Boerboom, and P. G. Kistemaker, *Int. J. Mass Spectrom. Ion Phys.*, **47**, 7 (1983).
83. K. L. Busch, S. E. Unger, A. Vincze, R. G. Cooks, and T. Keough, *J. Am. Chem. Soc.*, **104**, 1507 (1982).
84. R. Stoll and F. W. Röllgen, *Org. Mass Spectrom.*, **14**, 642 (1979).
85. E. D. Hardin and M. L. Vestal, *Anal. Chem.*, **53**, 1492 (1981).
86. E. D. Hardin, T. P. Fan, C. R. Blakley, and M. L. Vestal, *Anal. Chem.*, **56**, 2 (1984).
87. R. J. Perchalski, R. A. Yost, and B. M. Wilder, *Anal. Chem.*, **55**, 2002 (1983).
88. D. A. McCrery, E. B. Ledford, Jr., and M. L. Gross, *Anal. Chem.*, **54**, 1435 (1982).
89. K. L. Busch and R. G. Cooks, *Science*, **218**, 247 (1982).
90. G. D. Daves, Jr., *Acc. Chem. Res.*, **12**, 359 (1979).
91. A. Eicke, W. Sichtermann, and A. Benninghoven, *Org. Mass Spectrom.*, **15**, 289 (1980).
92. B. Schueler and F. R. Krueger, *Org. Mass Spectrom.*, **14**, 439 (1979); *Org. Mass Spectrom.*, **15**, 295 (1980); *Adv. Mass Spectrom.*, **8A**, 918 (1980).
93. B. Schueler, P. Feigl, F. R. Krueger, and F. Hillenkamp, *Org. Mass Spectrom.*, **16**, 502 (1981).
94. H. J. Heinen, *Int. J. Mass Spectrom. Ion Phys.*, **38**, 309 (1981).
95. P. G. Kistemaker, M. M. J. Lens, G. J. Q. van der Peyl, and A. J. H. Boerboom, *Adv. Mass Spectrom.*, **8A**, 928 (1980).
96. G. J. Q. van der Peyl, K. Isa, J. Haverkamp, and P. G. Kistemaker, *Int. J. Mass Spectrom. Ion Phys.*, **47**, 11 (1983).
97. K.-D. Kupka, F. Hillenkamp, and Ch. Schiller, *Adv. Mass Spectrom.*, **8A**, 935 (1980).
98. P. K. Dutta and Y. Talmi, *Anal. Chim. Acta*, **132**, 111 (1981).
99. G. J. Q. van der Peyl, J. Haverkamp, and P. G. Kistemaker, *Int. J. Mass Spectrom. Ion Phys.*, **42**, 125 (1982).
100. T. A. Dang, R. J. Day, and D. M. Hercules, *Anal. Chem.*, **56**, 866 (1984).
101. K. Balasanmugam and D. M. Hercules, *Anal. Chem.*, **55**, 145 (1983).
102. R. J. Cotter, *Anal. Chem.*, **53**, 719 (1981).
103. R. J. Cotter and A. L. Yergey, *J. Am. Chem. Soc.*, **103**, 1596 (1981).
104. G. J. Q. van der Peyl, K. Isa, J. Haverkamp, and P. G. Kistemaker, *Org. Mass Spectrom.*, **16**, 416 (1981).
105. R. B. Van Breeman, M. Snow, and R. J. Cotter, *Int. J. Mass Spectrom. Ion Phys.*, **49**, 35 (1983).

106. F. Heresch, E. R. Schmid, and J. F. K. Huber, *Anal. Chem.*, **52**, 1803 (1980).

107. R. A. Fletcher, I. Chabay, D. A. Weitz, and J. C. Chung, *Chem. Phys. Lett.*, **104**, 615 (1984).

108. S. W. Graham, P. Dowd, and D. M. Hercules, *Anal. Chem.*, **54**, 649 (1982).

109. J. A. Gardella, Jr., D. M. Hercules, and H. J. Heinen, *Spectrosc. Lett.*, **13**, 347 (1980).

110. J. A. Gardella, Jr., and D. M. Hercules, *Fresenius Z. Anal. Chem.*, **308**, 297 (1981).

111. S. W. Graham and D. M. Hercules, *Spectrosc. Lett.*, **15**, 1 (1982).

112. J. K. De Waele, E. F. Vansant, and F. C. Adams, *Mikrochim. Acta*, **3**, 367 (1983).

113. J. K. De Waele, P. Van Espen, E. F. Vansant, and F. C. Adams, *Anal. Chem.*, **55**, 671 (1983); *Int. J. Mass Spectrom. Ion Phys.*, **46**, 515 (1983).

114. J. K. De Waele, J. J. Gybels, E. F. Vansant, and F. C. Adams, *Anal. Chem.*, **55**, 2255 (1983).

115. K. Balasanmugam and D. M. Hercules, *Microbeam Analysis—1982*, **17**, 389 (1982).

116. P. K. Dutta and Y. Talmi, *Fuel*, **61**, 1241 (1982).

117. N. E. Vanderborgh and C. E. Roland Jones, *Anal. Chem.*, **55**, 527 (1983).

118. R. J. Cotter, *Anal. Chem.*, **53**, 719 (1981).

119. J.-C. Tabet and R. J. Cotter, *Int. J. Mass Spectrom. Ion Phys.*, **53**, 151 (1983); *Int. J. Mass Spectrom. Ion Phys.*, **54**, 151 (1983).

120. K. Balasanmugam, S. K. Viswanadham, and D. M. Hercules, *Anal. Chem.*, **55**, 2424 (1983).

121. L. Ramaley, M. E. Nearing, M.-A. Vaughan, R. G. Ackman, and W. D. Jamieson, *Anal. Chem.*, **55**, 2285 (1983).

122. H.-R. Schulten, W. D. Lehmann, and D. Haaks, *Org. Mass Spectrom.*, **13**, 361 (1978).

123. H.-R. Schulten, P. B. Morkhouse, and R. Müller, *Anal. Chem.*, **54**, 654 (1982).

124. H.-R. Schulten, B. Bohl, U. Bahr, R. Müller, and R. Palavinskas, *Int. J. Mass Spectrom. Ion Phys.*, **38**, 281 (1981).

125. C. Y. Ng, "Molecular Beam Photoionization Studies of Molecules and Clusters," in I. Prigogine and S. A. Rice, Eds., *Advances in Chemical Physics*, Vol. 52, Wiley, New York, 1983, p. 263.

126. J. Berkowitz, *Photoabsorption, Photoionization, and Photoelectron Spectroscopy*, Academic Press, New York, 1979.

127. HNU Systems, Inc., Newton Upper Falls, MA.

128. G. S. Hurst, M. G. Payne, M. H. Nayfeh, J. P. Judish, and E. B. Wagner, *Phys. Rev. Lett.*, **35**, 82 (1975).

129. G. S. Hurst, M. H. Nayfeh, and J. P. Young, *Appl. Phys. Lett.*, **30**, 229 (1977); *Phys. Rev. A*, **15**, 2283 (1977).

130. G. S. Hurst, *J. Chem. Educ.*, **59**, 895 (1982); *Anal. Chem.*, **53**, 1448A (1981).

131. G. S. Hurst, M. G. Payne, S. D. Kramer, and J. P. Young, *Rev. Mod. Phys.*, **51**, 767 (1979); *Anal. Chem.*, **51**, 1050A (1979).

132. V. I. Balikin, G. I. Belov, V. S. Letokhov, and V. I. Mishin, *Usp. Fiz. Nauk*, **132**, 293 (1980).

133. D. W. Beekman, T. A. Callcott, S. D. Kramer, E. T. Arakawa, G. S. Hurst, and E. Nussbaum, *Int. J. Mass Spectrom. Ion Phys.*, **34**, 89 (1980).

134. C. H. Chen, G. S. Hurst, and M. G. Payne, *Chem. Phys. Lett.*, **75**, 473 (1980).

135. C. H. Chen, S. D. Kramer, S. L. Allman, and G. S. Hurst, *Appl. Phys. Lett.*, **44**, 640 (1984).

136. J. P. Young and D. L. Donohue, *Anal. Chem.*, **55**, 88 (1983).

137. D. L. Donohue, D. H. Smith, J. P. Young, H. S. McKown, and C. A. Pritchard, *Anal. Chem.*, **56**, 379 (1984).

138. C. M. Miller and N. S. Nogar, *Anal. Chem.*, **55**, 1606 (1983).

139. C. M. Miller, N. S. Nogar, A. J. Gancarz, and W. R. Shields, *Anal. Chem.*, **54**, 2377 (1982).

140. P. J. Savickas, K. R. Hess, R. K. Marcus, and W. W. Harrison, *Anal. Chem.*, **56**, 817 (1984).

141. J. D. Fassett, L. J. Moore, R. W. Shidaler, and J. C. Travis, *Anal. Chem.*, **56**, 203 (1984).

142. S. Mayo, T. B. Lucatorto, and G. Luther, *Anal. Chem.*, **54**, 553 (1982).

143. P. M. Johnson, M. R. Berman, and D. Zakheim, *J. Chem. Phys.*, **62**, 2500 (1975).

144. P. M. Johnson, *J. Chem. Phys.*, **62**, 4562 (1975); **64**, 4143, 4638 (1976).

145. G. Petty, C. Tai, and F. W. Dalby, *Phys. Rev. Lett.*, **34**, 1207 (1975).

146. M. Klewer, M. J. M. Beerlage, J. Los, and M. J. Van der Wiel, *J. Phys. B*, **10**, 2809 (1977).

147. L. Zandee, R. B. Bernstein, and D. A. Lichtin, *J. Chem. Phys.*, **69**, 3427 (1978).

148. L. Zandee and R. B. Bernstein, *J. Chem. Phys.*, **70**, 2574 (1979).

149. V. Boesl, H. J. Neusser, and E. W. Schlag, *Z. Naturforsch. A*, **33**, 1546 (1978).

150. V. S. Antonov, I. N. Knyazev, V. S. Letokshov, V. M. Matiuk, V. G. Movshev, and V. K. Potapov, *Opt. Lett.*, **3**, 37 (1978).

151. S. Rockwood, J. P. Reilly, K. Hohla, and K. L. Kompa, *Opt. Commun.*, **28**, 175 (1979).

152. P. M. Johnson, *Acc. Chem. Res.*, **13**, 20 (1980).

153. P. M. Johnson and C. E. Otis, *Ann. Rev. Phys. Chem.*, **32**, 139 (1981).

154. V. S. Antonov and V. S. Letokhov, *Appl. Phys.*, **24**, 89 (1981).

155. V. S. Antonov, V. S. Letokhov, and A. N. Shibanov, *Usp. Fiz. Nauk.*, **142**, 177 (1984).

156. R. B. Bernstein, *J. Phys. Chem.*, **86**, 1178 (1982).

157. D. A. Lichtin, L. Zandee, and R. B. Bernstein, "Potential Analytical Aspects of Laser Multiphoton Ionization Mass Spectrometry," in G. M. Hieftje et al., Eds., *Lasers in Chemical Analysis*, Humana Press, Clifton, NJ, 1981, Chapter 6.

158. E. W. Schlag and H. J. Neusser, *Acc. Chem. Res.*, **16**, 355 (1983).

159. M. Seaver, J. W. Hudgens, and J. J. DeCorpo, *Int. J. Mass Spectrom. Ion Phys.*, **34**, 159 (1980).

160. M. P. Irion, W. D. Bowers, R. L. Hunter, F. S. Rowland, and R. T. McIver, Jr., *Chem. Phys. Lett.*, **93**, 375 (1982).

161. T. J. Carlin and B. S. Freiser, *Anal. Chem.*, **55**, 955 (1983).

162. Finnigan MAT, San Jose, CA.

163. J. H. Brophy and C. T. Rettner, *Chem. Phys. Lett.*, **67,** 351 (1979).

164. D. M. Lubman, *Laser Focus,* **20**(6), 110 (1984).

165. D. M. Lubman and M. N. Kronick, *Anal. Chem.*, **54,** 660 (1982).

166. R. E. Smalley, L. Wharton, and D. H. Levy, *Acc. Chem. Res.*, **10,** 139 (1977).

167. J. M. Hayes and G. J. Small, *Anal. Chem.*, **54,** 1202 (1982); **55,** 565A (1983).

168. J. P. Reilly, J. W. Chai, and R. B. Opsal, *1984 Pittsburgh Conf. Anal. Chem. Appl. Spectrosc.*, Atlantic City, NJ, Paper No. 588.

169. D. M. Lubman and M. N. Kronick, *Anal. Chem.*, **54,** 2289 (1982); **55,** 867, 1486 (1983).

170. D. M. Lubman and R. N. Zare, *Anal. Chem.*, **54,** 2117 (1982).

171. J. W. Chai and J. P. Reilly, *Opt. Commun.*, **49,** 59 (1984).

172. G. Rhodes, R. B. Opsal, J. T. Meek, and J. P. Reilly, *Anal. Chem.*, **55,** 280 (1983).

173. M. L. Vestal, *Mass Spectrom. Rev.*, **2,** 481 (1983).

174. P. J. Arpino and G. Guiochon, *Anal. Chem.*, **51,** 682A (1979); **54,** 2375 (1982).

CHAPTER

19

LASER ABLATION
FOR ATOMIC SPECTROSCOPY

EDWARD H. PIEPMEIER

Department of Chemistry
Oregon State University
Corvallis, Oregon

627

1. INTRODUCTION

The plasma produced by high irradiance of a focused laser beam has been used since 1962 (1) to help determine the concentrations of chemical elements in selected micron-diameter regions of many types of samples. A schematic of a laser microprobe is shown in Fig. 19.1. The first microprobe used a pulsed ruby laser; Nd^{3+}, CO_2, N_2, and dye lasers are now in use. Virtually any type of sample material, including gases, can be atomized and excited. The atoms and molecules in the plasma are spectroscopically observed by emission, absorption, or fluorescence. Cross excitation of the plasma by a spark or microwave discharge may be used to increase the emission intensity and provide emission lines with narrower spectral line profiles than are usually produced without cross excitation at atmospheric pressure (2, 3). Or, the vaporized material may be swept by a gas stream into another chamber (4), into a flame (5), or into an electrical plasma for spectroscopic observation (6–10). Alternatively, the vaporized material may be condensed onto an electrode and then analyzed by an arc or spark (11). The use of mass spectrometer detection is considered in Chapter 18.

In many cases only a single laser shot is needed to provide an analysis, and less than a microgram of sample is vaporized. Consequently, the laser microprobe technique is often considered as an essentially nondestructive analytical

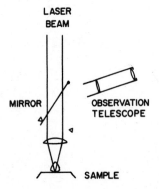

Fig. 19.1. Schematic of laser microprobe showing how sample can be positioned with the aid of a telescope.

technique. As such it is used to analyze precious archeological samples (12, 13) and forensic evidence (14, 15). Since the depth of a crater formed in a sample can be restricted to tens of micrometers or less, the laser microprobe is used to analyze successive layers of paintings to aid in their restorations (16). In biological and medical applications the laser microprobe is used to analyze various types of tissues (17–23) and even individual cells (24, 25). It finds a wide variety of applications in helping to identify inclusions in alloys, ceramics, plastics, polymers, and glasses (26–30). The identification of microscopic particles in air samples is aided by the use of a laser microprobe (31).

In addition to analyzing microscopic samples, the laser microprobe is used for the rapid analysis of bulk materials such as ceramics (28, 29), minerals (32–37), alloys (38–53), molten metals (7, 54), powders (44–46, 51, 55–58) and liquids (59, 60). Changes in chemical composition along a sample are determined by scanning the sampling region across the sample and recording the spectra of each sampling on sequential regions of a photographic plate (44, 61, 62). The thicknesses of coatings are also determined with a laser microprobe (5, 63).

The laser microprobe has unique capabilities that make it complementary to other microprobe techniques, such as the electron microprobe, the ion microprobe, and the controlled high-voltage spark. The other techniques require samples with minute to heavy electrical conduction, while the laser microprobe can atomize any type of material including electrical nonconductors. Uneven and large samples can be handled with the laser microprobe, and greater distances across a sample can be scanned than with the electron or ion microprobes. Perhaps one of the most important advantages is the very minimal (if any) sample preparation required for many applications of the laser microprobe. This frequently leads to a more rapid answer to a question than other techniques provide.

Accompanying the advantages of the laser microprobe are limitations, some of the most important of which are matrix effects. Causes for matrix effects and ways to minimize their influence will be considered in this chapter. However, even matrix effects can occasionally be used to advantage. For instance, one laser microprobe study (44) showed that the intensity of chromium lines from a steel sample is a measure of the strain in the sample, the result of a unique physical matrix effect.

The results from any particular laser microprobe system depend on many variables, including the characteristics of the laser beam, the way that the beam is focused on the sample, the interaction of the beam with the plasma above the sample surface, and the way in which the resulting spectra are observed. Cross excitation of the plasma adds other variables to the complex processes that occur. An understanding of the variables and their influences aids in the selection of an appropriate instrument, aids in the interpretation of analytical results, and is essential in the proper adjustment of the experimental variables if the best results are to be obtained. This chapter considers these variables from the time that the laser beam is generated until the spectra are recorded. A review of methods of improving analytical information is presented in a section on analytical applications. Finally, future needs and developments in the laser microprobe field are briefly discussed.

2. IMPORTANT CHARACTERISTICS OF LASER BEAMS

Characteristics of laser beams that significantly influence the analytical results obtained from a laser microprobe include the peak power and energy content per pulse, the temporal behavior of the pulses, the angular beam divergence and spatial intensity distribution in the beam, and the wavelength and spectral bandwidth of the beam. This section will consider these characteristics and how the focusing properties of the beam are affected. The next section will consider the interaction of the beam with the sample and the plume.

2.1. Irradiance

Vaporization and atomization of a sample requires that the energy be delivered to the sample at a rate that exceeds energy losses so that a high average energy density is achieved. Therefore, not only is the amount of energy absorbed important in determining the amount of material that is vaporized but the rate at which the energy is delivered to a specific sampling region, that is, the power density, is an important factor. The power density absorbed by the sample depends on the irradiance (W cm^{-2}) of the laser pulse at the surface and the reflectivity of the sample. The irradiance of the laser pulse also determines the fraction of laser energy absorbed by the plasma (or plume) that forms above surface of the sample. Consequently, the beam irradiance influences the energy density of the plume, which in turn influences many factors including the fraction of vapor atomized in the hot plume, the material transport properties (and thereby concentration gradients) within the plume, and the spectral excitation and line profile characteristics of the analytical spectral lines.

2.2. Temporal Behavior

The temporal behavior of a laser beam is important in determining the power density at the sample. There are several different types of pulsed modes of operation. The CO_2 and Nd:YAG lasers may be operated in a continuous mode or in a repetitively Q-switched mode at repetition rates exceeding 100 Hz. A flashlamp-pumped dye laser produces a pulse that is typically 0.4–1 μs long.

a.

b.

Fig. 19.2. Diagram showing how the intensity envelope of the spikes in a normal free-running laser pulse (*b*) generally follows the shape of the flashlamp pump pulse (*a*).

The ruby and Nd-glass lasers operate in what is known as a *normal* or free-running mode when the laser is simply pumped with a flashlamp and allowed to lase whenever the threshold condition for lasing happens to be exceeded. Normal pulses last from 0.1 to 1 ms depending on the flashlamp and usually consist of many microsecond-duration pulses, sometimes called spikes, unevenly spaced in time. The spikes usually are not reproducible in energy or time, Fig. 19.2.

Q-switched pulses of 10–100-ns duration can be reproducibly generated if they do not contain multiple spikes, which are difficult to control. Q-switched pulse trains consisting of many picosecond mode-locked pulses may or may not produce significantly different results compared to ordinary Q-switched pulses. Their influence has received little experimental study in laser microprobe work. Theoretical estimates of the influence of picosecond mode-locked pulses could be made if the response time of the irradiated sample were known. However, unexpected nonlinear responses to pulses with the large peak power densities of mode-locked pulses might produce as yet unknown behavior. Since subpicosecond pulses deposit their energy in a time comparable to a lattice vibration, unique analytical results might be possible.

2.3. Focal Characteristics

When the laser beam is focused onto a sample, the irradiance at the surface of the sample depends on the focal properties of the beam. Off-axis mode structure increases the divergence angle of the beam. Consequently, obtaining the smallest focal spots requires suppression of off-axis modes. Also, with multimode operation the mode structure of the beam may change so that one pulse is focused differently than the next. The spatial mode structure may even change during a pulse (63, 64).

The best laser pulse for analytical determinations is one with well-defined modes, ideally a single mode. A beam with a Gaussian distribution can be focused to a spot of the minimum size theoretically possible. When a Gaussian beam is focused with an aberration-free lens of focal length L and radius a, a Gaussian distribution of power density is produced at the focal plane. The radius R_0 at which the power density of the focal spot is $1/e^2$ times the maximum value at the center of the focused Gaussian beam is given by (65)

$$R_0 = \frac{\lambda L}{\pi R} \qquad (19.1)$$

which shows that the spot size is directly proportional to the focal length or the f/number $(L/2R)$ of the lens. The power density at the focal spot is increased by the factor $(a/R_0)^2$ compared to the power density of the unfocused beam of radius R. Since the factor $L/\pi R$ may be about unity, Eq. (19.1) is a basis for the common statement that the focal spot size may be as small as about one wavelength.

Along the optical axis, the power density falls to one-half of its peak value at distances $\pm R_0 L/R$ away from the focal plane. This is an indication of the depth of focus.

The focal properties of a laser beam with a uniform power density distribution are also of interest in helping to evaluate the power density at a focal spot. A Gaussian beam that is passed through an aperture considerably smaller than the diameter of the beam approximates this case. When a circular aperture of radius a is uniformly illuminated, the resulting beam diverges with an angle equal to $1.22\lambda/2R$ because of diffraction. When focused, the beam produces an image with a power density that follows the familiar Airy function, which has a central maximum surrounded by circular fringes (66). If the first minimum in the fringe pattern is used to define the radius of the focal spot, then the spot has a radius

$$R_s = \frac{1.22\lambda L}{2R} = \theta L \qquad (19.2)$$

where θ is the divergence angle of the beam in radians. This is an arbitrary choice for the radius, which may be misleading since the power densities in the circular fringes may be high enough to vaporize target material. Equation (19.2) is commonly used to estimate the minimum spot size of a focused laser beam from a knowledge of the beam divergence angle and the focal length of a lens. The depth of focus is indicated by the distance $\pm 1.45 R_s L/R$ at which the power density along the optical axis drops to one-half its maximum value.

The spatial distribution of a pulsed laser is often a rather complex pattern because of inhomogeneities in the laser cavity and mirror imperfections. The distribution of the power density in the focal spot of a multiple-mode beam is approximately a small image of the power density in the original beam.

When the pump power of a laser is increased, most of the increased laser power may go into weaker off-axis modes or into generating additional off-axis modes. Therefore, if the weaker modes located between the stronger modes become more intense, the power densities in the most intense regions of the focal spot may not increase significantly, although the average power density of the overall focal spot may increase. The appearance of additional off-axis modes increases the divergence angle of the beam and increases the spot size, so that even the average power density of the focal spot may not increase as much as expected, if at all.

Many methods may be used to reduce the number of off-axis modes in pulsed lasers, and thereby improve their focal properties (67–70). Improved selection occurs if the distance between the cavity mirrors is increased or the cross section of the cavity is decreased. An important variable is the Fresnel number, which is equal to $a^2/L\lambda$, where a is the cavity radius, L is the length of the cavity, and the units of all quantities are the same (e.g., centimeters). For good axial mode selection, the Fresnel number should be less than 10. The divergence of a plane mirror ruby laser was reduced to 0.3 mrad, an almost diffraction-limited beam, by increasing the cavity length to 5 m and adding a limiting aperture of 3 mm diameter (71). This length might be made more practical by folding the cavity with mirrors into five 1-m lengths or ten 0.5-m lengths if the mirrors did not introduce additional aberrations.

Mode selection is also improved by locating a small aperture at the position where the laser beam has been narrowed down by a lens or spherical mirror system. A good example is the placement of two confocal lenses in a plane mirror cavity, Fig. 19.3. The aperture is placed at the common focal plane between the lenses, where the beam is very narrow. The best size of the aperture can be estimated from calculations (72) that show for an ideal laser cavity that the best rejection of the TEM_{10} mode (the strongest other than the desired TEM_{00} mode) occurs when the radius of the aperture is $0.27L'\lambda/a$, $0.30L'\lambda/a$, $0.32L'\lambda/a$, and $0.35L'\lambda/a$, respectively, for cavity Fresnel numbers of 2.5, 5, 10, and 20. The distance L' between the confocal lenses is used to calculate the Fresnel number in this case. Smaller than optimum apertures cause poorer selection and greater diffraction losses for the axial modes. Larger apertures rapidly cause poorer rejection of the off-axis mode, and when the aperture becomes 30% larger than optimum, the rejection is worse than if no aperture were used at all. With proper choice of the aperture size and careful alignment, this arrangement can select specific modes even with laser materials that have relatively poor homogeneity. *Although the total power output of a laser may be reduced by mode selecting devices, the power per mode may be increased several orders of magnitude, resulting in much higher irradiances.*

The power density at the aperture must not be high enough to cause break-

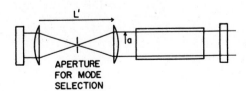

APERTURE
FOR MODE
SELECTION

Fig. 19.3. Diagram showing how two lenses and a small aperture hole are placed within a laser cavity to select the TEM_{00} mode by blocking higher-order modes.

down of the atmosphere or cause deterioration of the aperture material. Apertures made from diamonds are most resistant to destruction and are commercially available. The lenses should be corrected for spherical aberration, and their surfaces should be of the same quality as the laser cavity mirror surfaces (typically deviations of less than 0.1 wavelength from an ideal surface for a solid-state laser).

Even for an ideal laser beam, the irradiance pattern at the focal region is distorted when spherical aberrations are present in the focusing system. Interference effects occur because of the phase differences between various parts of the wave front arriving at the focal region. Spherical aberrations cause the interference patterns, and therefore the irradiance pattern, to be highly complex. A measure of spherical aberration is the maximum displacement D, expressed in units of wavelength, that occurs between the ideal and real wave fronts (66). For a plano-convex lens with its flat surface toward the focal region for minimum spherical aberration, the displacement varies as the fourth power of the radius from the optical axis and inversely as the third power of the focal length of the lens. The maximum displacement D_{max} occurs for rays farthest from the optical axis. Values for the focal length of a lens with a refractive index of 1.5 that will ensure a value of $D_{max} < 1$ wavelength for ruby and Nd-glass lasers are shown in Table 19.1.

Maps showing isophotes (lines of constant irradiance) have been published for calculated irradiance patterns for varying magnitudes of spherical aberration (74). An example is shown in Fig. 19.4. The size of the focal region increases in size as the aberration increases. The irradiance distribution along the optical axis (75) has been calculated and photographed for spherical aberration displacements from 0.3 to 164 wavelengths. The spherical aberration was increased by increasing the diameter of the beam. The maximum irradiance increases as the beam diameter increases until $D_{max} = 2$. For $D_{max} > 2$, the maximum irradiance does not change, but the focal region extends farther along the optical axis in the direction of the lens.

Table 19.1. Minimum Focal Length, for a Lens with $n = 1.5$, Required to Give a Spherical Aberration Displacement of Less Than One Wavelength at the Focal Plane[a]

Beam Diameter (cm)	Minimum Focal Length (cm)	
	$\lambda = 694$ nm	$\lambda = 441.06$ nm
0.2	0.85	0.75
0.5	3.0	2.5
1.0	7.2	6.3

[a] From ref 73.

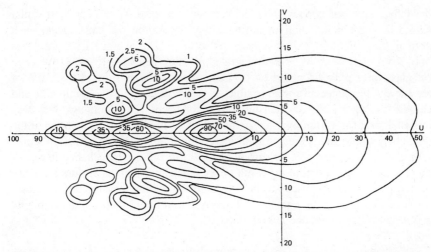

Fig. 19.4. Lines of constant intensity in the focal region of a $f/5$ simple lens with spherical faces. The intensity is normalized to 100% at the principal focal point. The axes are expressed in universal variables for light of wavelength λ, and a lens with a diameter to focal length ratio of a/f. The distances in real space, z, along the optical axis and, y, perpendicular to the optical axis are related to the plotted coordinates by the transformations $U = (2\pi/\lambda)(a/f)^2z$ and $V = (2\pi/\lambda)(a/f)y$; reproduced from ref. 74c by permission.

3. PRODUCTION OF AN ANALYTICAL PLASMA

The focused laser beam of a laser microprobe not only vaporizes and atomizes a selected region of the sample but its energy is also absorbed by the sample and atmospheric plasma that forms above the surface of the sample, thereby helping to maintain and excite the analyte atoms. This section considers the processes that occur when a pulsed laser beam interacts with an analyte sample and resulting plasma.

3.1. Direct Laser Interactions with the Sample

3.1.1. Reflection

Reflection of part of the energy of an incident laser beam is an important consideration in determining the fraction of laser energy absorbed by a sample, especially at low irradiances, during the beginning a laser pulse. For metals a model based on the Drude–Zener free-electron theory predicts a linear decrease in reflectivity of the metal as its temperature increases (76). Consequently, more laser energy would be expected to be absorbed by a hot, laser-heated metal surface than that predicted using room temperature reflectivity. Experimentally, the reflectivities of copper and steel samples decrease to 25% of their initial values during the first half of a single 1-μs relaxation oscillation spike in a normal (free-running) Nd-glass laser pulse (77). The average irradiance was only 7.5×10^3 W cm^{-2}. Time-integrated measurements for 15-ns, Q-switched Nd-glass laser pulses focused on tin and copper samples showed that 50% of the energy was reflected for an irradiance of 10^8 W cm^{-2} (78). For ebonite and carbon less than 20% of the energy was reflected. For aluminum 80% was reflected, and for Teflon over 95% was reflected. In all cases the fraction of the total energy reflected decreased with increasing irradiance, and at 10^{10} W cm^{-2} ranged from less than 5% for carbon to 25% for Teflon. The change in reflectivity may be due in part to the result of phase changes that occur during intense heating, which will be discussed below. In any case these measurements indicate that laser energy can be coupled effectively into a target that is initially highly reflective, if the irradiance is high enough.

3.1.2. Photoelectrons

Before the laser irradiance reaches the level where a sample begins to vaporize, photoelectrons may be ejected by multiphoton absorptions. For Q-switched pulses the affect may be relatively weak compared to thermionic emission of electrons, to be discussed shortly, but the photoelectric effect is assumed to play an important role in the generation of the plume plasma using a 5-ns, 5-mJ, 10-Hz N$_2$ laser (79). Photoelectric electrons are coincident with the laser pulse and do not involve a thermally related delay. Photoelectrons are emitted by the simultaneous absorption of n photons when the photoelectric work function of the sample is less than n times the energy of one laser photon. A Nd-glass laser has 1.17-eV photons, a ruby laser 1.78-eV photons, a Rhodamine 6G dye laser about 2.1-eV photons, and a N$_2$ laser 3.68-eV photons. The work functions of most elements are less than 5 eV, so that only a few photons are needed for each electron. Two- and three-photon photoelectric effects have been observed and studied (80–84). The electron current density increases as the nth power of the irradiance. Irradiances as low as 10 W cm^{-2} produce measurable currents with electron multiplier amplification (80). Irradiances up to 0.4 MW cm^{-2} produce distinguishable two-photon photoelectrons from a Au sample with ruby laser pulses having a 20-ns half-width (85). At higher irradiances the electron pulse is delayed by several nanoseconds and is produced mainly by thermionic emission.

3.1.3. Thermionic Emission

Thermionic emission can be produced by free-running laser pulses as well as Q-switched pulses at irradiances lower than those that destroy the surface. The only requirement is that the surface temperature be raised sufficiently to allow significant escape of thermal electrons. The classical Richardson equation relates the current density to the temperature of a surface:

$$j = AT^2\exp\frac{-\phi}{kT} \tag{19.3}$$

where j is the current density (A cm^{-2}), ϕ the work function in electron volts, T the absolute temperature in Kelvin, k is Boltzmann's constant (in eV K^{-1}), and A is a proportionality constant equal to 60 A cm^{-2} for many metals (65, p. 132). Current densities obtained from calculated surface temperatures caused by free-running (86) and Q-switched (87) laser pulses agree quite well with measured current densities for irradiances low enough to prevent surface destruction.

Thermionic emission of positive ions has been observed at irradiances low enough to prevent visible cratering of the surface (88–93) and comparable to irradiances used to observe thermionic electrons. The ion density is given by the Richardson–Smith equation (94), identical in form to Eq. (19.3) with ϕ replaced by the positive ion work function. Observed ions include those from species that were absorbed on the surface, as well as those from the bulk material (92, 93). The largest signals from a W surface were caused by Na$^+$ and K$^+$, and included H$^+$, C$^+$, H$_2$O$^+$, CO$^+$, CO$_2{}^+$, and W$^+$ (93).

The ejection of neutral molecules, including desorbed gases, has also been observed at laser irradiances low enough to prevent visible cratering (95–97). Thermionic emissions of ions and molecules are related by the Langmuir–Saha equation (98)

$$\frac{j_+}{j_0} = \frac{g_+}{g_0}\exp\frac{\phi - I}{kT} \tag{19.4}$$

provided equilibrium conditions exist about the surface (a condition that may not be reached during laser interactions). In Eq. (19.4), j_+ and j_0 are the positive ion and neutral particle fluxes leaving the surface at temperature T, g_+ and g_0 are the statistical weights of the ionic and neutral states, and I and ϕ are the ionization potential and electron work functions, respectively.

Thermionic ejection of electrons, ions, and neutral molecules show that important species are ejected from the surface of the sample prior to the onset of melting and major vaporization of the sample material during a laser pulse. These species add to the complexity of the laser interaction with the surface. For instance, as the free electrons absorb more laser energy (inverse bremsstrahlung) they may become energetic enough to initiate a breakdown of the gas above the sample surface. A breakdown at a sufficiently high irradiance then initiates a radiation-supported atmospheric shock wave, which precedes the expansion into the atmosphere of the major portion of the vaporized sample (99, 100).

3.1.4. Sample Heating

Vaporization and atomization of a major portion of the sample material is preceded by heating of the sample, which leads to phase changes and ejection of material away from the surface. Heating of conducting solids by laser pulses, even as short as Q-switched pulses, is essentially thermal in nature. Laser photons are absorbed by electrons in the conduction band of the material, and the energy is given up by collisions with other electrons and lattice phonons, that is, converted into heat. The time between collisions is on the order of 10^{-13} s, which is a very short period of time compared to a 10^{-8}-s Q-switched laser pulse. Consequently, the absorbed laser energy is effectively transferred instantaneously to the solid as heat. Heat transfer equations have been used to predict the temperatures at various depths in the sample for cases where a phase change does not occur in the material (65, pp. 68, 102). The irradiances for such studies (typically below 10^7 W cm^{-2}) are below those used in a laser microprobe, but the results are recognized as limiting cases and are useful in helping to understand the extent of thermal conduction that occurs during Q-switched and free-running laser pulses for higher irradiances.

The absorption coefficient for metals is high enough (10^5–10^6 cm^{-1}) that the laser energy can be considered to be deposited at the surface, and then transferred into the metal by conduction. For a Q-switched laser pulse with 20-ns width at half height and a peak power density sufficient to cause an absorbed photon flux density of 1.6×10^7 W cm^{-2}, the predicted temperature rise at 10 ns is predicted to be 330 K at the surface of a copper sample (101). The temperature rise is almost the same down to a depth of 10^{-4} cm, and then drops rapidly to half its surface value at twice that depth. The surface temperature is near its maximum value of 500 K by 20 ns. The temperature rise never exceeds 250 K (half the maximum surface value) below a depth of 4×10^{-4} cm. The maximum temperature below a depth of 3×10^{-4} cm occurs after the laser pulse has ended.

Therefore, the highest temperature rises caused by the heat deposited by a Q-switched laser pulse in a surface layer with a thickness on the order of 10^{-5}–10^{-6} cm (for absorption coefficients of 10^5–10^6 cm^{-1}) reach to a depth on the order of only 10^{-4} cm. This region is relatively shallow even for a focal spot diameter as small as 10 μm (10^{-3} cm). Large temperature rises caused by heat conduction parallel to the surface of the material also reach out to about 10^{-4} cm from the edge of the focal spot. The total affected area is therefore not much larger in diameter than the focal spot itself; heat conduction parallel to the surface appears to be relatively insignificant.

Longer laser pulses, such as free-running laser pulses lasting on the order of

10^{-3} s, allow greater depths to be reached by the highest temperatures, and conduction in directions parallel to the surface becomes significant (102). Depths of the same magnitude as the diameter of the focused laser beam reach significant temperatures (65, p. 78).

3.1.5. Phase Changes

When the temperatures are high enough to cause a phase change in the material, some of the incident energy is used to produce the phase change. A one-dimensional model that includes surface heating and the presence of a solid–liquid boundary is presented in Masters (103). The thermal conductivity and specific heats of the fluid phase differ from those of the solid phase. For metals, the thermal conductivity of the liquid phase is only about half that of the solid (104). The results of using some of the incident energy as latent heat to produce a phase change, and a reduced thermal conductivity of the new phase, are shallower penetration of the highest temperatures into the material and a higher concentration of the energy nearer the surface. A case that assumes no phase change, therefore, provides a limiting case for the penetration that occurs.

When temperatures near the boiling point are reached, a significant part of the material begins to vaporize. For a metal the latent heat of vaporization now consumes even more of the incident energy than the latent heat of fusion. For a free-running laser pulse with relatively low average laser flux densities, the amount of material vaporized depends more on the thermal conductivity of the material than the latent heat of vaporization (105). For materials with lower thermal conductivities, the absorbed heat remains nearer the focal region, causing higher temperatures to be reached and greater amounts of material to be vaporized, than for materials that conduct heat away more quickly. At higher laser flux densities, the heat is supplied too quickly to be conducted away; a larger fraction of the heat is used to vaporize the material and the latent heat of vaporization becomes a more important factor in determining the amount of material vaporized during a laser pulse. Experimental results (105) show less Cu than Type 18-8 stainless steel vaporized by a free-running laser pulse at low laser output, while more Cu than stainless steel is vaporized at higher laser energies. The latent heat of vaporization begins to become more important than thermal conductivity when the absorbed flux density F exceeds the value given approximately by

$$F > 2 L_v \rho k^{1/2} t^{1/2} \tag{19.5}$$

where L_v is the latent heat of vaporization per unit mass, ρ is the density of the material, k is the thermal diffusivity, and t is the laser pulse length (65, p. 100). The heat capacity is an important variable in determining the time t_v that it takes the surface to reach the vaporization temperature, or boiling point. This time can be estimated from a one-dimensional heat flow equation (106), assuming a constant laser flux density F absorbed at the surface:

$$t_v = \frac{\pi K \rho c}{4F^2} \Delta T \tag{19.6}$$

In this equation K is the thermal conductivity, c the heat capacity per unit mass, and ΔT the temperature rise necessary to reach the vaporization temperature. After the material reaches a state of steady vaporization, the rate of vaporization V (g cm^{-2} s^{-1}) can be estimated by

$$V = \frac{F}{c \Delta T + L_v} \tag{19.7}$$

The first term in the denominator represents the heat used to raise the temperature of an amount of the material to the vaporization temperature, and the second term the heat required to vaporize that amount. Experimental results for a 0.7-ms free-running laser pulse with average energy densities in the range of 2–16 kJ cm^{-2} show a range of more than three decades in the amount of material vaporized from various metal samples (107), the amount varying with the time it takes to reach the vaporization temperature, as well as the rate of vaporization.

3.1.6. Other Variables

Other variables besides sample reflectivity, thermal conductivity, vaporization temperature, heat capacity, and latent heats of fusion and vaporization are important in determining the total amount of material removed from the sample. For instance, particles of molten metal can be photographed as they are ejected from the focal region, and cooled deposits are commonly observed at some distance from the resulting crater. The details of the ejection of molten material are difficult to formulate into a quantitative theoretical description. Microcraters have been observed within an area that was shielded from a laser beam by a 0.1-mm steel blade in contact with a steel sample (108). The microcraters, observed up to 150 μm from the main crater, were apparently caused by volcanic-type ejection from superheated centers in material heated by thermal conduction. Pressure may build up in the central region of the main crater because of thermal expansion, internal boiling, or recoil (109) caused by rapid vaporization. The fluctuations in temperature caused by the temporal and spatial mode structure of the beam add instability to the dynamics of the fluid and add to the difficulty in developing a quantitative theory.

When the incident flux density reaches values that can be obtained by Q-switched or pulsed dye lasers (in excess of 10^{-8} W cm^{-2}), the ejected vapor, which becomes ionized, begins to absorb some of the incident laser energy. At atmospheric pressure, an atmospheric shock wave may precede the ejected material, absorbing a large fraction of the laser energy before the energy reaches the sample material (99, 100). Expansion of the hot sample plasma across the surface causes additional vaporization, and increases the diameter of the sample spot. The energy density of the surface plasma is high enough that reradiation of energy to the surface is significant.

The flux density can be so high that the vaporized material may not be able to escape faster than new material is vaporized. This results in a higher pressure and an increase in the effective vaporization temperature. One theoretical model (110) predicts that these effects become important for absorbed flux densities above 10^8 W cm^{-2}. In another study, which attempted to predict the amount of material ejected by a Q-switched laser pulse, it was assumed that the material reached the pressure and temperature of its critical point before vaporization occurred (101). Measured crater depths for 5 metals agreed within a factor of two of the calculated depth reached by the critical temperature for this model. It was suggested that most of the material below the depth reached by the critical temperature did not vaporize because of the high pressures exerted upon it by the vaporizing material above it. The details of the processes that actually occur in any given case remain in question, however, because other experiments using graphite (111) indicate that little or no such superheating occurs. Evidence for the existence of high pressure, however, has been indicated by surface displacements around the edges of craters in Teflon (2).

The amount of material vaporized with a Q-switched or dye laser pulse is up to several orders of magnitude less than for a longer free-running laser pulse of the same energy (43, 101). This is because of the much shorter time allowed for conduction of heat into the sample; the energy for the shorter pulses is therefore expended in processes near the surface rather than in vaporizing material at greater depths. Crater depths range from 1 to 10 μm for a Q-switched pulse, and the amount of material ejected (although it may not all be vaporized) is generally at and below the microgram level (100). The craters in metals show a splash pattern that indicates that molten material is present (2, 99, 112). Measurements of crater cross sections show that a large fraction of the molten material is redeposited as a rim around the crater, extending above the original surface (2). Measurements using alloy samples suggest that there is even time during a Q-switched pulse for unequal partitioning of the elements between the vapor and liquid phases (100, 112). The need to obtain reproducible partitioning makes control of the temporal shape and irradiance of the laser pulse even more critical in obtaining reproducible analytical results.

3.2. Plume Production

Attention will now be turned to the region above the surface of the sample into which the sample vapor is ejected. This region is called a plume, spark, or microplasma and is the region that provides analytically useful spectra.

3.2.1. Sample Plasma

Early in a laser pulse, electrons and ions have been released from the surface by photoelectric and thermionic processes. The free electrons above the surface gain energy by absorbing photons from the laser beam. Although this initial plasma may trigger the generation of more plasma when the sample is in a gaseous atmosphere, its influence may in fact be quickly overshadowed by a more massive plasma that is formed by thermal ionization of the vaporizing bulk material of the sample (113). For a free-running laser pulse, the sample material at the surface is not heated much above its boiling temperature. Observed velocities of the vaporizing material are on the order of 10^4 cm^{-1}, consistent with velocities of atoms at normal boiling temperatures (114, 115). However, high-speed framing camera studies (116) indicate that an observable plasma forms near the surface of an aluminum sample prior to the major sample ejection steps, even for a free-running laser pulse with irradiances of 7–8 MW cm^{-2}. At the much higher irradiances caused by a Q-switched pulse, the velocities of the sample species reach above 10^6 cm s^{-1}, corresponding to particles with energies far above their normal boiling temperatures (117, 118). X-rays and many highly ionized species have been spectrally observed, indicating that energies of ten to hundreds of electron volts are reached (119–128). Interferometric measurements (129, 130) indicate the presence of high concentrations of electrons, which can absorb photon energy by the inverse bremsstrahlung process (a free-free transition of an electron in the field of an ion). Streak camera measurements (131) show that the velocity of the luminous front of the plasma material in a vacuum actually increases during a Q-switched pulse, indicating that the plasma gains its energy from the laser beam. Individual particles of high energy apparently result from the highly energetic plasma generated by inverse bremsstrahlung absorption of the high power density of the laser beam.

3.2.2. Atmospheric Plasma and Shock Waves

When the sample is immersed in an atmosphere, an atmospheric plasma may occur prior to a plasma of sample material. For Q-switched pulses an atmospheric plasma is a major spectral phenomenon and propagates rapidly through the atmosphere along the direction of the incident laser beam. Time-resolved spectral observations show that the atmospheric plasma moves ahead of the majority of the sample material (99). When the beam strikes the surface at an angle, the atmospheric plasma also propagates in the direction of the reflected laser beam, while the sample vapor is ejected perpendicular to the surface of the sample.

The front of the atmospheric plasma region propagates with supersonic velocity through the atmosphere, indicating the presence of a shock wave (100). When a disturbance of this type moves with supersonic velocity, gas dynamic theory predicts that a wave front forms. The wave front is a discontinuity, ahead of which is found undisturbed gas and immediately behind which there is a region of relatively high mass density, energy density, and momentum, compared to the undisturbed gas. This region is energetic enough to be a luminous plasma, which emits an undesirable spectral continuum background. Farther behind the wave front the mass density becomes less than in the undisturbed gas. The wave front and its energetic discontinuities propagate through the gas, and there is little if any net flow of the gas ahead of the wave front. The shock wave leaves behind it a hot region into which some of the sample vapor enters.

The energetic shock wave derives the power needed to maintain and propagate it by absorbing photons from the laser beam via inverse bremsstrahlung in the plasma region behind its wave front. The absorption of laser energy near the front of the rapidly moving (10^7 cm s^{-1}), radiation-supported shock wave can prevent a large fraction of the laser energy from reaching the sample surface that lies behind the shock wave (99, 100). The influence of the radiation-supported shock wave can be reduced by reducing the atmospheric pressure (100). At atmospheric pressure the influence can be reduced by reducing the instantaneous irradiance (and thereby the power density input) below the level necessary to sustain the shock wave.

The laser wavelength also has a significant influence on the shock wave. The inverse bremsstrahlung absorption coefficient decreases as the square of the wavelength of the absorbed photons for high temperatures (132) and approximately as the cube of the wavelength for low temperatures (133). The absorption coefficient for a dye laser at 0.56 μm is less than for the ruby (0.6943 μm), Nd-glass (1.06 μm), or CO_2 (10.6 μm) lasers. A flashlamp-pumped Rhodamine 6G dye laser with its shorter wavelength and longer (1 μs) pulses of 100 mJ essentially eliminates the highly absorbing radiation-supported shock wave at atmospheric pressure caused by a Q-switched Nd-glass laser pulse with the same energy (134).

3.2.3. Fluid Dynamic Effects

Ejection of the sample material into the plume involves fluid dynamic processes in addition to simple vaporization. High-speed framing camera studies (116) of the plume caused by a free-running laser pulse focused on an aluminum sample show the ejection of particles and what is apparently a cylindrical sheath of molten metal that rises from the circumference of the crater. After the beginning of the laser pulse, a heated spot appears on the sample surface, followed by formation of a plasma that initially expands with a velocity of about 10^4 cm s^{-1}. This event is followed by a fine spray of particles having a velocity of 4 $\times$ 10^{-3} cm s^{-1}. Above 0.6 mm, horizontal striate layers of high intensity are formed. About 375 μs after the laser pulse begins, the cylindrical sheath is noticed. As it rises above the surface with a velocity of 10^3 cm s^{-1} it contracts to a diameter of 0.2 mm and interrupts the incident laser beam at a height of 1.6 mm above the surface. Absorption of the laser beam energy causes the molten metal to disintegrate into an excited plasma and particles in chaotic motion. The sheath then becomes detached from the surface in a spiral motion and is soon followed by another sheath that rises from the surface. These events demonstrate the complex and rather chaotic nature of the formation of a plume for a typically long free-running laser pulse having many spikes.

Similar photographs taken by the same authors for a 25–30-μs multispike Q-switched pulse show the formation of a plasma that initially expands with horizontal and vertical velocities of 5 $\times$ 10^4 and 3 $\times$ 10^4 cm s^{-1}, respectively. The average irradiance caused by each individual spike was 55 MW cm^{-2}. Expansion of the plasma across the surface leads to its extension over the nearby edge and somewhat down the side of the sample. During the laser pulse, a tapered cone of plasma in the center of the plume is more intense than the surrounding region. After the pulse the plume intensity weakens, becomes more heterogeneous, and divides into a surface component and an elevated component. For a 50-ns single-spike Q-switched laser pulse with an irradiance of 4200 MW cm^{-2}, a spherically shaped plasma is observed to expand at an initial rate greater than 10^6 cm s^{-1}. The plasma is visible for about 20 μs and "during this period it remained relatively motionless, resembling a nebula suspended above the surface of the sample." No particle ejection was observed.

A dye laser with a 0.1-J pulse in 800 ns produces from a sodium sample a plume of absorbing sodium atoms that has a mushroom shape, which breaks up into several ball-shaped regions as it moves away from the surface (135). The absorbing atoms were detected by atomic absorption by forming the plume between an expanded laser source and a vidicon. The atoms are visible for over 0.8 ms.

Time-integrated photographs of Q-switched laser plumes from metal samples in our laboratory have at times shown streaks that are attributed to ejected hot particles. Craters (2, 21, 99) and adjacent areas show streaks that appear to be caused by ejected particles or streams of molten metal, Fig. 19.5. Significant differences in the appearances of craters are caused by changes in focal conditions (2, 46). Apparently the extent of particle ejection depends in part on the

 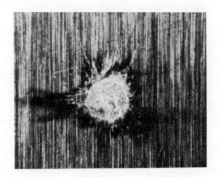

a.   **b.**

Fig. 19.5. Photomicrographs of craters formed in an aluminum alloy by (a) a single-spike Q-switched laser pulse and (b) a double-spike Q-switched laser pulse, showing streaks apparently caused by ejected particles or streams of molten metal. Dark stains may be caused by oxides or nitrides; reproduced from ref. (99) with permission.

ways that the fluid dynamic forces are influenced by the spatial pattern of the irradiance at the focal spot (as well as the temporal shape of the laser beam) as it differs from shot to shot and from one laser instrument to the next. Even on the short submicrosecond time scale of a Q-switched pulse, it appears that macroscopic hydrodynamic forces are at work along with submicroscopic thermal vaporization processes to determine the way in which the sample material enters the plume.

3.3. Plume Spectroscopy

Spectral properties of a laser plume help to determine detection limits, spectral interferences, and the shape of the analytical curve. Radiation from a free-running laser plume consists of pulses that are related to individual laser spikes (27, 116, 136). Resonance lines are often self-reversed, indicating the presence of a central region of excited atoms surrounded by a region of relatively unexcited atoms that absorbs almost all of the photons in the center of a spectral emission line profile. The line emission is accompanied by a continuum background emission that also may be absorbed at the spectral line centers by the relatively unexcited outer regions of the plume.

3.3.1. Background Continuum

For a free-running laser pulse with an irradiance per spike averaging 7–8 MW cm^{-2}, background continuum is concentrated in the center of the plume (116) and corresponds in time to individual laser spikes. The line intensity does not increase to the same extent for similar laser spikes, probably because of non-uniform absorption of the laser spikes in the rapidly changing plume. All resonance lines are broad and asymmetrically self-reversed because of spectral line shifts that accompany pressure broadening. After about 400 μs, emission terminates even though the laser spikes continue. The termination occurs about the same time as a molten metal sheath (described in Section 3.2.3.) rises above the surface and contracts to block the laser beam from the sample surface.

Q-switched and multispike Q-switched laser pulses produce a more intense continuum background than lower irradiance free-running laser pulses because more of the laser energy is used to heat the plasma above the sample surface, and relatively less goes to vaporize the sample material. Continuum is emitted primarily from two locations, one near the crater, and the other from a region extending several millimeters above the surface in the direction of the laser beam. At low atmospheric pressures, the crater region is the dominant source of continuum (100), indicating that the source of emission from this region is a plasma of sample material. Spatial masking of the continuum background allows the line-to-background ratio to be improved because line emission from sample material species occurs from a larger area than the continuum (100, 137). Another way to spatially improve the line-to-background ratio is to incline the laser beam at an angle to the surface (99). Photographs of the plume when the surface is at a 45° angle to the beam show an atmospheric continuum in the direction of the incident and reflected beams, while the sample material appears to be ejected approximately perpendicular to the surface. The region perpendicular to the surface would be expected to contain most of the sample species, and observation of this region would produce a better line-to-background ratio than observations of a region to one side of a beam that was perpendicular to the surface.

3.3.2. Time-Resolved Spectra

Time resolution improves the line-to-background ratio (138, 139). The most intense part of the continuum lasts for a relatively short time, only during the laser pulse and several tenths of a microsecond later (99, 100). Line emission from sample material lasts for several microseconds after the continuum emission (2, 99, 100, 116). Time-resolved observations (116) of Al (I), Al (II), and Al (III) lines show that the peak intensity of the doubly ionized atom (III) is reached sooner (0.3 μs after the laser pulse) than the peak intensity of the singly ionized atom (1.8 μs). This suggests that recombination of ions and electrons plays an important role in the production of excited neutral atoms. Time-resolved studies (99) using multispike Q-switched pulses show an intensification of Al (I), Al (II), and Al (III) lines 0.6 mm above the surface when the second laser spike reheats the plasma. However, at 1.2 mm above the surface only the ion lines are intensified during the second laser spike, indicating a relative absence of neutral Al atoms. It seems unlikely that recombination alone would produce so many neutral atoms at 0.6 mm above the surface and relatively few if any at 1.2 mm. The difference is due more likely to the later arrival of atoms from the sample. Reheating of the plasma by the second laser spike also causes intensification of the 399.5-nm N (II) line at 1.2 mm height, but not at 0.6 mm, indicating that the atmospheric species are being replaced by the Al atoms. Time and spatially resolved fluorescence observations also indicate that the rising plume material first replaces the atmospheric species and then mixes with them (140).

3.3.3. Atmospheric Influences

The spectral characteristics of the plume are influenced by density changes of the atmosphere in which the sample is immersed (79, 136). Spatially resolved spectra of a Q-switched laser plume (100) at lower pressures (10–100 torr) and room temperature show an increase by a factor of 2 to 3 in height of the region where line emission of sample species is observed. Similar results are obtained using a flashlamp-pumped dye laser pulse with the same energy (134). In addition to emission observations of excited atoms, spatially resolved atomic absorption and atomic fluorescence measurements (134, 135, 140–142) of ground-state atoms in dye laser plumes also show an increase in the expanse of the plume at lower atmospheric densities. Collisions with atmospheric species are therefore important in determining how far the sample vapor is transported before condensation or chemical recombination causes depletion of its free atoms.

The chemical composition of the atmosphere influences the free atom concentration of the sample species in the plume. However, time and spatially resolved fluorescence studies of Al and AlO in a laser plume formed from an aluminum alloy sample in an oxygen atmosphere show that a reaction between the aluminum atoms and oxygen takes place in the outer regions of the plume before complete mixing of the plume and the atmosphere has occurred (140). Mixing is still taking place after 85 μs at a pressure of 150 torr. Observations of the central region of the plume would minimize the chemical influences of the atmosphere upon analytical results.

3.3.4. Line Broadening and Shifting

Emission lines from the hottest regions of Q-switched laser plumes are usually relatively broad because of pressure and Doppler broadening. Line shifts accompany pressure broadening, and net Doppler shifts may also be present (2, 99). Self-absorption adds to observed broadening when the product of concentration and pathlength of absorbing species along the optical viewing path is high enough. Self-reversal is often present for resonance lines of major constituents because of their high concentration in the cooler region of the plume where there is a decreasing gradient of the fraction of excited species in the direction of the viewing system. Linewidths of several tenths of a nanometer are common when the region of observation is limited to the intense plasma regions near the surface and along the laser beam. Linewidths become narrower as the region of observation is moved slightly above the surface and off to one side of the axis of the laser beam (100). Lines become narrower with time, as the plasma cools (99, 120).

Narrower lines also result when the atmospheric pressure at room temperature is reduced (100, 136, 79). Time-resolved measurements at a pressure of several torr (134) and at 100 torr (142) show that linewidths late in the plume are comparable to the linewidths of hollow cathode lamps, where the main source of line broadening is only Doppler broadening.

3.3.5. Cross Excitation

Cross excitation of a laser plume with another excitation source, such as a high-voltage spark (1, 27), microwave discharge (6, 9, 143) or radio-frequency discharge (144), tends to eliminate self-reversal and commonly produces narrower lines (2). Experimentally the total line profile appears to consist of the line profile of the spark, superimposed upon the profile emitted during the early stages of the plume. Emission caused by cross excitation may be more intense and last much longer than emission from the plume alone, so that the final result is a spectrum that mainly has the characteristics of the cross-excitation discharge.

For cross excitation by a spark, a spark is usually struck between two pointed electrodes spaced from 0.2 to 2 mm apart and located a similar distance above the sample surface. A 1–25-μF capacitor charged to a voltage ranging from 0.4 to 4 kV is connected to the electrodes. A spark occurs when ions from the plume trigger the discharge. When a Q-switched laser pulse is used, the residual inductance of the capacitor and the circuit may be all that is used in order to keep the lifetime of the spark relatively short. It is undesirable to continue the spark after the sample species have left the observation region because emission from the spark alone only adds to the total background signal.

When a free-running laser pulse is used, an inductance of 30–1000 μH may be placed in series with the capacitor to extend the spark oscillation for several hundred microseconds. In this case the emission spectra generally increase and decrease with oscillations of the spark (145). An increase in capacitance increases the intensity of the spectrum, including unwanted cyanogen bands. An increase in inductance decreases the ratio of ion to atom line intensities. An increase in initial capacitor voltage increases the ratio of ion to atom line intensities.

Time-resolved spectral observations of a Q-switched laser plume cross excited with a relatively short spark (with unspecified temporal characteristics) show the onset of spark-intensified lines 0.5 μs following the laser pulse (2). The 358-nm Al (II) ion line dips to zero between plume excitation and spark reexcitation, and then reaches its maximum value at 2 μs. The 308-nm Al (I) line behaves as a continuation of the intense background continuum pulse and begins its drop toward zero intensity at 6 μs. The region where plume excitation ceases and spark excitation begins is not distinguishable. The 248-nm C line appears at about 0.4 μs, peaks at about 1.5 μs, and slowly decreases in intensity for more than 20 μs. Cyanogen bands around 388 nm show a rapid increase 0.2 μs after the appearance of the 248-nm C line. Several peaks are present at 4-μs intervals. The 399-nm N emission slowly rises after the laser is fired and then increases sharply at about 5 μs. Many peaks then occur about 1–1.5 μs apart. These results suggest that time resolution could be used to improve line-to-background intensity ratios. For sets of single-laser plumes, the relative standard deviations using photoelectric detection and a gated integrator were at best 4.3% and more often between 7 and 20% without cross excitation. With cross excitation the relative standard deviation rose to 25–45%.

The 40-ns half-width of a typical Q-switched laser pulse and the rapid changes in intensities that occur in less than a few tenths of a microsecond indicate that the laser and spark should be synchronized to each other to within a jitter of less than a few nanoseconds to obtain good reproducibility. The use of an externally triggered spark, synchronized to the laser has improved the results for some elements (146–148).

4. OBSERVING THE PLUME

4.1. Illumination of the Spectrometer/Spectrograph

A spectrometer or spectrograph with an efficient photon throughput is desirable when analyses are made with single-laser plumes (in contrast to superimposing the results from many shots). Consequently, a small $f/$ number (the ratio of focal length to aperture diameter) and efficient optical components are both required. The optical aperture of the spectrometer/spectrograph should be filled by the external optics that focus the plume onto the entrance slit; that is, the $f/$ number of the external optical system should be as small or smaller than the $f/$ number of the spectrometer/spectrograph. Of the various types of optical systems that have been used to illuminate the entrance slit, the over-and-under mirror system using two concave mirrors that fold the optical axis into a Z shape seems to be the best for photoelectric detection, where uniform illumination of the spectral image in the focal plane of the spectrometer is not particularly important. When mirrors are used, chromatic aberration is absent, an important consideration in multielement analysis where many wavelengths are used. The mirror position can be adjusted to compensate for coma and astigmatism of a spectrograph/spectrometer (149). The photon throughput of the two-surface mirror system is comparable to that of other mirror and lens systems.

When photographic recording is used for quantitative analysis, uniformly illuminated spectral images may be desired for easier determination of the optical densities of the images on the photographic emulsion. To obtain a uniformly illuminated slit image, other optical systems are used, which usually result in a decrease in photon throughput (150). The decrease is unimportant if there are enough photons from the background emission to make the background emission visible on the photographic emulsion.

Spatial discrimination may be desirable to increase the line-to-background ratios (99). With the over-and-under mirror system, a point source of light is focused to a vertical line image (sagittal focus) at the entrance slit of the spectrometer/spectrograph and to a horizontal line image (tangential focus) a short distance in front of the entrance slit (toward the light source). Consequently, the width of the vertical entrance slit defines the horizontal dimension of the region of the plume that is observed. The vertical dimension is determined by the width of a horizontal slit placed in front of the entrance slit at the tangential focus of the mirror system. When the external mirrors have been positioned to compensate for the astigmatism of a spectrograph, the vertical position along the spectral line image in the focal plane of the spectrograph corresponds to the vertical position in the plume. Spatially resolved information is therefore present in the spectral line images, and horizontal apertures can be used in the focal plane, or in the densitometer used to read the film, to obtain vertical spatial discrimination.

Alignment of the optical system is critical because of the relatively small few-millimeter size of the laser plume. Reverse illumination of the spectrometer/spectrograph provides a good way to check the alignment of the entire optical system. A diverging light source, such as a tungsten bulb, is placed in or near the focal plane of the spectrometer/spectrograph (in a region where green light would be diffracted) and positioned to fill the aperture of the spectrometer/spectrograph. The light coming from the entrance slit should be centered on the external mirrors. If any light falls off the edges of either mirror, then the mirror system is not fully illuminating the aperture of the spectrograph/spectrometer.

A He–Ne laser could be used to locate the optical axis. However, its beam would have to be made divergent (e.g., by scattering) to fill the aperture of the optical system, in order to be able to see if the external mirrors were completely filling the aperture of the spectrometer/spectrograph.

The sagittal focus located in front of the entrance slit of a reverse-illuminated spectrometer/spectrograph can be located by placing a horizontal slit in the focal plane and reverse illuminating the spectrometer/spectrograph through the horizontal slit. This locates the position where the tangential (horizontal) focus of the over-and-under external mirror system must be located to compensate for astigmatism. Reverse illumination through the horizontal slit in the focal plane of the spectrograph will produce a small square image at the point where the plume is ordinarily located when the external mirror system is properly aligned to compensate for astigmatism. The dimensions of the image will be defined by the entrance slit width and the width of the horizontal slit in the focal plane.

4.2. Photoelectric Detection

Photomultiplier tubes used to detect the spectral light pulses should have low-inductance capacitors of $0.1-1-\mu F$ capacitance connected across the four or five dynode resistors that are closest to the anode, to supply the relatively large signal current pulses without significantly changing the dynode voltages. Gated integrators can be used to discriminate against the initial background continuum pulse or oscillatory background signals that occur when an oscillating spark is used for cross excitation. Time gating can be done by pulsing the dynode chain of the photomultiplier tube, or by switching the charge from the photomultiplier anode to an integrating capacitor for the desired time interval (using field-effect transistors or integrated circuit analog gates). Alternatively, a boxcar integrator can sample the voltage signal across a resistor connected between the photomultiplier anode and ground or the voltage signal from a current-to-voltage converter operational amplifier circuit. If a *voltage* signal is produced and then sampled, the response time of the anode circuit caused by the resistor and any capacitance in parallel with the resistor must be two to three times less than the duration of the sampling gate to obtain the desired time resolution (151). For instance, if a 30-cm length of coaxial cable (capacitance 13–30 pF) were connected across the anode resistor, the resistance should be less than 10 kΩ to provide better than 1-μs resolution.

4.3. Atomic Absorption

Atoms in the plume can be observed by absorption as well as by emission. Flashlamps, pulsed hollow cathode lamps, and lasers can be used as the primary light source. Flashlamps require a high-resolution spectrograph/spectrometer, and time-resolved photographic (152) and photoelectric detection (153) have been used. Pulsed hollow cathode lamps with pulses longer than the plume duration can be used as primary light sources with low-resolution spectrometers (42, 45, 154). Figure 19.6 shows typical atomic absorption signals for four elements in a Q-switched laser plume. The pulse duration and repetition rate of the hollow cathode lamp must be chosen to minimize spectral line broadening (155–157). Guidelines for pulse control are presented in Piepmeier and de Galan (157).

Shot noise is a major source of noise. A low-pass RC filter can be used to reduce the signal fluctuations caused by shot noise, but a balance must be chosen between reducing fluctuations caused by shot noise and loss of temporal information about the changing atomic population (42). Noise limits the concentration range over which the peak value of absorbance can be used as the analytical signal. The concentration range for an analytical working curve can be increased to higher values by using the width of the transient absorption signal at half height as a measure of concentration (158).

Pulsed hollow cathode lamps with pulses much shorter than the duration of a laser plume can be used (134). When the lamp pulse is shorter than the time it takes for significant changes in absorption to occur, the photoelectric signal can be integrated over the entire lamp pulse without distorting the results (151). Short lamp pulses with higher currents can provide more photons than longer pulses to reduce shot noise. They also help to avoid spectral line broadening that occurs during longer pulses that allow the concentration of absorbing atoms to build up in the lamp (causing line broadening). Intense pulses as short as 1 μs have been generated (134).

A narrow-band tunable dye laser with 1-μs pulses has been used as a primary light source for atomic absorption measurements in a plume (135, 140, 142).

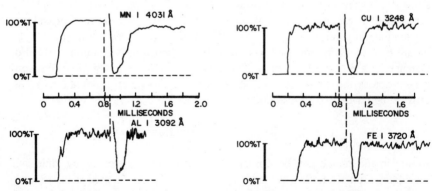

Fig. 19.6. Atomic absorption signals for four elements in a Q-switched laser plume using hollow cathode lamps as the primary sources. Fluctuations in the 100% T signal are caused by shot noise. The vertical dashed line is the time of laser firing; reproduced from ref. (42) by permission.

The intensity is high enough to make shot noise negligible, and a photodiode rather than a photomultiplier may be used as the detector.

4.4. Atomic Fluorescence

Atomic fluorescence with a tunable dye laser primary source has detected 10^9 atoms of Li in a sample of X-ray film (159) and 10^9 atoms of Cr in flour (160). By working in a vacuum (10^{-3}–10^{-4} torr), chemical matrix effects in powdered samples are minimized (160), and a linear relationship between fluorescence signal and concentration is obtained. This linearity is a significant characteristic of this method because, although emission and absorption log-log working curves are linear, their slopes are usually less than unity, indicating that the analytical signal is nonlinearly related to concentration.

5. ANALYTICAL APPLICATIONS

Laser microprobes have been used to determine over 60 elements in many types of samples. These include light elements such as lithium and boron that require special detectors when determined with an electron microprobe. Relative precision with solid samples is comparable to that of emission spectrochemical analysis with a dc arc (55, 161). Absolute detection limits (27, 42) are several orders of magnitude lower than for emission spectrochemical analysis with a dc arc or spark and approach those of atomic absorption with an electrothermal atomizer (graphite furnance). Relative detection limits of the dc arc technique are better by one or two orders of magnitude, although the amount of sample vaporized (10 mg) is typically 10^3 or 10^4 more than for a laser microprobe.

Sample preparation for a laser microprobe is usually relatively simple. A considerable amount of time can often be saved since in many cases no preparation is needed. Therefore, errors caused by loss of a constituent or contamination with foreign elements from reagents, mortars, and auxiliary electrodes (unless spark cross excitation is used) can be avoided. Smaller samples are mounted on a microscope stage and very carefully positioned at the focal spot of the microprobe. Large samples, such as antique ceramics, may require replacing the standard microscope stage with a suitable stand. The stand should have micrometer adjustments to make it easy to position selected submillimeter regions of the sample at the focal spot.

Focal spots less than a micrometer in diameter (approaching the diffraction limit) can be obtained with a single-mode laser and a corrected short focal length lens system. However, relatively low energy pulses are required to restrict the diameter of the sampled region to this range (25) since diffraction patterns may also cause sampling outside the central focal region (58). Typically, the sampled region is in the 30–300-μm range. When using spark cross excitation to increase the emission signal, care must be exercised because the spark may hit the sample surface, further expanding the sampled region (25, 27, 34, 162). A thin plastic mask is used to minimize spark sampling of this type (32).

5.1. Matrix Effects

As mentioned in Section 3.1, thermal properties of the sample play important roles in determining the amount of material vaporized. For instance, in one study low-boiling sample materials produced emission signals too weak to accurately measure unless spark cross excitation was used to enhance the signal (41). The low signals are attributed to the laser pulse producing mainly small droplets and only a relatively few free atoms for the low-boiling materials. Attempts to correlate simple mathematical functions of thermal properties of the sample material with crater volumes have produced only fair correlations (2, 163). The amount of material sampled also depends on other properties of the material discussed in Section 3.1.

Consequently, changes in matrix effects (changes in the analytical signal caused by changing from one type of sample material, or matrix, to another) must be considered when trying quantitatively to determine amounts or concentrations of elements in a sample. An accurate way to compensate for matrix effects is to prepare an analytical working curve from a series of standard samples that are identical to the unknown sample except for the concentration of the analyte. The analyte concentrations of the standards should fall on both sides of the analyte concentration in the unknown sample. The unknown concentration is found by interpolation. (Extrapolation away from either end of a working curve is dangerous because the shape of the curve is not known in the unmeasured regions.) Ideally, one of the standards should be identical to the sample, but this is not generally practical for solid samples.

Quite often standards may not be available, and the preparation of accurate standards for a special type of sample is usually a difficult and time-consuming task. Studies have been made to find ways to partially compensate for matrix effects. In one study (46) the crater volumes were measured optically with a metallograph, a process requiring a few minutes for each crater. A log-log plot of spectral line relative intensities versus amount of analyte lost from the sample (calculated from the crater volume and known analyte concentration) produced a line with a correlation coefficient of 0.98 (unity is ideal) for copper in widely differing alloys, with base materials of Al, Pb, Zn, Sn, and Fe. Crater volumes ranged over two orders of magnitude. For a dc arc analysis using the same samples, the points were more widely scattered, producing a correlation coefficient of only 0.84. Therefore, the matrix effect was less for the laser technique than for the dc arc. Although considerable improvement was obtained by measuring crater volumes, the residual point scatter of the log-log plot showed that complete compensation for matrix effects could not be achieved by normalizing the intensities to the crater volumes.

5.2. Internal Reference

The internal reference (or internal standard) technique can be used to help compensate for some (but not all) types of matrix effects as well as changes in other experimental variables. To use this method, the analyte line intensity (or other quantity proportional to concentration) is replaced by the *ratio* of the analyte line intensity to an appropriate internal reference line intensity. Assuming that the intensities of the analyte and the internal standard lines are influenced by exactly the same factor when an experimental condition (other than analyte concentration) is changed, their ratio remains constant and dependent only on analyte concentration. The success of the internal reference technique depends on how well the experimental results adhere to the underlying assumption. In some cases the internal reference method could produce poorer results than if it were not used at all. For solid samples the main matrix material is usually used as the internal reference since its concentration is usually known and often changes relatively little from one sample to the next. The internal reference method is widely used, especially in emission spectrochemical methods of analysis to compensate for such things as dc arc wander, minor temperature fluctuations, and sample flow rates in solution techniques. It is also commonly used in laser microprobe analysis when a proper internal reference of known concentration can be identified.

Spectral lines for the internal reference are usually chosen so that their upper energy levels are close to the upper energy levels of the analyte lines. This criteria assumes that the emitting vapor is close to spectral or thermal equilibrium. In a dynamic laser plume, however, this may not be the case, and processes other than thermal collisions (e.g., pumping from higher excited states) may dominate the excitation mechanisms for some lines. Therefore, an arbitrary internal reference line and an analyte line may not behave in the same manner when uncontrolled experimental conditions vary; compensation for matrix effects may not then occur.

The internal reference method was used to compensate for one matrix effect discovered while using low-index crystallographic plane surfaces of single-crystal samples and a multispike Q-switched laser microprobe (38). Crater volumes for single crystals were independent of crystal orientation for crystals with cubic symmetry. These results are expected since the reflectivity and thermal properties are isotropic in cubic crystals. However, for the same experiments, higher emission signals with better precision were observed when crystal planes of higher atomic packing densities were parallel to the surface. The masses ejected were the same for any orientation of a given metal. Momentum measurements of the ejected material showed that the momentum in the direction perpendicular to the surface was higher for crystal planes with higher atomic packing densities. Since the mass of the ejected material was the same for all crystal orientations, the higher momentum indicated a narrower plume with preferred velocities perpendicular to the surface. A narrower plume provides more atoms in the center of a cross-excitation spark, causing the higher emission intensities to be observed. The dependence of velocity direction upon crystal orientation may be partly caused by the anisotropy of the elasticity coefficients of cubic metals. Similar anisotropic effects are well known for other microprobe techniques.

For analytical purposes the anisotropic effect was reduced below the level of measurement precision for a Cu–Al alloy by using the internal reference technique. These results indicate that the intensities of the analyte and matrix (internal reference) lines were influenced to about the same extent by a change in crystal orientation. The results are important for other types of metallic samples because they behave as single crystals when the laser beam is focused only upon a single grain. Even when the metal grain size is much smaller than the focal spot, the metal usually has a texture where certain crystal plane orientations are preferred.

A study of the precision of laser energy, crater volume, and line intensities indicates that the internal reference method is useful for polycrystalline samples (32). A nonlinear relationship between laser energy and observed line intensities occurred for a multispike Q-switched laser that had a relative standard deviation in laser energy of 0.02. The relative standard deviations of the resulting crater volumes and line intensities were 0.07 and 0.12, respectively. This is attributed to a nonlinear dependence on laser energy of the amount of laser energy absorbed in the plume. The relative standard deviation of the *ratio* of pairs of line intensities was only 0.05, compared to 0.12 for the line intensities alone, indicating that the use of an internal standard would improve analytical results.

Another study showed that some line intensities decreased by a factor of 2 for an eightfold increase in grain size of the same material (44). In this case, however, the ratio of two lines did not improve the results, emphasizing that care must be exercised when choosing lines for the internal reference technique.

5.3. Powdered Samples

Matrix effects are reduced and the preparation of standards made easier by dispersing the sample into a fine powder or solution. When the sample is not a powder, making it into a powder compromises one of the advantages of the laser microprobe technique—ease of sample preparation and freedom from contamination. However, matrix effects were less when a laser microprobe with spark cross excitation was used than when a dc arc was used for the same powdered samples (55). For the laser microprobe technique, powders were hand packed rather than pressed into pellets. Matrix powders that were studied included BaF_2, Al_2O_3, Yb_2O_3, and WO_3. The samples were synthesized by thoroughly mixing and grinding an ethanol slurry of appropriate powders having particles sizes of 5 μm or less. After drying, the resulting cake was dry ground and then hand packed to a smooth surface for the laser microprobe. Unpacked powders scattered, and pressed pellets were not sampled reproducibly, compared to the hand-packed surfaces. Detection limits were comparable to those of the dc arc, and the analyte intensity precision of the microprobe was better than that of the arc. Atomic fluorescence detection in a vacuum shows similar freedom from matrix effects for powdered samples (160).

5.4. Liquid Samples

Several methods are used to determine the concentrations of elements in solutions. In one case the solution is drawn into a capillary of Teflon (sold as "spaghetti" insulation for B&S Gauge 28 wire). The walls of the capillary reduce splattering of the sample. The laser is focused on the liquid inside the tip of the capillary. A few particles of graphite are floated on the liquid surface if the liquid is transparent. Results of a qualitative analysis were reported (60).

Solutions have been evaporated to dryness in small quantities (40 μL) onto plastic coverslips (for microscope slides) (164). Solutions have also been mixed with gelatin, frozen, microtomed, and mounted on slides (137). Molten metal has been directly used as a sample, with the expressed purpose of showing how its chemical composition could be determined when a sample is in a hostile environment such as a furnace (7, 54).

Samples that are difficult to dissolve in ordinary solvents may sometimes be fused into a glass matrix at high temperature. The glass is used directly as the sample for the laser microprobe (33, 165).

5.5. Improving Precision

The relative standard deviation of an analysis for relatively homogeneous samples with sufficient surface area to accept many laser shots has been improved to below 0.01 by integrating the results of many laser shots (40, 41). The sample is automatically moved between laser shots to expose a new part of the surface for each shot. Cross excitation is not required, except for samples with low boiling points. Cross excitation with a spark improved the detection limits but degraded the relative precision by a factor of 3. Opaque samples with high boiling points (e.g., ceramics and silicon) behaved like metals. Transparent samples with high boiling points (e.g., glasses) required relatively high irradiances (5×10^{10} W cm^{-2}) for reliable vaporization and good precision, consistent with the requirement of forming an absorbing plasma at the surface early in the lifetime of the laser pulse to absorb the laser energy and vaporize the sample.

5.6. Relative Concentrations

Although the accurate determination of analyte concentration is a desirable goal, it is worthwhile to recognize that absolute concentrations are not always required in practice for a useful analysis of some samples. For instance, a qualitative analysis that identifies the major component or components of a sample, such as an inclusion in an alloy or glass, may be all that is necessary to help solve the problem at hand (30, 61). Relative concentration profiles can be obtained without the use of external standards, when the sample matrix changes relatively little from point to point (44, 61, 62). Changes in the analytical signal caused by changes in the matrix as the microprobe is scanned across a sample may be as important in helping to solve a problem as would be changes in analyte concentrations.

5.7. Summary of Applications

To give a brief indication of the broad scope of laser microprobe analyses, Table 19.2 lists many of the types of samples that have been reported in the literature, along with the elements that have been either qualitatively or quantitatively determined for a particular sample. In some references mention is made only of the types of samples used, and the analyte elements were not specified. A review and discussion of some types of problems that laser microprobe analyses have helped to solve (up to 1973) is presented in a monograph by Moenke and Moenke (145). Commercial instruments have been produced by Jarrell-Ash (Waltham, MA) VEB Carl Zeiss (Jena, Germany), LOMO (Leningrad), and Japan Electric Optics Laboratory (Akishima-she, Tokyo).

6. FUTURE NEEDS AND DEVELOPMENTS

In the excitement that followed the discovery of the laser, it was found that it was relatively easy to focus almost any laser pulse upon a sample and obtain an emission spectrum. Unfortunately, the laser pulses were usually not reproducible, and accurate analytical results were not automatically obtained, for reasons only later to be discovered. A review of the processes that occur when a laser plume is produced clearly indicates the need for more than a laser that generates pulses of reproducible energy. Because the events that take place are

Table 19.2. Elements Determined in Various Samples by Laser Microprobes

Sample	Elements	Reference
Metals		
Ag	C, Pb	153
Low-alloy steel	Cu, Fe, Mn, Ni, Si	27
Iron alloys	Cr, Ni, W, Si, Mn, Al	50
Steel	Ni, Cr, Mn, Fe, Si, Al, Ti, V, P, C	51, 53, 167, 170, 171
Inclusions in steel	Mn, Si, Fe, Al, Ti, Ca, Zr, Mg, Na, K	29, 61
NBS 1174A	Cu	46
1209604 steel (Armco)	Cu, Sn	46
AISI 4340 steel	Mn, Ni, Cr	61
Stainless steel	Cr	31, 44
Molten & solid stainless steel	Cr, Ni	54
Al alloy	Mn, Cu, Fe	31, 42
Al	Cu	116
Al–Mn alloy (Aluman)	Al, Mn	44
Al alloy standards	Cu, Sn	46
8% Al–Mg alloy	Al, Zn, Mn, Cu, Si, Fe, Ni, Pb, Sn	58
Duralumin	Mn, Mg, Cu, Si	172
Cu–Al alloys	Al, Cu	38
Copper alloys		31, 173
Copper	C, Al, Ni, Zn, Pb	153
Brass	Zn, Cu	52, 112
Bronze	Al	53
Alloy of 61.5% Cu/35.3% Zn	Zn, Cu	2
Mn	C, Fe, Zn	153
Pb alloys	Cu, Ag, Sn	46
Si	Al, Ca	40, 41
Zn alloys	Cu	46
Cr	C, Fe	153
Sn alloys	Cu	46
Titanium alloy	Al	53
U	Fe	44
Plutonium		174
Zircaloy-2	Fe	44
Zn		27
Zr	Y, La	39
Powders		
Powdered glass	Cu, Ag, Mn	45
Powders of BaF$_2$, Al$_2$O$_3$, Yb$_2$O$_3$, WO$_3$	Co, Cr, Fe, Mn, Mo, Zr, Fe, Mn, Ni, Sn	55

Table 19.2. (*Continued*)

Sample	Elements	Reference
Powders		
UO$_2$		44
Talc		27
Jarrell-Ash 367-60 Spectro Quality powder standards	Ag, Al, As, Au, B, Ba, Be, Ca, Co, Cr, Cu, Fe, Mg, Mn, Mo, Ni, Pb, Sb, Si, Sn, Ti, V	31, 58
Loose soil		58
Al$_2$O$_3$	Y, Eu, La, Yb	40, 41, 56
Al$_2$O$_3$–CaO	Ni, Cu, Co	57
Al$_2$O$_3$–MgO	Ni, Cu, Co	57
Al$_2$O$_3$–SiO$_2$	Ni, Cu, Co	57
Enamel powdered	Mn, Fe, Si, Al, Ti	51
Graphite	Cu, Ca, Ag	153
Y$_2$O$_3$	La, Eu	56
La$_2$O$_3$	Y	56
NaCl	Sc, Y, La, Pr, Nd, Sm, Eu, Gd, Tb, Dy, Ho, Er, Tm, Yb, Lu	56
NaCl, NaNO$_3$, NaClO$_4$, Na$_2$SO$_4$, KH$_2$PO$_4$	Y, La, Yb	56
Silicate Minerals		
Amphibole	Mn, Al, Si, Fe	36
Basalt	Cu, Ag, Mn, Zn	154
Biotite	Ca, Mg, Cu, Si, Al, Na, Ti, Ni, Sr, Cr, Mn	33
Clinopyroxene	Si, Ca, Mg, Fe, Al, Ti	35
Diabase	Cu, Ag, Mn, Zn	154
Enstatite (in a meteorite)	Si, Ti, Mg, Fe, Al	35
Garnet	Fe, Mg, Mn, Ca	165
Granite	Cu, Ag, Mn, Zn	154
Granodiorite & basalt (dissolved in glass)	Fe, Mg, Cr, V, Ca, Cu, Ti, Ni, Si	32
Orthopyroxene	Si, Ca, Fe, Mg, Ti, Al, Cr, Mn	35
Plagioclase	Mg, Ca, Si, Al, B, Fe, Cu, Na, Ti, Ni, Sr, Cr, Mn	35, 33
Potassium Felspar	Si, Na, K, Al, Mg, Ca, Fe, Mn, Ti	35
Quartz	Ca, Mg, Fe, Cu, Si, Al, Na, Ti, Ni, Sr, Cr, Mn	33
Trachyte	Cu, Ag, Mn, Zn	154
Ore Minerals		
Arsenopyrile	Fe, As, Ni, Cr, Mg	35
Bornite	Fe, Cu, Cr, Mg, Si, Ca, Ag	35

Table 19.2 *(Continued)*

Sample	Elements	Reference
Ore Minerals		
Chalcopyrite	Fe, Cu, Al, Cr, Mg, Ca	35
Chromite	Fe, Cr, Ti, Al, Mn, V	35
Galena	Pb, Fe, Ag, Si, Mg, Bi, Ca	35
Magnetite	Fe, Mg, Si, Mn, Ca	35
Niccolite	As, Ni, Mg, Fe	35
Pyrite	Fe, Ni, Mg, Ca, Cu, Co, Ag, Zn, Pb, Ti	35, 37
Pyrrhotite	Fe, Ni, Mg, Ca, Cr, Ag	35
Siegenite	Ni, Fe, Cu, Zn, Co, Mg, Cu, Zn	35
Sphalerite	Zn, Fe, Si, Pb, Cd, Cu, Si, Mg, Pb, Ca	35, 37
Sulfide ores	Ni, Co	37
Biological Samples		
Liver, kidney, heart sections	P, Mg, Fe, Cu	23
Albumin	Ca, Mg	20
Blood serum	Ca, Mg	20
Mouse strain L fibroblasts	Au	22
Pancrease sections	B, C, Mg, Ca, Mn, Fe, Zn, Cu, Ti, Al, Cr	18
Other		
Wear metal oil	Al, Cr, Cu, Fe, Mg, Ni, Sn, Pb, Ag	60
Grease		60
Polyethylene		40, 41
Glass defects		28
Glass	Fe, Cu	172
Paint chips and surfaces		60
Metal filings		60
Putty		60
Lipstick		60
Slag particles	Na, K, Ca, Mg, Al, Si	61
Wolframite	Mn, Fe	167
Alrborr particles		30
Air	S, halogens	175
Gelatin	P, Mg, Fe, Cu, Li, Zn, Hg, Pb	20, 23, 137
Meteorite		176
Alumina substances		177
Gunshot residue		15

dynamic and localized, reproducible analytical results also depend on a reproducible temporal shape for the laser pulse, and reproducible spatial mode structure. These are stringent requirements even for present-day lasers, which have not yet fully utilized control and feedback techniques to obtain reproducible energy, temporal shape, and spatial mode structure; indeed, the discovery and development of some of these techniques still lies in the future.

The short pulse of a N_2 laser has produced a linear analytical working curve when the sample is at low pressure. The use of picosecond lasers should also be investigated because the duration of the pulse is comparable to the lattice vibration times of solids and may provide as yet unknown advantages.

Future improvements may also lie in controlling the laser pulse to conform to any desired temporal shape. For instance, it may be best to have a pulse with a very high initial irradiance to generate a small surface plasma, especially for initially transparent samples. If high irradiance were continued throughout most of the pulse, the surface plasma would initiate a radiation-supported shock wave that would propagate into the atmosphere and isolate some of the laser energy from the sample surface. If, however, the irradiance of the pulse were reduced as soon as the small surface plasma formed, most of the laser energy might be transferred to the sample. Versatile control of the pulse shape would allow a balance to be achieved between laser energy absorption in the atmosphere (possibly desirable for sample excitation) and absorption in the sample.

A few studies have shown some of the influences that atmospheric density has upon the analytical results. A delay in observing the analytical signal until the plume has had a chance to more fully develop, to become more homogeneous, and to interact with the atmosphere may lead to improved analytical results. On the other hand, plume expansion during the delay, and some atmospheric interactions, may detract from the analytical signal. Clearly, further studies are needed to show how the atmosphere should be manipulated to improve the desired analytical results. Such studies would be relatively easy to make since the atmosphere is readily varied. However, they would necessarily be complicated by studies of how the laser energy is partitioned between the sample and the atmosphere, and of how to distinguish between chemical and thermal effects. They might also include studies of what happens when the sample material rises into an atmosphere that has been significantly heated by the laser pulse.

Most cross excitation, when needed, has been done with an electric spark that ignites when charged particles in the laser plume pass between the electrodes. A laser-produced, radiation-supported shock wave in the atmosphere could initiate the spark before the analyte has had a chance to reach the gap, and unnecessary background radiation would be produced. A few studies (146–148, 166, 167) have used a spark that is electronically ignited at a predetermined time after the laser pulse when the analyte has reached the spark gap. More work is needed in this area, including studies of controlled waveform sparks (160).

Laser plume spectroscopy has been at the mercy of nonlinear and seemingly unpredictable fluid dynamic processes that help to determine the ways that sample ejection occurs, and, to some extent, the shape of the plume. The discovery of ways to bring these processes under control would aid the reproducibility of analytical results. Although this may seem to be a difficult task, it is encouraging to recall that the notoriously random electric spark was brought under spatial control (169).

Although a wealth of information is available about processes that occur during a laser microprobe analysis, it is not clear that what has been observed by one worker is totally applicable to the laser microprobe studies of another. Some of the important processes in the sampling and plume development steps vary dramatically with laser beam irradiance. The pattern of laser irradiance in the focal region is not uniform and depends on the spatial mode structure of the beam, which may change from pulse to pulse, and upon the type of focusing system and its aberrations. Rarely do two pulsed lasers have the same spatial mode structure, and the mode structure of any given laser changes from one laser cavity alignment condition to another. Mode structure and/or the irradiance pattern at the focal region have been only occasionally included in literature reports. This has impeded a clearer understanding of laser microprobe processes. Improvements in pulsed laser technology are needed to ensure the spatial quality of the beam. Even when this is done, however, the need will continue for accurately determining and reporting the spatial mode structure of the beam for each measurement and the aberrations of the focusing system, along with an accurate determination of beam energy and temporal shape, if a unified and quantitative understanding of the processes that are being studied is to be achieved.

As with other microprobe techniques, the need will continue for accurate analytical standard samples that are homogeneous throughout the small sampled region. Standards of a wide variety of materials are needed to encompass the broad applicability of the laser microprobe technique. Better ways to correct for matrix effects are needed when standards are not available.

When viewing the needs of a field such as this, it is easy to lose sight of many successes that have been achieved. Accurate analytical results have been obtained in laser microprobe measurements. Ways to compensate for matrix effects have been found. Some accurate analytical standards do exist. And the solutions to many problems have been aided by qualitative or relative analyses using a laser microprobe without analytical standards. As applications continue to be reported in the literature, the laser microprobe is finding its place in the analytical world, helping the analyst to solve unique and difficult analytical problems.

ACKNOWLEDGMENT

The author thanks the National Science Foundation for its grant support of the research that helped him to better understand the laser microprobe (Grant Numbers GP-11121, GP-28069, CHE-7305031, CHE-7901759).

BIBLIOGRAPHY

Charschan, S. S., *Lasers in Industry*, Van Nostrand Reinhold, New York, 1972.
Moenke, H., and L. Moenke-Blankenburg, *Laser Micro-Spectrochemical Analysis*, Eng. trans. by R. Auerbach, Crane, Russak & Co., New York, 1973.
Ready, J. F., *Effects of High Power Laser Radiation*, Academic Press, New York, 1971.
Ross, D., Lasers, *Light Amplifiers and Oscillators*, Eng. trans. ed. by O. S. Heavens, Academic Press, New York, 1969.
Schafer, F. P., *Dye Lasers*, Vol. 1, *Topics in Applied Physics*, Springer-Verlag, New York, 1975.

REFERENCES

1. F. Brech, *Appl. Spectrosc.*, **16,** 59 (1962).
2. C. D. Allemand, *Spectrochim. Acta*, **27B,** 185 (1972).
3. A. V. Karyakin, M. V. Akhmanova, and V. A. Kaigorodov, *Zh. Analit. Khim.*, **20,** 145 (1965).
4. T. Ishizuka, Y. Uwamino, and H. Sunahara, *Anal. Chem.*, **49,** 1339 (1977).
5. T. Ka'ntor, L. Bezúr, E. Pungor, P. Fodor, J. Nagy-Balogh, and Gy. Heincz, *Spectrochim. Acta*, **34B,** 341 (1979).
6. T. Ishizuka and Y. Uwamino, *Anal. Chem.*, **52,** 125 (1980).
7. D. A. Cremer, H. C. Dilworth, C. Griggs, F. A. Archuleta, D. Basinger, W. C. Danen, and L. S. Blair, Chemistry Division Semiannual Review, Oct. 1983–Mar. 1984, Los Alamos National Laboratory, Los Alamos, New Mexico, p. 78.
8. W. Schrön, *Analytiktreffen-Atomspektroskopie*, Finsterbergen, GDR, 1978.
9. F. Leis and K. Laqua, *Spectrochim. Acta*, **33B,** 727 (1975).
10. F. N. Abercrombie, M. D. Silvester, A. D. Murray, and A. R. Barringer, *Application of Inductively Coupled Plasma to Emission Spectroscopy*, R. M. Barnes Ed., Franklin Institute Press, Philadelphia, 1978, p. 121.
11. A. A. Boitsov and Kh. I. Zil'bershtein, *Spectrochim. Acta*, **36B,** 1201 (1981).
12. Jarrell-Ash Co., Waltham, Mass., *Spectrum Scanner*, **21,** Nos. 3 and 4 (1966).
13. W. J. Young, Abstract 64, 10th National Meeting of the Society for Applied Spectroscopy, St. Louis, MO, Oct. 1971.
14. E. Rudolph, "Nachweis vom Fremdmetallabrieb auf Stahlwerkzeugen mit Hilfe der Laser-Mikrospektralanalyse," Thesis, Köthen, 1969.
15. R. C. Sullivan, C. Pompa, L. V. Sabatino, and J. J. Horan, *J. Forensic Sci.*, **19,** 486 (1974).

16. A. Pterakeiv, A. Samow, and G. Dimitrov, *Jena Rev.*, **16**, 250 (1971).
17. R. C. Rosan, M. K. Healy, and W. F. McNary, *Science,* **142**, 236 (1963).
18. R. Galabova, A. Petrakiev, A. Paneva, and P. Petkov, *Acta Histochem.*, **54**, 66 (1975).
19. D. Glick and K. W. Marich, *Clin. Chem.*, **21**, 1238 (1975).
20. K. W. Marich, "Laser Microprobe Emission Spectroscopy Biomedical Applications," in T. Hall, P. Echlin, and R. Kaufmann, Eds., *Microprobe Analysis as Applied to Cells and Tissues,* Academic Press, New York, 1974.
21. D. B. Sherman, M. P. Ruben, H. M. Goldman, and F. Brech, *Ann. N.Y. Acad. Sci.*, **122**, 767 (1965).
22. M. M. Herman, K. G. Bensch, K. W. Marich, and D. Glick, *Exp. Mol. Path.*, **16**, 186 (1972).
23. K. W. Marich, W. J. Trytl, J. G. Hawley, N. A. Peppers, R. E. Myers, and D. Glick, *J. Phys. E*, **7**, 830 (1974).
24. R. Kaufmann, F. Hillenkamp, and E. Remy, *Microscopica Acta,* **73**, 1 (1972).
25. N. A. Peppers, E. J. Scribner, L. E. Alterton, R. C. Honey, E. S. Beatrice, I. Harding-Barlow, R. C. Rosan, and D. Glick, *Anal. Chem.*, **40**, 1178 (1968).
26. N. V. Korolev and G. A. Faivilevich, *Zavodsk. Lab.*, **30**, 557 (1964).
27. S. D. Rasberry, B. F. Scribner, and M. Margoshes, *Appl. Opt.*, **6**, 1 (1967), 87 (1967).
28. J. R. Ryan, E. Ruh, and C. B. Clark, *Am. Ceram. Soc. Bull.*, **45**, 260 (1966).
29. J. Debras-Gueden and N. Liodec, *Compt. Rend.*, **257**, 3336 (1963).
30. J. R. Ryan and J. L. Cunningham, *Metal Progr.*, **90**, 100 (Dec. 1966).
31. H. Uchida, F. Adachi, O. Mori, and R. Negishi, *Bunseki Kagaku,* **24**, 325 (1975).
32. K. G. Snetsinger and K. Keil, *Am. Mineral.*, **52**, 1842 (1967).
33. M. Yoshida and M. Murota, *J. Geosci. (Osaka)*, **19**, 81 (1975).
34. M. Yoshida, H. Hirano, and M. Murota, *J. Geosci. (Osaka)*, **18**, 73 (1974).
35. J. A. Maxwell, *Can. Mineral.*, **7**, 727 (1963).
36. A. V. Karyakin, V. A. Kaigorodov, and M. V. Akhmanova, *Zh. Prikl. Spektrosk.*, **2**, 364 (1965).
37. N. I. Eremin, *Mineral Mag.*, **40**, 312 (1975).
38. R. Kirchheim, U. Nagorny, K. Maier, and G. Tölg, *Anal. Chem.*, **48**, 1505 (1976).
39. E. P. Krivchikova and N. M. Vasil'eva, *Zh. Anal. Khim.*, **28**, 822 (1973).
40. A. Felske, W.-D. Hagenah, and K. Laqua, *Spectrochim. Acta,* **27B**, 1 (1972).
41. A. Felske, W.-D. Hagenah, and K. Laqua, *Spectrochim. Acta,* **27B**, 295 (1972).
42. D. E. Osten and E. H. Piepmeier, *Appl. Spectrosc.*, **27**, 165 (1973).
43. H. Klocke, *Spectrochim. Acta,* **24B**, 263 (1969).
44. E. Cerrai and R. Trucco, *Energ. Nucl. (Milan)*, **15**, 581 (1968).
45. E. K. Vul'fson, A. V. Karyakin, and A. I. Shidlovskii, *Zavodskaya Lab.*, **40**, 1134 (1974) (Eng. trans.).
46. K. L. Morton, J. D. Nohe, and B. S. Madsen, *Appl. Spectrosc.*, **27**, 109, (1973).
47. M. A. Krishtal, Yu. I. Davydov, and V. D. Korvachev, *Zavodsk. Lab.*, **30**, 950 (1964).
48. E. F. Runge, F. R. Bryan, and R. W. Minck, *Can. Spectrosc.*, **9**, 80 (1964).
49. E. F. Runge, R. W. Minck, and F. R. Bryan, *Spectrochim. Acta,* **20**, 733 (1964).

50. E. P. Krivachikova, *Zh. Anal. Khim.*, **30**, 187 (1975).
51. M. L. Pelukh and A. A. Yankovskii, *Zh. Prikl. Spektrosk.*, **21**, 208 (1974).
52. S. A. Mikhnov, V. V. Panteleev, V. S. Strizhnev, and A. A. Yankovskii, *Zh. Prikl. Spektrosk.*, **17**, 394 (1972).
53. G. I. Nilolaev and V. I. Podgornaya, *Zh. Prikl. Specktrosk.*, **16**, 911 (1972).
54. E. F. Runge, S. Bonfiglio, and F. R. Bryan, *Spectrochim. Acta*, **22**, 1678 (1966).
55. A. B. Whitehead and H. H. Heady, *Appl. Spectrosc.*, **22**, 7 (1968).
56. T. Ishizuka, *Anal. Chem.*, **45**, 538 (1973).
57. A. V. Karyakin and V. A. Kaigorodov, *Zh. Anal. Khim.*, **22**, 504 (1967).
58. M. Birnbaum, *J. Appl. Phys.*, **36**, 3688 (1965).
59. R. C. Rosan, *Appl. Spectrosc.*, **19**, 97 (1965).
60. S. F. Bosen, A. Pudzianowski, and G. Dragutinovich, *J. Forensic Sci.*, **19**, 357 (1974).
61. J. T. McCormack, *Metals Rev.*, **38**, 6 (Mar. 1965).
62. M. Montanarini and J. Steffen, *IEEE J. Quantum Electronics*, **QE-12**, 126 (1976).
63. E. S. Dayhoff and B. Kessler, *Appl. Opt.*, **1**, 339 (1962).
64. E. S. Dayhoff, *Proc. IRE*, **50**, 1684 (1962).
65. J. F. Ready, *Effects of High-Power Laser Radiation*, Academic Press, New York, 1971.
66. M. Born and E. Wolf, *Principles of Optics*, Macmillan (Pergamon), New York, 1964, p. 435.
67. D. Röss, *Lasers, Light Amplifiers and Oscillators*, O. S. Heavens, Ed., Academic Press, New York, 1969, p. 405.
68. J. A. Baker and C. W. Peters, *Appl. Opt.*, **1**, 674 (1962).
69. G. Magyar, *Rev. Sci. Instr.*, **38**, 517 (1967).
70. W. F. Hagen, *J. Appl. Phys.*, **40**, 511 (1969).
71. D. Röss, *Frequenz*, **17**, 233 (1963).
72. T. Li, *Bell Syst. Tech. J.*, **42**, 2609 (1963).
73. C. G. Morgan, *Rep. Prog. Phys.*, **38**, 621 (1975).
74. (a) L. R. Evans and C. G. Morgan, *Nature (Lond.)*, **219**, 712 (1968); (b) *Phys. Med. Biol.*, **14**, 205 (1969); (c) *Phys. Rev. Lett.*, **22**, 1099 (1969).
75. J. M. Aaron, C. L. M. Ireland, and C. G. Morgan, *J. Phys. D: Appl. Phys.*, **7**, 1907 (1974).
76. M. N. Libenson, G. S. Romanov, and Ya. A. Imas, *Zh. Tekh. Fiz.*, **38**, 1116 (1968) [*Sov. Phys.-Tech. Phys.*, **13**, 925 (1969)].
77. A. M. Bonch-Bruevich, S. E. Potanov, and Ya. A, Imas, *Zh. Tekh. Fiz.*, **38**, (1968) [*Sov. Phys.-Tech. Phys.*, **13**, 640 (1968)].
78. N. G. Basov, V. A. Boiko, O. N. Krokhin, O. G. Semeno, and G. V. Sklizkov, *Zh. Tekh. Fiz.*, **38**, 1973 (1968) [*Sov. Phys.-Tech. Phys.*, **13**, 1581 (1969)].
79. K. Kagawa and S. Yokoi, *Spectrochim. Acta*, **37B**, 789 (1982).
80. H. Sonnenberg, H. Heffner, and W. Spicer, *Appl. Phys. Lett.*, **5**, 95 (1964).
81. I. Adawi, *Phys. Rev.*, **134**, A788 (1964).
82. E. M. Logothetis and P. L. Hartman, *Phys. Rev. Lett.*, **18**, (1967).
83. F. Shiga and S. Imamura, *Appl. Phys. Lett.*, **13**, 257 (1968).
84. M. C. Teich and G. J. Wolga, *Phys. Rev.*, **171**, 809 (1968).
85. Gy. Farkas, S. Kertész, and Zs. Náray, *Phys. Lett.*, **28A**, 190 (1968).

86. C. M. Verber and A. H. Adelman, *J. Appl. Phys.*, **36,** 1522 (1965).
87. J. F. Ready, *Phys. Rev.*, **137,** A620 (1965).
88. J. K. Cobb and J. J. Muray, *Brit. J. Appl. Phys.*, **16,** 271 (1965).
89. S. H. Khan, F. A. Richards, and D. Walsh, *IEEE J. Quantum Electron.*, **QE-1,** 359 (1965).
90. M. Iannuzzi and R. Williamson, *Nuovo Cimento,* **36,** 1130 (1965).
91. D. M. Stevenson, *Proc. IEEE,* **54,** 1471 (1966).
92. E. Bernal G., J. F. Ready, and L. P. Levine, *Phys. Lett.,* **19,** 645 (1966).
93. E. Bernal G., J. F. Ready, and L. P. Levine, *IEEE J. Quantum Electron.,* **QE-2,** 480 (1966).
94. L. P. Smith, *Phys. Rev.,* **35,** 381 (1930).
95. L. P. Levine, J. F. Ready, and E. Bernal G., *J. Appl. Phys.,* **38,** 331 (1967).
96. L. P. Levine, J. F. Ready, and E. Bernal G., *IEEE J. Quantum Electron.,* **QE-4,** 18 (1968).
97. J. Berkowitz and W. A. Chupka, *J. Chem. Phys.,* **40,** 2735 (1964).
98. K. H. Kingdon and I. Langmuir, *Phys. Rev.,* **21,** 380 (1923).
99. E. H. Piepmeier and H. V. Malmstadt, *Anal. Chem.,* **41,** 700 (1969).
100. E. H. Piepmeier and D. E. Osten, *Appl. Spectrosc.,* **25,** 642 (1971).
101. J. F. Ready, *J. Appl. Phys.,* **36,** 462 (1965).
102. J. L. Deming, J. H. Weber, and L. C. Tao, *AIChE J.,* **15,** 501 (1969).
103. J. I. Masters, *J. Appl. Phys.,* **27,** 477 (1956).
104. *American Institute of Physics Handbook,* 2nd ed., McGraw-Hill, New York, 1963.
105. T. Kato and T. Yamaguchi, *NEC Res. Dev.,* **12,** 57 (Oct. 1968).
106. H. G. Landau, *Quart. J. Appl. Math.,* **8,** 81 (1950).
107. V. B. Braginskii, I. I. Minakova, and V. N. Rudenko, *Zh. Tekh. Fiz.* **37,** 1045 (1967) [*Sov. Phys.-Tech. Phys.,* **12,** 753 (1967)].
108. K. Vogel and P. Backlund, *J. Appl. Phys.,* **36,** 3697 (1965).
109. F. Neuman, *Appl. Phys. Lett.,* **4,** 167 (1964).
110. S. I. Anisimov, A. M. Bonch-Bruevich, M. A. El'yashevich, Ya. A. Imas, N. A. Pavlenko, and G. S. Romanov, *Zh. Tekh. Fiz.,* **36,** 1273 (1966) [*Sov. Phys.-Tech. Phys.,* **11,** 945 (1967)].
111. H. Weichel and P. V. Avizonis, *Appl. Phys. Lett.,* **9,** 334 (1966).
112. J. M. Baldwin, *Appl Spectrosc.,* **24,** 429 (1970).
113. R. E. Honig, *Appl. Phys. Lett.,* **3,** 8 (1963).
114. J. A. Howe and T. V. Molloy, *J. Appl. Phys.,* **35,** 2265 (1964).
115. J. A. Howe, *J. Chem. Phys.,* **39,** 1362 (1963).
116. R. H. Scott and A. Strasheim, *Spectrochim. Acta,* **25B,** 311 (1970).
117. J. F. Ready, *Appl. Phys. Lett.,* **3,** 11 (1963).
118. E. W. Sucov, J. L. Pack, A. V. Phelps, and A. G. Engelhardt, *Phys. Fluids,* **10,** 2035 (1967).
119. N. G. Basov, V. A. Boiko, Yu. P. Voinov, E. Ya. Kononov, S. L. Mandel'shtam, and G. V. Sklizkov, *JETP Lett.,* **5,** 141 (1967).
120. E. Archbold, D. W. Harper, and T. P. Hughes, *Brit. J. Appl. Phys.,* **15,** 1321 (1964).
121. D. D. Burgess, B. C. Fawcett, and N. J. Peacock, *Proc. Phys. Soc.,* **92,** 805 (1967).

122. P. Dhez, P. Jaegle, S. Leach, and M. Velghe, *J. Appl. Phys.*, **40**, 2545 (1969).
123. B. C. Fawcett and N. J. Peacock, *Proc. Phys. Soc.*, **91**, 973 (1967).
124. B. C. Fawcett, D. D. Burgess, and N. J. Peacock, *Proc. Phys. Soc.*, **91**, 970 (1967).
125. C. DeMichelis and S. A. Ramsden, *Phys. Lett.*, **25A**, 162 (1967).
126. P. Langer, G. Tonon, F. Floux, and A. Ducauze, *IEEE J. Quantum Electron.*, **QE-2**, 499 (1966).
127. J. L. Bobin, F. Floux, P. Langer, and H. Pignerol, *Phys. Lett.*, **28A**, 398 (1968).
128. N. G. Basov, V. A. Boiko, V. A. Gribkov, S. M. Zakharov, O. N. Krovhin, and G. V. Sklizkov, *JETP Lett.*, **9**, 315 (1969).
129. C. W. Bruce, J. Deacon, and D. F. Vonderhaar, *Appl. Phys. Lett.*, **9**, 164 (1966).
130. C. D. David, *Appl. Phys. Lett.*, **11**, 394 (1967).
131. H. Weichel and P. V. Avizonis, *Appl. Phys. Lett.*, **9**, 334 (1966).
132. Ya. B. Zel'dovich and Yu. P. Paizer, *Physics of Shock Waves and High Temperature Hydrodynamic Phenomena*, Vol. 1, Eng. transl. W. D. Hayes and R. F. Probstein, Eds., Academic Press, New York, 1967.
133. L. Spitzer, *Physics of Fully Ionized Gases*, Wiley-Interscience, New York, 1956.
134. R. M. Manabe and E. H. Piepmeier, *Anal. Chem.*, **51**, 2066 (1979).
135. L. E. Steenhoek and E. S. Yeung, *Anal. Chem.*, **53**, 528 (1981).
136. W. Börgershausen and R. Vesper, *Spectrochim. Acta*, **24B**, 103 (1969).
137. W. J. Treytl, K. W. Marich, and D. Glick, *Anal. Chem.*, **47**, 1275 (1975).
138. W. J. Treytl, J. B. Orenberg, K. W. Marich, and D. Glick, *Appl. Spec.*, **25**, 376 (1971).
139. W. J. Treytl, J. B. Orenberg, K. W. Marich, A. J. Saffir, and D. Glick, *Anal. Chem.*, **44**, 1903 (1972).
140. G. J. Beenen and E. H. Piepmeier, *Appl. Spectrosc.*, **38**, 851 (1984).
141. A. L. Lewis II and E. H. Piepmeier, *Appl. Spectrosc.*, **37**, 523 (1983).
142. W. C. Pesklak, "Laser Induced Nonresonance Atomic Fluorescence Observations In an Analytical Laser Microprobe Plume," Master's Thesis, Oregon State University, Corvallis, Oregon, 1984.
143. W. D. Hagenah, K. Laqua, and F. Leis, *Colloq. Spectrosc. Int. (Proc.) 17th*, **2**, 491 (1973).
144. H. U. Möde, W. D. Hagenah, and K. Laqua, *Colloq. Spectrosc. Int. (Proc.) 15th*, **4**, 383 (1969).
145. H. Moenke and L. Moenke-Blankenburg, *Laser Micro-Spectrochemical Analysis*, R. Auerbach (trans.), Crane, Russak, New York, 1973.
146. L. Moenke-Blankenburg, J. Mohr, and W. Quillfeldt, *Mikrochim. Acta Suppl.*, **4**, 229 (1970).
147. R. Klockenkämper and K. Laqua, *Spectrochim. Acta*, **32B**, 207 (1977).
148. N. V. Korolev, V. V. Ryukhin, and G. B. Lodin, *J. Appl Spectrosc.*, **19**, 838 (1975).
149. S. A. Goldstein and J. P. Walters, *Spectrochim. Acta*, **31B**, 201 (1976), 295 (1976).
150. K. Laqua, "Analytical Spectroscopy Using Laser Atomizers," in N. Omenetto, Ed., *Analytical Laser Spectroscopy*, Wiley, New York, 1979.
151. E. H. Piepmeier and L. de Galan, *Spectrochim. Acta*, **31B**, 163 (1976).

152. E. H. Piepmeier, Doctoral Thesis, University of Illinoise, Urbana IL, 1966.

153. V. G. Mossotti, K. Laqua, and W.-D. Hagenah, *Spectrochim. Acta,* **23,** 197 (1967).

154. E. K. Vul'fson, A. V. Karyakin, and A. I. Shidlovskii, *Zh. Anal. Khim.,* **26,** 1253 (1973).

155. G. J. DeJong and E. H. Piepmeier, *Anal. Chem.,* **46,** 318 (1974).

156. G. J. DeJong and E. H. Piepmeier, *Spectrochim. Acta,* **29B,** 159 (1974).

157. E. H. Piepmeier and L. de Galan, *Spectrochim. Acta,* **30B,** 263, (1975).

158. J. P. Matousek and B. J. Orr, *Spectrochim. Acta,* **31B,** 475 (1976).

159. A. L. Lewis II, G. J. Beenen, J. W. Hosch, and E. H. Piepmeier, *Appl. Spectrosc.,* **37,** 263 (1983).

160. H. S. Kwong and R. M. Measures, *Anal. Chem.,* **51,** 428 (1979).

161. A. V. Karyakin, M. V. Achmanova, and V. A. Kaigorodov, "Possibilities for the Use of a Laser in the Spectrochemical Analysis of Pure Substances," in *Proc. XII Coll. Spec. Int., Exeter,* Adam Hilger Ltd., 1965, p. 353.

162. T. Yamane and S. Matsushita, *Spectrochim. Acta,* **27B,** 27 (1972).

163. J. D. Nohe and K. L. Morton, *W. Elec. Eng. (USA),* **14,** 34 (Oct. 1970).

164. K. W. Marich, P. W. Carr, W. J. Treytl, and D. Glick, *Anal. Chem.,* **42,** 1775 (1970).

165. W. H. Blackburn, Y. J. A. Pelletier, and W. H. Dennen, *Appl. Spectrosc.,* **22,** 278 (1968).

166. L. Moenke-Blankenburg, H. Moenke, J. Mohr, W. Quillfeldt, K. Wiegand, W. Grassme, and W. Schrön, *Spectrochim. Acta,* **30B,** 227 (1975).

167. N. V. Korolev, V. V. Ryukhin, and G. B. Lodin, *Zh. Prikl. Spektrosk.,* **19,** 21 (1973).

168. J. P. Walters, *Anal. Chem.,* **40,** 1672 (1968).

169. J. P. Walters, *Appl. Spectrosc.,* **26,** 323 (1972).

170. A. V. Karyakin, M. V. Akhmanova, and V. A. Kaigorodov, *Zh. Anal. Khim.,* **23,** 1610 (1968).

171. Yu. M. Buravlev, I. I. Morokhovskaya, V. N. Murav'ev, and B. P. Nadezhda, *Zh. Anal. Khim.,* **27,** 2337 (1972).

172. K. I. Taganov and L. M. Fainberg, *Zh Prikl. Spektrosk.,* **20,** 571 (1974).

173. V. V. Panteleev and A. A. Yankovskii, *Zh. Prikl. Spektrosk.,* **3,** 96 (1965).

174. H. N. Barton, *Appl. Spectrosc.,* **23,** 519 (1969).

175. T. Yamane, *Bunko Kenkyu,* **22,** 321 (1973).

176. D. Dimov, N. Nikolov, G. Dimitrov, and A. Petraknev, *Annu. Univ. Sofia Fac. Phys. (Bulgaria),* **64–65,** 477 (1970–72).

177. D. R. Watson, J. E. Davis, T. H. Briggs, and D. W. Ports, *Am. Ceram. Soc. Bull. (USA),* **52,** 632 (Aug. 1973), abstract only.

INDEX